MW00980642

Dynamic Energy Budgets in Biological Systems

Dynamic Energy Budgets in Biological Systems

Theory and Applications in Ecotoxicology

S.A.L.M. Kooijman

Professor of Theoretical Biology
Vrije Universiteit,
Amsterdam

Published by the Press Syndicate of the University of Cambridge
The Pitt Building, Trumpington Street, Cambridge CB2 1RP
40 West 20th Street, New York, NY 10011–4211, USA
10 Stamford Road, Oakleigh, Melbourne 3166, Australia

© Cambridge University Press 1993

First published 1993

Printed in Great Britain at the University Press, Cambridge

British Library Cataloguing in publication data available

Library of Congress cataloguing in publication data available

ISBN 0 521 45223 6 hardback

Contents

Preface

In 1978 Thea Adema asked me to develop a statistical methodology for screening toxicants for their effects on daphnid reproduction. I observed that large daphnids tended to have bigger litters than small ones and this led me to realize that reproduction cannot be modelled without including variables such as growth, feeding, food quality and so on. Since then I have found myself working on a theory of Dynamic Energy Budgets (DEB), which has rapidly covered more ground. Ten years ago, I would not have seen any connection between topics such as feeding of daphnids, embryo development of birds and the behaviour of recycling fermenters. Now, I recognize the intimate relationship between these and many other phenomena and the fundamental role of surface/volume ratios and energy reserves. Although new relationships continue to fit into this theory, the time seems ripe to collect the results into a book and to reveal new, exciting forms of coherence in biology.

DEB theory is central to eco-energetics, which is the study of the mechanisms involved in the acquisition and use of energy by individuals; this includes the many consequences of the mechanisms for physiological organization and population and ecosystem dynamics. The related field of bioenergetics focuses on molecular aspects and metabolic pathways in a thermodynamic setting. Although the first and second laws of thermodynamics are frequently used in eco-energetics, thermodynamics is not used to derive rate equations, as is usual, for example, in non-equilibrium thermodynamics. One of the reasons is that the behaviour of individuals cannot be traced back to a restricted number of biochemical reactions. This difference in approach blocks possible cross-fertilizations between levels of organization. This barrier is particularly difficult to break down because eco-energetics usually deals with individuals in a static sense; an individual of a given size allocates energy to different purposes in measured percentages. This tradition hampers links with physiological processes. DEB theory, in contrast, treats individuals as non-linear, dynamic systems. This process oriented approach has firm physiological roots and at the same time it provides a sound basis for population dynamics theories, as will be demonstrated in this book. The hope is that DEB theory will contribute to the cross-fertilization of the different species of energetics.

I like my job very much as it offers good opportunities to enjoy the diversity of life during spare time hikes. Many of my fellow biologists stress the interesting differences between species to such an extent that the properties they have in common remain largely hidden. I believe that this obscures the way in which a particular species deviates from the common pattern, and the causes of deviations, and urges me to stress phenomena that species seem to have in common. I fully understand the problem of being overwhelmed by the diversity of life, but I think that reactions of ecstasy, apathy or complaint hardly

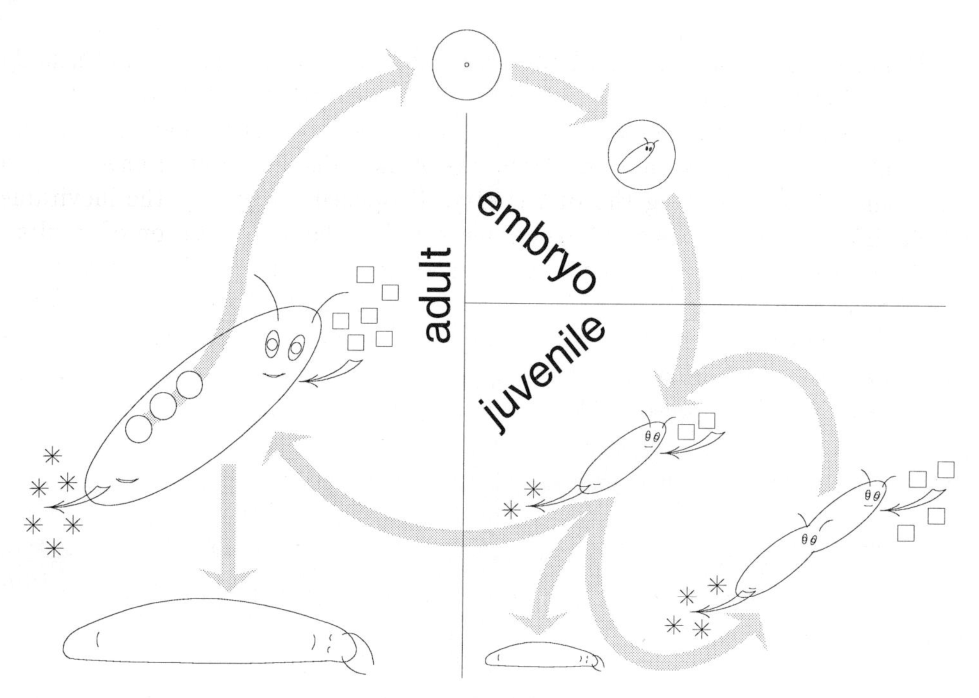

Dynamic Energy Budget theory aims to quantify the energetics of heterotrophs as it changes during life history. The key processes are feeding, digestion, storage, maintenance, growth, development, reproduction and aging. The theory amounts to a set of simple rules, summarized in table 4.1 on page {116}, and a wealth of consequences for physiological organization and population dynamics. Although some of the far reaching consequences turn out to be rather complex, the theory is simple, with only one parameter per key process. Intra- and inter-specific body size scaling relationships form the core of the theory and include dividing organisms, such as microbes, by conceiving them as juveniles.

contribute to insight. This book explores to what extent a theory that is not species-specific can be used to understand observations and experimental results, and it culminates in a derivation of body size scaling relationships for life history traits without using empirical arguments.

DEB theory is quantitative, so it involves mathematics; I feel no need to apologize for this, although I realize that this may be an obstacle for many biologists. My hope is that an emphasis on concepts, rather than mathematical technicalities, avoidance of jargon as much as possible and a glossary will reduce communication problems. Only in some parts of the chapter on population dynamics may the mathematics used be called 'advanced', the remainder being elementary. The text is meant for scientists and mathematicians with a broad interest in fundamental and applied quantitative problems in biology.

The *aim* is summarized in the diagram on this page. The primary aim is not to describe energy uptake phenomena and energy use in as much detail as feasible, but to evaluate consequences of simple mechanisms that are not species-specific. The inclusion or exclusion of material in the book was judged on its relevance with respect to a set of mechanisms that appeared to be tightly interlocked. This book, therefore, does not review all that is relevant to energetics. It does, however, include some topics that are not usually encountered in texts on energetics, because DEB theory appears to imply predictions for the topics. Discrepancies between predictions and actual behaviour of particular species will, hopefully, stimulate a guided search for explanations of these discrepancies. I have learned to appreciate this while developing the DEB theory. It opened my eyes to the inevitable preconceptions involved in the design of experiments and in the interpretation of results.

The emphasis is on mechanisms. This implies a radical rejection of the standard application of allometric equations, which I consider to be a blind alley that prevents understanding. Although it has never been my objective to glue existing ideas and models together into one consistent framework, many aspects and special cases of the DEB theory turned out to be identical to classic models, for example:

author	year	page	model
Descartes	1638	{24}	logarithmic spirals in shells
Arrhenius	1889	{44}	temperature dependence of physiological rates
Huxley	1891	{252}	allometric growth of body parts
Pütter	1920	{81}	von Bertalanffy growth of individuals
Pearl	1927	{174}	logistic population growth
Fisher & Tippitt	1928	{108}	Weibull aging
Emerson	1950	{145}	linear growth of colonies of bacteria on plates
Huggett & Widdas	1951	{89}	foetal growth
Best	1955	{141}	diffusion limitation of uptake
Smith	1957	{103}	embryonic respiration
Leudeking & Piret	1959	{190}	microbial product formation
Holling	1959	{63}	hyperbolic functional response
Marr & Pirt	1962	{162}	maintenance in yields of biomass
Droop	1973	{162}	reserve (cell quota) dynamics
Rahn & Ar	1974	{235}	water loss in bird eggs
Hungate	1975	{249}	digestion

The DEB theory not only shows how and why these models are related, it also specifies the conditions under which these models might be realistic, and it extends the scope from the thermodynamics of subcellular processes to population dynamics.

Discussion will be restricted to heterotrophic systems, those that show either a close coupling between energy and nutrients (i.e. building blocks) or do not suffer from nutrient limitation. Autotrophic systems are characterized by the decoupling of energy and nutrient uptake and are not discussed here.

Potential *practical applications* are to be found in the control and optimization of biological production processes. In my department, for example, we use DEB theory in research on reducing sludge production in sewage treatment plants and optimizing microbial product formation. Other potential applications are to be found in medicine. Ecotoxicological applications will be discussed in chapter 8.

Book organization

The first two chapters are introductory.

Chapter 1 gives the historical setting and some philosophical, methodological and technical background. Many discussions with colleagues about the way particular observations fit or do not fit into a theory rapidly evolved into ones about the philosophical principles of biological theories in general. These discussions frequently related to the problem of the extent to which biological theories that are not species-specific, are possible. This chapter, and indeed the whole book, introduces the idea that the value of a theory is in its usefulness and therefore a theory must be coupled to a purpose. I have written a section on the position I take in these matters and insert throughout the book many remarks on aspects of modelling and testability to point to fundamental problems in practical work. Chapter 1 sets out the context within which DEB theory, developed in subsequent chapters, has a meaning.

Chapter 2 introduces some basic concepts that are pertinent to the organization level of the individual. The concept 'system' is introduced and the state variables body size and energy reserves are identified to be of primary importance. The relationship between different measures for body size and energy are discussed; these are rather subtle due to the recognition of storage materials. A variant of the principle of 'homeostasis' is given to accommodate 'storage'. Effects of temperature on physiological rates are presented and the notion of life stages is discussed. Chapter 2 also introduces vital concepts for the development of DEB theory, and paves the way for testing theory against experimental data, which will occur during the development of the DEB theory.

The next two chapters develop the DEB theory. Chapter 3 describes processes of energy uptake and energy use by individuals in all life stages, which together form the DEB theory. This set of processes gives a complete specification of the transformation of food into biomass and will later be used to analyze consequences and implications. The discussion includes the processes of development and aging because of their relevance for energetics. Special attention is given to the process of aging because most other energetics theories select age as a primary state variable; the focus is therefore on the coupling between aging and energetics. Chapter 3 provides the meat of the DEB theory.

Chapter 4 summarizes the DEB theory in a dimensionless form and lists the 12 assumptions on which the theory is based. It is shown that some assumptions of simple mechanisms that are difficult to test directly can be replaced by a mathematically equivalent set that is easier to test, but has a more complex relationship with mechanisms. Implications for developing individuals are analyzed and some observations that do not seem to fit into the theory at first glance are discussed. The auxiliary purpose of these analyses is to give examples of DEB theory application and to show how it can be used to improve biological insight.

Chapters 5, 6 and 7 deal with applications to other levels of biological organization and to inter-species comparisons. The consequences, implications and usefulness of the DEB model are evaluated.

Chapter 5 develops population consequences. It starts with some well known standard population dynamic theories and introduces the DEB machinery step by step. The population, after its introduction as a collection of individuals, is considered as a new entity in terms of systems analysis with its own relationships between input, output and state. These new relationships are expressed in terms of those for individuals. The coupling between mass and energy fluxes at population level is studied as an important application of this point of view. Chapter 5 ends with an analysis of the behaviour of food chains and explains why long chains cannot exist. Applications to communities are discussed very briefly.

Chapter 6 compares the energetics of different species and studies some of the evolutionary implications. DEB theory also relates to parameter values. Chapter 6 shows how, for a wide variety of biological variables, body size scaling relationships can be derived rather than established empirically. This approach to body size scaling relationships is fundamentally different from that of existing studies. The chapter also compares different life history strategies and gives some speculations about the origin of life.

Chapter 7 explores consequences of DEB theory for several suborganismal phenomena: details of the digestion process, the organization of protein synthesis and the growth of parts of organisms. The purpose is to show that a model at the level of the individual can help when modelling at the suborganismal level, and even at the molecular level in some cases. The relationship between the energy allocation rule of DEB theory and the allometric growth of body parts is demonstrated to make the link to classical theory for this topic.

The last chapter, chapter 8, illustrates applications of the theory in ecotoxicology. The purpose is twofold: to show that fundamental science can really contribute to the solution of practical problems, and to show that applied research can contribute to fundamental science. The chapter focuses on realistic models for uptake and elimination of xenobiotics, which account for body size and lipid fraction, the coupling of kinetics with lethal and sublethal effects and the evaluation of the consequences of effects of toxicants on individuals for the population level.

The logical structure of the chapters is indicated in the diagram (right). The level of biological organization is indicated on the horizontal axis. The best reading sequence is from top to bottom, but a first quick glance through the section on notation and symbols, page {326}, can save time and annoyance.

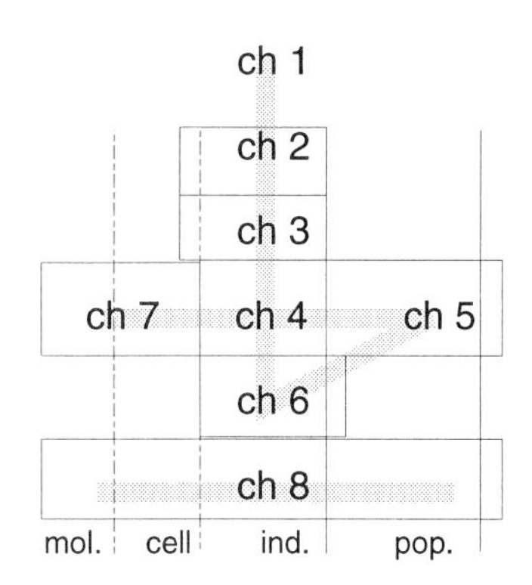

Acknowledgements

Many people have contributed to this book in different ways. I would specifically like to thank the following persons.

Anneke de Ruiter, Trudy Bakker, Hanneke Kauffman, Eric Evers and Harry Oldersma have done a lot of experiments with daphnids and rotifers at the TNO laboratories, and provided the essential experimental material on energetics and toxicology for the development of DEB theory. At a later stage Cor Zonneveld and Arjan de Visser contributed to the theory by experimental and theoretical studies on pond snails, helping to fit the model to embryo development data. Nelly van der Hoeven and Lisette Enserink carried out most useful experimental work on the energetics and population dynamics of daphnids in relation to the DEB theory, while Bob Kooi, Hans Metz, Odo Diekmann, Henk Heijmans, André de Roos and Horst Thieme contributed significantly to the mathematical aspects of DEB-structured populations. Rob van Haren and Hans Schepers worked on mussel energetics in relation to the accumulation elimination behaviour of xenobiotics. Sigrid Bestebroer, Paul Bruijn, Christa Ratsak and Erik Muller worked on various DEB aspects of sewage water treatment. Paul Hanegraaf studied the coupling between energy and mass fluxes. Rienk-Jan Bijlsma is working on extension of the theory to include plants. Various students have done excellent work on specific details: Wyanda Yap, Karin Maarsen, Rik Schoemaker, Arianne van der Berg, Karen Karsten, Marinus Stulp, Bert van der Werf.

I gained a lot from productive collaboration and discussions with Ad Stouthamer, Henk van Verseveld, Hans de Hollander, Arthur Koch (microbiological physiology), Dick Eikelboom, Arnbjørn Hanstveit (applied microbiology), Nico de With, Andries ter Maat ((neuro)physiology), Nico van Straalen, Ger Ernsting (ecology), Jacques Bedaux (statistics), Leo and Henk Hueck, Thea Adema, Kees Kersting (ecotoxicology), Jan Parmentier (chemistry), Wim van der Steen (methodology), Odo Diekmann (permanent skeletons of isomorphs), Schelten Elgersma (Taylor expansion incubation time) and last but not least Roger Nisbet and Tom Hallam (energetics, population dynamics).

The text of the book has been improved considerably by critical comments from a number of readers, particularly Ger Ernsting, Wout Slob, Miranda Aldham-Breary, Tom Hallam, Karen Karsten, Henk Hueck, Andree de Roos, Bob Kooi, Martin Boer, Wim van der Steen, Emilia Persoon, Gabriëlle van Diepen, Ad van Dommelen and Mies Dronkert. Present behind all aspects of the 15 years of work on the theory is the critical interest of Truus Meijer, whose loving patience is unprecedented. The significance of her contribution is beyond words.

S.A.L.M. Kooijman
Professor of Theoretical Biology
Vrije Universiteit
de Boelelaan 1087
1081 HV
Amsterdam

July 1993

Chapter 1

Energetics and models

This introductory chapter presents some general background to theoretical work in energetics. I start with an observation that feeds the hope that it is possible to have a theory that is not species-specific, something that is by no means obvious in view of the diversity of life! A brief historical setting follows giving the roots of some general concepts that are basic to Dynamical Energy Budget (DEB) theory. I will try to explain why the application of allometry restricts the usefulness of almost all existing theories on energetics. This explanation is embedded in considerations concerning philosophy and modelling strategy to give the context of the DEB theory.

1.1 Energy and mass fluxes

1.1.1 Hope for generality

Growth curves are relatively easy to produce and this may explain the fact that the literature is full of them. Yet they remain fascinating. When environmental conditions, including temperature and food availability, are constant and the diet is adequate, organisms ranging from yeasts to vertebrates follow, with astonishing accuracy, the same growth pattern as that illustrated in figure 1.1. This is amazing because different species have totally different systems for regulating growth. Some species, such as daphnids, start to invest at a certain moment during growth a considerable amount of energy in reproduction. Even this does not seem to affect the growth curve. So one wonders how the result can be so similar time and again. Is it all coincidence resulting from a variety of different causes, or do species have something in common despite their differences? Are these curves really similar, or is the resemblance a superficial one?

These questions led me on a breathtaking hike into many corners of biological territory. They became an entertaining puzzle: Is it possible to construct a set of simple rules, based on mechanisms for the uptake and use of material by individuals, that is consistent with what has been measured? The early writers made a most useful start: growth results from processes of build-up and break-down. Break-down has something to do with making energy and elementary compounds available, so how are they replenished? What processes determine digestion and feeding? What determines food availability? Build-up results in

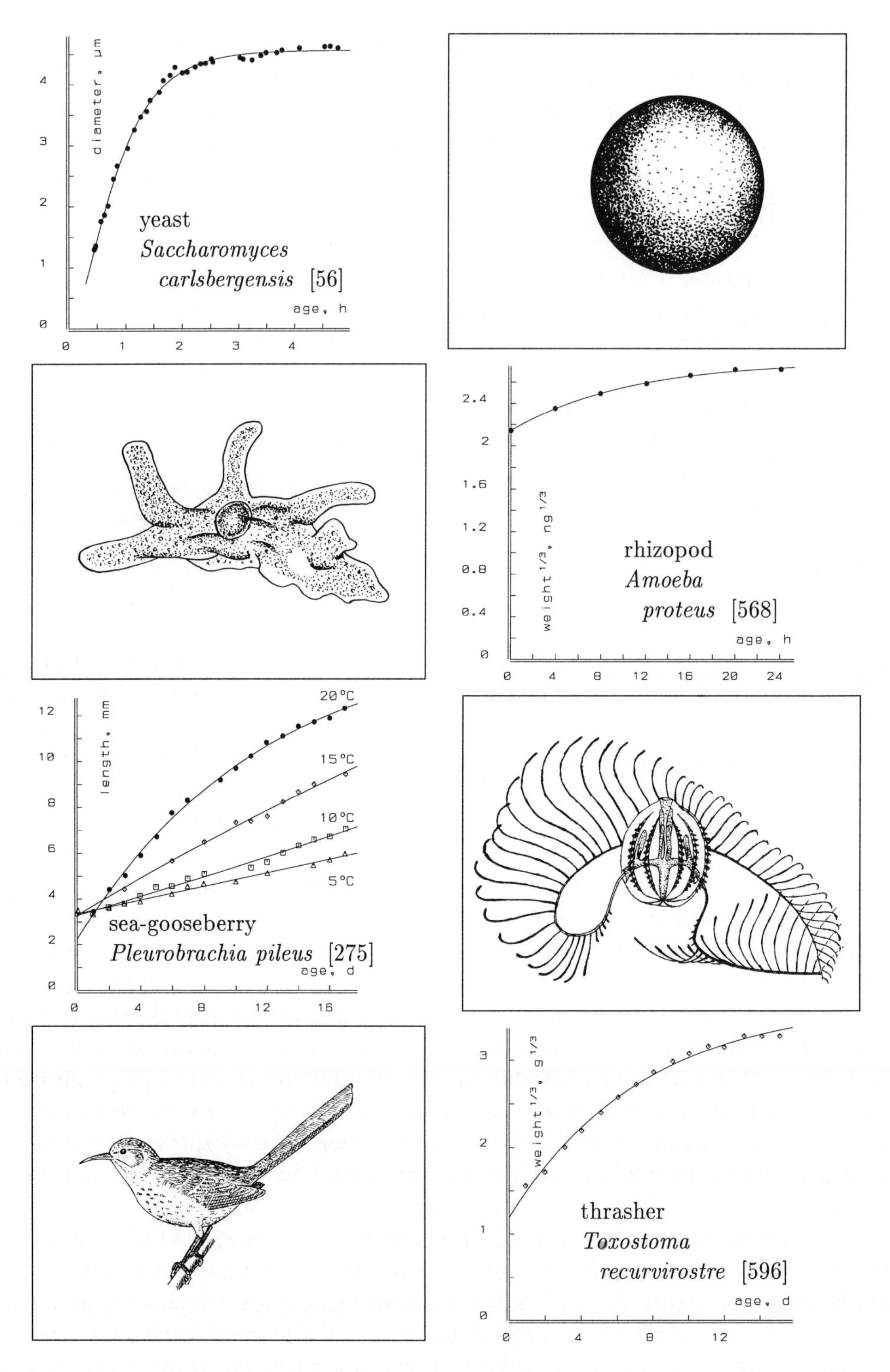

Figure 1.1: These growth curves all have the shape $L(t) = L_\infty - (L_\infty - L_0)\exp\{-\dot{\gamma}t\}$, while the organisms differ considerably in their growth regulating systems. How is this possible? Data sources are indicated by entry numbers of bibliography.)

size increase, and so affects feeding, but offspring are produced as well. This obviously affects food availability. Where does maintenance fit in? Why should there be any maintenance at all? What is the role of age? This is just a sample of the questions that should be addressed to give a satisfactory explanation of a growth curve.

1.1.2 Historical setting

Many of these questions are far from new. Boyle, Hooke and Mayow in the 17th-century were among the first to relate respiration to combustion, according to McNab [472]. The first measurements of the rate of animal heat production are from Crawford in 1779, Lavoisier and de Laplace in 1780 aimed at relating it to oxygen consumption and carbon dioxide production [472]. Interest in how metabolic rate, measured as oxygen consumption rate, depends on body size goes back at least as far as the work of Sarrus and Rameau [628] in 1839. They were the first to find rates proportional to surface area for warm-blooded animals [66]. Later this became known as the Rubner's surface law [614]. Pütter [574] used it in a model for the growth of individuals in 1920. He saw growth as the difference between build-up and break-down. The processes of build-up, which later became known as anabolic processes, were linked directly to the metabolic rate, which was assumed to follow the surface law. The processes of break-down, now known as catabolic processes, were assumed to proceed at a constant rate per unit of volume. Volume was thought to be proportional to weight. The growth rate then results a weighted difference between surface area and volume. The casual way Wallace mentioned this idea in a note to Poulton (appendix 3 in [219]), suggests that its roots go back to before 1865. The resulting growth curve is presented in figure 1.1. The fact that Pütter applied the model to fish, whereas the surface law was based on work with warm-blooded animals, generated a lot of criticism.

More data were generated with improved methods of measurement; invertebrates were also covered. Kleiber [390] found in 1932 that metabolic rates are proportional to weight to the power $\frac{3}{4}$ and this became known as Kleibers law. Extensive studies undertaken by Brody [97] confirmed this proportionality. Von Bertalanffy [66] saw anabolic and catabolic rates as special cases of the allometric relationship, i.e. a relationship of the type $y = \alpha x^\beta$, where y is a variable and x is usually body weight. He viewed this as a simplified approximation which could be applied to almost all types of metabolic rates, including the anabolic and the catabolic, but the constant β varied somewhat. It depends on tissue, physiological condition and experimental procedure. The growth curve proved to be rather insensitive to changes in β for catabolism, so, like Pütter, von Bertalanffy took the value 1 and classified species on the basis of β for anabolism. The surface law is just one of the possibilities.

Although von Bertalanffy [65] was the genius behind the ideas of general systems theory, he never included the feeding process in his ideas about growth. I still do not know why, because mass balance equations are now always bracketed together with systems. I think that the use of allometric equations, which are a step away from a mechanistic explanation towards a meaningless empirical regression, act as an obstacle to new ideas in metabolic control. I will explain this in later sections. The idea of allometry goes back to Snell [672] in 1891 and, following the work of Huxley [345], it became widely known. Both Huxley and von Bertalanffy were well aware of the problems connected with allometric equations, and used them as first approximations. Now, a century later, it is hard to find a study that involves body size and does not use them.

Zeuten [787] was the first to point to the necessity of distinguishing between size differences within a species and between species. The differences in body size within a species, as measured in one individual at different points during development, are treated here as an integral part of the processes of growth and development. Those between species are discussed in a separate chapter on parameter values, where I will show that body size scaling relationships can be deduced without any empirical arguments.

1.1.3 Energetics

The problem that everything depends on everything else is a hard one in biology, anything left out may prove to be essential in the end. If one includes as much as possible one loses an intellectual grasp of the problem. The art is to leave out as much as possible and still keep the essence. Here, I will focus the discussion on an abstract quantity, called energy, rather than a selection of the many thousands of possible compounds usually found in organisms. Any selection of compounds would exclude others, so what is the role of the ones that have been left out? Jeong *et al.* [357] made an heroic attempt to model the compound-based physiology of *Bacillus* and introduced more than 200 parameters. Many compounds, however, have not yet been identified and the quantities and dynamics of most compounds are largely unknown. Moreover, organisms such as yeasts and vertebrates differ in their main compounds. So tracking down compounds does not seem a promising route to an understanding of the similarity in growth curves. A better route would seem to be the use of the concept of energy, meaning something like 'the ability to do work', which primarily consists of driving chemical reactions against the direction of their thermodynamic decay. The term was first proposed by Thomas Young in 1807, according to Blaxter [73]. Energy is stored in a collection of organic compounds. So a full explanation cannot do without mass fluxes, as I will explain on {41}.

The general idea is that at least some phenomena can be understood on the assumption that elementary compounds are available in sufficient amounts, given a certain availability of energy. I assume, therefore, a close coupling between energy and material flows. Many aspects of this basic assumption will be discussed, and certainly examples exist where it does not make sense.

The reasoning boils down to the following: Food is conceived as material that bears energy. It is partially converted into energy upon entry into the individual (i.e. through the outer membrane in bacteria, membrane of feeding vacuole in ciliates, gut wall in animals; I am not using the term 'conversion' in the sense of nuclear physics but in a conceptual sense). Energy can be reconverted into material constituting the individual. These conversions come with an overhead cost to be paid, and rules for conversion can be derived, while the first and second laws of thermodynamics are observed. The material aspect of energy will be discussed, of course, but it is important to realize here that there is a close link with material flows. For example, proteins in food are first decomposed into amino acids, and amino acids are polymerized to proteins again. A similar process applies to carbohydrates and lipids, which together with proteins constitute the main materials of life. The decomposition of many types of source materials into a limited number of types of central metabolites before polymerization into biomass is known as the 'funnel' concept. The rich diversity of the catabolic machinery, especially among the prokaryotes, and the poor diversity of the anabolic machinery was already recognized by Kluyver in 1926.

The role of energy in cellular metabolism, in particular the generation and use of ATP, is the main focus of bioenergetics [496]. This compound is called the energy currency of the cell. Together with NADPH, which provides reducing power, it drives the anabolic processes. Compounds involved in the decomposition processes are important for the cell in two respects: through the production of ATP from ADP, which is produced in anabolic processes, and through the production of elementary compounds as a substrate for anabolic processes [324]. The final stages of the catabolic processing of lipids, carbohydrates and proteins all make use of same cellular machinery: the Krebs cycle. To some extent, these substrates can substitute for each other with respect to the energy demands of the cell. The cell chooses between the different substrates on the basis of their availability and its need of particular substrates in anabolic processes.

After this introduction, it perhaps comes as a surprise that ATP is not the main focus in eco-energetics. This is because ATP itself does not play a leading role in energy fluxes. It has a role similar to that of money in your purse, while your bank account governs your financial status. A typical bacterial cell has about 5×10^6 ATP molecules, which is just enough for 2 seconds of biosynthetic work [431]. The mean lifetime of an ATP molecule is about 0.3 seconds [289]. The cell has to make sure that the adenylate energy charge ($\frac{1}{2}$ ADP +ATP) (AMP+ ADP+ATP)$^{-1}$ remains fairly constant (usually around 0.9, but this matter is not settled yet). It does so by coupling endergonic (energy requiring) and exergonic (energy releasing) reactions. If the energy charge is reduced, the energy yield of the reaction ATP→ADP+P declines rapidly. The situation where the energy charge as well as the concentration of AMP+ADP+ ATP remain constant relates to the concept of homeostasis, {38}. Cells keep their purses well filled, which makes the dynamics of the purse contents less interesting. ATP is part of the machinery by which energy is harvested or mobilized.

The development of the chemiosmotic theory for the molecular mechanism of ATP generation has boosted biochemical research in cellular energetics and it is now a central issue in all texts on molecular biology [507], although competing theories exist [434]. The role of membrane-bound enzymes in the synthesis and membrane-mediated transport of compounds gradually became important. The link between activity coupled to a surface area and volume-based bulk (substrate, product) is a cornerstone in the DEB theory for the uptake and use of energy. It is through the ratio between membrane surface area and cell volume, that body size exercises its influence on cellular processes.

Photoautotrophic plants form a major group of organisms where a close coupling between energy and material flows does not exist; they derive energy from light, but obtain water, carbon dioxide and nutrients independently. Moreover, they are extremely adaptable in the relative size of organs, they require specific partitioning of body size to get access to the nutrient and light uptake potential and to the maintenance costs, and the fact that they are site-bound complicates population dynamics considerably. Thus they are not considered here, but I hope to deal with them in the future. Extensive literature exists on the coupling between energy and nutrient flows in algae. This coupling is usually modelled by writing the population growth rate as a product of terms for different limiting factors. Such a presentation does not have a mechanistic counterpart at the individual level and is, therefore, incompatible with my main objective. Although this book is about chemoheterotrophs only, this does not mean that the present theory has nothing useful to say about plants. The way energy reserves are treated here is closely related to the concept of cell quota, as introduced by Droop [184]. Cell quota are intracellular pools of nutrients, such as phosphate, nitrate etc. in algae. Droop proposed a model to describe pool size in the equilibrium of chemostats. This idea has been extended by Nyholm [515,516] to include transients. This model turns out to be a special case of the DEB theory. An important difference between nutrients and energy is that maintenance, using energy, is not required for nutrients. No strict classification of resource uptakes is possible, but the focus in this book is on those situations where it is sensible to study just one commodity: energy.

Bacteria also take energy and nutrients independently and are thus left out. Chemoheterotrophic bacteria, i.e. bacteria that degrade organic compounds to gain energy, are an exception under special conditions. The conditions are that they have unlimited access to nutrients and are limited in the uptake of energy and I will include these bacteria under these conditions.

I will assume that all essential compounds can either be obtained from food, or are unrestrictedly available from the environment. Some rotifers, for example, take up vitamins directly from the environment, independent of their algal food; the assumption being that these vitamins are always available.

1.1.4 Population energetics

The strategy chosen makes it possible to deal indirectly with questions that relate to chemical conversions, see {192}, and preserves a relative simplicity which allows penetration into the population level. The idea is this: If a population consists of individuals that take up and use energy in a particular way, how will the population to which they belong behave in a given environment? If populations are tied up in food chains or webs, how will these structures change dynamically? What new phenomena play a role at the population level, when compared with the individual level?

Except for work in the tradition of mathematical demography on which modern age-structured population dynamic theory is based [125], most literature on population dynamics up to a few years ago dealt with unstructured populations, i.e. populations that can be characterized by the number of individuals only. So all individuals are treated as identical, they are merely counted. This also applies to microbiological literature, which basically deals with microbial populations and not with individual cells. This has always struck me as most unrealistic, because individuals have to develop before they can produce offspring. The impact of a neonate on food supplies is very different from that of an adult. In the chapter on population dynamics, {171}, I will show that neonates producing new neonates can dominate the dynamics of unstructured populations. This absurdity makes one wonder to what extent unstructured population models have something useful to say about real populations. Many modern views in ecology, e.g. concerning the relationship between stability and diversity, are based on models for unstructured populations. I will use arguments from energetics to structure populations, i.e. to distinguish different individuals. This, however, complicates population dynamics considerably, and the first question to be addressed is: does this increase in complexity balance the gain in realism? I know only one route to an answer: try it and see!

1.2 The art of modelling

1.2.1 Strategies

Before I start to develop a theory for energetics, I think it is important to explain my ideas about theories and models in general. It is certainly possible that you may disagree with part of what follows, and it is helpful to know exactly where the disagreement lies. The source of a disagreement is frequently at a point other than where it first became apparent. I started this chapter pointing to growth curves as an example, because they feed the hope that it is possible to build a quantitative theory that is not species-specific. My primary interest, however, is not limited to growth curves, it is less concrete. How do phenomena operating at different levels of organization relate to each other and how can these relationships be used to cross-fertilize different biological specializations?

Let me state first that I do not believe in the existence of objective science. The type of questions we pose, the type of observations we make, bear witness to our preconceptions. There is no way to get rid of them. There is nothing wrong with this, but we should be aware of it. When we look around us we actually see mirrors of our ideas. We can try to change ourselves on the basis of what we see, but we cannot do without the projections we impose on reality. Observations and statements span the full range from facts via interpretation to abstract ideas. The more abstract the idea, the more important the mirror effect. Let me give an example of something that is not very abstract. I spend a long day looking for a particular plant species. At the end of the day luck strikes, I find a specimen. Then I return home, using the same path, and shame, oh shame, this species turns out to be quite abundant. What makes matters worse, I am quite experienced in this type of activity. So, if someone maintains that he would not miss the plants, I am inclined to think that he is simply not able to criticize his own methodology. (This example is used in the hope that this book helps to develop a search image for valuable biological observations that would go unnoticed otherwise.)

I do not believe in the existence of one truth, one reality. If such a 'truth' existed, it would have so many partially overlapping aspects, that it would be impossible to grasp them all simultaneously and recognize that there is just one truth. A consequence of this point of view is that I do not accept a classification of theories into 'true' and 'false' ones. In connection with this, I regard the traditional concepts of verification and falsification as applied to theories as meaningless. I also think that theories are always idealizations, so when we look hard, it must be possible to detect differences between theory-based predictions and observations. Therefore, I have taught myself to live happily with the knowledge that, if there is only one reality and if theories can only be classified into 'true' or 'false' ones, all of them will be classified as 'false'. As it is not possible to have the concept 'a bit true', believers in one reality do not seem very practical to me. Perhaps you judge this as cynical, but I do not see myself as a cynic. Discussions suggest that colleagues with a quantitative interest are more likely to share this point of view than those with a qualitative interest.

I classify theories on the basis of their usefulness. This classification is sensitive to the specification of a purpose and to a 'state of the art'. Theories can be most useful to detect relationships between variables, but can lose their usefulness when the state of the art develops. Theories can be useful for one purpose, but totally useless for another. When theories produce predictions that deviate strongly from observations, they are likely to be classified as useless, so I do not think that this pragmatism poses a threat to science in the eyes of the apostles of Popper. Although it is satisfying to have no difference between prediction and observation, small differences do not necessarily make a theory useless. It all depends on the amount of difference and on the purpose one has. A 'realistic' description then just means that observations and descriptions do not differ much. There will always be the possibility that a well fitting description rests on arguments that will prove to be not realistic in the end. Perhaps you think that this is trivial, but I doubt it. Take for instance goodness of fit tests in statistics and how they are applied, e.g. in ecological journals. The outcome of the test itself is not instructive, for the reasons given. It would be instructive, however, to have a measure for the difference between prediction and observation that allows one to judge the usefulness of the theory. Such measures should, therefore, depend on the theory and the purposes one has; it would be a coincidence to find them in a general text on statistics.

The sequence, 'idea, hypothesis, theory, law' is commonly thought to reflect an increasing degree of reliability. I grant that some ideas have been tested more extensively than others and may be, therefore, more valuable for further developments. Since I deny the existence of a totally reliable proposition, because I do not accept the concept 'true', I can only use this sequence to reflect an increasing degree of usefulness. It is, however, hard and probably impossible to quantify this on an absolute scale, so I treat the terms in this sequence more or less as synonyms. Each idea should be judged separately on its merits.

Mathematical models are statements written in a certain language. Mathematics as a language is most useful for formulating quantitative relationships. Therefore, quantitative theories usually take the form of mathematical models. This does not imply that all models are theories. It all depends on the ideas behind the model. Ideally a model results, mathematically, from a list of assumptions. When model predictions agree with observations in a test, this supports the assumptions, i.e. it gives no reason to change them and it gives reason to use them for the time being. As explained on {14}, the amount of support such a test gives is highly sensitive to the model structure. If possible, the assumptions should be tested one by one. From a strict point of view, it would then no longer be necessary to test the model. Practice, however, teaches us to be less strict. I am inclined to identify assumptions with theories.

A statement that is frequently heard from people with a distaste for models, is: 'a model is not more than you put into it'. Done in the proper way, this is absolutely right and it is the single most important aspect of the use of models. As this book illustrates, assumptions, summarized in table 4.1, have far reaching consequences, that cannot be revealed without the use of models. Put into other words: any mathematical statement is either wrong or follows from assumptions. Few people throw mathematics away for this reason.

The problem that everything depends on everything else in biology has strong implications for models that represent theories. When y depends on x, it is usually not hard to formulate a set of assumptions, which imply a model that describes the relationship with acceptable accuracy. This also holds for a relationship between y and z. When more and more relationships are involved, the cumulative list of assumptions tends to grow and it becomes increasingly difficult to keep them consistent. This holds especially when the same variables occur in different relationships. It is sometimes far from easy to test the consistency of a set of assumptions. For example: when a sink of material and/or energy in the maintenance process is assumed for individuals, it is no longer possible to assume a constant conversion of prey biomass into predator biomass at the population level. It takes a few steps to see why; this will be explained in the section on yield, {181}.

Complexity concerning the number of variables is a major trap in model building. This trap became visible with the introduction of computers, because they removed the technical and practical limitations for the inclusion of variables. Each relationship, each parameter in a relationship comes with an uncertainty, frequently an enormous one, in biology. With considerable labour, it is usually possible to trim computer output to an acceptable fit with a given set of observations. This, however, gives minimal support for the realism of the whole, which turns simulation results into a most unreliable tool, e.g. for making predictions in other situations. A model of the energetics of individuals can easily become too complex for use in population dynamics. If it is too simple, many phenomena at the individual level will not fit in. It will be difficult then to combine realism at the individual level and coherence between levels of organization. The need for compromise, which is not typical for energetics, makes modelling an art, with subjective flavours.

The only solution to the trap of complexity is the use of nested modules. Sets of closely interacting objects are isolated from their environment and combined into a new object, a module, with simplified rules for input-output relationships. This strategy is basic to all science. A chemist does not wait for the particle physicist to finish his job, though the behaviour of the elementary particles determines the properties of atoms and molecules taken as units by the chemist. The same applies to the ecologist who does not wait for the physiologist. The existence of different specializations testifies to the relative success of the modular approach and still amazes me. The recently proposed hierarchy theory in ecology [14,520], does basically the same within that specialization.

The problems that come with defining modules are obvious, especially when they are rather abstract. The first problem is that it is always possible to group objects in different ways to form new objects which then makes them uncomparable. The problem would be easy if we could agree about the exact nature of the basic objects, but life is not that simple. The second problem with modules lies in the simplification of the input-output relationships. An approximation that works well in one circumstance can be inadequate in another. When different approximations are used for different circumstances, and this is done for several modules in a system, the behaviour of the system can easily become erratic and no longer contribute insight into the behaviour of the real thing. The principle of reduction in science relates to the attempt to explain phenomena in terms of the smallest feasible objects. I subscribe to a weaker principle: that of coherence. This aims to relate the behaviour of modules to that of their components while preserving consistency.

If we accept community ecology as a feasible science, I see two research strategies for riding this horse. The first one is to accept that species differ considerably in the way they take up and use resources. This would mean modelling the energetics of each species, stripping the model of most of its details in various ways, and then trying to determine the common features in population dynamics that these simplified models and the full model produce. I do not share the hope that different traits of individuals will indeed result in similar dynamics of populations. The second strategy, which is followed here, is to try to capture the diversity of the energetics of the different species into one model with different parameter values and build theories for the parameter values. The simplification step before the assemblage of populations into a community remains necessary.

1.2.2 Systems

The DEB theory is built on dynamic systems. The idea behind the concept of a system is simple in principle, but in practice, as in energetics, some general modelling problems arise that can best be discussed here. A system is based on the idea of state variables, which are supposed to specify completely the state of the system at a given moment. Completeness is essential. The next step is to specify how the state variables change with time as a function of a number of inputs and each other. The specification of change of state variables usually takes the form of a set of differential equations, which have parameters, i.e. constants that are considered to have some fixed value in the simplest case. Usually this specification also includes a number of outputs.

Parameters are typically constant, but sometimes the values change with time. This can be described by a function of time, which again has parameters that are now considered to be constant. Physiological rates, for instance, depend on temperature. Parameters that have the interpretation of a rate are, therefore, constant as long as the temperature does not change. If the temperature changes, so do the rate parameters which then become functions of time. As a side product of metabolism, heat is generated. In ectotherms, i.e. animals that do not heat their body to a constant high temperature, heat production is low, due to their usually low body temperature. The body temperature usually follows that of the environment, and can thus be treated as a time function. The situation is more complex in developing birds, which make the transition to the endothermic state some days after hatching when they invest energy to keep body temperature at a fixed high level. As the temperature of the hatchling is high due to breeding, metabolism is high and so is heat production as a side product. On top of this, the bird starts to invest extra energy in heating. Here, the state variables of the system interfere with the environment, but not via input; this requires that body temperature is considered as an additional state variable.

The choice of the state variables is the most crucial step in the definition of a system. It is usually a lot easier to compare and test alternative formulations for the change of state variables, than different choices of state variables. Models with different sets of state variables are hardly comparable. The variables that are easy to measure or the ones that will be used to test the model are not always the variables that should appear as state variables. An example is metabolic rate, which is measured as the respiration rate, i.e. oxygen consumption rate or carbon dioxide production rate. Metabolic rate is not chosen as a primary variable or parameter in the DEB theory. It only has the role of a derived variable, which is nonetheless important. This point will doubtlessly generate controversy. I will divide the metabolic rate into its different components, each of which follows simple rules. The sum of these components is then likely to behave in a less simple way in non-linear models. The same holds for, for example, dry weights, which I will decompose into structural biomass and reserve materials. A direct consequence of such decompositions is that experimental results that only include composite variables are difficult to interpret. For mechanistic models, it is essential to use variables that are the most natural players in the game. The relationship between these variables and those to be measured is the next problem to be solved, once the model is formulated, cf. {36}.

Thermodynamics makes a most useful distinction between intensive variables, which are independent of size, such as temperature, concentration, density, pressure, viscosity, molar volume, molar heat capacity, etc., and extensive variables, which depend on size, such as mass, heat capacity, volume. Extensive variables can sometimes be added in a meaningful way if they have the same dimension, but intensive variables can not. Concentrations, for example, can only be added when they relate to the same volume. Then they can be treated as masses, i.e. extensive variables. When the volume changes, as body volumes do, we face the basic problem that concentrations are the most natural choice for dealing with mechanisms, while we need masses, i.e. absolute values to make use of conservation laws. This is one of the reasons why one needs a bit of training in the application of the chain rule for differentiation.

1.2.3 Physical dimensions

A few remarks on physical dimensions are needed here, because a test for dimensions is such a useful tool in the process of modelling. Only a few texts deal adequately with them.

Models which do not have a match of dimensions over '=' signs are meaningless. This does not imply that models that treat dimension well are necessarily useful models. The elementary rules are simple: addition and subtraction is only meaningful if the dimensions are the same, but the addition or subtraction of variables with the same dimensions is not always meaningful. Meaning depends on interpretation. Multiplication and division of variables correspond with multiplication and division of dimensions. The simplification of the dimension, however, should be treated with care. A dimension that occurs in both the numerator and the denominator in a ratio does not cancel automatically. A handy rule of thumb is that such dimensions only cancel if the sum of the variables to which they belong can play a meaningful role in the theory. The interpretation of the variable and its role in the theory always remains attached to dimensions. So the dimension of the biomass density in the environment expressed on the basis of volume, is cubed length (of biomass) per cubed length (of environment); it is not dimensionless. This argument is sometimes quite subtle. The dimension of the total number of females a male butterfly meets during its lifetime is number (of females) per number (of males), as long as males and females are treated as different categories. If it is meaningful for the theory to express the number of males as a fraction of the total number of animals, the ratio becomes dimensionless.

The connection between a model and its interpretation gets lost if it contains transcendental functions of variables that are not dimensionless. Transcendental functions, such as logarithm, exponent, sinus, frequently occur in models. pH is an example, where a logarithm is taken of a variable with dimension number per cubed length ($\ln\{\#l^{-3}\}$). When it is used to specify environmental conditions, no problems arise, it just functions as a label. However, if it plays a quantitative role, we must ensure that the dimensions are cancelled correctly. For example, take the difference between two pH-values. This difference is dimensionless: $\text{pH}_1 - \text{pH}_2 = \ln\{\#_1 l^{-3}\} - \ln\{\#_2 l^{-3}\} = \ln\{\#_1 l^{-3} \#_2^{-1} l^3\}$. In linear multivariate models in ecology, the pH sometimes appears together with other environmental variables, such as temperature, in a weighted sum. Here dimension rules are violated and the connection between the model and its interpretation is lost.

Another example is the Arrhenius relationship, cf. {44} where the logarithm of a rate is linear in the inverse of the absolute temperature: $\ln \dot{v}(T) = \alpha - \beta T^{-1}$, where $\dot{v}$ is a rate, T the absolute temperature and α and β are regression coefficients. At first sight, this model seems to violate the dimension rule for transcendental functions. However, it can also be presented as $\dot{v}(T) = \dot{v}(\infty)\exp\{T_A T^{-1}\}$, where T_A is a parameter with dimension temperature and $\dot{v}(\infty)$ is the rate at very high temperatures. In this presentation, no dimension problem arises. So, it is not always easy to decide whether a model suffers from dimension problems.

A further example is the allometric function: $\ln y(x) = \alpha + \beta \ln x$, or $y(x) = \alpha x^\beta$, where y is some variable and x has the interpretation of body weight. At first sight, this model also seems to violate the dimension rule for transcendental functions. Huxley introduced it as a solution of the differential equation $\frac{dy}{dx} = \beta\frac{y}{x}$, cf. {252}. This equation, however, does not suffer from dimensional problems, neither does its solution $y(x) = y(x_1)(\frac{x}{x_1})^\beta$. This function has three rather than two parameters. It can be reduced to two parameters for dimensionless variables only. The crucial point is that in most body size scaling relationships, a natural reference value x_1 does not exist for weights. The choice is arbitrary. This differs from the Arrhenius example, where the choice for the unit of temperature does not influence the relationship. The allometric function violates the dimension rule for transcendental functions and should, therefore, not be used in models that represent theories. Models that violate dimension rules are bound to be purely empirical. Although this has been stated by many authors, the use of allometric functions is so widespread in energetics that it almost seems obligatory.

Many authors who use allometric functions are well aware of this problem. In discussions, they argue that they just give a description that does not pretend to be explanatory. However, they frequently use it in models that claim to be explanatory at another point. For me, this is walking in marshy country, which is why I have been explicit in my point of view on theories, where there is no useful role for allometric functions. I accept that they offer a description that is sparse in parameters and frequently accurate. I also understand the satisfaction that a log-log plot can give by the optical reduction of the frequently huge scatter. I think, however, that they are an obstacle to understanding what is going on. I will show that energetics is in no need of allometric functions and that they are at the root of many problems. One problem is that as soon as two groups of species are found to differ in the scaling parameter β, they can no longer be compared on the basis of their parameter values, because the dimensions of the parameter α differ. (The dimensions of α will even have a statistical uncertainty.) This seems most paradoxical to me, because many authors use allometric functions specifically for the purpose of comparing species.

The comparison of different systems that share common principles can be a most powerful tool in biology. I give two examples, which will be discussed later, {81,85}.

Individuals of some species, such as humans, loose their ability to grow. Cartilage tissue is replaced by bone, which makes further growth impossible. Is this the reason for the cessation of growth? This question cannot be answered by studying these species, because they stop growing and also change cartilage to bone. The answer should be 'no', I think, because it is possible to formulate a model for growth that applies to these species as well as to species that continue to grow, such as fish. Growth in mammals ceases even if they would not lose the ability to grow, and cartilage is replaced, possibly to obtain a mechanically better structure.

Another example is the egg shell of birds, which limits the diffusion of oxygen and, therefore, the development of the embryo, according to some authors [579]. A frequently used argument is the strong negative correlation between diffusion rates across the egg shell and diffusion resistance, when different egg sizes are compared, ranging from kolibries to ostriches. Again I think that the shell does not limit the development of the embryo, because it is possible to formulate a model for embryo development that applies to birds as well as to animals that do not have egg shells. The physical properties of the egg shell are well adapted to the needs of the embryo, which causes the observed correlation.

The crux of the argument is that the same model applies to different systems and that the systems can be compared on the basis of their parameters. If one or more parameters cannot be compared for different species, because they have different dimensions, a most useful type of argument is lost and this is why allometric functions spoil the argument.

I shall frequently use dimensionless variables, rather than the original ones which bear dimensions. Although this procedure is standard in the analysis of properties of models, my experience is that many biologists are annoyed by it. I will, therefore, explain briefly the rationale behind this usage.

The first reason for working with dimensionless variables is to simplify the model and get rid of as many parameters as possible. This makes the structure of the model better visible, and, of course, is essential for understanding the range of possible behaviours of the model when the parameter values are changed. The actual values of parameters are known with a usually high degree of uncertainty and they can vary a lot.

The second reason is to detect the parameter combinations that can actually be estimated on the basis of a given set of observations. In the model $y(x) = y(x_1)(\frac{x}{x_1})^\beta$, the parameters x_1, $y(x_1)$ and β cannot be estimated at the same time from a set of observations $\{x_i, y_i\}$, no matter how extensive the set is. When all parameter values are wanted, we need different, rather than more, observations. In many cases, knowledge about the values of all parameters is not necessary for the use of the model. One intriguing aspect is that it is not only impossible, but it is also not necessary to know the value of any parameter that has energy in its dimension, when the purpose is to test the energy-based model against observations that do not contain energies. On dimensional grounds, it is obvious that all estimatable parameters are composed of ratios or products of parameters that contain energy in their dimension, such that the energy dimension drops out. This holds for all models that treat physical dimensions well, irrespective of their realism. Some remarks on the ability to test a model must be made in this context.

The most rigorous way to test a model is to test all assumptions one by one. This is usually impossible, but this does not turn the model into a useless one. A weaker test can be based on consequences of the model. It always remains possible that different sets of assumptions have exactly or practically the same consequences. It is, therefore, a weaker test. One can reduce the collection of different sets of assumptions by testing more consequences in different circumstances. When a model survives all these tests, it means an increase in usefulness for further developments. If a model does not have consequences that can be tested, it is simply useless and does not deserve discussion. It is important to realize, however, that testability comes in gradations.

The third reason for working with dimensionless variables is that numerical methods for integration and parameter estimation usually involve appropriate choices of step lengths, norm values and the like. When the step length is not dimensionless, it is dependent on the units of measurement in which the parameter are expressed, which is most inconvenient.

1.2.4 Statistics

The amount of support which a successful test of a model gives depends on the model structure and has an odd relationship with the ability to estimate parameters: the better one can estimate parameters, the less support a successful test of a model gives. This is a rather technical, but vital point in work with models. I will try to make this clear with a simple model that relates y to x, and which has a few parameters, to be estimated on the basis of a given set of observations $\{x_i, y_i\}$. We make a graph of the model for a given interval of the argument x, and get a set of curves if we choose the different values of the parameters between realistic boundaries. Two extremes could occur, with all possibilities in between:

- The curves have widely different shapes, together filling the whole x, y-rectangular plot. In this case one particular curve will probably match the plotted observations, which determines the parameters in an accurate way, but a close match gives little support for the model; if the observations are totally different, another curve, with different parameter values will have a close match.

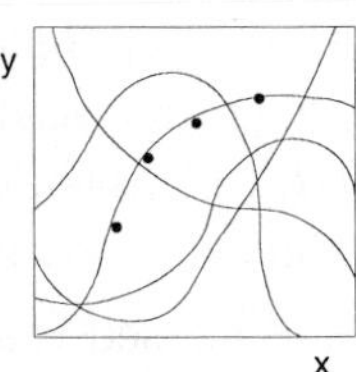

- The curves all have the same shape and are close together in the x, y-rectangular plot. If there is a close match with the observations, this gives substantial support for the model, but the parameter values are not well determined by the observations. All curves with different parameter values fit well.

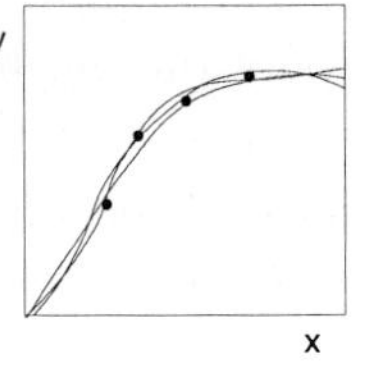

The polynomial is an example of the first category, the model for embryonic growth on {83} is an example of the second. The choice of the structure of the model is of course not free; it is dictated by the assumptions. I mention this problem to show that testability is a property of the theory and that nice statistical properties can combine with nasty theoretical ones and vice versa. It is essential to make this distinction.

The properties of parameter estimates also depend on the way the parameters are introduced. In the regression of y on x, the estimators for parameters a and b in the relationship $y = x^2(a + bx)$ are strongly negatively correlated when in the observations $\{x_i, y_i\}_{i=1}^n$ all $x_i > 0$; the mathematically totally equivalent relationship $y = x^2(c + b(x - \sum_i x_i^3 / \sum_i x_i^2))$ suffers much less from this problem. Replacement of the original parameters by appropriately chosen compound parameters can also reduce correlations between parameter estimates.

An increase in the number of parameters usually allows models to assume any shape in a graph. This is closely connected with the structural property of models just mentioned. So a successful test against a set of observations gives little support for such a model, unless the set includes many variables as well. A fair comparison of models should be based on the number of parameters per variable that is described, not the absolute number.

Observations show scatter, which reveals itself if one variable is plotted against another. It is such an intrinsic property of biological observations that deterministic models should be considered as incomplete models. Only complete models, i.e. models that describe observations which show scatter, can be tested. The standard way to complete deterministic models is to add 'measurement error'. The definition of a measurement error is that, if the measurements are repeated frequently enough, the error will disappear in the mean of these observations. Such models are called regression models: $\underline{y}_i(x_i) = f(x_i|\text{pars}) + \underline{\epsilon}_i$. They are characterized by a deterministic part, here symbolized with the function f, plus a stochastic part, $\underline{\epsilon}$. The latter term is usually assumed to follow a normal probability density, with mean 0 and a fixed variance, which is one of the parameters of the model. The interpretation of scatter as measurement error originates from physics. It is usually not realistic in biology, where many variables can be measured accurately in comparison with the amount of scatter. The observations just happen to differ from model expectations. When the scatter is large, the model is useless, despite its goodness of fit as a stochastic model. A realistic way of dealing with scatter is far from easy and usually gives rise to highly complicated models. Modellers are frequently forced to compromise between realism and mathematical over-simplicity. This further degrades the strict application of goodness of fit tests for models with unrealistic stochastic components.

For lack of better ready-to-use alternatives, the tests against observations in this book will be based mainly on the regression method. This is most unsatisfactory, but such is life. I will, however, discuss two alternatives: individuals with stochastic inputs, {121,257}, and individuals that have different parameter values, {112,210}; the motivation is that behavioural components of the feeding process are notoriously erratic, thus contributing significantly to the scatter, and individuals tend to deviate from each other in their input/output behaviour. Observations from a single individual usually have less scatter than those from different ones. The mathematics behind these alternatives is quite tedious, so I rely mainly on computer simulation studies.

I give estimates in this book for standard deviations for many parameter values that are obtained from experimental results, to indicate accuracy. I follow this standard procedure with some hesitation on two grounds. The first reason to doubt the usefulness is that the value of the standard deviation is rather sensitive to the stochastic part of the model, which might not be very realistic, as discussed. The second reason is that such standard deviations do not account for correlations between parameters. A small standard deviation for a parameter, therefore, does not necessarily mean that such a parameter is known accurately, an error that is easy to make.

Chapter 2

Individuals

From a systems analysis point of view, individuals are special because at this organization level it is relatively easy to make mass balances. This is important, because the conservation law for mass and energy is one of the few hard laws available in biology. At the cellular and at the population level it is much more difficult to measure and model mass and energy flows. It will be argued on {245} that life started as an individual in evolutionary history rather than as a particular compound, such as RNA. The individual is seen as an entity separated from the environment by physical barriers. Discussion should, therefore, start at the level of the individual.

While developing the DEB theory in the next chapter, I will present many tests against experimental data. These tests require careful interpretation of data that makes use of the material presented in this chapter, which introduces some general concepts that relate to individuals.

2.1 Input/output relationships

Any systems model relates inputs to a system with outputs of that system, as a function of its state. Although many formulations suggest that the output is the result of the state of the system and its input, this directional causality is, in fact, a matter of subjective interpretation. Input, state and output display simultaneous behaviour, without an objective, directional causality. The DEB model for uptake and use of energy in terms of input/output descriptions is neutral with respect to the interpretation in terms of 'supply' and 'demand'. With the 'supply' interpretation, I mean that the lead is in the feeding process, which offers an energy input to the individual. The available energy flows towards different destinations, more or less as water flows through a river delta. With the 'demand' interpretation, I mean that the lead is in some process using energy, such as maintenance and/or growth, which requires some energy intake. Food searching behaviour is then subjected to regulation processes in the sense that an animal eats what it needs. I think that in practice species span the whole range from 'supply' to 'demand' systems. A sea-anemone, for example, is a 'supply' type of animal. It is extremely flexible in terms of growth and shrinkage, which depend on feeding conditions. It can survive a broad range of food densities. Birds are examples of a 'demand' system and they can only survive at relatively high food densities. The range of possible growth curves is thus much more restricted.

Even in the 'supply' case, growth may be regulated carefully by hormonal control systems. Growth should not proceed at a rate beyond the possibility of mobilizing energy and elementary compounds necessary to build new structures. Models that describe growth as a result of hormonal regulation should deal with the problem of what determines hormone levels. The answer invokes the individual level. The conceptual role of hormones is linked to the similarity of growth patterns despite the diversity of regulating systems. In the DEB theory, messengers such as hormones are part of the physiological machinery by which an organism regulates its growth. Their functional aspects can only be understood from other variables and compounds.

Balance equations are extremely useful for the specification of constraints for the simultaneous behaviour of input, state and output of systems. The problem of unnoticed sources of sinks can only be circumvented by precise book-keeping. The possibility of being able to formulate balance equations will turn out to be the most useful aspect of the abstract quantity 'energy', cf. {41}. The conservation law for energy was originally formulated by von Mayer [461] in 1842. Precursors of the principle of conservation of energy go back as far as Leibnitz in 1693 [118]. This law is known today as the first law of thermodynamics. The law of conservation of mass was first described in a paper by Lavoisier in 1789.

2.2 State variables

Many models for growth have age as a state variable. Age itself has excellent properties as a measuring-tape, because it has a relatively well defined starting point (here taken to be the start of embryogenesis and not birth, i.e. the transition from the embryonic state to the juvenile one). It can also be measured accurately. Some well-studied species only thrive on abundant food supply, which results in well-defined and repeatable size-age curves. This has motivated a description of growth in terms of age, where food is considered as an environmental variable, like temperature, rather than a description in terms of input/output relationships and energy allocation rules.

One frequently applied model was proposed by Gompertz in 1825:

$$W(t) = W_\infty \left(W_0/W_\infty\right)^{\exp\{-\dot{\gamma}t\}}$$

where $W(t)$ is the weight, usually the wet weight, of an individual of age t and $\dot{\gamma}$ the Gompertz growth rate. The individual grows from weight W_0 asymptotically to weight W_∞. This is essentially an age-based model, which becomes visible from a comparison of alternative ways to express it as a differential equation: $\frac{d}{dt}\ln W = -\dot{\gamma}\ln\frac{W}{W_\infty}$ or $\frac{d^2}{dt^2}\ln W = -\dot{\gamma}\frac{d}{dt}\ln W$. The first equation states that the weight-specific growth rate decreases proportionally to the logarithm of weight as a fraction of ultimate weight. (Note that the notation $\frac{d}{dt}\ln W$ suggests a dimension problem, because it looks as if the argument of a transcendental function is not dimensionless. Its mathematically equivalent notation $W^{-1}\frac{d}{dt}W$, shows that no dimension problem exists here.) It is hard to put a mechanism behind this relationship. The second equation states that the change in weight-specific

Figure 2.1: These talking gouramis, *Trichopsis vittatus*, come from the same brood and therefore have the same age. They also grew up in the same aquarium. The size difference resulted from competition for a limited amount of food chunks, which amplified tiny initial size differences. This illustrates that age cannot serve as a satisfactory basis for the description of growth

growth rate decreases proportionally with the growth rate, which can be linked to a simple aging mechanism where the ability to grow fades according to a first order process. In the situation of abundant food, this model usually gives an acceptable fit. The problems with this model and similar ones become apparent when growth has been measured at different food availabilities.

Figure 2.1 shows two fish from the same brood, which have lived in the same 5 litre aquarium. Their huge size difference shows that age-based growth models are bound to fail. The mechanism behind the size difference in this case is the way of feeding, which involved a limited number of relatively big food chunks for the whole brood. Initially, the size differences were very small, but the largest animal always took priority over its smaller siblings, which amplified the size differences. Similar results apply to prokaryotes, which have a poor control over age-at-division at constant substrate density, but a high control over size-at-division [398].

Apart from empirical reasons for rejecting age as a state variable for the description of growth, it cannot play the role of an explanatory variable from a physical point of view. Something that proceeds with age, such as damage caused by free radicals, cf. {106}, can play that role. One will need an auxiliary model to show in detail how such a variable depends on age. One of the problems with the Gompertz model and related ones is that growth does not result from a difference between an uptake and a usage term. It is formulated as an intrinsic property of the organism. The environment can only affect growth via the parameter values.

When feeding is conceived as input of energy, size must be one of the state variables. A large individual eats much more than a small one, so it is hard to imagine a realistic model for growth that does not have size as one of the state variables; however, many quantities can be taken to measure size. Examples are volume, wet weight, dry weight, ash-free dry weight, amount of carbon or energy etc. Originally I thought that, to some extent, they were more or less exchangeable, depending on the species. Now, I am convinced that volume is the only natural choice to measure size in the context of the present theory, where surface areas play such an important role. A volume (organism), living in another volume (environment), is bound to communicate with it over a surface area. The DEB theory makes use of the interpretation of the ratio of size and surface area in terms of length. Weights remain of considerable practical interest for several reasons, the most important ones relating to the implementation of mass balances. The relationships between size measures will be discussed in the next section.

Size alone is not sufficient to describe the process of energy uptake and cannot be used with any degree of accuracy. Energy reserves should be considered as well, even in the most simple models. There are several reasons for this.

The first one is the existence of maintenance, i.e. a continuous drain of energy necessary to keep the body going. Feeding on particles, even if these particles are molecules, implies that there are periods in which no particles arrive. The capacity of a digestive system cannot realistically be made big enough to smooth out the discrete arrival process, in order to 'pay' the steady costs for maintenance. Other costs are paid as well in the absence of any food input. Spectacular examples of prolonged action without food intake are the European, North American and New Zealand eels *Anguilla*, which cease feeding at a certain moment. Their alimentary canal even degenerates, prior to the 3000 km long journey to their breeding grounds, where they spawn. The male emperor penguin *Aptenodytes forsteri* breeds its egg in Antarctic midwinter for two months and feeds the newly hatched chick with milky secretions from the stomach without access to food. The male loses some 40% of its body weight before assistance from the female arrives.

The second argument for including storage is that individuals react slowly to changes in their feeding conditions. Again, this cannot be described realistically with the digestive system as a buffer because its relaxation time is too short.

The third argument is that well-fed individuals happen to have a different (chemical) body composition than individuals in poor feeding conditions. The type of difference depends on the species, as will be discussed later. Originally I thought that, as long as food density is constant, one can do without storage. This is why the first version of the DEB model [416], did not have energy storage. However, when growth at different food densities is compared and storage levels depend on food density, one should include storage even under these simple conditions.

Size and stored energy should play a role in even the simplest model for the uptake and use of energy. Several other state variables, such as the content of the digestive system, energy density of the blood etc., will be necessary to describe the finer details of some physiological processes, but they need not play a significant role at the population level. For the purpose of the analysis of population dynamics and the contribution of aging therein, it makes sense to introduce age as an auxiliary third state variable. It also proves

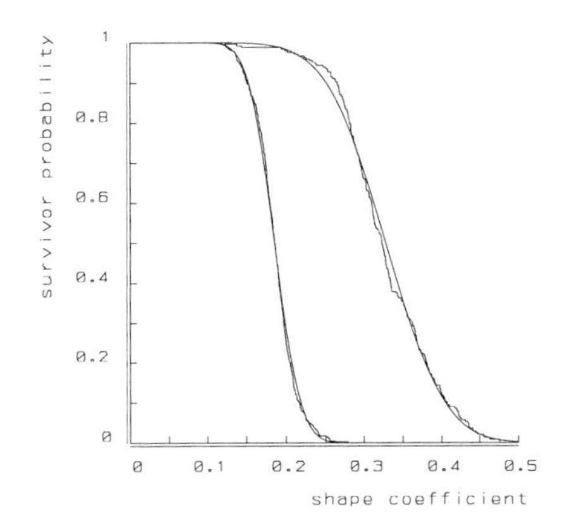

Figure 2.2: The sample survivor function (see glossary) of shape coefficients for European birds (left) and Neotropical mammals (right). The lengths include the tail for the birds, but not for mammals. Data from Bergmann and Helb [58] and Emmons and Feer [203]. The fitted survivor functions are those of the normal distribution.

necessary to distinguish life stages to catch qualitative differences in energetics, cf. {49}.

2.3 Size and shape

2.3.1 Length/surface area/volume relationships

The shape that organisms can take resists any accurate description when different species are compared. For an understanding of energetics two aspects of size and shape are relevant, as will be explained later: surface areas for acquisition processes and volumes for maintenance processes. Shape defines how these measures relate to each other. Measurements of lengths and weights are usually easy to obtain in a non-destructive way, so the practical problem has to be solved of how these measurements are related to surface areas and volume.

Volume is rather difficult to measure for some species. As a first crude approximation, wet weights, W_w, i.e. the weight of a living organism without adhering water, can be converted to volumes, V, by division through a fixed specific density $[d_w]$, which is close to 1 g cm^{-3}. So $W_w = [d_w]V$ or $[W_w] = [d_w]$, where $[d_w]$ is here taken to be a (fixed) parameter. If an organism does not change its shape during development, an appropriately chosen length measure, L, can be used to obtain its volume. The length is multiplied by a fixed dimensionless shape coefficient d_m and the result is raised to the third power. So $V = (d_m L)^3$. The shape coefficient, defined as volume$^{1/3}$ length^{-1}, is specific for the particular way the length measure has been chosen. Thus the inclusion or exclusion of a tail in the length of an organism results in different shape coefficients. A simple way to obtain an approximate value for the shape coefficient belonging to length measure L is on the basis of the relationship $d_m = (\frac{W_w}{[d_w]})^{1/3} L^{-1}$.

The following considerations help in getting acquainted with the shape coefficient. For a sphere of diameter L and volume $L^3\pi/6$, the shape coefficient is 0.806 with respect to the diameter. For a cube with edge L, the shape coefficient takes the value 1, with respect to this edge. The shape coefficient for a cylinder with length L and diameter L_ϕ is $(\frac{\pi}{4})^{1/3}(L/L_\phi)^{-2/3}$ with respect to the length.

Table 2.1: The means and coefficients of variation of shape coefficients of European birds and mammals and Neotropical mammals.

Taxon	source	number	mean	cv	mean	cv
			tail included		tail excluded	
European birds	[203,105]	418	0.186	0.14		
European mammals	[94]	128	0.233	0.27	0.335	0.28
Neotrop. mammals	[58]	246	0.211	0.41	0.328	0.18

The shapes of organisms can be compared in a crude way on the basis of shape coefficients. Figure 2.2 shows the distributions of shape coefficients among European birds and Neotropical mammals, they fit the normal distribution closely. Summarizing statistics are given in table 2.1, which includes European mammals as well. Some interesting conclusions can be drawn from the comparison of shape coefficients. They have an amazingly small coefficient of variation, especially in birds including sphere-like wrens and stick-like flamingos; which probably relates to constraints for flight. Mammals have somewhat larger shape coefficients than birds, because they tend to be more spherical and this possibly relates to differences in mechanics. The larger coefficient of variation indicates that the constraints are perhaps less stringent than for birds. The spherical shape is more efficient for energetics because cooling is proportional to surface area and a sphere has the smallest surface area/volume ratio, namely $6/L_\phi$. When the tail is included in the length, European mammals have somewhat larger shape coefficients than Neotropical mammals, but the difference is absent when the tail is excluded. Neotropical mammals tend to have longer tails, which is probably due to the fact that most of them are tree dwellers. The temperature differences between Europe and the Neotropics do not result in mammals in Europe being more spherical to reduce cooling.

These considerations should not obscure the practical purpose of shape coefficients: to convert shape-specific length measures to volumetric lengths, i.e. cubic roots of volumes. Each parameter that has length in its dimensions is sensitive to the way that lengths have been measured (in- or excluding extremities, etc.). As long as the comparison is made between bodies of the same shape, there is no need for concern, but as soon as different shapes are compared, it is essential to convert length to volumetric length, the rationale being that a comparison made on the basis of unit volumes of organisms is made on the basis of cells.

2.3.2 Isomorphism

Isomorphism is an important property which applies to the majority of species on earth. It refers to conservation of shape as an individual grows in size. The shape can be any shape and the comparison is only between shapes that a single individual takes during its development. Two bodies of a different size are isomorphic if it is possible to transform one body into the other by a simple geometric scaling in three dimensional space: scaling involves only multiplication, translation and rotation. This implies, as Archimedes already knew, that if two bodies have the same shape and if a particular length takes value L_1 and L_2 in the different bodies, the ratio of their surface areas is $(L_1/L_2)^2$ and that of their volumes $(L_1/L_2)^3$, irrespective of their actual shape. It is, therefore, possible to make assertions about the surface area and the volume of the body relative to some standard, on the basis of length only. One only needs to measure the surface area or volume if absolute values are required. This property will be used extensively.

The significance of the relationship between length, surface area and volume for isomorphs does not show up in the first place, in the context of practical measurement, but for the body itself.

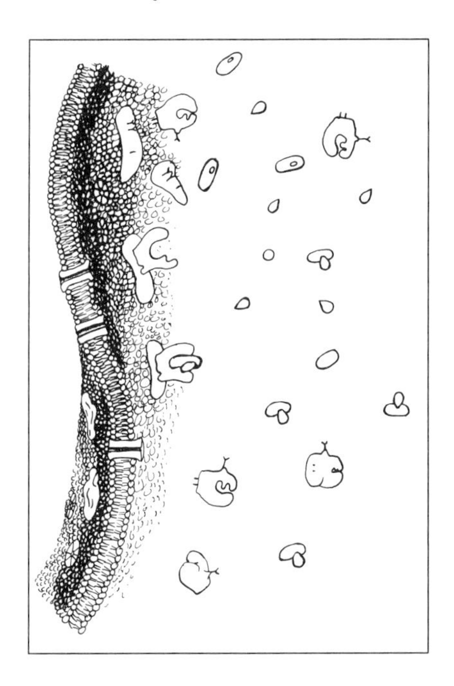

These relationships play an essential role in the communication between the extensive variable body size and intensive variables such as concentrations of compounds and reaction rates between compounds. Secreting organs 'know' their volume relative to body volume by the build up of the concentration of their products in the body. Each cell in the body 'knows' its volume by the ratio between its volume and the surface area of its membranes. One mechanism is that most enzymes only function if bound to a membrane, with their substrates and products in the cell volume as illustrated. The functional aspect is that the production of enzymes is a relatively slow process, a handicap if a particular transformation needs to be accelerated rapidly. Most enzymes can be conceived of as fluffy, free floating structures, with performance depending on the shape of the outer surface of the molecule and the electrical charge distribution over it.
If bound to a membrane, the outer shape of the enzyme changes into the shape required for the catalysis of the reaction specific to the enzyme. Membranes thus play a central role in cellular physiology [249,292,763]. Many pathways require a series of transformations and so involve a number of enzymes. The binding sites of these enzymes on the membrane are close to each other, so that the product of one reaction is not dispersed in the cytosol before being processed further. The product is just handed over to the neighbouring enzyme in a process called piping. Interplay between surface areas and volumes is basic to life, not only at the level of the individual, but also at the molecular level.

Most species are approximately isomorphic. It is not difficult to imagine the physiological significance of this. Process regulating substances in the body tend to have a short lifetime to cope with changes, so such substances have to be produced continuously. If some organ secretes at a rate proportional to its volume (i.e. number of cells), isomorphism will result in a constant concentration of the substance in the body. The way the substance exercises its influence does not have to change with changing body volume in order to obtain the same effect in isomorphs.

Exoskeletons

Isomorphism itself poses no constraints on shape, but if organisms have a permanent exoskeleton, then stringent constraints on shape exist and as most animals with a permanent exoskeleton actually meet these constraints, it is helpful to work them out. This subsection can be skipped without loss of continuity.

A grasshopper remains isomorphic and has an exoskeleton, but it grows by moulting, thus the exoskeleton is not permanent and isomorphism poses no constraints in this case. The same holds for an organism which resembles a sphere, such as a sea urchin; it cannot have a permanent (rigid) exoskeleton, because the curvature of its surface changes during growth. A cylindrical organism that grows in length only, is not isomorphic. A cylindrical organism that grows isometrically has only its caps as a permanent exoskeleton; thus this includes only the caps, i.e. two growing disks separated by a growing distance. The permanent exoskeleton generally represents a (curved) surface in three dimensional space, which can be described in a simple way using logarithmic spirals. The idea of the logarithmic spiral or *spira mirabilis* (in the plane) goes back to Descartes' studies of *Nautilus* in 1638 and to Bernoulli in 1692. The function has been used by Thompson [713], Rudwick [616,617] and Raup [584,585] to describe the shape of brachiopods, ammonites and other molluscs. I will rephrase their work in modern mathematical terms and extend the idea a bit.

A natural starting point for a description of the isomorphic permanent exoskeleton is the mouthcurve. This is a closed curve in three dimensional space that describes the 'opening' of the permanent exoskeleton (shell). This is where the skeleton synthesizing tissue is found. The development of the exoskeleton can, in most cases, be retraced in time to an infinitesimally small beginning, giving the permanent exoskeleton just the one 'opening'. This method avoids the problem of the specification of the shape of an invisibly small object. To follow the mouth curve back in its development, we introduce a dummy variable l, which has the value 0 for the present mouth curve and $-\infty$ at the start of development. By placing the start of development at the origin, the test on isomorphism of the developing exoskeleton is reduced to mapping one exoskeleton to another by multiplication and rotation only (so no translation). We can always orient the exoskeleton such that the rotation is around the x-axis. Let $\mathbf{R}(l)$ denote the rotation matrix

$$\mathbf{R}(l) = \begin{pmatrix} 1 & 0 & 0 \\ 0 & \cos l & \sin l \\ 0 & -\sin l & \cos l \end{pmatrix}$$

The closed mouthcurve $\mathbf{m}$ at an arbitrary value for the dummy variable l, can be described by

$$\mathbf{m}(l) = c^{l/2\pi}\mathbf{R}(-l)\mathbf{m}(0) \tag{2.1}$$

where c is a constant describing how fast the mouth curve reduces in size when the exoskeleton rotates over an angle 2π. If c is very large, it means that the exoskeleton does not rotate during its reduction in size. Size reduction relates in a special way to the rotation rate to ensure (self) isomorphism. It follows from the requirement that for any two points $\mathbf{m}_0$ and $\mathbf{m}_1$ on the mouthcurve, the distance $\|\mathbf{m}_1(l+h) - \mathbf{m}_0(l)\|$ depends on l in a way that does not involve the particular choice of points. The rotation matrix is here evaluated at argument $-l$, because most gastropods form left handed coils. For right handed coiling l, rather than $-l$, should be used. The mouth curve, together with the parameter c determine the shape of the exoskeleton.

An arbitrary point on the mouth curve will describe a logarithmic spiral to the origin. To visualize this, it helps to realize that a simple function such as the standard circle is given by $\mathbf{f}(l) = (\sin l, \cos l)$, where the dummy variable l takes values between $-\infty$ and ∞. A graphical representation can be obtained by plotting $\sin l$ against $\cos l$. Similarly, the logarithmic spiral with the vertex at the origin through the point $\mathbf{m}(0) \equiv (m_1, 0, m_3)$ is given by

$$\mathbf{f}(l) = c^{l/2\pi}(m_1, m_3 \sin -l, m_3 \cos -l) \tag{2.2}$$

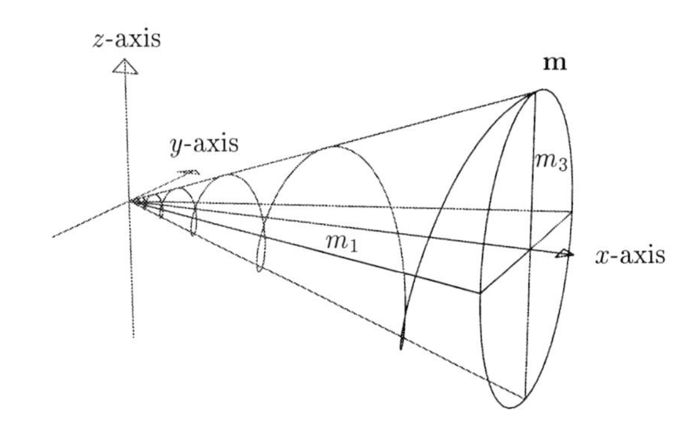

It lies on a cone around the x-axis with vertex at the origin, and tangent m_3/m_1 of the diverging angle with respect to the x-axis. For increasing l, the normalized direction vector of the spiral from the vertex, $(m_1, m_3 \sin -l, m_3 \cos -l)/\|\mathbf{m}\|$, with $\|\mathbf{m}\| = \sqrt{m_1^2 + m_3^2}$, describes a circle in the y,z-plane at x-value $m_1/\|\mathbf{m}\|$.

Until now, no explicit reference to time has been made. If the length measure of the animal follows a von Bertalanffy growth pattern, i.e. $1 - \exp\{-\dot{\gamma}t\}$ for $t \in (0, \infty)$, cf. {81}, the relationship $c^{l/2\pi} = 1 - \exp\{-\dot{\gamma}t\}$ results. So, $l = \frac{2\pi}{\ln c} \ln\{1 - \exp\{-\dot{\gamma}t\}\}$. I will argue on {81} that this is realistic when food density and temperature remain constant. In winter, when growth ceases in the temperate regions and calcification partially continues in molluscs, a thickening of the shell occurs, which is visible as a ridge ringing the shell. If the gradual transitions between the seasons can be neglected, these ridges will be found at $l = \frac{2\pi}{\ln c} \ln\{1 - \exp\{-\dot{\gamma}i\}\}$, $i = 1, 2, 3, ..$, when the unit of time is one growth season. In principle, this offers the possibility of determining the von Bertalanffy growth rate $\dot{\gamma}$ from a single shell found on the sea shore.

The mouth curve in living animals with a permanent exoskeleton frequently lies more or less in a plane, which reduces the specification of the three dimensional mouth curve to a two dimensional one, plus the specification of the plane of the mouth curve, which involves two extra parameters. The exoskeleton can always be oriented such that the plane of the mouth curve is perpendicular to the x,y-plane and the mouth opening is facing negative y-values.

Let $\mathbf{p} \equiv (p_1, p_2, 0)$ denote a point in the plane of the mouth curve, such that this plane is perpendicular to the vector $\mathbf{p}$ and $p_2 \leq 0$. (Remember that the axis of the spiral is the x-axis with the vertex at the origin so that the orientation of the exoskeleton is now completely fixed.) The mouth curve $\mathbf{n}$ in the plane is now measured using the point $\mathbf{p}$ as origin. If the mouth curve is exactly in a plane, a series of two coordinates suffice to describe the exoskeleton together with c, p_1 and p_2.

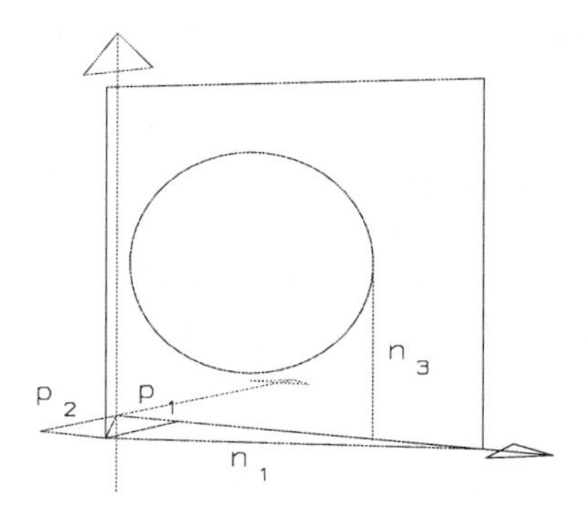

If it is not exactly in a plane, we can interpret the plane as a regression plane and still use three coordinates, where the y-values are taken to be small. The relationship between $\mathbf{n}$ measured in the coordinate system with the plane of the mouth curve as x, z-plane and $\mathbf{p}$ as origin with the original three dimensional mouth curve $\mathbf{m}$ is:

$$\mathbf{m} = \mathbf{p} + \begin{pmatrix} -p_2/||\mathbf{p}|| & -p_1/||\mathbf{p}|| & 0 \\ p_1/||\mathbf{p}|| & -p_2/||\mathbf{p}|| & 0 \\ 0 & 0 & 1 \end{pmatrix} \mathbf{n} \tag{2.3}$$

More specifically, if the mouth curve is a circle with radius r and the centre point at $(q_1, 0, q_3)$, we get $\mathbf{n}(\phi) = (q_1 + r \sin \phi, 0, q_3 + r \cos \phi)$, for an arbitrary value of ϕ between 0 and 2π. This dummy variable just scans the circle. The 6 parameters c, p_1, p_2, q_1, q_2 and r completely fix both shape and size of all isomorphic exoskeletons with circular mouth curves. If only the shape is of interest, we can choose r as the unit of distance, which leaves 5 free parameters for a full specification.

This class of morphs is too wide because it includes physically impossible shapes. The orientation of the mouth curve should be such that a mouth opening results and the shape may not 'bite' itself when walking along the spiral. This constraint can be translated into the constraint that the intersections of the exoskeleton with the x, z-plane should not intersect each other. The intersections of the mouth curve with the x, z-plane are easy to construct, given points on the mouth curve. When the point $\mathbf{m}_1 \equiv (m_1, m_2, m_3)$ on the mouth curve $\mathbf{m}(0)$ spirals its way back to the vertex, it intersects the x, z-plane at $c^{l_i/2\pi}\mathbf{R}(l_i)\mathbf{m}_1$, with $l_i = i\pi - \arctan m_2/m_3$ for $i = 0, -1, -2, \cdots$.

The distinction Raup [584] made between a generating curve and a biological one is purely arbitrary and has neither biological nor geometric meaning; Raup raises the problem that realistic values for the parameters he uses to characterize shape tend to cluster around certain values. Schindel [634] correctly pointed out that this depends on the particular way of defining parameters, and he used the intersection of (2.1) with the x, z-plane to characterize shape and showed that realistic values for parameters of this curve did not cluster. Any parameterization, however, is arbitrary unless it follows the growth mechanism. This shape of permanent exoskeletons is dealt with here to show that the shape is a result of the isomorphic constraint.

Nautilus has a fixed number of septa per revolution. This is to be expected as it makes a septum as soon as the end chamber in which it lives exceeds a given proportion of its body size. (The fact that the septa in subsequent revolutions frequently make contact implies that *Nautilus* somehow knows the number π.) These septa cause the shell to be no longer isomorphic in the strict sense, but to be what can be called periodically isomorphic, by which I mean that isomorphism no longer holds for any two values of l, but for values that differ by a certain amount. Many gastropods are sculptured at the outer surface of their shell; this sculpture is formed by the mantle curling around the shell edge. The distance from the shell edge and the height of the sculpture relates to the actual body size, the result being a shell that is also periodically isomorphic. Sculpture patterns that do not follow the mouth curve, but follow the logarithmic spirals, do not degrade isomorphism. Some shells of fully grown ammonites and gastropods have a last convolution that deviates in shape from the previous ones, showing a change in physiology related to life stage; this will be discussed later, {83,150}.

Most shapes are simple and correspond to special cases where the mouth curve lies in a plane. For $p_1 = 0$, the mouth curve lies in a plane parallel to the x, z-plane; shapes such as *Planorbis* and *Nautilus* result if the mouth curve is symmetrical around the x, y-plane. A growing sheet is obtained when $p_1 \to 0$ and $p_2 = 0$ so that the mouth curve lies in the y, z-plane. Age ridges can still show logarithmic spirals (in the plane), depending on the value of c. Figure 2.3 gives a sample of possible shapes. Although the shell of *Spirula* is internal rather than external, this does not spoil the argument.

From an abstract point of view, the closed mouth curve can secrete exoskeletons to either side and no formal restrictions exist for the parameters describing their surfaces. (The biological reality is that two mouth curves are lined up and can be moved apart to let the animal interact with the environment.) Animals such as bivalves have two logarithmic spirals sharing the same mouth-curve, one turns clockwise, one anti-clockwise. Many gastropods also have a second exoskeleton, the plane-like operculum, which is so small that it easily escapes notice. Gastropods of the genera *Berthelinia*, *Julia* and *Midorigai* have two valves, much like the bivalva. As illustrated in figure 2.4, more complex shape are possible when the mouth curve is branched.

2.3.3 Changing shapes

Huxley [345] described how certain parts of the body can change in size relative to the whole body, cf. {252}. He used allometric functions to describe this change and pointed to the problem that if some parts change in an allometric way, other parts can not. From an energetics point of view, the change in relative size of some extremities is not very important. The total volume is of interest because of maintenance processes, and certain surface areas for acquisition processes. The fact that wings, for instance, have a delayed development in birds is of little relevance to whole body growth. The basic problem is in the relationship between the size measure and the volume that has to be maintained. Reserve materials allocated to reproduction contribute substantially to the trunk length of the larvacean *Oikopleura*, but do not require maintenance. I will show on {147} how such lengths can be used to study growth and reproduction investment simultaneously.

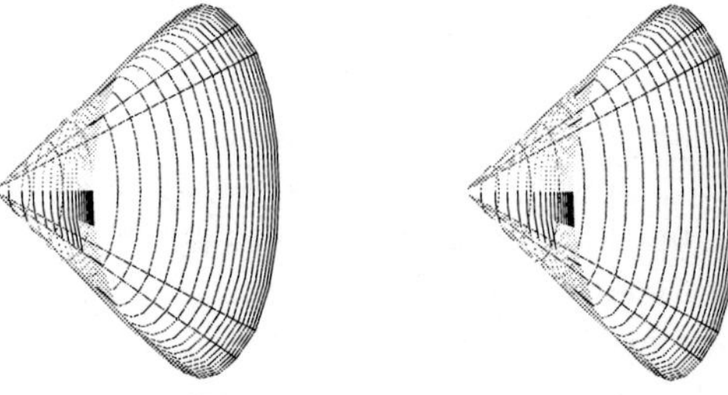

Patella, $c \to \infty$, $p_2 = 0$

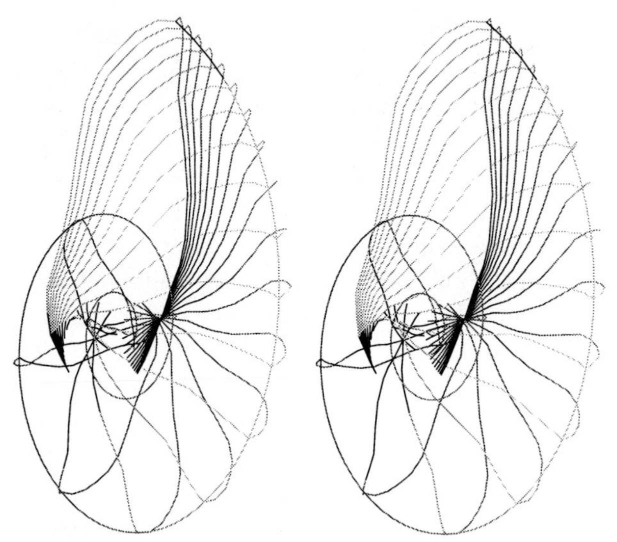

Nautilus, $c = 3$, $p_1 = 0$, $p_2 \to 0$

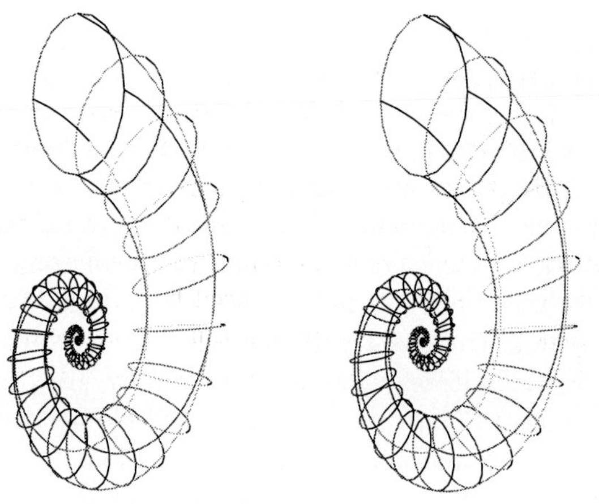

Spirula, $c = 5$, $p_1 = 0$, $p_2 \to 0$

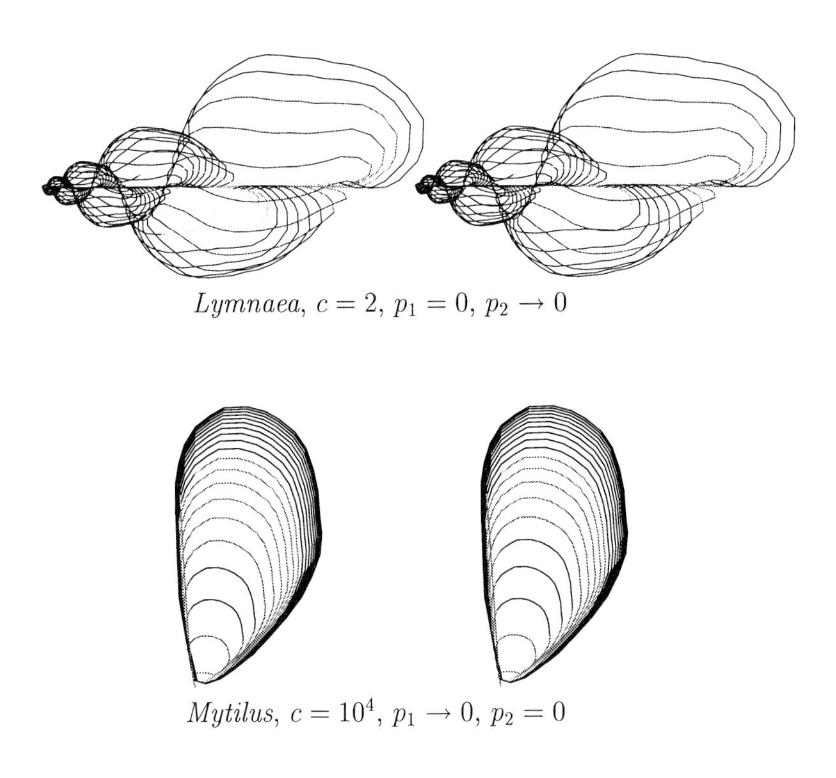

Lymnaea, $c = 2$, $p_1 = 0$, $p_2 \to 0$

Mytilus, $c = 10^4$, $p_1 \to 0$, $p_2 = 0$

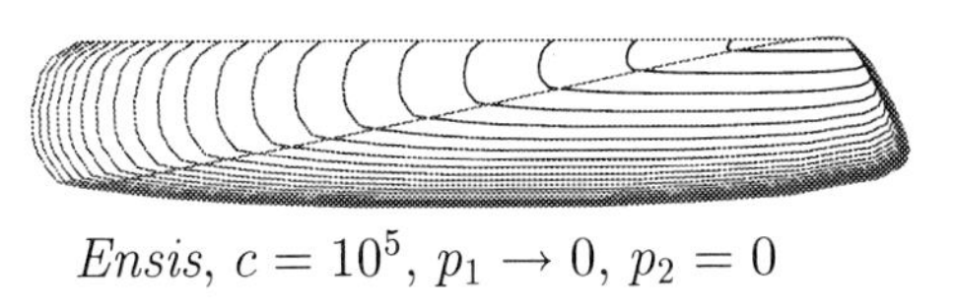

Ensis, $c = 10^5$, $p_1 \to 0$, $p_2 = 0$

Figure 2.3: A sample of possible shapes of isomorphs with permanent exoskeletons. The mouth curves are shown at equal steps for the dummy argument (*Lymnaea*, *Spirula*) or for time. Illuminate well and evenly to obtain the stereo effect. Hold your head about 50 cm from the page with the axis that connects your eyes exactly parelell to that for the figures. Do not focus at first on the page but on an imaginary point far behind the page. Try to merge both middle images of the four you should see this way. Then focus on the merged image. If this fails, try stereo glasses. If the grey is in front, rather than at the background, you are looking with your right eye to the left picture. Prevent this with a sheet of paper placed between your eyes and the page. About 10% of people actually look with one eye only and thus fail to see depth. If necessary, test this by raising one finger in front of your nose and counting the number of raised fingers that you see while focusing at infinity.

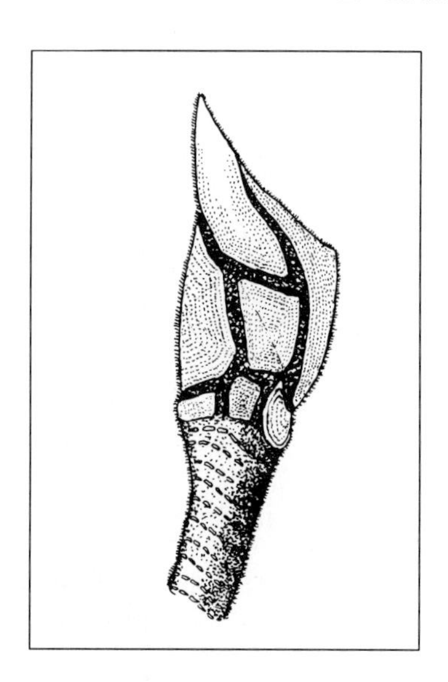

Figure 2.4: The goose barnacle (*Scalpellum scalpellum*) has an exoskeleton with a large number of components, each belonging to the family (2.1); it is an example of a branched mouth curve. Tetrahedrons provide an example of permanent exoskeletons with three branching points in the mouth curve and cubes with eight. If the (branched) mouth curve is a globular network, the exoskeleton can even resemble a sphere.

Some species such as echinoderms and some insects change shape over different life-stages. Some of these changes do not give problems because food intake is sometimes restricted to one stage only. If the shape changes considerably during development, and if volume has been chosen as the basis for size comparisons, the surface area related processes should be corrected for these changes in shape. A convenient way to do this is by means of the dimensionless shape correction function $\mathcal{M}(V)$, which stands for the actual surface area relative to the isomorphic one for a body with volume V, where a particular shape has been chosen as the reference. The derivation of this function will be illustrated for two important examples that will occur throughout the book: filamentous hyphae of fungi and rod-shaped bacteria. These organisms are both very important from a biological point of view and they serve to illustrate the important notion of 0D- and 1D-isomorphs.

If a filament can be conceived as a cylinder with variable length, and thus variable volume V, but a fixed diameter L_ϕ, its surface area equals $A(V) = 4VL_\phi^{-1}$ if the caps are excluded. Suppose now that the cylinder grows isomorphically from the start. The surface area of the isomorphic cylinder, i.e. a cylinder that has a diameter proportional to its length, is proportional to $V^{2/3}$. The constant of proportionally depends on the choice of a reference volume, say V_d. The isomorphic surface area is thus $A_d(V/V_d)^{2/3}$, where A_d denotes the surface area of a cylinder with volume V_d. So the shape correction function for filaments becomes

$$\mathcal{M}(V) = \frac{4VL_\phi^{-1}}{4V_dL_\phi^{-1}(V/V_d)^{2/3}} = (V/V_d)^{1/3} \tag{2.4}$$

It is not essential that the cross section through a filament is circular, it can be any shape, as long as it does not change during growth.
The important aspect is that growth is isomorphic in one direction. So it must be possible to the orient the body such that the direction of growth is along the x-axis, while no growth occurs along the y- and z-axes. The different body sizes can be obtained by multiplication of the x-axis by some scalar l. By doing so, both the surface area and the volume are

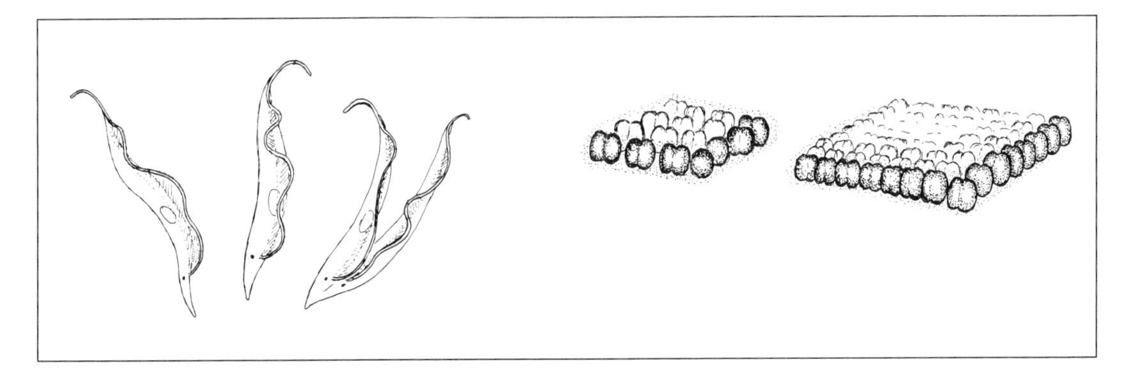

Figure 2.5: Left: The kinetoplastid *Trypanosoma* does not grow in the longitudinal direction, along which it divides. The change in shape is like a 2D-isomorph. Right: The blue-green bacterial colony *Merismopedia* is only one cell layer thick. Although this sheet also grows in two dimensions, it is a 1D-isomorph. The arrangement of the cells requires an almost perfect synchronization of the cell cycli.

multiplied by l, so $A = A_d V/V_d$, if the surface area at $V = 0$ is negligibly small. Division by the isomorphic surface area $A_d(V/V_d)^{2/3}$ results in the shape correction function for filaments. Filaments are therefore 1D-isomorphs, while the isomorphs of the previous subsection are in fact 3D-isomorphs.

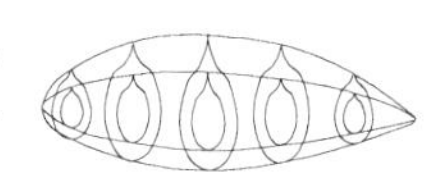

Several unicellulars divide longitudinally, such as members of several classes of the phylum *Zoomastigina* (*Opalinita*, *Retortamonadida*, *Choanomastigotes*, *Kinetoplastida*) and some filamentous bacteria (spirochetes [333]). The notorius *Trypanosoma*, which cause sleeping sickness, are among these unicellulars; see figure 2.5. Some are filamentous, but do in some respects just the opposite of the above mentioned filaments; they grow in diameter rather than length. To illustrate the concept of the shape correction function, I will derive here the shape correction function for this growth pattern, assuming that growth perpendicular to the longitudinal axis of the body is isomorphic and that no growth occurs in the longitudinal direction.

Thus it is possible to orient the body with its axis of no growth along the x-axis and multiply both the y- and z-axes by some scalar to obtain different body sizes. There are no restrictions in shape, as long as growth in the y, z-plane is isomorphic and for each x, y-value there are a limited number of z-values. The body need not be rotationally symmetric, it can taper towards its ends. Multiplying the y- and z-axis by some value l results in a multiplication of the surface area of the body by l and of the volume by l^2. (To see this, one should realize that surface area can be written as $A = \int_0^{L_d} L_c(x)\,dx$, where $L_c(x)$ denotes the circumference of the cross section through the body at x, and L_d the length of the body in the longitudinal direction. Multiplying the y- and z-axes by some value l results in a multiplication of $L_c(x)$ with l, while L_d remains untouched. Likewise, volume can be written as $V = \int_0^{L_d} A_c(x)\,dx$, where $A_c(x)$ denotes the surface area of the cross section at x, which is multiplied by l^2.) For some reference volume V_d for $l = 1$, we have thus $l = \sqrt{V/V_d}$, or $A = A_d\sqrt{V/V_d}$, where A_d denotes the surface area of a body with volume V_d. The surface area of a 3D-isomorph is $A_d(V/V_d)^{2/3}$, so that the shape correction function for 2D-isomorphs is

$$\mathcal{M}(V) = A_d\sqrt{V/V_d}(A_d(V/V_d)^{2/3})^{-1} = (V_d/V)^{1/6} \tag{2.5}$$

This correction function for 2D-isomorphs *de*creases with the cubic root of a length measure, while for 1D-isomorphs (filaments) it *in*creases with a length measure.

A subtlety of this reasoning can be illustrated by sheets, i.e. flat bodies that only grow in two dimensions, with a constant, but small, height. The archaebacterium *Methanoplanus* fits this description.

Several colonies, such as the sulphur bacterium *Thiopedia*, the blue-green bacterium *Merismopedia* and the green alga *Pediastrum* also fall into this category; see figure 2.5. How sheets grow in two dimensions does not matter: they may change wildly in shape during growth. Height must be small to neglect the contribution of the sides to the total surface area. The surface area of the sheet relates to its volume as $A(V) = 2VL_h^{-1}$, where L_h denotes the height of the sheet and the factor 2 accounts for the upper and lower surface area of the sheet. Division by the isomorphic surface area $A(V_d)(V/V_d)^{2/3}$ gives $\mathcal{M}(V) = (V/V_d)^{1/3}$, as for filaments, i.e. 1D-isomorphs. This may come as a surprise since sheets have much in common with 2D-isomorphs, where height L_h plays the same role as longitudinal length L_d. The important difference is that for $x = 0$ and $x = L_h$, there are infinitely many z-values for appropriately chosen y-values, a case that has been excluded for 2D-isomorphs. This suggests an obvious route for mixtures of 1D- and 2D-isomorphs: thick sheets that grow isomorphically in two dimensions. So the upper and lower surface areas behave as a 1D-isomorph, while the sides behave as a 2D-isomorph.

This conclusion invites a examination of the contribution of the caps in filaments. This can best be done via the introduction of biofilms, conceived as super-organisms, which resemble sheets, but in some ways they do the opposite; they grow in height rather than in the direction of the sheet, but the increase in surface area is negligibly small. A biofilm on a plane can be conceived formally as a 0D-isomorph. Its surface area is just A_d, while that of a 3D-isomorph is still $A_d(V/V_d)^{2/3}$, which leads to the shape correction function for 0D-isomorphs

$$\mathcal{M}(V) = (V_d/V)^{2/3} \tag{2.6}$$

Films relate to sheets as 1D-isomorphs relate to 2D-isomorphs. Films frequently occur in combination with 1D-isomorphs, as will be shown.

Cooper [137] argues that at constant substrate density *Escherichia* grows in length only, while the diameter-length ratio at division remains constant for different substrate densities.

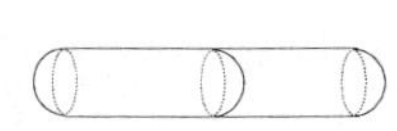

(For the use of the term 'density', see the remark under (3.3).) This mode of growth and division is typical for most rod-shaped bacteria, and most bacteria are rod-shaped. Shape and volume at division, at a given substrate density, are selected as a reference. The cell then has, say, length L_d, diameter δL_d, surface area A_d and volume V_d. The fraction δ is known as the aspect ratio of a cylinder. The index d will be used to indicate length, surface area and volume at division at a given substrate density. The shape of the rod shaped bacterium is idealized by a cylinder with hemispheres at both ends and, in contrast to a filament, the caps are now included. Length at division is $L_d = \left(\frac{4V_d}{(1-\delta/3)\delta^2\pi}\right)^{1/3}$, making length $L = \frac{\delta}{3}\left(\frac{4V_d}{(1-\delta/3)\delta^2\pi}\right)^{1/3} + \frac{4V}{\pi\delta^2}\left(\frac{(1-\delta/3)\delta^2\pi}{4V_d}\right)^{2/3}$. Surface area becomes $A = L_d^2\frac{\pi}{3}\delta^2 + \frac{4V}{\delta L_d}$. The surface area of an isomorphically growing rod equals $A_d(V/V_d)^{2/3}$. The shape correction function is the ratio of these surface area's. If volume, rather than length, is used as an argument the sought, dimensionless, correction function becomes

$$\mathcal{M}(V) = \frac{\delta}{3}\left(\frac{V_d}{V}\right)^{2/3} + \left(1 - \frac{\delta}{3}\right)\left(\frac{V}{V_d}\right)^{1/3} \tag{2.7}$$

When $\delta = 0.6$, the shape just after division is a sphere as in cocci, so this is the upper boundary for aspect ratio δ. This value is obtained by equating the volume of a cylinder to that of a sphere with the same diameter. When $\delta \to 0$, the shape tends to a filament.

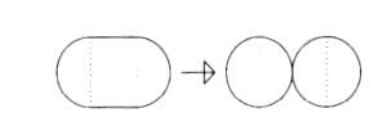

The shape correction function for rods can now be conceived as a weighted sum of those for a 0D- and a 1D-isomorph, with a simple geometric interpretation of the weight coefficients. A cylinder with blunt caps has the shape correction function

$$\mathcal{M}(V) = \frac{\delta}{\delta + 2}\left(\frac{V_d}{V}\right)^{2/3} + \frac{2}{\delta + 2}\left(\frac{V}{V_d}\right)^{1/3} \tag{2.8}$$

which is again a weighted sum of correction functions for 0D- and 1D-isomorphs. For the aspect ratio $\delta \to \infty$, the shape can become arbitrary close to a 0D-isomorph. The exact geometry of the caps is thus of less importance for surface area/volume relationships. Rods are examples of static mixtures of a 0D- and a 1D-isomorph, i.e. the weight coefficients do not depend on volume. Crusts are examples of dynamic mixtures and will be discussed on {145}.

The table right summarizes the shape correction functions for isomorphs of different dimensions. The power of the scaled volumes has an odd relationship with the dimension of isomorphy. Mixtures of 1D- and 0D- or 2D-isomorphs can resemble 3D-isomorphs, depending the weight coefficients and the range of values for the scaled volume.

Dim	$\mathcal{M}(V)$
0	$(V/V_d)^{-2/3}$
1	$(V/V_d)^{1/3}$
2	$(V/V_d)^{-1/6}$
3	$(V/V_d)^0$

2.3.4 Weight/volume relationships

In the discussion about shape coefficients, {21}, the crude relationship $W_w = [d_w]V$ was used to relate wet weight to structural biovolume. This mapping in fact assumes homeostasis, see {38}, without a decomposition of the organism into a structural and a storage component. Almost all of the literature is based on this relationship or the similar one for dry weights: $W_d = [d_d]V$.

For some purposes in energetics, such relationships between volumes and weights are far too crude. One needs a more refined definition of size to distinguish structural body volume from energy reserves. The necessity of making this distinction originates, among other things, from the quantification of metabolic costs. These costs are not paid for reserve materials; this is most obvious for freshly laid bird eggs. Such eggs are composed almost entirely of reserve materials and use practically no oxygen, as will be discussed later, {84,103}.

Some convenient size measures, such as weight, suffer more from the contribution of reserves than others. For example, energy allocated to reproduction, but temporarily stored in a buffer, will contribute to weight, but not to structural body volume. Energy reserves replace water in many aquatic species [553,788], but in the human species, for instance, energy reserves are often (painfully) visible. Energy reserves generally contribute more to dry weight than to wet weight [248]. While wet weight is usually easier to measure and can be obtained in a non-destructive way, dry weight has a closer link to chemical composition and mass balance implementations. I will show on {192} how to separate structural body mass from reserves and determine the relative abundances of the main elements for both categories on the basis of dry weight.

The relationship between wet weight W_w and dry weight W_d and structural body volume, V, non-allocated energy reserves E, and energy reserves allocated to reproduction $E_{\dot{R}}$ is

$$W_w = [d_{wv}]V + [d_{we}](E + E_{\dot{R}})/[E_m] \tag{2.9}$$

$$W_d = [d_{dv}]V + [d_{de}](E + E_{\dot{R}})/[E_m] \tag{2.10}$$

where $[E_m]$ denotes the maximum non-allocated reserve energy density as discussed in the next chapter. Its occurrence here is just to obtain the dimension weight volume^{-1} for the density $[d_{we}]$ and it is part of the tactic to avoid measurement of energies if not strictly required. If food is *ad libitum*, the energy reserve E will be found to evolve to $[E_m]V$ in the DEB theory, so that the energy reserves will then contribute $[d_{we}]V$ to wet weight. Under this condition weight is thus proportional to volume, apart from the possible contribution of reserves allocated to reproduction. If energy reserves replace water and the specific density of the energy reserves equals that of water, we have $[d_{we}] = 0$. If their specific density is less than that of water because of a high lipid content, for instance, $[d_{we}]$ can be negative if the reserves still replace water. The conversion coefficients $[d_{**}]$ have fixed values, due to homeostasis for the structural biomass and the reserves, see {38}.

Although this relationship between weight and structural biovolume is more accurate than a mere proportionality, it is by no means 'exact' and it depends again on the species. The gut contents of earthworms, shell of molluscs, exoskeleton of crustaceans and calcareous skeleton of corals do not require maintenance and for this reason they should be excluded from biovolume and weight. In the finer details, all species pose specific problems for the interpretation of size measurement. The contribution of inorganic salts to the dry weight of small marine invertebrates is frequently substantial. Figure 2.6 illustrates the interpretation problem in the measurement of ash-free dry weight in relation to length in

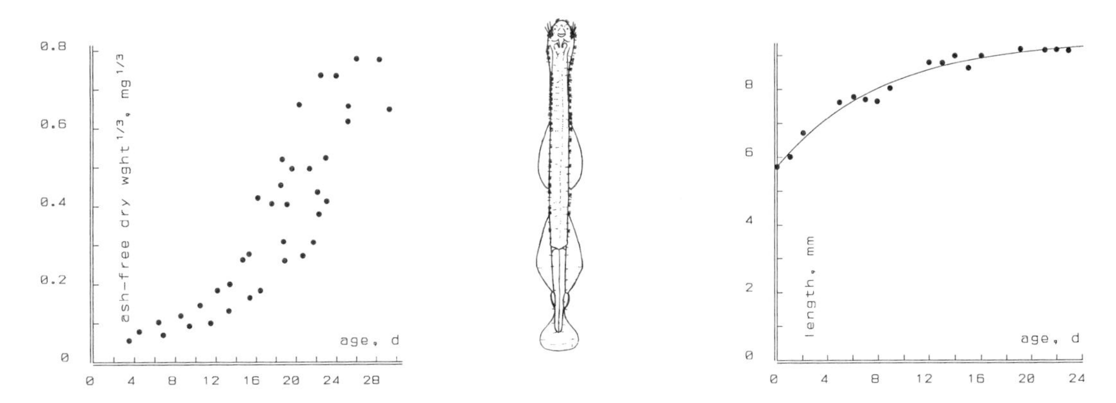

Figure 2.6: The ash-free dry weight and the length of the cheatognat *Sagitta hispida*. Data from Reeve [589,588]. The curve through the lengths is $L(t) = L_\infty - (L_\infty - L_0)\exp\{-\dot{\gamma}t\}$.

cheatognats. Length measurements follow the expected growth pattern closely at abundant food, while the description of weight seems to require an *ad hoc* reasoning. Although quickly said, this is an important argument in the use of measurements within a theoretical context: if an explanation that is not species-specific competes with a specific one, the first explanation should be preferred if the arguments are otherwise equally convincing. Since energy reserves contribute to weight and are sensitive to feeding conditions, weights usually show much more scatter, in comparison to length measurements. This is illustrated in Figure 2.7.

The determination of the size of an embryo is complicated by the extensive system of membranes the embryo develops in order to mobilize stored energy and materials and the decrease in water content during development [766]. In some species, the embryo can be separated from 'external' yolk. As long as external yolk is abundant, the energy reserves of the embryo without that yolk, if present at all, will on the basis of DEB theory turn out to be a fixed fraction of wet and dry weight, so that the embryo volume is proportional to weight. Uncertainty about the proportionality factor will hamper the comparison of parameter values between the embryonic stage and the post-embryonic one.

The aqueous fraction of an organism is of importance in relation to the kinetics of toxicants. The aqueous weight is the difference between wet weight and dry weight, so $W_a = W_w - W_d$. It can be written as $W_a = [d_{wa}]V$, for

$$[d_{wa}] = [d_{wv}] - [d_{dv}] + ([d_{we}] - [d_{de}])\frac{E + E_{\dot{R}}}{[E_m]V} \tag{2.11}$$

The contribution of the last term, which stands for that of the reserves to the volume-water weight conversion, is probably small in most cases. The volume occupied by water is $V_a = W_a/d_a$, where d_a stands for the specific density of water, which is close to 1 g cm^{-3}. The aqueous fraction of structural body volume is thus $V_a/V = [d_{wa}]/d_a$ and typically takes values between 0.7 and 0.9.

It is possible to use variations in weight relative to some measure of length to indicate

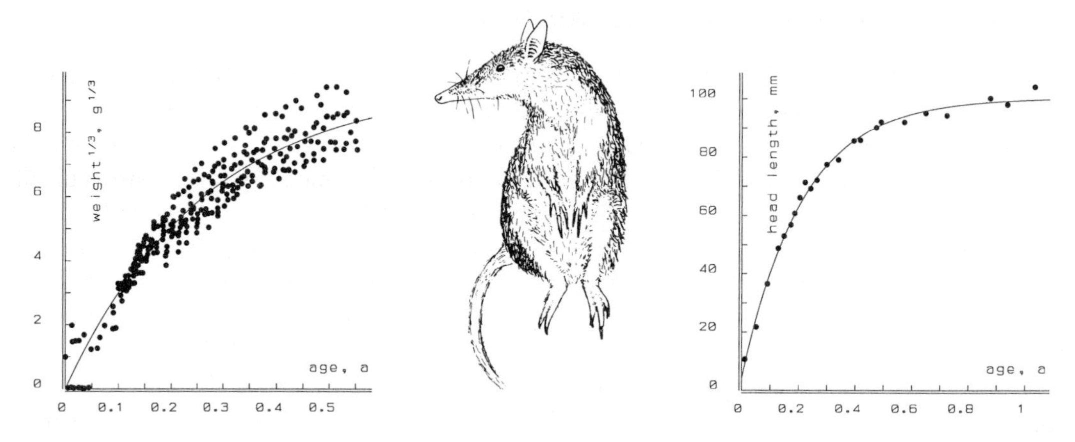

Figure 2.7: The weight to the power 1/3 and the head length of the long-nosed bandicoot *Perameles nasuta*. Data from Lyne [444]. The curves are again $L(t) = L_\infty - (L_\infty - L_0)\exp\{-\dot{\gamma}t\}$.

variations in energy reserves. This has been done for birds [552], and fish. There exists a series of coefficients to indicate the nutritional condition of fish, e.g. (weight in g)×(length in cm)$^{-1}$, which is sometimes used with a factor 1, 10 or 100. It is known as the condition factor, Hile's formula or the ponderal index [7,240,318,343].

The problems energy reserves pose for the finer details of the definition of size are not restricted to weights. They also affect the relationship between total volume and structural volume, in a way comparable to wet weights for species that do not interchange water for energy reserves. I see structural volume and energy reserves as rather abstract quantities that define the state of an organism. DEB theory specifies how the behaviour of the organism depends on its state. In addition to that, we need theories that relate these abstract quantities to things we can measure in order to substantiate the claim that DEB theory is about the living world. This subsection presents such an auxiliary theory of how weights (things we can measure) relate to the (abstract) state variables structural volume and energy reserves. This mapping rests on a key concept of DEB theory: homeostasis. An intimate relationship thus exists between the (DEB) theory that is based upon abstract quantities and the auxiliary theory that relates abstract quantities to measurements. Later, {103,192}, I will show that this relationship is even more intimate for the auxiliary theory that relates respiration measurements to (abstract) energy fluxes. The reason for discussing the relationship between weight and structural volume and energy reserves in a chapter prior to the developing the DEB theory is to stress the distinction between 'core' theory of abstract variables and auxiliary theories of relationships with measurements. If the core theory is no longer useful, the auxiliary theories should be thrown away automatically. However, it is possible to change to other auxiliary theories, without changing the core theory.

The relationship between volume, reserves and mass will be worked out further in the section on mass-energy coupling, {192}.

2.3.5 C-mole/volume relationships

The microbiological tradition is to express the relative abundances of the elements hydrogen, oxygen and nitrogen in dry biomass relative to that of carbon and to conceive the combined compound so expressed as a kind of abstract 'molecule' that can be counted and written as $CH_{n_{HW}}O_{n_{OW}}N_{n_{NW}}$. For each C-atom in dry biomass, there are typically $n_{HW} = 1.8$ H-atoms, $n_{OW} = 0.5$ O-atoms and $n_{NW} = 0.2$ N-atoms for a randomly chosen micro-organism [608]. This gives a 'molecular weight' of $w_W = 24.6$ g mol^{-1}, which can be used to convert dry weights into what are called 'C-moles'. The relative abundances of elements in biomass-derived sediments largely remain unaltered on a geological time scale, apart from the excretion of water. In the geological literature the Redfield ratio C:N:P = 105:15:1 is popular [587], or for silica bearing organisms such as diatoms, radiolarians, silico-flagelates and (some) sponges C:Si:N:P = 105:40:15:1. This literature usually excludes hydrogen and oxygen, because their abundances in biomass-derived sediments change considerably during geological time. Other bulk elements in organisms are S, Cl, Na, Mg, K and Ca, while some 14 other trace elements play an essential role, as reviewed by Fraústo da Silva and Williams [231]. The ash that remains when dry biomass is burnt away is rich in these elements. Since ash-weight is typically some 5% of dry weight, I will here include the first four most abundant elements only, but the inclusion of more elements is straightforward. As stated before, some taxa require special attention on this point.

As for weight densities, the chemical composition of biomass cannot be taken constant for most purposes in energetics, {192}. If a 'molecule' of structural biomass is denoted by $CH_{n_{HV}}O_{n_{OV}}N_{n_{NV}}$ and of energy reserves by $CH_{n_{HE}}O_{n_{OE}}N_{n_{NE}}$, the relative abundances in dry biomass are for $[E] \equiv E/V$ given by

$$n_{*W} = \frac{n_{*V}[d_{mv}] + n_{*E}[d_{me}][E]/[E_m]}{[d_{mv}] + [d_{me}][E]/[E_m]} \tag{2.12}$$

where $*$ stands for H, O or N and $[d_{mv}]$ and $[d_{me}]$ denote the conversion coefficients from structural biovolume and energy volume to C-mole. These conversion coefficients have simple relationships with those from volume to dry weight, because the 'molar weight' of structural biovolume and energy reserves are given by

$$\begin{aligned} w_V &\simeq [d_{dv}]/[d_{mv}] = 12 + n_{HV} + 16n_{OV} + 14n_{NV} \quad \text{gram mol}^{-1} \\ w_E &\simeq [d_{de}]/[d_{me}] = 12 + n_{HE} + 16n_{OE} + 14n_{NE} \quad \text{gram mol}^{-1} \end{aligned}$$

since the contribution of the other elements to dry weight is negligibly small. The problem of uncovering the relative abundances n_{*V} and n_{*E} from measurements of n_{*W}, will be discussed on {192}.

As is standard in the microbiological literature, the concept of C-mole will be extended to (simple) substrates, the difference from an ordinary mole being that it always has at most 1 C-atom. For reasons of consistency of notation, substrate density X_0 will be expressed on a volume per volume basis, while $d_{mx}X_0$ gives substrate as C-mole per volume. The same strategy will be used for products that are produced by micro-organisms; $[d_{mp}]$ converts volume of product into mole of product.

2.4 Homeostasis

The compounds that cells use to drive metabolism require enzymes for their chemical transformation. Compounds that react spontaneously are excluded. In this way cells achieve full control over all transformations, because they synthesize enzymes, consisting of protein, themselves. No reaction runs without the assistance of enzymes. The properties of enzymes depend on their micro-environment. So homeostasis, i.e. a constant chemical composition, is essential for full control. Changes in the environment in terms of resource availability, both spatial and temporal, require the formation of reserve pools to ensure a continuous supply of essential compounds for metabolism. This implies a deviation from homeostasis. The cells solution to this problem is to make use of polymers that are not soluble. In this way these reserves do not change the osmotic value. In many cases cells encapsulate the polymers in membranes, to reduce interference even further, at the same time increasing access, as many cellular activities are membrane bound.

Reserve materials can be distinguished from materials of the structural biomass by a change in relative abundance if resource levels change. This defining property breaks down in case of extreme starvation when structural materials are degraded as well as when reserves are exhausted. An example of this is the break down of muscle tissue in mammals such as ourselves, which must be considered as structural material. The distinction between reserves and structural materials is meant to accommodate the fact that some materials are more mobile than others. DEB theory builds on a two-way classification and in fact assumes homeostasis implicitly for structural biomass and reserve separately, via two other assumptions. Homeostasis is assumed for the structural materials because the volume-specific energy costs for growth are assumed to be constant, as explained in the next chapter. The assumption that the energy content of reserve materials is just proportional to the amount of reserve material, without any labels relating to their composition, in fact implies the assumption of homeostasis for reserve materials as well, and because the amount of reserves can change relative to the structural materials, the chemical composition of the whole body can change. That is, it can change in a particular way. This is a consequence of the choice of energy as a state variable rather than the complete catalogue of all compounds.

Storage and structural compounds differ in the way in which they are non-permanent in organisms. Storage materials are continuously used and replenished, while structural materials, and in particular proteins, are subjected to continuous degradation and reconstruction. Most proteins (enzymes) have a fragile, tertiary structure, which results in very short mean, functional lifetimes. Energy costs for protein turnover are included in maintenance costs. The DEB model assumes no maintenance for energy reserves.

The two-way classification of compounds into permanent (structural biomass) and dynamic (reserves) will doubtlessly prove to be too simplistic on biochemical grounds. It is, however, a considerable improvement on the one-way classification, which is standard at present, in the field considered in this book. The consequences of a two-way classification for the interpretation of measurements and for the evaluation of population dynamical properties are complicated enough.

The reserve dynamics within the DEB model will work out such that homeostasis applies for the whole organism (including structural biomass and reserves) from birth to death, if food density does not change and reserves are at equilibrium. Realistic or not, any attempt to deviate from this property will soon break down with insurmountable problems of tying measurements of body size to the abstract variables structural biomass and reserves. This, of course, would degrade the testability of such a theory and so its usefulness.

2.4.1 Storage materials

Storage material can be classified into several categories; see table 2.2. These categories do not point to separate dynamics. Carbohydrates can be transformed into fats, for instance. Most compounds have a dual function as a reserve pool for both energy and elementary compounds for anabolic processes. For example, proteins stores supply energy, amino acids and nitrogen. Ribosomal RNA (rRNA) catalyzes protein synthesis. In rapidly growing cells such as those of bacteria in rich media, rRNA makes up to 80% of the dry weight, while the relative abundance in slowly growing cells is much less. For this reason, it should be included in the storage material. I will show how this point of view leads to realistic descriptions for peptide elongation rates, {250}, and growth rate related changes in the relative abundance of nitrogen, {192}. There is no requirement that storage compounds be inert.

Waxes can be transformed into fats (triglycerides) and play a role in buoyancy e.g. of zooplankton in the sea [55]. By increasing their fat/wax ratio, they can ascend to the surface layers, which offer different food types (phytoplankton), temperatures and currents. Since surface layers frequently flow in directions other than deeper ones, they can travel the earth by just changing their fat/wax ratio and stepping from one current into another. Wax ester biosynthesis may provide a mechanism for rapidly elaborating lipid stores from amino acid precursors [627].

Unsaturated lipids, which have one or more double bonds in the hydrocarbon chain, are particularly abundant in cold water species, compared with saturated lipids. This possibly represents a homeo-viscous adaptation [654].

The amount of storage materials depends on the feeding conditions in the (recent) past, cf. {72}. Storage density, i.e. the amount of storage material per unit volume of structural biomass, tends to be proportional to the volumetric length for different species, if conditions of food (substrate) abundance are compared, as explained on {218} and tested empirically on {224}. This means that the maximum storage density of bacteria is small. Under conditions of nitrogen limitation for instance, bacteria can become loaded with energy storage materials such as polyphosphate or polyhydroxybutyrate, depending on the species. This property is used in biological plastic production and phosphate removal from sewage water. Intracellular lipids can accumulate up to some 70% of the cell dry weight in oleaginous yeasts, such as *Apiotrichum* [582,785]. This property is used in the industrial production of lipids. The excess storage is due to the uncoupling of energy and mass fluxes in bacteria and these conditions have been excluded from the present analysis. Only situations of energy limitation are dealt with.

Table 2.2: Some frequently used storage materials in heterotrophs.

phosphates	
pyrophosphate	bacteria
polyphosphate	bacteria (*Azotobacter*)
carbohydrates	
β-1,3-glucans	
leucosin	*Chrysomonadida, Prymnesiida*
chrysolaminarin	*Chrysomonadida*
paramylon	*Euglenida*
α-1,4-glucans	
starch	*Cryptomonadida, Dinoflagellida, Volvocida*
glycogen	blue green bacteria, protozoa, yeasts, molluscs
amylopectin	*Eucoccidiida, Trichotomatida, Entodiniomorphida*
trehalose	fungi, yeasts
lipoids	
poly β hydroxybutyrate	bacteria
triglyceride	oleaginous yeasts, most heterotrophs
wax	sea water animals
proteins	most heterotrophs
ovalbumin	egg-white protein
casein	milk protein (mammals)
ferritin	iron storage in spleen (mammals)
cyanophycine	blue green bacteria
phyocyanin	blue green bacteria
ribosomal RNA	all organisms

Figure 2.8: Some storage deposits are really eye-catching.

2.4.2 Storage deposits

Lipids, in vertebrates, are stored in cell vacuoles in specialized adipose tissue, which occurs in rather well defined surface areas of the body. The cells themselves are part of the structural biomass, but the contents of the vacuole is part of the reserves. In molluscs specialized glycogen storage cells are found in the foot [308]. The areas for storage deposits are usually found scattered over the body and therefore appear to be an integral part of the structural body mass, unless super-abundant; see figure 2.8. The occurrence of massive deposits is usually in preparation for a poor feeding season. The rodent *Glis glis* is called the 'edible doormouse', because of its excessive lipid deposits just prior to dormancy, {131}. (Stewed in honey and wine, doormice were a gourmet meal for the ancient Romans.)

In most invertebrate groups, storage deposits do not occur in specialized tissues, but only in the cells themselves at an amount that relates to requirements. So reproductive organs tend to be rich in storage products. The mesoglea of sea anemones, for instance, has mobile cells that are rich in glycogen and lipid, called 'glycocytes', which migrate to sites of demand during gametogenesis and directly transfer the stored materials to e.g. developing oocytes [654]. Glycogen that is stored for long-term typically occurs in rosettes and for short-term in particles [322,654].

2.5 Energy

Energy fluxes through living systems are difficult to measure and even more difficult to interpret. Let me mention briefly some of the problems.

Although it is possible to measure the thermodynamic energy content of food through complete combustion, this only shows that the organism cannot gain more energy from food, since combustion is not complete. Food is degraded to a variety of elementary compounds, some of which are used for anabolism. Another problem is that of digestion efficiency. The difference between the energy content of food and faeces is just an upper boundary for the influx, because there are energy losses in the digestion process. Part of this difference is never actually used by the organism, but is used by e.g. the gut flora. Another part becomes lost by enhanced respiration coupled to digestion, especially of proteins, called 'specific dynamic action' or 'heat increment of feeding'.

Growth involves an energy investment, which is partially preserved in new biomass. On top of the energy content of the newly formed biomass, energy has been invested to give it its structure. Part of this energy is lost during growth and can be measured as dissipating heat. This heat can be considered as an overhead of the growth process. The energy that is fixed in new biomass is partly present as energy bearing compounds. Cells are highly structured objects and the information contained in their structure is not measured by bomb calometry.

Thermodynamics of irreversible or nonequilibrium processes offers a framework to pinpoint the problem, cf. [281,421] for instance. While bomb calometry measures enthalpy, Gibbs free energy is the more useful concept to quantify the energy performance of individuals. Enthalpy and Gibbs free energy are coupled via the concept of entropy: the enthalpy of a system equals its Gibbs free energy minus the entropy times the absolute temperature. This basic relationship was formulated by Gibbs in 1878. The direct quantification of entropy requires the complete specification of the biochemical machinery, which is exactly what we try to avoid. (Dörr [177], for instance, gives an entropy reduction of 0.05 eV $\simeq 5$ kJ mol^{-1} associated with the spatial fixation of one single amino acid group of a chain molecule at 25 °C.) Gibbs relationship can be used to measure entropy indirectly in simple systems such as micro-organisms growing on well defined substrates via enthalpy and free energy. Since such free energies for micro-organisms are measured at the population level, a detailed discussion is postponed till {201}. Although this discussion opens the way to determine the entropy of living systems, I did not yet attemp to obtain numerical estimates, unfortunately. Existing ideas still range from entropy values larger than that of substrate (succinic acid) [42] to very low values [434].

All these problems about the measurement and interpretation of energy hamper direct experimental testing of assumptions about energy flows. It is possible, however, to circumvent this problem to some extent in a stunningly simple way: by not measuring energies! By refraining from direct testing of assumptions about energies, one would think that theories about energy fluxes are not testable and, therefore, useless. The consequences of such assumptions for quantities that do not represent energies, however, are testable. Many testable consequences are presented and actually tested in this book. Tests on consequences of assumptions on energy are weaker tests, which becomes apparent as soon as one or more consequences are found to be not realistic enough. It can be quite a puzzle to identify which of the assumptions about energy is the least useful one. The procedure, however, allows one to include overheads in parameter values without the obligation to take the complete machinery apart for all species.

Despite the difficulties in interpretation of energy fluxes, many attempts have been made to measure them. A relatively successful method is through the measurement of respiration. One such empirical relationship is given by Brafield and Llewellyn [85] for aquatic animals:

heat loss in J = (11.16 mg O_2 cons.) + (2.62 mg CO_2 prod.) − (9.41 mg NH_3 prod.)
Blaxter [73] advises replacement of the last term by −(5.93 mg N) − (3.39 mg CH_4) for mammals and by −(1.2 mg N) for birds. The justification of these conversions to energy rests on the idea of homeostasis. This makes the relationship between energy fluxes and gas exchange to some extent species-specific. The ratio between carbon dioxide production and oxygen consumption on a molar basis is known as the respiration quotient (RQ). Complete combustion of fat gives a respiration coefficient of 0.71, starch gives 1.0 and meat protein gives 0.82 [324]. The respiration coefficient thus gives (partial) information about the compounds that are combusted. If the composition of the combusted material remains the same, so that the respiration coefficient is constant, the oxygen consumption rate is proportional to the energy used.

Von Bertalanffy [64] related the respiration rate to the rate of anabolism. I cannot follow this reasoning. At first sight, synthesis processes are reducing by nature, which makes catabolism a better candidate for seeking a relationship with respiration. In the standard static budget studies, respiration rates are identified with routine metabolic costs. Routine metabolic costs are a lump sum including the maintenance of concentration gradients across membranes, protein turnover, regulation, transport (blood circulation, muscle tonus), and an average level of movement. The Scope For Growth (SFG) concept rests on this identification. The idea behind this concept is that energy contained in faeces and the energy equivalent of respiration are subtracted from energy derived from food, the remainder being available for growth [46]. In the DEB model, where energy derived from food is added to the reserves, the most natural candidate for a relationship with respiration is the rate at which the reserves are used.

This interpretation is also not completely free of problems, even if the respiration measurements are done on animals that are not digesting at the time. Some of the energy used from the reserves is not lost, but is fixed in the structural biovolume. This introduces some double counting. However, it seems realistic to assume that this flow is small in comparison to the overheads of the anabolic processes. This is a rather crucial point in the interpretation of respiration rates. Although respiration rates are measured over short periods (typically a couple of minutes) and the actual growth of the body is absolutely negligible, the energy invested in the growth process is by no means negligibly small. Parry [531] estimates the cost of growth between 17 and 29% of the metabolism of an 'average' ectotherm population. The respiration rate includes routine metabolic costs as well as costs for growth [619]. This interpretation is, therefore, incompatible with the SFG concept. Since the DEB model does not use respiration rates as a primary variable, the interpretation problems concerning respiration rates only play a role in testing the model.

In the next chapter, I will argue that routine metabolic costs are proportional to structural biovolume, {76}, heating costs to surface area, {78}, and growth costs to volume increase, {80}. I will show that these assumptions result in a respiration rate that is a weighted sum of surface area and volume in steady state conditions for the reserves. This is, for all practical purposes, numerically indistinguishable from the well known Kleibers rule, that takes respiration to be proportional to weight to the power 0.75 or length to the power 9/4; see figure 2.9. There are three major improvements in comparison to Kleibers rule: this model does not suffer from dimensional problems, it provides an explanation

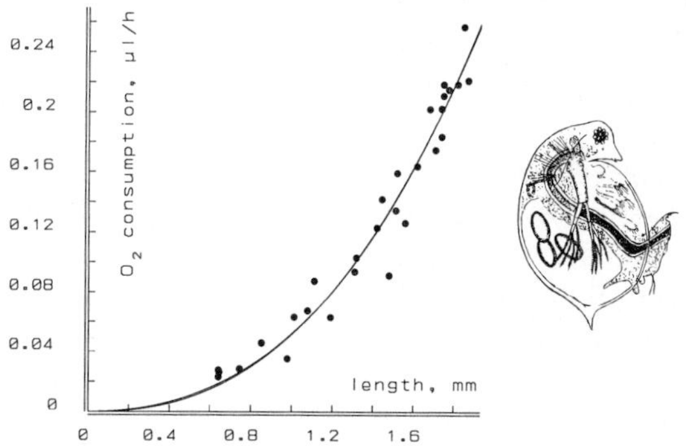

Figure 2.9: The respiration rate of *Daphnia pulex* with few eggs at 20 °C as a function of length. Data from Richman [595]. The DEB model based curve $0.0336L^2 + 0.01845L^3$ as well as the standard allometric curve $0.0516L^{2.437}$ have been plotted on top of each other, but they are so similar that this is hardly visible. Looking hard, you will notice that the line width varies a little.

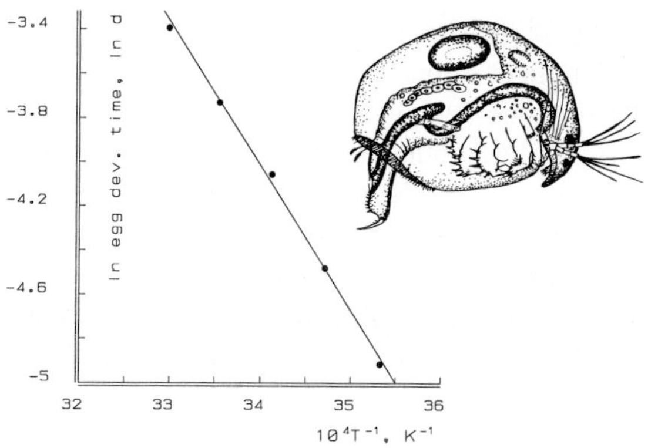

Figure 2.10: The Arrhenius plot for the development time for eggs of the waterflea *Chydorus sphaericus*, i.e. the time between egg laying and hatching. Data from Meyers [482].

rather than a description and it accommodates species that deviate from Kleibers rule, such as endotherms. This will be discussed later in somewhat more detail, {103}.

2.6 Temperature

All physiological rates depend on the temperature of the body. For a species-specific range of temperatures, the description proposed by Arrhenius in 1889, see e.g. [260], usually fits well:

$$\dot{k}(T) = \dot{k}(T_1) \exp\left\{\frac{T_A}{T_1} - \frac{T_A}{T}\right\} \tag{2.13}$$

where T is the absolute temperature (in Kelvin), T_1 a chosen reference temperature, T_A a parameter known as the Arrhenius temperature and $\dot{k}$ a (physiological) reaction rate. So, when $\ln \dot{k}$ is plotted against T^{-1}, a straight line results with slope T_A; see figure 2.10.

Arrhenius based this formulation on the van't Hoff equation for the temperature coefficient of the equilibrium constant and amounts to $\dot{k}(T) = \dot{k}(\infty) \exp\{\frac{-E_a}{RT}\}$, where $\dot{k}(\infty)$ is known as the frequency factor, R is the gas constant 8.31441 $\mathrm{J\,K^{-1}\,mol^{-1}}$, and E_a is called the activation energy. Justification rests on the collision frequency which obeys the law of mass action, i.e. it is proportional to the product of the concentrations of the reactants. The Boltzmann factor $\exp\{\frac{-E_a}{RT}\}$ stands for the fraction of molecules that manage to obtain

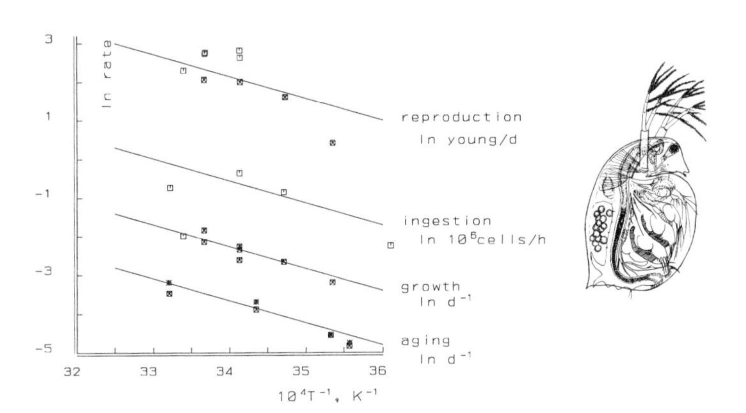

Figure 2.11: The Arrhenius plot for reproduction, ingestion, von Bertalanffy growth and Weibull aging of *Daphnia magna*; from [415]. The Arrhenius temperature is 6400 K. ◇ males, □ females. Food: the algae *Scenedesmus subspicatus* (open symbols) or *Chlorella pyrenoidosa* (filled symbols). The ingestion and reproduction rates refer to 4 mm individuals.

the critical energy E_a to react.

Eyring [260] studied the thermodynamical basis of the Arrhenius relationship in more detail. He came to the conclusion that this relationship is approximate for bimolecular reactions in the gas phase. His absolute rate theory for chemical reactions proposes a more accurate description where the reaction rate is proportional to the absolute temperature times the Boltzmann factor. This description, however, is still approximate [260,320].

The step from a single reaction between two types of particles in the gas phase to physiological rates where many compounds are involved and gas kinetics do not apply is, of course, enormous. If, however, each reaction depended in a different way on temperature, cells would have a hard time coordinating the different processes when the temperature fluctuated. The Arrhenius relationship seems to describe the effect of temperature on physiological rates with acceptable accuracy in the range of relevant temperatures. Due to the somewhat nebulous application of thermodynamics to describe how physiological rates depend on temperature, I prefer to work with the Arrhenius temperature, rather than the activation energy. I even refrain from the improvement offered by Eyring's theory, because the small correction does not balance the increase in complexity of the interpretation of the parameters for biological applications.

Figure 2.11 shows that the Arrhenius temperatures for different rates in a single species are practically the same, which again points to the regulation problem an individual would experience, if they were different. Obviously, animals cannot respire more without eating more.

In chemistry, activation energy is known to differ widely between different reactions. Processes such as the incorporation of ^{14}C-leucine into protein by membrane-bound rat-liver ribosomes have an activation energy of 180 kJ mol^{-1} in the range 8-20 °C and 67 kJ mol^{-1} in the range 22–37 °C. The difference is due to a phase transition of the membrane lipids, [723] after [10]. Many biochemical reactions seem to have an activation energy in this range [680]. This supports the idea that the value of activation energy is a constraint for functional enzymes in cells.

Table 2.3 gives Arrhenius temperatures for several species. The mean Arrhenius temperature, T_A, is somewhere between 10000 and 12500 K, which is consistent for the embryo development of 35 species [790] and the von Bertalanffy growth of 250 species [410]. The value is in the upper range of values usually applied. This is due to the fact that many experiments do not allow for an adaptation period. The problem is that many enzymes are changed a little when temperature changes. This takes time, depending on species and body size. Without an adaptation period, the performance of enzymes adapted to one temperature is measured at another temperature, which usually results in a lower Arrhenius temperature being measured.

At low temperatures, the actual rate of interest is usually lower than expected on the basis of (2.13). If the organism survives, it usually remains in a kind of resting phase, until the temperature comes up again. For many sea water species, this lower boundary is between 0 and 10 °C, but for terrestrial species it can be much higher; caterpillars of the large-blue butterfly *Maculinea rebeli*, for instance, cease growth below 14 °C [200]. The lower boundary of the temperature tolerance range frequently sets boundaries for geographical distribution. Reef building corals only occur in waters where the temperature never drops below 18 °C.

At high temperatures, the organism usually dies. At 27 °C, *Daphnia magna* grows very fast, but at 29 °C, it dies almost instantaneously. The tolerance range is sharply defined at the upper boundary. Nisbet [509] gives upper temperature limits for 46 species of protozoa, ranging from 33 to 58 °C. The width of the tolerance range depends on the species; many endotherms have an extremely small one. Thermophilic bacteria and organisms living in deep ocean thermal vents thrive at temperatures of 100 °C or more.

Sharpe [644,651] proposed a quantitative formulation for the reduction of rates at low and high temperatures, on the basis of the idea that the rate is controlled by an enzyme that has an inactive configuration at low and high temperatures. The reaction to these two inactive configurations is taken to be reversible with rates depending on temperature in the same way as the reaction that is catalized by the enzyme, however the Arrhenius temperatures might differ. This means that the reaction rate has to be multiplied by the fraction of enzyme that is in its active state, which is assumed to be at its equilibrium value. This fraction turns out to be

$$\left(1 + \exp\left\{\frac{T_{A,L}}{T} - \frac{T_{A,L}}{T_L}\right\} + \exp\left\{\frac{T_{A,H}}{T_H} - \frac{T_{A,H}}{T}\right\}\right)^{-1} \tag{2.14}$$

where T_L and T_H relate to the lower and upper boundaries of the tolerance range and $T_{A,L}$ and $T_{A,H}$ are the Arrhenius temperatures for the rate of decrease at both boundaries. All are taken to be positive and all have dimension temperature. We usually find $T_{A,H} \gg T_{A,L}$. The fraction of enzyme that is active is close to 1 between T_L and T_H for realistic values of these four temperatures.

Many extinctions are thought to be related to changes in temperature. This is the conclusion of an extensive study by Prothero, Berggren and others [571] on the change in fauna during the middle-late Eocene (40–41 Ma ago). This can most easily be understood if the ambient temperature makes excursions outside the tolerance range of a species. If a leading species in a food chain is a victim, many species that depend on it will follow. The wide variety of indirect effects of changes in temperature complicate a detailed analysis of climate-related changes in faunas. Grant and Porter [270] discuss in more detail the

Table 2.3: Arrhenius temperatures as calculated from literature data on the growth of ectothermic organisms. The values for the mouse cells are obtained from Pirt [557]. The other values were obtained using linear regressions.

species	range (°C)	T_A (K)	type of data	source
Escherichia coli	23–37	6590	pop. growth	[491]
Escherichia coli	26–37	5031	pop. growth	[350]
Escherichia coli	12–26	14388	pop. growth	[350]
Psychrophilic pseudomonad	12–30	6339	pop. growth	[350]
Psychrophilic pseudomonad	2–12	11973	pop. growth	[350]
Klebsiella aerogenes	20–40	7159	pop. growth	[722]
Aspergillus nidulans	20–37	7043	pop. growth	[724]
9 species of algae	13.5–39	6842	pop. growth	[264]
mouse tissue cells	31–38	13834	pop. growth	[751]
Nais variabilis	14–29	9380	pop. growth	[366]
Pleurobrachia pileus	5–20	10000	Bert. growth	[275]
Mya arenaria	7–15	13000	Bert. growth	[18]
Daphnia magna	10–26.5	6400	Bert. growth	[410]
Ceriodaphnia reticulata	20–26.5	6400	Bert. growth	[410]
Calliopius laeviusculus	6.5–15	11400	Bert. growth	[151]
Perna canaliculus	7–17	5530	lin. growth	[317]
Mytilus edulis	6.5–18	8460	lin. growth larvae	[679]
Cardium edule & C.glaucum	10–30	8400	lin. growth larvae	[385]
Scophthalmus maximum	8–15	15000	lin. growth larvae	[360]
25 species of fish	6–29	11190	embryonic period	[467]
Brachionus calyciflorus	15–25	7800	embryonic period	[282]
Chydorus sphaericus	10–30	6600	embryonic period	[482]
Canthocampus staphylinus	3–12	10000	embryonic period	[629]
Moraria mrazeki	7–16.2	13000	embryonic period	[629]

Table 2.4: The von Bertalanffy growth rate for the waterfleas *Ceriodaphnia reticulata* and *Daphnia magna*, reared at different temperatures in the laboratory both having abundant food. The length at birth is 0.3 and 0.8 mm respectively.

	Ceriodaphnia reticulata				*Daphnia magna*			
temp	growth rate	s.d.	ultimate length	s.d.	growth rate	s.d.	ultimate length	s.d
°C	a^{-1}	a^{-1}	mm	mm	a^{-1}	a^{-1}	mm	mm
10					15.3	1.4	4.16	0.16
15	20.4	4.0	1.14	0.11	25.9	1.3	4.27	0.06
20	49.3	3.3	1.04	0.09	38.7	2.2	4.44	0.09
24	57.3	2.6	1.06	0.01	44.5	1.8	4.51	0.06
26.5	74.1	4.4	0.95	0.02	53.3	2.2	4.29	0.06

geographic limitations for lizards set by temperature, if feeding during daytime is only possible when temperature is in the tolerance range, which leads to constraints on ectotherm energy budgets.

As a first approximation it is realistic to assume that all physiological rates are affected by temperature, so that a change in temperature amounts to a simple transformation of time. Accelerations, such as the aging acceleration that will be introduced on {107}, must thus be corrected for temperature differences by application of the squared factor, so $\ddot{k}(T) = \ddot{k}(T_1)\exp\{-2T_A(T_1^{-1} - T^{-1})\}$. It will be argued, {81}, that ultimate size results from a ratio of two rates, so it should not depend on the temperature if all rates are affected in the same way. Table 2.4 confirms this for two species of daphnids cultured under well standardized conditions and abundant food [410]. This is consistent with the observation by Beverton, see appendix to [126], that the walleye *Stizostedion vitreum* matures at 2 years at the southern end of its range in Texas and at 7 or 8 years in northern Canada, while the size at maturation of this fish is the same throughout its range.

Ultimate sizes are, however, frequently found to decrease with increasing temperature. The reason is usually that the feeding rate increases with temperature, so at higher temperatures, food supplies are likely to become limited, which reduces ultimate size. I will discuss this phenomenon in more detail in relation to the Bergmann rule, {132}. For a study of the effects of temperature on size, it is essential to test for the equality of food density. This requires special precautions.

A common way to correct for temperature differences in physiology is on the basis of Q_{10} values, known as van't Hoff coefficients. The Q_{10} is the factor that should be applied to rates for every 10 °C increase in temperature: $\dot{k}(T) = \dot{k}(T_1)Q_{10}^{(T-T_1)/10}$. The relationship with the Arrhenius temperature is thus $Q_{10} = \exp\{\frac{10T_A}{TT_1}\}$. Because the range of relevant temperatures is only from about 0 to 40 °C, the two ways to correct for temperature differences are indistinguishable for practical purposes. If the reference temperature is 20 °C, or $T_1 = 293$ K, Q_{10} varies from 3.49 to 2.98 over the full temperature range for $T_A = 10000$ K.

2.7 Life-stages

Three life-stages are to be distinguished: embryo, juvenile and adult. The triggers for transition from one stage to another and details of the different stages will be discussed later, {97}. This section serves to introduce the stages.

The first stage is the embryonic one, which is defined as a state early in the development of the individual, when no food is ingested. The embryo relies on stored energy supplies. Freshly laid eggs consist, almost entirely, of stored energy, and for all practical purposes, the initial volume of the embryo can realistically be assumed to be negligibly small. At this stage it hardly respires, i.e. it uses no oxygen and does not produce carbon dioxide. (The shells of bird eggs initially produce a little carbon dioxide [77,294].) In many species, this is a resting stage. Although the egg exchanges gas and water with the environment, it is a rather closed system. Foetal development represents an exception, where the mother provides the embryo with reserve material, such as in the placentals and some species of velvet worm *Peripatus*. Complicated intermediates between reproduction by eggs and foetuses exist in fishes [781,782], reptiles and amphibians [71,555,684]. The evolutionary transition from egg to foetus occured many times independently. From the viewpoint of energetics, foetuses are embryos because they are not taking food. The digestive system is not functional and the embryo does not have a direct impact on food supplies in an ecological sense. The crucial difference from an energetics point of view is the supply of energy to the embryo. In lecithotrophic species, nutrients are provided by the yolk of the ovum, whereas in matrotrophic species nutrients are provided by the mother as the foetus grows, not just in vitellogenesis. The fact that eggs are kept in the body (viviparity) or deposited into the environment (oviparity) is of no importance. (The difference is of importance in a wider evolutionary setting, of course.) As in eggs, a number of species of mammal have a developmental delay just after fertilization, called diapause [656].

The second stage is the juvenile one, in which food is taken but as yet resources are not allocated to the reproductive process. In some species, the developing juvenile takes a sequence of types of food or sizes of food particles. Most herbivores, for instance, initially require protein rich diets which provide nitrogen for growth, cf. {60}. Some species, such as *Oikopleura*, seem to skip the juvenile stage. It does not feed as a larva, a condition known as lecithotrophic, and it starts allocating energy to reproduction at the same moment as feeding. The larva is a morphologically defined stage, rather than an energy defined stage. If the larva feeds, it is here treated as a juvenile, if not, it is considered to be an embryo. So, the tadpole of the mouth-breeding frog *Rheobatrachus*, which develops into a frog within the stomach of the parent, should for energy purposes be classified as an embryo, because it does not feed. Parthenogenetic aphids have a spectacular mode of reproduction: embryos producing new embryos [383] cf. {171}. Since aphids are oviviparous, females carry daughters and grand daughters at the same time. The juvenile stage is lacking and the embryo stage overlaps with the adult one.

The word mammal refers to the fact that the young usually receive milk from the mother during the first stage after birth, called the baby stage. The length of the baby stage varies considerably. If adequate food is available, the guinea-pig *Cavia* can do without milk [656]. At weaning the young experience a dramatic change in diet and frequently the growth rate drops substantially. Few biochemical conversions are required for milk to become building blocks for new tissue. The baby, therefore, represents a transition stage between embryo and juvenile. The baby stage relates to the diet in the first instance, cf. {60}, and not directly to a stage in energetic development, such as embryo and juvenile. This can best be illustrated by the stoat *Mustela erminea*. Although blind for some 35–45 days, the female offspring reaches sexual maturity when only 42–56 days of age, before they are weaned. Copulation occurs whilst they are still in the nest [384,656].

Asexually propagating unicellulars take food from their environment, though they do not reproduce in a way comparable to the production of eggs or young by most multicellulars. For this reason, I treat them as juveniles in this energetics classification of stages. Although I realize that this does not fit into standard biological nomenclature, it is a logical consequence of the present delineations. I do not know of better terms to indicate energy defined stages, which points to the absence of literature dealing with the individual-based energetics of both micro-organisms and multicellulars. This book will show that both groups share enough features to try to place them into one theoretical framework. Some multicellulars, such as some annelids and triclads, propagate also by division. Some of them sport sexual reproduction as well, causing the distinction between both groups to become less sharp and the present approach perhaps more amenable. Some authors think that ciliates stem from multicellulars that have lost their cellular boundaries. This feature is standard in fungi and acellular slime molds. Some bacteria have multicellular tendencies [650]. So no sharp separation exists between unicellulars and multicellulars.

The eukaryotic cell cycle is usually partitioned into the interphase and the mitotic phase, which is here taken to be infinitesimally short. The interphase is further partitioned into the first gapphase, the synthesis phase (of DNA) and the second gapphase. Most cell components are made continuously through the interphase, so that this distinction is less relevant for energetics. The second gapphase is usually negligibly short in prokaryotes. Since the synthesis phase is initiated upon exceeding a certain cell size, size at division depends on growth conditions and affects the population growth rate. These phenomena will be discussed in detail.

Holo-metabolic insects are unique in having a pupal stage between the juvenile and adult one. It closely resembles the embryonic stage from an energetics point of view, cf. {151}. Pupae do not take food and start the synthesis of (adult) tissue from tiny imaginal disks. A comparable situation occurs in phyla such as echinoderms, bryozoans, sipunculans and echiurans, where the adult stage develops from a few undifferentiated cells of the morphologically totally different larva. Williamson [771] gave intriguing arguments for interpreting this transition, called cataclysmic metamorphosis, as evidence that the larval stage has been acquired later in phylogeny from, sometimes, unrelated taxa. In some cases, the larval tissues are resorbed, so converted to storage materials, in other cases the new stage develops independently. When *Luidia sarsi* steps off its bipinnaria larva as a tiny starfish, the relatively large larva swims actively for another 3 months, [702] in [771]. Some jelly fishes (Scyphomedusae) alternate between an asexual stage, small sessile polyps, and a sexual stage, large free swimming medusae. Many parasitic trematods push this alternation of generations into the extreme. From an energetic perspective, the sequence embryo, juvenile is followed by a new sequence, embryo, juvenile, adult, with different values for energy parameters for the two sequences. The coupling between parameter values is discussed on {217}.

The third stage is the adult one, which allocates energy to the reproduction process. The switch from the juvenile to the adult stage, puberty, is here taken to be infinitesimally short. The actual length differs from species to species and behavioural changes are also involved. The energy flow to reproduction is continuous and usually quite slow, while reproduction itself is almost instantaneous. This can be modelled by the introduction of a buffer, which is emptied or partly emptied upon reproduction. The energy flow in females is usually larger than that in males, and differs considerably from species to species.

Most animal taxa have two sexes, male and female, but even within a set of related taxa, an amazing variety of implementations can occur. Some species of mollusc and annelid for instance, are hermaphrodite, being male and female at the same time; some species of fish and shrimp for instance, are male during one part of their life and female during another part; some have very similar sexes while other species show substantial differences between male and female; see figure 2.12. The male can be bigger than the female, as in many mammals, especially sea elephants, or the reverse can occur, as in spiders. Males of some fish, rotifers and some echiurans are very tiny, compared to the female, and parasitize in or on the female or do not feed at all. The latter group combines the embryo stage with the adult one, not unlike aphids. As will be explained in the chapter on the comparison of species on {217}, differences in ultimate size reflect differences in values for energy parameters. Parameter values, however, are tied to each other, because it is not possible to grow rapidly without eating a lot (in the long run). Differences in energy budgets between sexes are here treated in the same way as differences between species.

In some species a senile stage exists, where reproduction diminishes or even ceases. This relates to the process of aging and is discussed on {105}. An argument is presented for why this stage cannot be considered as a natural next stage within the context of DEB theory.

The summary of the nomenclature used here reads:

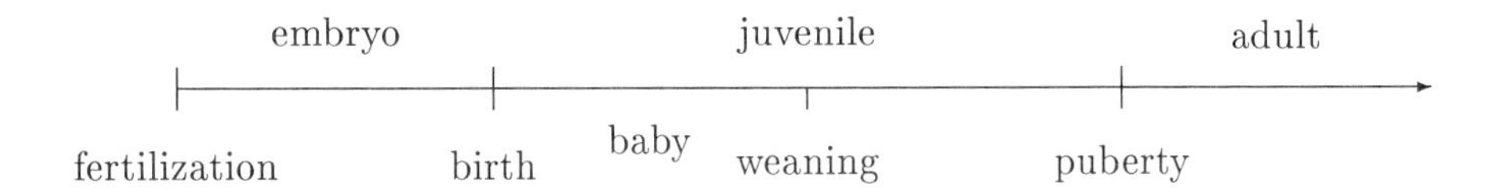

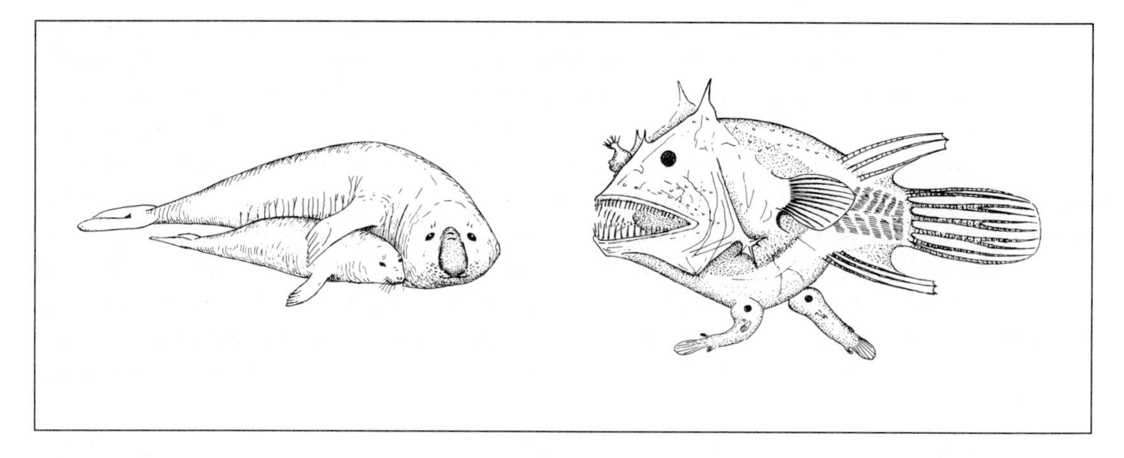

Figure 2.12: Sexual dimorphy can be extreme. The male of the southern sea elephant *Mirounga leonina* is ten times as heavy as the female, while the parasitic males of the angler fish *Haplophryne mollis* are just pustules on female's belly.

Chapter 3

Energy acquisition and use

This chapter discusses the mechanistic basis of different processes which together constitute the Dynamic Energy Budget (DEB) model. The next chapter will summarize and simplify the model and evaluate consequences at the individual level. Tests against experimental data are presented during the discussion to examine the realism of the model formulations, and also to develop a feeling for the numerical behaviour of the model elements. The next chapter presents additional tests that involve combinations of processes. The sequential nature of human language does not do justice to the many interrelationships of the processes. These interrelationships are what makes the DEB model more than just a collection of independent sub-models. I have chosen here to follow the fate of food, ending up with production processes and aging. This order fits 'supply' systems, but for 'demand' systems, another order may be more natural. The relationships between the different processes is schematically summarized in figure 3.1.

The details and logic of the energy flows will be discussed in this chapter, and a brief introduction will be given below.

Food is ingested by an animal, transformed into faeces and egested. Energy derived from food is taken up via the blood, which has a low capacity for energy but a high transportation rate. Blood exchanges energy with the storage, and delivers energy to somatic and reproductive tissues. A fixed part, κ, of the utilization rate, i.e. the energy delivered by the blood, is used for (somatic) maintenance plus growth, the rest for development and/or reproduction. The decision rule for this fork is called the κ-rule. Maintenance has priority over growth, so growth ceases if all energy available for maintenance plus growth is used for maintenance. Energy used for development in embryos and juveniles is similarly partitioned into maintenance of a certain degree of maturation and an increase in the degree of maturity. The energy spent on increasing the degree of maturity in juveniles is allocated to reproduction in adults.

Substrate is taken up and processed by unicellulars (including prokaryotes) in a way conceptually comparable to food by animals, although defecation and utilization share partly the same machinery to mobilize energy. The coupling between mass and energy fluxes, particularly relevant to micro-organisms, is discussed on {192}.

Figure 3.1: Energy fluxes through a heterotroph. The rounded boxes indicate sources or sinks. Rates 3, 7, 8, 9 and 10 also contribute a bit to heating, but this is not indicated in order to simplify the scheme.

1 ingestion (uptake)	2 defecation	3 assimilation	4 demobilization
5 mobilization	6 utilization	7 maintenance	8 maturation maintenance
9 growth investment	10 maturation	11 reproduction	12 heating (endotherms only)

multicellular

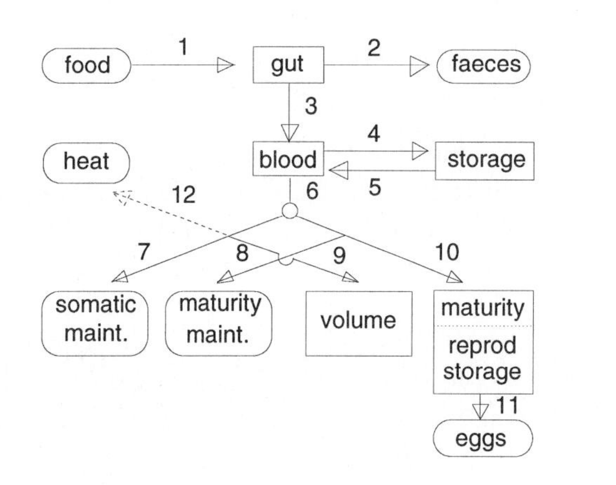

unicellular

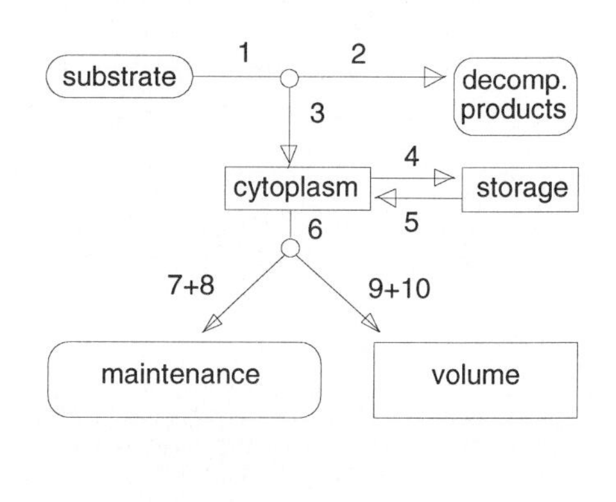

3.1 Feeding

Feeding is part of the behavioural repertoire and, therefore, notoriously erratic compared with other processes involved in energetics. The three main factors that determine feeding rates are body size, food availability and temperature. If different types of food are available, many factors determine preferences, e.g. relative abundances, size and searching patterns, which relate to experience and nutritional aspects. For some species it is sensible to express food availability per surface area of environment, for others food per volume makes more sense, and intermediates also exist. Body size of the organism and spatial heterogeneity of the environment hold the keys to the classification. Food availability for krill, which feed on algae, is best expressed in terms of biomass or biovolume per volume of water, because this links up with processes that determine filtering rates. The spatial scale at which algal densities differ is large with respect to the body size of the krill. Baleen whales, which feed on krill, are intermediate between surface and volume feeders because some dive below the top layer, where most algae and krill are located, and sweep the entire column to the surface; so it does not matter where the krill is in the column. Cows and lions are typically surface feeders and food availability is most appropriately expressed in terms of biomass per surface area.

These considerations refer to the relevance of the dimensions of the environment for feeding, be it surface or volume. The next section discusses the relevance of the size of the organism for feeding. The significance of food density returns in the section on functional response.

3.1.1 Feeding methods

The methods organisms utilize to get their meal are numerous; some sit and wait for the food to pass by, others search actively. Figure 3.2 illustrates a small sample of methods, roughly classified with respect to active movements by prey and predator. The food items can be very small with respect to the body size of the individual and rather evenly distributed over the environment, or it can occur in a few big chunks. This section mentions briefly some feeding strategies and explains why feeding rates tend to be proportional to the surface area, when a small individual is compared to a large one of the same species. (Comparisons between species will be made in a separate chapter, {217}.)

Bacteria, floating freely in water, are transported even by the smallest current, which implies that the current relative to the cell wall is effectively nil. Thus bacteria must obtain substrates through diffusion, {141}, or attach to hard surfaces (films) or each other (flocs) to profit from convection, which can be a much faster process. Some species develop more flagellae at low substrate densities, which probably reduces diffusion limitation (Dijkhuizen, pers. com.). Uptake rate is directly proportional to surface area, when the carriers that bind substrate and transport it into the cell have a constant frequency per unit surface area of the cell membrane [5,114]. *Arthrobacter* changes from a rod shape into a small coccus at low substrate densities to improve its surface area to volume ratio. Caulobacters do the same by enhancing the development of stalks under those conditions [560].

Some fungi, slime molds and bacteria glide over or through the substrate releasing enzymes and collecting elementary compounds via diffusion. Upon arrival at the cell surface, the compounds are taken up actively. The bakers' yeast *Saccharomyces cerevisiae* typically lives as a free floating, budding unicellular, but under nitrogen starvation it can switch to a filamentous multicellular phase, which can penetrate solids [329]. Many protozoans engulf particles (a process known as phagocytosis) with their outer membrane (again a surface), encapsulate them into a feeding vacuole and digest them via fusion with bodies that contain enzymes (lysosomes). Such organisms are usually also able to take up dissolved organic material, which is much easier to quantify. In giant cells, such as the Antarctic foraminiferan *Notodendrodes*, the uptake rate can be measured directly and is found to be proportional to surface area [161]. Ciliates use a specialized part of their surface for feeding, which is called the 'cytostome'; isomorphic growth here makes feeding rate proportional to surface area again.

Marine polychaetes, sea anemones, sea lilies and other species that feed on blind prey are rather apathetic. Sea lilies simply orient their arms perpendicular to an existing current (if mild) at an exposed edge of a reef and take small zooplankters by grasping them one by one with many tiny feet. The arms form a rather closed fan in mild currents, so the active

Figure 3.2: A small sample of feeding methods classified with respect to the moving activities of prey and predator.

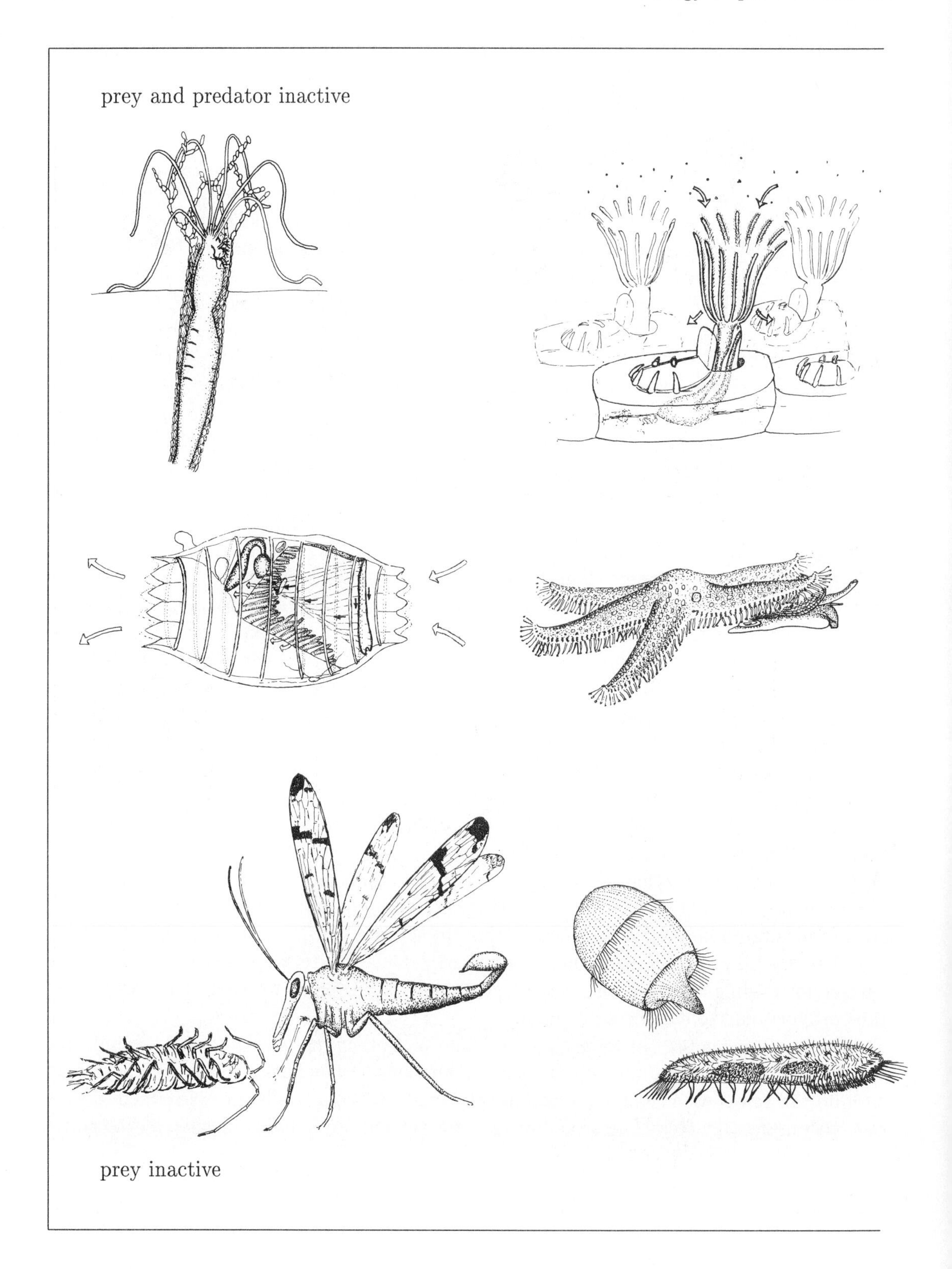
prey and predator inactive
prey inactive

predator inactive
prey and predator active

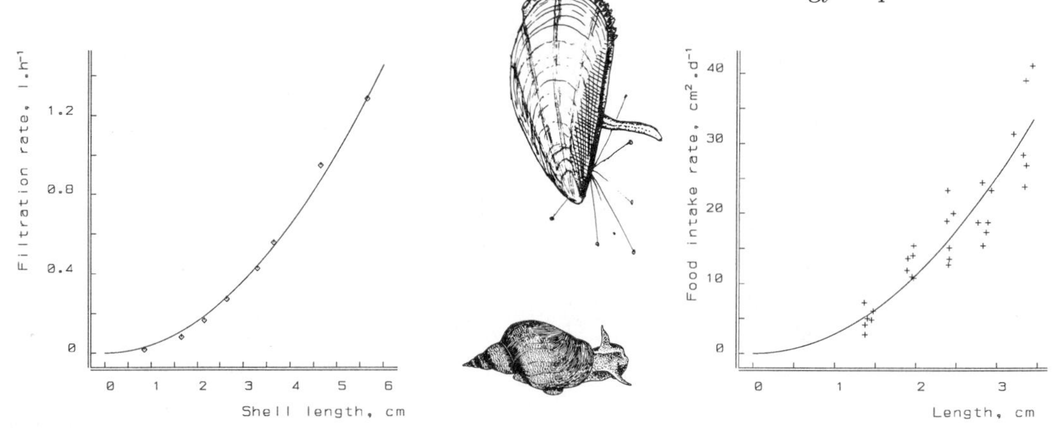

Figure 3.3: Filtration rate as function of shell length, L, of the blue mussel *Mytilus edulis* at constant food density (40×10^6 cells l^{-1} *Dunaliella marina*) at 12 °C. Data from Winter [774]. The least squares fitted curve is $\{\dot{F}\}L^2$, with $\{\dot{F}\}= 0.041$ (s.d. 6.75×10^{-4}) $\mathrm{l\,h^{-1}\,cm^{-2}}$.

Figure 3.4: Lettuce intake as a function of shell length, L, in the pond snail *Lymnaea stagnalis* at 20 °C [788]. The weighted least squares fitted curve is $\{\dot{I}\}L^2$, with $\{\dot{I}\} = 2.81$ (s.d. 0.093) $\mathrm{cm^2\,d^{-1}\,cm^{-2}}$.

area is proportional to the surface area of the animal. Sea-gooseberries stick plankters to the side branches of their two tentacles using cells which are among the most complex in the animal kingdom. Since the length of the side branches as well as the tentacles are proportional to the length of the animal, the encounter probability is proportional to a surface area.

Filter feeders, such as daphnids, copepods and larvaceans, generate water currents of a strength that is proportional to their surface area [100], because the flapping frequency of their limbs or tails is about the same for small and large individuals [565], and the current is proportional to the surface area of these extremities. (Allometric regressions of currents gave a proportionality with length to the power 1.74 [90], or 1.77 [196] in daphnids. In view of the scatter, they are in good agreement to a proportionality with squared length.) The ingestion rate is proportional the current, so to squared length. Allometric regressions of ingestion rates resulted in a proportionality with length to the power 2.2 [468], 1 [566], 2.4–3 [163], 2.4 [529] in daphnids. This wide range of values illustrates the limited degree of replicatability of these type of measurements. This is partly due to the inherent variability of the feeding process, and partly to the technical complications of measurement. Feeding rate depends on food density, as will be discussed, {63}, while most measurement methods make use of changes of food densities so that the feeding rate changes during measurement. Figure 3.11 illustrates results obtained with an advanced technique that circumvents this problem [209].

The details of the filtering process differ from group to group. Larvaceans are filterers in the strict sense, they remove the big particles first with a coarse filter and collect the small ones with a fine mesh. The collected particles are transported to the mouth in a mucous stream generated by a special organ, the endostyle. Copepods take their minute food particles out of the water, one by one with grasping movements [732]. Daphnids exploit centrifugal force and collect them in a groove. Ciliates, bryozoans, brachiopods, bivalves and ascidians generate currents not by flapping extremities, but by beating cilia on part of their surface area. The ciliated part is a fixed portion of the total surface area [227], and this again results in a filtering rate proportional to squared length; see figure 3.3.

Some surface feeding animals, such as crab spiders, trapdoor spiders, mantis, scorpion fish and frogs, lay an ambush for their prey, who will be snatched as soon as they arrive within reach, i.e. within a distance that is proportional to the length of a leg or jaw or tongue. The catching probability is proportional to the surface area of the predatory isomorphs. When aiming at prey having rather keen eye sight, they must hide or apply camouflage.

Many animals search actively for their meal, be it plant or animal, dead or alive. The standard cruising rate of surface feeders tends to be proportional to their length, because the energy investment in movement as part of the maintenance costs tends to be proportional to volume, while the energy costs for transport are proportional to surface area; see {63}. Proportionality of cruising rate to length also occurs if limb movement frequency is more or less constant [570]. The width of the path searched for food by cows or snails is proportional to length if head movements perpendicular to the walking direction scale isomorphically. So feeding rate is again proportional to surface area, which is illustrated in figure 3.4 for the pond snail.

The duration of a dive for the sperm whale *Physeter macrocephalus*, which primarily feeds on squid, is proportional to its length, as is well known to the whalers [752]. This can be understood, since respiration rate of this endotherm is about proportional to surface area, as I will argue on {103}, and the amount of reserve oxygen proportional to volume on the basis of a homeostasis argument. It is not really obvious how this translates into feeding rate, if at all; large individuals tend to feed on large prey, which tend occur less frequently than small prey and depends on depth. Moreover, time investment in hunting can depend on size as well. If the daily swimming distance during hunting would to be independent of size, the searched water volume is about proportional to surface area for a volume feeder such as the sperm whale. If the total volume of squid per volume of water is about constant, this would imply that feeding rate is about proportional to surface area.

The amount of food parent birds feed per nestling relates to the requirements of the nestling, which is proportional to surface area; figure 3.5 illustrates this for chickadees. This is only possible if the nestlings can make their needs clear to the parents, by crying louder.

Catching devices, such as spider or pteropod webs and larvacean filter houses [13], have effective surface areas that are proportional to the surface area of the owner.

All these different feeding processes relate to surface areas in comparisons between different body sizes within a species at a constant low food density. At high food densities, the encounter probabilities are no longer rate limiting, but digestion and other food processing activities involving other surface areas, for example the mouth opening and the gut wall. The gradual switch in the leading processes becomes apparent in the functional response, i.e. the ingestion rate as function of food density, {63}.

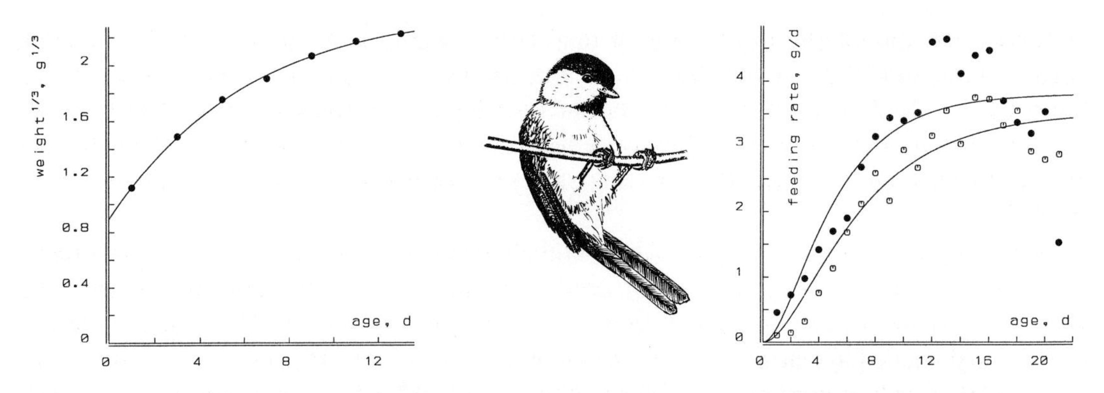

Figure 3.5: The von Bertalanffy growth curve applies to the black-capped chickadee, *Parus atricapillus* (left figure, data from Kluyver [392,671]. Brood size was a modest 5.) The amount of food fed per male (•) or female (∘) nestling in the closely related mountain chickadee, *P. gambeli*, is proportional to weight$^{2/3}$ (right figure), as might be expected for individuals that grow in a von Bertalanffy way. Data from Grundel [277,671]. The last five data points were not included in the fit, because of transition to independent food gathering behaviour.

3.1.2 Selection

Details of growth and reproduction patterns can only be understood in relation to selection of food items and choice of diet. The reverse relationship holds as well, especially for 'demand' systems. I will, therefore, mention some aspects briefly.

Many species change their diet during development in relation to their shifting needs with an emphasis on protein synthesis during the juvenile period and on maintenance during the adult one. Mammals live on milk during the baby stage, cf. {50}. The male emperor penguin *Aptenodytes* and mouth-brooding frog *Rhinoderma darwinii* provide their young initially with a secretions from the stomach. Plant eating ducks live on insects during the first period after hatching. The first hatching tadpoles of the alpine salamander *Salamandra atra* live on their siblings inside the mother, where they are also supported by blood from her reproductive organs, and the 1–4 winners leave the mother when fully developed. The same type of prenatal cannibalism seems to occur in the coelacanth *Latimeria* [715], and several sharks (sand tiger sharks *Odontaspidae*, mackerel sharks *Lamnidae*, thresher sharks *Alopiidae* [581]). Some species of poison dart frog *Dendrobates* feed their offspring with unfertilized eggs in the water-filled leaf axils of bromeliads, high up in the trees [187,188]. Many juvenile holo-metabolic insects live on different types of food than adults. Many wasps, for instance, are carnivorous when juvenile, while they feed on nectar as adults. Prickleback fish change from being carnivorous to being herbivorous at some stage during development [167].

Some species select for different food items in different seasons apart from changes in relative abundances of the different food sources. This is because of the tight coupling between feeding and digestion. The bearded tit *Panurus biarmicus* is a spectacular example; it lives on the seed of *Typha* and *Phragmites* from September to March and on insects in summer [676,754]. This change in diet comes with an adaptation of the stomach which is much more muscular in winter when it contains stones to grind the seeds. Once converted to summer conditions, the bearded tit is unable to survive on seeds. The example is remarkable because the bearded tit stays in the same habitat over the seasons. Many temperate birds change habitats over the seasons. Divers, for instance, inhabit fresh water tundra lakes during the breeding season and the open ocean during winter. Such species also change prey, of course, but the change is usually not as drastic as the one from insects to seeds.

When offered different food items, individuals can select for type and size. Shelbourne [652] reports that the mean length of *Oikopleura* eaten by plaice larvae increased with the size of the larvae. Copepods appear to select the larger algal cells [695]. Daphnids do not collect very small particles, < 0.9 μm cross-section [266], or large ones, > 27 and > 71 μm, the latter values were measured for daphnids of length 1 and 3 mm respectively [112]. Kersting and Holterman [381] found no size-selectivity between 15–105 (and probably 165) μm^3 for daphnids. Selection is rarely found in daphnids [601], or in mussels [226,767].

The relationship between feeding rates and diet composition gives a clue as to which processes actually set the upper limits to the ingestion rate. An indication that the maximum ingestion rate is determined by the digestion rate comes from the observation that the maximum ingestion rate of copepods feeding on diatoms expressed as amount of carbon is independent of the size of the diatom cells, provided that the chemical composition of the cells is similar [235]. The maximum ingestion rate is inversely related to protein, nitrogen and carbon content fed to the copepod *Acartia tonsa* [338]. The observation that the maximum ingestion rate is independent of cell size on the basis of ingested volume [247], points to the capacity of gut volume as the limiting factor.

These remarks should make it clear that the quantitative details of the feeding process cannot be understood without some understanding of the fate of the food. This involves the digestion process in the first place, but a whole sequence of other processes follow. Regulation of (maximum) ingestion depends by definition on the need in 'demand' systems, which is especially easy to observe in species that lose the ability to grow, such as birds and mammals. Temporarily elevated food intake can be observed in birds preparing for migration or reproduction or in mammals preparing for hibernation or in pregnant mammals [731]. For simplicity's sake, these phenomena will not be modelled explicitly.

Prokaryotes show a diversity and adaptability of metabolic pathways which is huge in comparison to that of eukaryotes. Many bacteria, for example, are able to synthesize all the amino acids they require, but will only do so if they are not available from the environment. The fungus *Aspergillus niger* only feeds on cellulose if no compounds are available that are easier to decompose. Another example is growth on glucose limited media. Figure 3.6 illustrates that prolonged exposure to limiting amounts of glucose eventually results in substantially improved uptake of glucose from the environment. The difference can amount to a factor of 1000. The outer membrane is adapted to this specialized task and may jeopardize a rapid change to other substrates. This adaptation process takes many cell division cycles, as is obvious from the measurement of population growth rates, which itself takes quite a few division cycles.

The relationship between food quality and physiological performance is taken up again in the discussion on food intake reconstructions {137} and on dissipating heat {201}.

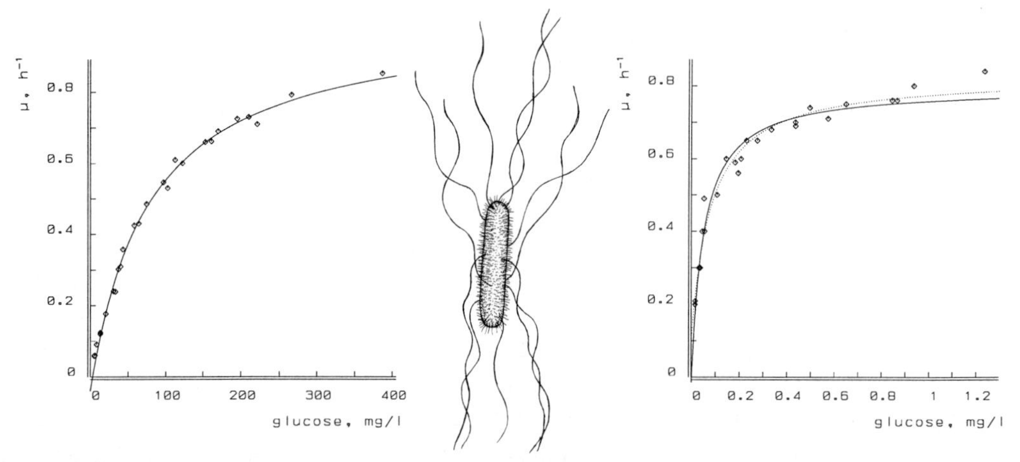

Figure 3.6: The population growth rate of *Escherichia coli* on glucose limited media. Schulze and Lipe's culture [645], left, had been exposed to glucose limitation just prior to the experiment, while that of Senn [648], right, had been pre-adapted for a period of three months.

3.1.3 Feeding and movement costs

As feeding methods are rather species-specific, costs for feeding will also be species-specific, if they contribute substantially to the energy budget. I will argue here that costs for feeding and movements that are part of the routine repertoire are usually insignificant with respect to the total energy budget. For this reason this subsection does not do justice to the voluminous amount of work that has been done on the energetics of movements, a field that is of considerable interest in other contexts. Alexander [9] has recently given a most readable and entertaining introduction to the subject of energetics and biomechanics of animal movement. Differences in respiration between active and non-active individuals give a measure for the energy costs of activity. The resting metabolic rate is a measure that excludes active movement. The standard or basal metabolic rate includes a low level of movement only. The field metabolic rate is the daily energy expenditure for free ranging individuals. Karasov [371] found that the field metabolic rate is about twice the standard metabolic rate for several species of mammal, and that the costs for locomotion ranges from 2–15% of the field metabolic rate. Mammals are amoung the more active species. The respiration rate associated with filtering in animals such as larvaceans and ascidians was found to be less than 2% of the total oxygen consumption [223]. The circumstance that energy investment into feeding is generally small, makes it unattractive to introduce many parameters to describe this investment. Feeding costs can be accommodated in two ways within the DEB theory without introduction of new parameters, and this subsection aims to explore to what extent this accommodation is realistic.

The first way is when the feeding costs are proportional to the feeding rate. They then show up as a reduction of the energy gain per unit of food. One can, however, argue that feeding costs per unit of food should increase with decreasing food density, because of the increased effort to extract it from the environment. This type of costs can only be accommodated without complicating the model structure if these costs cancel against an increased digestion efficiency, due to the increased gut residence time, cf. {247}.

The second way to accommodate feeding costs without complicating the model structure is when the feeding costs are independent of the feeding rate and proportional to body volume. They then show up as part of the maintenance costs, cf. {76}. This argument can be used to understand that feeding rates for some species tend to be proportional to surface area if transportation costs are also proportional to surface area, so that cruising rate is proportional to length, {59}. In this case feeding costs can be combined with costs for other types of movement that are part of the routine repertoire. A fixed (but generally small) fraction of the maintenance costs then relates to movement.

Schmidt-Nielsen [638] calculated 0.65 ml O_2 cm^{-2} km^{-1} to be the surface area-specific transportation costs for swimming salmon, on the basis of Brett's work [91]. (He found that transportation costs are proportional to weight to the power 0.746, but respiration was not linear with speed. No check was made for anaerobic metabolism of the salmon. Schmidt-Nielsen obtained, for a variety of fish, a power of 0.7, but 0.67 also fits well.) Fedak and Seeherman [213] found that the surface area-specific transportation costs for walking birds, mammals and lizards tend to be about 5.39 ml O_2 cm^{-2} $km^{-1} \simeq 0.03$ Wh cm^{-2} km^{-1}. (They actually report that transportation costs are proportional to weight to the power 0.72 as the best fitting allometric relationship, but the scatter is such that 0.67 fits as well.) This is consistent with data from Taylor *et al.* [703] and implies that the costs for swimming are some 12% of the costs for running. Their data also indicate that the costs for flying are between swimming and running and amount to some 1.87 ml O_2 cm^{-2} km^{-1}.

The energy costs of swimming are frequently taken to be proportional to squared speed on sound mechanical grounds [422], which questions the usefulness of the above mentioned costs and comparisons because the costs of transportation become dependent on speed. If the inter-species relationship that speed scales with the square root of volumetric length, see {223}, also applies to inter-species comparisons, the transportation costs are proportional to volume if the travelling time is independent of size.

The energy required for walking and running is found to be proportional to velocity for a wide diversity of terrestrial animals including mammals, birds, lizards, amphibians, crustaceans and insects [244]. This is quite a relief, as otherwise temperature would be a significant variable, to mention just one problem, affecting rates in a different way and making movements a complicated variable to handle at the population and the community level; the energy costs for walking or running a certain distance are independent of speed and just proportional to distance.

3.1.4 Functional response

The feeding or ingestion rate, $\dot{I}$, of an organism as a function of food or substrate density, X, expressed as number of items per surface area or volume, is described well by the hyperbolic functional response

$$\dot{I} = f\dot{I}_m \quad \text{with } f \equiv \frac{X}{K+X} \tag{3.1}$$

where K is known as the saturation coefficient or Michaelis constant, i.e. the density at which food intake is half the maximum value, and $\dot{I}_m$ the maximum ingestion rate. This

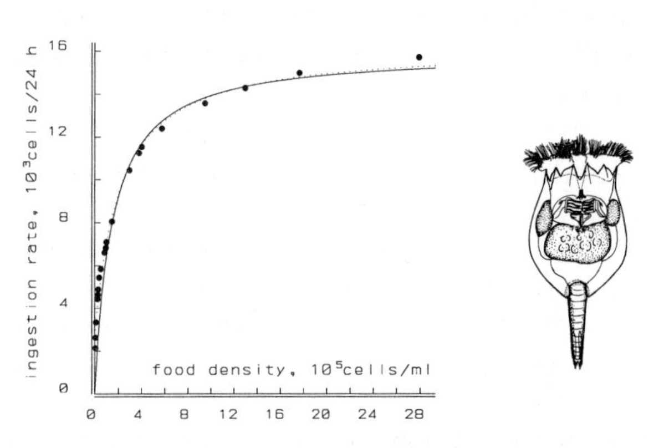

Figure 3.7: The ingestion rate, $\dot{I}$ of an individual (female) rotifer *Brachionus rubens*, feeding on the green alga *Chlorella* as a function of food density, X, at 20 °C. Data from Pilarska [554]. The curve is the hyperbola $\dot{I} = \dot{I}_m \frac{X}{K+X}$, with $\dot{I}_m =$ 15.97 (s.d. 0.81) 10^5cells d^{-1} and $K =$ 1.47 (s.d. 0.26) 10^5cells d^{-1}. The stippled curve allows for an additive error in the measurement of the algal density of 34750 cells ml^{-1}.

functional response has been proposed by Holling [332] as type II, and is illustrated in figure 3.7. It applies to ciliates feeding on organic particles (phagocytosis), algae filtering daphnids, mantis catching flies, substrate uptake by bacteria, or the enzyme mediated transformation of substrates. Although these processes differ considerably in detail, some common principle gives rise to the hyperbolic function. This can be explained on the basis of a simple model for feeding, that will be generalized subsequently.

Suppose that the handling of a particle takes a certain time τ and that particles arriving during handling are ignored. ('Handling' is used here in a wide sense, for feeding animals it might refer to the act of catching and eating as well as to decomposing the particles in the gut or the transfer of products across the gut wall.) Suppose further, that particles do not interfere with each other. So the number of particles arriving in a unit of time is Poisson distributed with a parameter proportional to the particle density, X, say $\dot{F}X$. Here $\dot{F}$ relates to a filtering rate or the speed of an animal relative to prey particles, a rate that is taken to depend on mean particle density only, and not on particle density at a particular moment. The time between subsequent arrivals, $\underline{t}_i$, is then exponentially distributed, with mean $(\dot{F}X)^{-1}$. The time between the end of a handling period and the next arrival is again exponentially distributed with mean $(\dot{F}X)^{-1}$. (To see this, one should make use of a defining property for an exponentially distributed variable $\underline{y}$, that $\underline{y}$ and $\underline{y}|\,\underline{y} > y$ are identically distributed, i.e. $\phi_{\underline{y}|\,\underline{y}>y}(t+y) = \phi_{\underline{y}}(t)$.) The time required to eat N particles is thus given by $\underline{t} = N\tau + \sum_{i=1}^{N} \underline{t}_i$ if one starts observations at a randomly chosen arrival of a particle. The mean ingestion rate, $\dot{I} = N/\mathcal{E}\underline{t}$, is thus $\dot{I} = (\tau + (\dot{F}X)^{-1})^{-1} = \tau^{-1}X((\tau\dot{F})^{-1} + X)^{-1}$, which is hyperbolic in the density X. The saturation coefficient is inverse to the product of the handling time and the filtering (or searching) rate, i.e. $K = (\tau\dot{F})^{-1}$. The maximum ingestion rate is inverse to the handling time. (The ingestion rate is here taken to be the ratio of a fixed number of particles eaten and the measured time it takes the animal to do this. If the feeding period is fixed, rather than the number of particles eaten, the mean ingestion rate might, in principle, deviate from the hyperbolic function. Moreover, we make sure that the particle density does not change during the observation period.)

This derivation can be generalized in different ways without changing the model. Each arriving particle can have an attribute that stands for the catching probability. The i-th particle has some fixed probability p_i of being caught upon encounting an animal, if the animal is not busy handling particles, and probability 0 if it is. It is not essential that the handling time is the same for all particles; this can be conceived as a second attribute attached to each particle, but it must be independent of food density. The condition of zero catching probability when the animal is busy can be relaxed. Metz and van Batenburg [476,477] and Heijmans [301], tied catching probability to satiation, which is thought to relate to gut content in mantis. An essential condition for hyperbolic functional responses is that catching probability equals zero if satiation (gut content) is maximal.

Another generalization is from one server, i.e. the individual handling the particles, to a large but fixed number of identical servers handling particles simultaneously, but without interfering with each other. The term 'server' stems from an extensive theory of applied probability calculus, known as queueing theory, that deals with this type of problem. See for instance [625,662]. Think of a server as an active site (enzyme molecules) in a membrane, of particles as substrate molecules and of catching as adsorption. If θ stands for the fraction of busy servers, then the change of this fraction due to arrivals is given by $\frac{d}{dt}\theta = \dot{k}_a X(1-\theta)$, where the adsorption rate constant, $\dot{k}_a$, plays exactly the same role as the filtering or searching rate $\dot{F}$. The change of the fraction of busy servers due to termination of handling is proportional to the number of busy servers, so it is given by $\frac{d}{dt}\theta = \dot{k}_d\theta$, where the desorption rate constant, $\dot{k}_d$, is just inverse to the mean handling time τ. In equilibrium, the fraction θ does not change, so $\dot{k}_a X(1-\theta) = \dot{k}_d\theta$, or $\theta = X/(K+X)$, with $K = \dot{k}_d/\dot{k}_a$. We assume here that the absorption and desorption process is rapid with respect to the changes in the particle density X.

The fraction of occupied sites as a function of the density of adsorbable particles (i.e. partial pressure in gas), is called the adsorption isotherm in physical chemistry. If the sites operate independently, as here, and so give rise to a hyperbolic function, this isotherm is called the Langmuir isotherm [24]. The adsorption rate of particles is found easily by substituting the Langmuir isotherm into the change of the busy fraction of servers:

$$\dot{k}_a X(1-\theta) = \dot{k}_a X\left(1 - \frac{X}{K+X}\right) = \frac{\dot{k}_d X}{K+X}$$

So the adsorption rate depends hyperbolically on the particle density in equilibrium. The saturation coefficient has the interpretation of the ratio of the desorption and the adsorption rate constants and the maximum adsorption rate of particles equals the number of servers times the desorption rate. If the desorbed particles are transformed with respect to the adsorbed ones, the process stands for an enzyme mediated transformation of substrate into product. The simple kinetics discussed here are called Michaelis–Menten kinetics. The condition of constant particle density can be somewhat relaxed; if the total number of particles, N, is really large with respect to the number of servers (a condition formulated by Briggs and Haldane [92]), or if the rate of product formation $\dot{k}_d$ is really small (a condition formulated by Michaelis and Menten [483]), or if K times the number of servers is really small with respect to $(K+N)^2$, (a more general condition formulated by Segel [647]), the reaction still follows Michaelis–Menten kinetics.

Although the details of feeding and adsorption processes differ considerably, from a more abstract point of view the mechanisms are closely related. What is essential is that a busy period exists and that, if more servers are around, they operate identically and independently.

It is entirely possible that the hyperbolic response also arises from completely different mechanisms. A most interesting property of the hyperbolic function is that it is the only one with a finite number of parameters that maps into itself. For instance, an exponential function of an exponential function is not again an exponential function. A polynomial (of degree higher than one) of a polynomial is also a polynomial, but it is of an increasingly higher degree if the mapping is repeated over and over again. The hyperbolic function of a hyperbolic function is also a hyperbolic function. (Note that the linear response function is a special case of the hyperbolic one.) In a metabolic pathway each product serves as a substrate for the next step. Neither the cell nor the modeller needs to know the exact number of intermediate steps to relate the production rate to the original substrate density, if and only if the functional responses of the subsequent intermediate steps are of the hyperbolic type. If, during evolution, an extra step is inserted in a metabolic pathway the performance of the whole chain does not change in functional form. This is a crucial point because each pathway has to be integrated with other pathways to ensure the proper functioning of the individual as a whole. If an insert in a metabolic pathway simultaneously required a qualitative change in regulation at a higher level, the probability of its occurrence during the evolutionary process would be remote.

A most useful property of the hyperbolic functional response is that it has only two parameters which serve as simple scaling factors on the food density and ingestion rate axis. So if food density is expressed in terms of the saturation coefficient, and ingestion rate in terms of maximum ingestion rate, the functional response no longer has dimensions or parameters.

Filter feeders, such as rotifers, daphnids and mussels, reduce filtering rate with increasing food density [226,565,603,604], rather than maintaining a constant rate, which would imply the rejection of some food particles. They reduce the rate by such an amount that no rejection occurs due to handling (processing) of particles. If all incoming water is swept clear, the filtering rate is found from $\dot{F}(X) = \dot{I}/X$, which reaches a maximum if no food is around (temporarily), so that $\dot{F}_m = \{\dot{I}_m\}V^{2/3}/K$, and approaches zero for high food densities. The braces stand for 'surface area-specific', thus $\{\dot{I}_m\} \equiv \dot{I}_m V^{-2/3}$, stands for the maximum surface area-specific ingestion rate, which is considered as a parameter that depends on the composition of the diet. An alternative interpretation of the saturation coefficient in this case would be $K = \dot{I}_m/\dot{F}_m = \{\dot{I}_m\}/\{\dot{F}_m\}$, which is independent of the size of the animal, as long as only intraspecific comparisons are made. It combines the maximum capacity for food searching behaviour, only relevant at low food densities, with the maximum capacity for food processing, which is only relevant at high food densities.

Mean ingestion rate for an isomorph of volume V at food density X thus amounts to

$$\dot{I} = \{\dot{I}_m\} f V^{2/3} \text{ with } f \equiv \frac{X}{K+X} \tag{3.2}$$

where $\{\dot{I}_m\}$ stands for the maximum surface area-specific ingestion rate, expressed in volumetric length. When starved animals are fed, they often ingest at a higher rate for some time [753], but this is usually a fast process which will be neglected here. Starved daphnids for instance are able to fill their guts within 7.5 minutes [247].

The ingestion rate, or substrate uptake rate for filaments and rods are found from (3.2) by multiplication of $\{\dot{I}_m\}$ with the shape correction function (2.4) or (2.7), which leads to

$$\begin{aligned} \dot{I} &= [\dot{I}_m] f V && \text{for filaments} \\ \dot{I} &= [\dot{I}_m] f \left(\frac{\delta}{3} V_d + \left(1 - \frac{\delta}{3}\right) V\right) && \text{for rods} \end{aligned} \tag{3.3}$$

for $[\dot{I}_m] \equiv \{\dot{I}_m\} V_d^{-1/3}$. Since food for rods and filaments in cultures usually consists of a simple organic compound, it is standard to quantify uptake rate in gram or mole, rather than in volume as is done here. The choice of unit is free and to some extent arbitrary, the present one being motivated by the study of food chains, see {212}, where the conversion from food to structural biomass urges symmetry. Likewise, I will use the term 'substrate density', rather than 'substrate concentration' to stress the relationship with food density and to cover insoluble substrates as well.

An important source of deviations from the hyperbolic functional response will be discussed on {144}.

3.1.5 Food deposits and claims

Any description of the feeding process that is not species-specific can only be roughly approximative at best. In this subsection I want to point briefly to some important types of feeding behaviour that are likely to cause deviations from the hyperbolic functional response: stocking food and claiming resources via a territory. The importance of these types of behaviour is at the population level, where the effect is strongly stabilizing for two reasons. The first is that the predator lives on deposits if prey is rare, which lifts the pressure on the prey population under those conditions. The second one is that high prey densities in the good season do not directly result in an increase in predator density. This also reduces the predation pressure during the bleak seasons. Although the quantitative details will not be worked out here because of species-specificity, I want to point to this behaviour as an introduction to other smoothing phenomena that will be covered. The DEB model differs from almost all other models in dealing with such phenomena, so these remarks serve to point to the necessity of including smoothing phenomena in realistic models.

Many food deposits relate to survival during winter, frequently in combination with dormancy, cf. {131}. In the German, Dutch and Scandinavian languages, the word 'hamster' is the stem of a verb for stocking of food in preparation for adverse conditions. This rodent is famous for the huge piles of maize it stocks in autumn. The English language has selected the squirrel for this purpose. This type of behaviour is much more widespread, for example in jays; cf. figure 3.8.

Many species defend territories just prior to and during the reproductive season. Birds do it most loudly. The size of the territories depends on bird as well as food density. One of the obvious functions of this behaviour is to claim a sufficient amount of food for the peak demand when the young grow up. The behaviour of stocking and reclaiming of food typically fits 'demand' systems and is less likely to be found in 'supply' systems.

Figure 3.8: The great grey shrike *Lanius excubitor*, also known as the 'nine-killer' in Dutch, hoards throughout the year, possibly to guard against bad luck when hunting. Many other shrikes do this as well.

3.2 Digestion

Details of the digestion process are discussed on {247} because they do not bear directly on the specification of the DEB model. Logic of arguments requires, however, that some aspects of the digestion process should be discussed here.

3.2.1 Smoothing and satiation

The capacity of the stomach/gut volume is specific to a species. It depends strongly on type of food specialized on. Fish feeding on plankters, i.e. many small constantly available particles, have a low capacity, while fish such as the swallower, that feed on rare big chunks of food, have high capacities. It may wait weeks for new chunks of food; see figure 3.9. The stomach/gut volume, which is still 'environment' rather than animal, is used to smooth out fluctuations in nutritional input to the organism. Organisms attempt to run their metabolic processes under controlled and constant conditions. Food in the digestive tract and reserves inside the organism together make it possible for regulation mechanisms to ensure homeostasis. Growth, reproductive effort and the like do not depend directly on food availability but on the internal state of the organism. This even holds, to some extent, for those following the 'supply' strategy, where energy reserves are the key variable. These reserves rapidly follow the feeding conditions.

If the food in the stomach, X_s, follows a simple first order process, the change of stomach contents is

$$\frac{d}{dt}X_s = \{\dot{I}_m\}V^{2/3}\left(\frac{X}{K+X} - \frac{X_s}{[X_{sm}]V}\right) \tag{3.4}$$

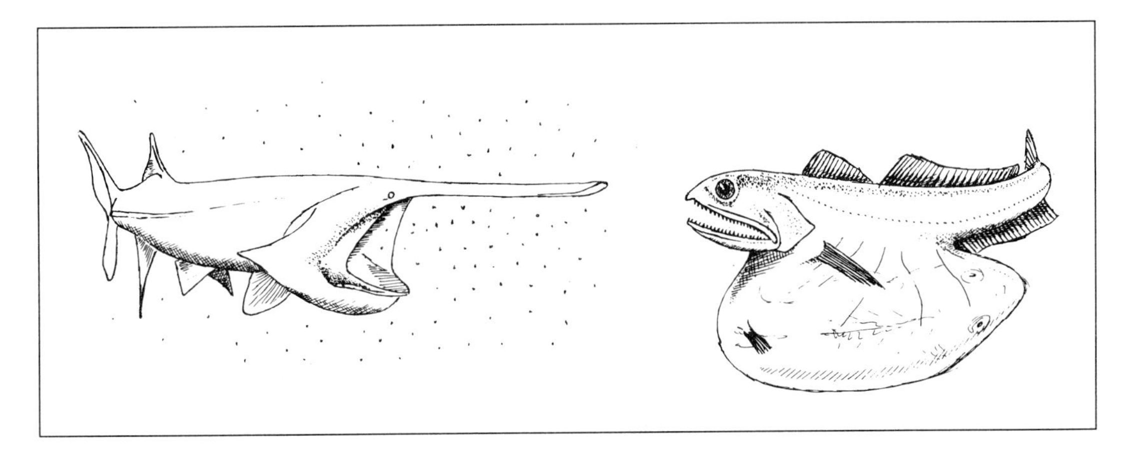

Figure 3.9: The 2 m paddlefish *Polyodon spathula* feeds on tiny plankters, while the 18 cm black swallower *Chiasmodon niger* can swallow fish bigger than itself. They illustrate extremes in buffer capacities of the stomach.

where $[X_{sm}]$ is the maximum food capacity density of the stomach. The derivation is as follows. A first order process here means that the change in stomach contents can be written as $\frac{d}{dt}X_s = \dot{I} - \dot{\alpha}X_s$, where the proportionality constant $\dot{\alpha}$ is independent of the input, given in (3.2). Since food density is the only variable in the input, $\dot{\alpha}$ must be independent of food density X, and thus of scaled functional response f. If food density is high, stomach content converges to its maximum capacity $\dot{I}_m/\dot{\alpha} = \{\dot{I}_m\}V^{2/3}\dot{\alpha}^{-1}$. The assumption of isomorphism implies that the maximum storage capacity of the stomach is proportional to the volume of the individual. This means that we can write it as $[X_{sm}]V$, where $[X_{sm}]$ is some constant, independent of food density and body volume. This allows one to express $\dot{\alpha}$ in terms of $[X_{sm}]$, which results in (3.4).

The mean residence time in the stomach is thus $t_s = V^{1/3}[X_{sm}]/\{\dot{I}_m\}$, and so it is proportional to length and independent of the ingestion rate. First order dynamics implies complete mixing of food particles in the stomach, which is unlikely if fermentation occurs. This is because the residence time of each particle is then exponentially distributed, so a fraction $1 - \exp\{-1\} = 0.63$ of the particles stays less time in the stomach than the mean residence time, and a fraction $1 - \exp\{-\frac{1}{2}\} = 0.39$ less than half the mean residence time. This means incomplete, as well as 'too complete', and thus wasteful fermentation.

The extreme opposite of complete mixing is plug flow, where the variation in residence times between the particles is nil in the ideal case. Pure plug flow is not an option for a stomach, because this excludes smoothing. These conflicting demands probably separated the tasks of smoothing for the stomach and digestion for the gut to some extent. Most vertebrates do little more than create an acid environment in the stomach to promote protein fermentation, while actual uptake is via the gut. Plug flow of food in the gut, X_g, can be described by

$$\frac{d}{dt}X_g(t) = t_s^{-1}(X_s(t) - X_s(t - t_g)) \tag{3.5}$$

where t_g denotes the gut residence time and t_s the mean stomach residence time. This equation follows directly from the principle of plug flow. The first term $t_s^{-1}X_s(t)$, stands for the influx from the stomach and follows from (3.4). The second one stands for the outflux, which equals the influx with a delay of t_g. Substitution of (3.4) and (3.2) gives $\frac{d}{dt}X_g(t) = \dot{I}(t) - \dot{I}(t - t_g) + \frac{d}{dt}X_s(t - t_g) - \frac{d}{dt}X_s(t)$. Since $0 \leq X_s \leq [X_{sm}]V$, $\frac{d}{dt}X_s \to 0$ if $[X_{sm}] \to 0$. So the dynamics of food in the gut reduces to $\frac{d}{dt}X_g(t) = \dot{I}(t) - \dot{I}(t - t_g)$ for animals without a stomach.

Some species feed in meals, rather than continuously, even if food is constantly available. They only feed when 'hungry' [178]. Stomach filling can be used to link feeding with satiation. From (3.4) it follows that the amount of food in the stomach tends to $X_s^* = f[X_{sm}]V$, if feeding is continuous and food density is constant. Suppose that feeding starts at a rate given by (3.2) as soon as food in the stomach is less than $x_{s0}X_s^*$, for some value of the dimensionless factor x_{s0} between 0 and 1, and feeding ceases as soon as food in the stomach exceeds $x_{s1}X_s^*$, for some value of $x_{s1} > x_{s0}$. The mean ingestion rate is still of the type (3.2), where $\{\dot{I}_m\}$ now has the interpretation of the *mean* maximum surface area-specific ingestion rate, not the one during feeding. A consequence of this on/off switching of the feeding behaviour is that the periods of feeding and fasting are proportional to a length measure. This matter is taken up again on {121}.

3.2.2 Gut residence time

The volume of the digestive tract is proportional to the whole body volume in strict isomorphs. This has been found for e.g. ruminant and nonruminant mammals [162] ($\simeq$ 11%) and for daphnids [209] ($\simeq$ 2.5% if the whole space in the carapace is included). If the animal keeps its gut filled to maximum capacity, $[X_{gm}]V$ say, and if the volume reduction due to digestion is not substantial, this gives a simple relationship between gut residence time of food particles t_g, ingestion rates $\dot{I}$, and body volume V:

$$t_g = [X_{gm}]V/\dot{I} = \frac{V^{1/3}[X_{gm}]}{f\{\dot{I}_m\}} \tag{3.6}$$

This is exactly what has been found for daphnids [209], see figure 3.10, and mussels [286]. Copepods [127] and carnivorous fish [358] seem to empty their gut at low food densities, which leads to a gut residence time of $V^{1/3}[X_{gm}]/\{\dot{I}_m\}$, if the throughput is at maximum rate.

Since ingestion rate, (3.2), is proportional to squared length, the gut residence time is proportional to length for isomorphs. For filaments such as worms, which have a fixed diameter, ingestion rate is proportional to cubed length, (3.3), so gut residence time is independent of body volume.

Daphnids are translucent, which offers the possibility of studying the progress of digestion as a function of body length.

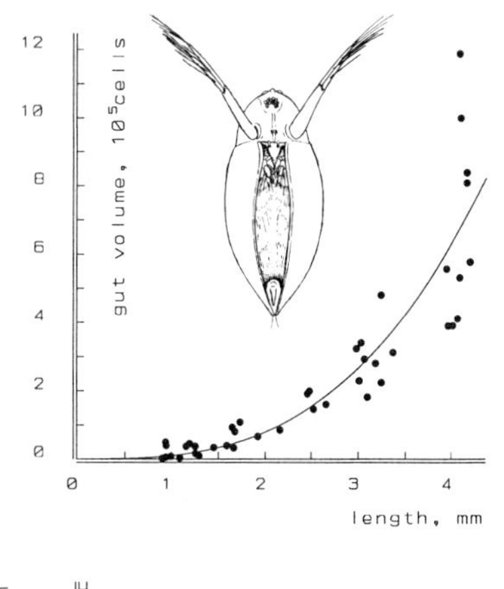

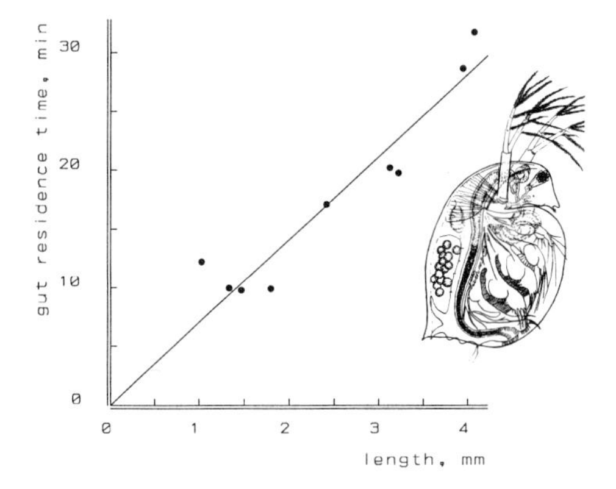

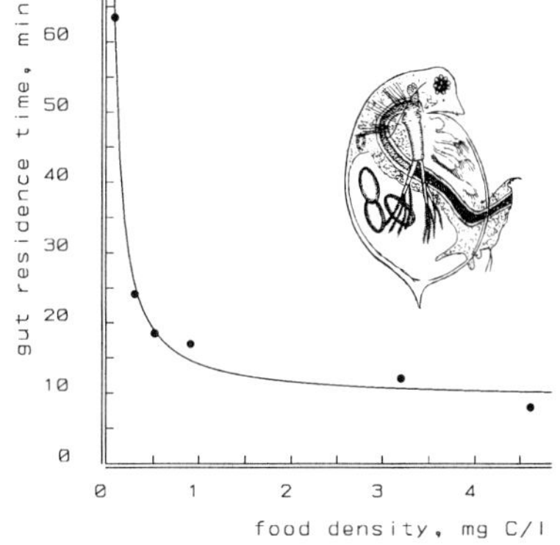

Figure 3.10: Gut volume is proportional to cubed length (right) and gut residence time is proportional to length (lower left), while the latter depends hyperbolically on food density (lower right), as illustrated for daphnids. The first two figures relate to *D. magna* feeding on the green alga *Scenedesmus* at 20 °C. Data from Evers and Kooijman [209]. The third one relates to *D. pulex* of 2 mm feeding on the diatom *Nitzschia actinastroides* at 15 °C. Data from Geller [247].

The photograph of *D. magna* on the right shows the sharp transition between the chlorophyll of the green algae and the brown-black digestion products, which is typical for high ingestion rates. The relative position of this transition point depends on the ingestion rate, but not on the body length. Even in this respect daphnids are isomorphic. At low ingestion rates, the gut looks brown from mouth to anus. The paired digestive caecum is clearly visible just behind the mouth.

3.3 Assimilation

In animal physiology it is standard to call the enthalpy of ingested food the 'gross' energy intake [441]. It is used to quantify the energy potential for the individual. In microbial physiology and biochemistry [37], the more appropriate free energy content of consumed substrate is used for the same purpose, cf. {201}. The difference obviously relates to the poor thermodynamical definition of food of complex chemical nature. The term 'digestible' energy is used for gross energy minus energy in faeces. Then comes 'metabolisable' energy, which is taken to be digestable energy minus energy in urine and in released methane gas, followed by 'net' energy, which is metabolisable energy minus energy lost in heat increment of feeding. The term 'assimilated' energy will here be the free energy intake minus free energy in faeces and in all losses in relation to digestion. The energy in urine is treated somewhat differently and tied to the process of maintenance, cf. {77}.

The assimilation efficiency of food is here taken to be independent of the feeding rate. This makes the assimilation rate proportional to the ingestion rate, which seems to be realistic, cf. figure 7.1. I will discuss later the consistency of this simple assumption with more detailed models for enzymatic digestion, {247}. The conversion efficiency from food into assimilated energy is written as $\{\dot{A}_m\}/\{\dot{I}_m\}$, where $\{\dot{A}_m\}$ is a diet-specific parameter standing for the maximum surface area-specific assimilation rate. This notation may seem clumsy, but the advantage is that the assimilated energy that comes in at food density X is now given by $\{\dot{A}_m\}fV^{2/3}$, where $f = X/(K+X)$ and V the body volume. It does not involve the parameter $\{\dot{I}_m\}$ in the notation, which turns out to be useful in the discussion of processes of energy allocation in the next few sections.

The conversion from substrate to energy in bacteria is substantially more efficient under aerobic (oxygen rich) conditions than under anaerobic ones, while metabolic costs are not affected by oxygen availability [417]. This means that the parameter $\{\dot{A}_m\}$ and not $\{\dot{I}_m\}$ is of direct relevance to the internal machinery, cf. {201}.

3.4 Storage dynamics

Energy crossing the gut wall enters the blood or body fluid and is usually circulated through the body rapidly. It therefore does not matter where in the gut uptake takes place. Residence time in the digestive tract is usually short compared to that in the energy reserves, which means that for most practical purposes, the effect of digestion can simply be summarized as a conversion of ingested food, $\dot{I} = \{\dot{I}_m\}fV^{2/3}$, into (assimilated) energy, $\dot{A} = \{\dot{A}_m\}fV^{2/3}$. Blood has a low uptake capacity for energy (or nutrient), but a high transportation rate; it is pumped through the body many times an hour. The changes of energy in blood, E_{bl}, and in reserves, E, are coupled by $\frac{d}{dt}E_{bl} = \dot{A} - \frac{d}{dt}E - \dot{C}$ where $\dot{C}$ denotes the energy consumed by the body tissues and is called the utilization or catabolic rate. The change of energy reserves can be positive or negative. Since the energy capacity of blood is small, the change of energy in blood cannot have a significant impact on the whole body. It therefore seems safe to assume that $\frac{d}{dt}E_{bl} \simeq 0$, which means that $\frac{d}{dt}E = \dot{A} - \dot{C}$ as a first approximation.

The reserve density, $[E] \equiv E/V$, is assumed to follow simple first order dynamics

$$\frac{d}{dt}[E] = \frac{\{\dot{A}_m\}}{V^{1/3}}\left(f - \frac{[E]}{[E_m]}\right) \tag{3.7}$$

where $[E_m]$ is the maximum energy reserve density. Its derivation is completely analogous to (3.4). A first order process for the reserve density means that it can be written as $\frac{d}{dt}[E] = \dot{A}/V - \dot{\alpha}[E]$, where the proportionality constant $\dot{\alpha}$ is independent of food density X. At high food density, the reserve density converges to its maximum $\dot{A}_m(V\dot{\alpha})^{-1} = \{\dot{A}_m\}V^{-1/3}\dot{\alpha}^{-1}$. Because of the homeostasis assumption for energy reserves, the maximum capacity must be independent of body volume, so it can be written as a constant $[E_m]$, independent of both food density and body volume. This allows one to express $\dot{\alpha}$ in terms of the maximum capacity $[E_m]$, which gives (3.7).

An essential difference between stomach and reserves dynamics is that the first is in absolute quantities, because it relates to bulk transport, while the latter is in densities because it relates to molecular phenomena. (One cannot simply divide by body volume in (3.4) to turn to densities because body volume depends on time. One should, therefore, correct for growth to observe the mass conservation law.) Note that the requirement of homeostasis for energy density overrules the interpretation of reserve dynamics in terms of a simple mechanism where reserve 'molecules' react with the catabolic machinery at a rate given by the law of mass action. (Due to the concept of homeostasis, the density of the catabolic machinery is constant.) The organism has to adjust the reaction rate between reserves and the catabolic machinery during growth to preserve homeostasis. These adjustments are small, as long as dilution of energy density by growth is small with respect to the use of energy, i.e. if $\frac{d}{dt}\ln V \ll V^{-1/3}\{\dot{A}_m\}[E_m]^{-1}$. In practice, this condition is always fulfilled, which is not surprising because growth can only be high if the use of energy is really high. This naive picture of the mechanism can be made much more realistic without disturbing the first order kinetics.

Since inflow of energy is over a surface area and use over a volume, use of energy density is inversely proportional to length. This too corresponds closely with processes at the molecular level. Since energy reserves should not interfere with osmolarity, they are formed from insoluble polymers which are frequently further separated from the body fluid by membranes and confined to particular surface areas of the body, both macroscopically, e.g. around the gut, and microscopically. The bigger the body, the less accessible the energy reserves expressed as density. Many consequences of these extremely simple dynamics for the reserves will be tested against observations in this book. Direct testing is hampered by the problem of measuring energy fluxes inside organisms. Tests on the basis of respiration rates are probably the most direct ones feasible. Some auxiliary theory has to be developed first.

A consequence of the assumption of a first order dynamics for energy reserves is that the utilization rate must obey

$$\dot{C} = \dot{A} - \frac{d}{dt}([E]V) = \dot{A} - V\frac{d}{dt}[E] - [E]\frac{d}{dt}V = [E](\dot{v}V^{2/3} - \frac{d}{dt}V) \tag{3.8}$$

where $\dot{v} \equiv \{\dot{A}_m\}/[E_m]$; as $\dot{v}$ will show up time and again, I have given it a name, *energy conductance*, as a result of one of many discussions with Roger Nisbet. Its dimension is length per time and stands for the ratio of the maximum surface area-specific assimilation rate and the maximum volume-specific reserve energy density. The inverse, $\dot{v}^{-1}$, has the interpretation of a resistance. It is remarkable that the biological use of conductance measures seems to be restricted to plant physiology [362,511]. An important property of utilization rate is that it does not depend directly on the assimilation rate and, therefore, not on food density. It only depends on the volume of the organism and energy reserve.

The storage residence time in (3.7) is thus $V^{1/3}[E_m]/\{\dot{A}_m\}$, which must be large with respect to that of the stomach, $V^{1/3}[X_{sm}]/\{\dot{I}_m\}$ and the gut, $V^{1/3}[X_{gm}]/\{\dot{I}_m\}$, to justify neglect of the smoothing effect of the digestive tract.

If the energy reserve capacity, $[E_m]$, is extremely small, the dynamics of the reserves degenerates to $[E] = f[E_m]$, while both $[E]$ and $[E_m]$ tend to 0. The utilization rate then becomes $\dot{C} = \{\dot{A}_m\}fV^{2/3}$. This case has been studied by Metz and Diekmann [479].

3.5 The κ-rule for allocation

Some animals, such as birds, reproduce after having obtained their final size. Others, such as daphnids continue growth after onset of reproduction. *Daphnia magna* starts reproducing at a length of 2.5 mm, while its ultimate size is 5 to 6 mm, if well-fed. This means an increase of well over a factor 8 in volume during the reproductive period. Figure 3.11 illustrates a basic problem for energy allocation rules that such animals pose. It becomes visible as soon as one realizes that a considerable amount of energy is invested in reproductive output. The volume of young produced exceeds $\frac{1}{4}$ of that of the mother each day, or 80% of the utilization rate [595]. The problem is that growth is not retarded in animals crossing the 2.5 mm barrier; they also do not feed much more but they simply follow the surface area rule with a fixed proportionality constant at constant food densities. It seems unlikely that they digest their food much more efficiently, so where does the energy allocated to reproduction come from?

A solution to this problem can be found in development. Juvenile animals have to mature and become more complex. They have to develop new organs and install regulation systems. Increase in size (somatic growth) of the adult does not include an increase in complexity. The energy no longer spent on development in adults is spent on reproduction. Growth continues smoothly at the transition from development to reproduction. This suggests the 'κ-rule' : a fixed proportion κ of energy utilized from the reserves is spent on growth plus maintenance, the remaining portion, $1 - \kappa$ on development plus reproduction. The background and rationale of the κ-rule is as follows.

At separated sites along the path that blood follows, somatic cells and ovary cells pick up energy. The only information the cells have is the energy content of the blood and body size, cf. {23}. They do not have information about each others activities in a direct way. This also holds for the mechanism by which energy is added to or taken from energy reserves. The organism only has information on the energy density of the blood, and on size, but not on which cells removed energy from the blood. This is why the parameter

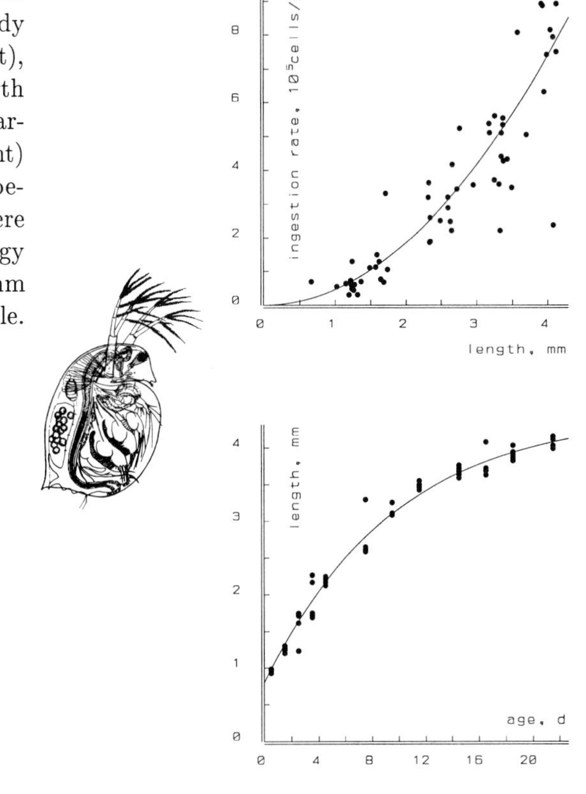

Figure 3.11: Ingestion in the waterflea *Daphnia magna* as a function of body length at 20 °C and abundant food (right), its reproduction (below) and body length (below right) as functions of age. Comparison of the quadratic feeding curve (right) and the von Bertalanffy growth curve (below right) leads to the question: where did the substantial reproductive energy come from in animals crossing the 2.5 mm barrier? The answer leads to the κ-rule. Original data and from [209].

κ does not show up in the dynamics of energy density. The activity of all carriers which remove energy from the body fluid and transport it across the cell membrane depends, in the same way, on the energy density of the fluid. Both somatic cells and ovary cells may use the same carriers, but the concentration in their membranes may differ so that κ may differ from the ratio of ovary and body weight. This concentration of active carriers is controlled, e.g. by hormones, and depends on age, size and environment. Once in a somatic cell, energy is first used for maintenance, the rest is used for growth. This makes maintenance and growth compete directly, while development and reproduction compete with growth plus maintenance at a higher level. The κ-rule makes growth and development parallel processes that interfere only indirectly, as has been discussed by Bernardo [60], for instance.

If conditions are poor, the system can block allocation to reproduction, while maintenance and growth continue to compete in the same way. This will be discussed further in the next chapter. I will show that Huxley's allometric model for relative growth closely links up with the κ-rule on {252}.

It is important to realize that although the fraction of utilized energy spent on maintenance plus growth remains constant, the absolute size of the flow tends to increase during development at constant food densities, as does the energy flow to maintenance plus growth.

The κ-rule solves quite a few problems from which other allocation rules suffer. Although it is generally true that reproduction is maximal when growth ceases, a simple allocation shift from growth to reproduction leaves similarity of growth between different sexes unexplained, since the reproductive effort of males is usually much less than that of females. The κ-rule implies that size control is the same for males and females and for organisms such as yeasts and ciliates which do not spend energy on reproduction, but do grow in a way that is comparable to species that reproduce; see figure 1.1. A strong support for the κ-rule comes from situations where the value for κ is changed to a new fixed value. Such a simple change affects reproduction as well as growth and so food intake in a very special way. Parasites such as the trematod *Schistosoma* in snails harvest all energy to reproduction and increase κ to maximize the energy flow they can consume, as will be discussed on {243}. Parasite induced gigantism, coupled to a reduction of the reproductive output, is also known from trematod infested chaetognats [499], for instance. The daily light cycle also affects the value for κ in snails; see {128}. The effect of some toxic compounds can be understood as an effect on κ, as will be discussed in the chapter on ecotoxicity, {282}. I will show how the κ-rule can be derived from a number of other assumptions that lend themselves to direct experimental testing on {119}.

3.6 Maintenance

Maintenance stands for a collection of processes necessary to 'stay alive'. More precisely, maintenance energy is defined as the (mean) energy requirement of an organism, excluding the production processes of growth, reproduction and development. Maintenance costs are species-specific and depend on the size of the organism and on body temperature. Maintenance processes include the maintenance of concentration gradients across membranes, the turnover of structural body proteins, a certain (mean) level of muscle tension and movement, and the (continuous) production of hairs, feathers, scales. I do not include heating of endotherms in maintenance for convenience, although it is a process necessary to stay alive and will be treated accordingly. As explained in the discussion on the κ-rule, {74}, development is excluded from maintenance, as it relates (partly) to a type of production process. The maintenance part of development is referred to as maturity maintenance, and will be discussed in the section on development, {97}. To distinguish maturation maintenance from other maintenance costs, the latter will be called costs for somatic maintenance, if necessary.

The notion of maintenance costs for advanced taxa is probably as old as man himself. Duclaux [186] was the first to recognize in 1898 that maintenance costs should be separated from production costs to understand the energetics of micro-organisms. The next reference to maintenance costs for micro-organisms stems from Sherris *et al.* [653] in 1957, in relation to motility. In the early 1960s maintenance costs for micro-organisms received considerable attention [310,380,455,466,533,556].

As is customary, I use the term 'metabolism' or 'respiration' to cover non-maintenance processes as well. The realization that respiration includes growth leads, I think, to the solution of a long standing problem: the acceptance that maintenance energy is proportional to biovolume, while metabolism or respiration is about proportional to volume to the power $\frac{3}{4}$. I will discuss this further in the section on respiration, {103}.

The idea that maintenance costs are proportional to biovolume is simple and rests on homeostasis: a metazoan of twice the volume of a conspecific has twice as many cells, which each use a fixed amount of energy for maintenance. A unicellular of twice the original volume has twice as many proteins to turn over. Bacteria, which grow in length only, have a surface area that is a linear function of cell volume. The energy spent on concentration gradients, which is coupled to membranes is, therefore, proportional to volume. Protein turnover seems to be low in prokaryotes [397]. Eukaryotic unicellular isomorphs are filled with membranes, and this ties the energy costs for concentration gradients to volume. (The argument for membrane-bound food uptake works out differently in isomorphs, because feeding involves only the outer membrane directly.) Working with mammals, Porter and Brand [567] argued that proton leak in mitochondria represents 25% of the basal respiration in isolated hepatocytes and may contribute significantly to the standard metabolic rate of the whole animal.

The energy costs for movement are also taken to be proportional to volume if averaged over a sufficiently long period. Costs for muscle tension in isomorphs are likely to be proportional to volume, because they involve a certain energy investment per unit volume of muscle. In the section on feeding, I discussed briefly the energy involved in movement, {63}, which has a standard level that includes feeding. This can safely be assumed to be a small fraction of the total maintenance costs. Sustained powered movement such as in migration requires special treatment. Such activities involve temporarily enhanced metabolism and feeding. The occasional burst of powered movement hardly contributes to the general level of maintenance energy requirements. Sustained voluntary powered movement seems to be restricted to humans and even this seems of little help in getting rid of weight!

Energy lost in excretion products is here included in maintenance costs, because the excretion of nitrogen is the most important component. This flux is tied to protein turnover, the costs of which are also included in maintenance. Products directly derived from food can also be excreted. These products are linked to the feeding process and should, therefore, show up in the value for $\{\dot{A}_m\}$. Such a partitioning of products complicates the analysis of excretion fluxes and the practical significance is limited because the energy flux involved in excretion is usually very small. Microbial product formation is discussed on {189}.

Maintenance costs are here taken to be independent of the growth rate. Tempest and Neijssel [707] argued that the concentration gradients of potassium and glutamic acid can involve a substantial energy requirement in prokaryotes. However, the concentrations of these compounds vary markedly with growth rate so that this energy drain is not taken to be part of maintenance here, but as part of the overhead costs of the growth process. The high costs for patassium gradients is at odds with Ling's association-induction hypothesis [434], which states a.o. that virtually all K^+ in living cells exists in an absorbed state. The mechanism is via a liquid crystal type of structure for the cytoplasm [102].

Some species have specific maintenance costs, such as daphnids which produce moults every other day at 20 °C. The synthesis of new moults occurs in the intermoult period and is a continuous and slow process. The moults tend to be thicker in the larger sizes. The exact costs are difficult to pin down, because some of the weight refers to inorganic compounds, which might be free of energy cost. Larvaceans produce new feeding houses every 2 hours at 23 °C [214], and this contributes substantially to organic matter fluxes in oceans [11,12,157]. These costs are taken to be proportional to volume. The inclusion of costs for moults and houses in maintenance costs is motivated by the observation that these rates do not depend on feeding rate [214,407], but only on temperature. Euryhaline fishes have to invest energy for osmoregulation in waters that are not iso-osmotic. The cichlid *Oreochromis niloticus* is iso-osmotic at 11.6 ppt and 29% of the respiration rate at 30 ppt can be linked to osmoregulation [780]. Similar results have been obtained for brook trout *Salvelinus fontinalis* [237].

The maintenance costs $\dot{M}$, are thus taken to be proportional to volume

$$\dot{M} = [\dot{M}]V \tag{3.9}$$

and the volume-specific costs for maintenance, $[\dot{M}]$, can be partitioned into a variety of processes that together are responsible for these costs.

As stated on {39}, no maintenance costs are paid over reserves. The empirical justification can most easily be illustrated by the absence of respiration in freshly laid eggs, which consist almost entirely of reserves; see figure 3.14. Note, however, that costs for turnover of reserves are covered by overheads in assimilation and utilization. Although the difference between turnover costs for reserves and structural biomass is subtle, eggs show that the turnover costs for reserves are not equivalent with maintenance for reserves, since they do not respire when freshly laid.

3.7 Homeothermy

Heat comes free as a side product of all uses of energy, cf. {201}. In ectotherms, this heat simply dissipates without increasing the body temperature above that of the environment to any noticeable amount as long as the temperature is sufficiently low. If the environmental temperature is high, as in incubated bird eggs just prior to hatching, metabolic rates are high as well, releasing a lot more free energy in the form of heat and increasing the body temperature even further, cf. {135}. This is called a positive feedback in cybernetics. The rate of heat dissipation obviously depends on the degree of insulation and is directly related to surface area. A small number of species, known as endotherms, use energy for the purpose of keeping their body temperature at a predetermined high level, 34 °C in monothremes, 37 °C in most mammals, 39 °C in non-passerine birds, 41 °C in passerine birds. Mammals and birds change from ectotherms to endotherms during the first few days of their juvenile stage. Some species temporarily return to the ectothermic state or partly so in the night (kolibries) or during hibernation (rodents, bats) or torpor (tenrecs, cf. {131}). Not all parts of the body are kept at the target temperature, especially not the extremities. The naked mole rat *Heterocephalus glaber* (see figure 3.12) has a body temperature that is almost equal to that of the environment [441] and actually behaves as an ectotherm. Huddling in the nest of this colonial species plays an important role in thermoregulation [778]. Many ectotherms can approach the state of homeothermy under favourable conditions by walking from shady to sunny places, and back, in an appropriate

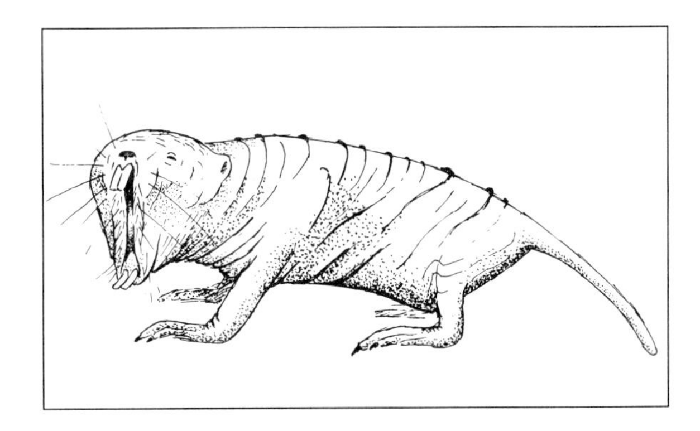

Figure 3.12: The naked mole rat *Heterocephalus glaber* (30 gram) is one of the few mammals that are essentially ectothermic. They live underground in colonies of some 60 individuals. The single breeding female suppresses reproductive development of all 'frequent working' females and of most 'infrequent working' females, a social system that reminds us of termites [442].

way. In an extensive study of 82 species of desert lizards from three continents, Pianka [551] found that body temperature T_b relates to ambient air temperature T_e as

$$T_b = 311.8 + (1 - \beta)(T_e - 311.8)$$

where β stands for the species-specific thermoregulatory capacity, spanning the full range from perfect regulation, $\beta = 1$ for active diurnal heliothermic species, to no regulation, $\beta = 0$ for nocturnal thigmothermic species. The target temperature of 311.8 K or 38.8 °C varied somewhat between the different sub-groups and is remarkably close to that of mammals. Other species can raise their temperature over 10 °C above that of the environment (bumble bees, moths, tuna fish, mackerel shark). These examples do make clear that energy investment into heating is species-specific and that the regulation of body temperature is a different problem.

The 'advantages' of homeothermy are that enzymes can be used that have a narrow tolerance range for temperatures and that activity can be maintained at a high level independent of environmental temperature. At low temperatures ectotherms are easy prey for endotherms. Development and reproduction are enhanced, which opens niches in areas with short growing seasons that are closed to ectotherms. The costs depend on the environmental temperature, insulation and body size. If temperature is high and/or insulation is excellent and/or body size is large, there may be hardly any additional costs for heating; the range of temperatures for which this applies is called the thermo-neutral zone. The costs for heating, $\dot{H}$, due to losses by convection or conduction can be written as

$$\dot{H} = \{\dot{H}\}V^{2/3} \qquad (3.10)$$

Heat loss is not only proportional to surface area but, according to Newton, also to the temperature difference between body and environment. This is incorporated in the concept of thermal conductance $\{\dot{H}\}/(T_e - T_b)$, where T_e and T_b denote the temperature of the environment and the body. It is about 5.43 $\mathrm{J\,cm^{-2}\,h^{-1}\,{}^\circ C^{-1}}$ in birds and 7.4-9.86 $\mathrm{J\,cm^{-2}\,h^{-1}\,{}^\circ C^{-1}}$ in mammals, as calculated from [311]. The unit $\mathrm{cm^{-2}}$ refers to volumetric squared length, not to real surface areas which involve shape. The values represent crude means in still air. The thermal conductance is roughly proportional to the square root of wind speed.

This is a simplified presentation. Birds and mammals moult at least twice a year, to replace their hair and feathers which suffer from wear, and change the thick winter coat for the thin summer one. Cat owners can easily observe that when their pet is sitting in the warm sun, it will pull its hair into tufts, especially behind the ears, to facilitate heat loss. Many species have control over blood flow through extremities to regulate temperature. People living in temperate regions are familiar with the change in the shape of birds in winter to almost perfect spheres. This increases insulation and generates heat from the associated tension of the feather muscles. These phenomena point to the variability of thermal conductance.

There are also other sources of heat exchange, through ingoing and outgoing radiation and cooling through evaporation. Radiation can be modulated by changes in colour, which chameleons and tree frogs apply to regulate body temperature [441]. Evaporation obviously depends on humidity and temperature. For animals that do not sweat, evaporation is tied to respiration and occurs via the lungs. Most non-sweaters pant when hot and lose heat by enhanced evaporation from the mouth cavity. A detailed discussion of heat balances would involve a considerable number of coefficients [492,677], and would obscure the main line of reasoning. I will, therefore, refrain from giving these details. It is important to realize that all these processes are proportional to surface area, and so affect the heating rate $\{\dot{H}\}$ and in particular its relationship with the temperature difference between body and environment.

3.8 Growth

Growth can now be derived on the basis of (3.8), (3.9), (3.10) and the κ-rule; see {74}. The κ-rule states that

$$\kappa\dot{C} = [G]\frac{d}{dt}V + \dot{M} + \dot{H} \tag{3.11}$$

where $[G]$ denotes the volume-specific costs for growth, which are taken as fixed in view of homeostasis of the structural biomass. These costs thus include all types of overhead costs, not just the costs for synthesis. There are no costs for heating for ectotherms, so $\dot{H} = 0$ for them. Substitution of (3.8), (3.9) and (3.10) gives

$$\frac{d}{dt}V = \dot{v}\frac{V^{2/3}[E]/[E_m] - V^{2/3}(V_h/V_m)^{1/3} - V/V_m^{1/3}}{[E]/[E_m] + g} \tag{3.12}$$

Note that growth does not depend on food density directly. It only depends on reserve density and body volume. The energy parameters combine in the compound parameters V_h, V_m, g and $\dot{v}$. The compound parameters will appear frequently in the sequel, so they are best introduced here. To aid memory, it is useful to give them names.

The *maintenance rate constant* $\dot{m} \equiv [\dot{M}]/[G]$ was introduced by Marr [455] and publicized by Pirt [556], and stands for the ratio of costs for maintenance and biovolume synthesis. It has dimension time^{-1}. It remains hidden here in the maximum volume V_m, but it will frequently play an independent role.

The quantity $g \equiv [G]/\kappa[E_m]$ is called the (energy) *investment ratio* and stands for the costs for new biovolume relative to the maximum potentially available energy for growth plus maintenance. It is dimensionless.

$V_m \equiv (\frac{\dot{v}}{g\dot{m}})^3 = (\kappa\{\dot{A}_m\}/[\dot{M}])^3$ stands for the *maximum volume* ectotherms can reach. (Endotherms cannot reach this volume because they loose energy through heating.) The comparison of species is based on this relationship between maximum volume and energy budget parameters and is the core of the relationship between body size and physiological variables together with the invariance property of the DEB model, to be discussed later, {217}.

The *heating volume* $V_h \equiv (\{\dot{H}\}/[\dot{M}])^3$ stands for the reduction in volume endotherms experience due to the energy costs for heating. It can be treated as a simple parameter as long as the environmental temperature remains constant. If the temperature changes slowly relative to the growth rate, the heating volume just a function of time. If environmental temperature changes rapidly, body temperature can be taken to be constant again while the effect contributes to the stochastic nature of the growth process, cf. {121}. Note that (3.12) shows that the existence of a heating volume is not an extra assumption, but a consequence of the volume-bound maintenance costs and the surface area-bound input and heating costs.

If food density X and, therefore, the scaled functional response f are constant, and if the initial energy density equals $[E] = f[E_m]$, energy density will not change. Volumetric length as a function of time since hatching where $V(0) \equiv V_b$, can then be solved from (3.12)

$$\frac{d}{dt}V^{1/3} = \frac{\dot{v}}{3(f+g)}\left(f - (V_h/V_m)^{1/3} - (V/V_m)^{1/3}\right) \tag{3.13}$$

$$V^{1/3}(t) = V_\infty^{1/3} - (V_\infty^{1/3} - V_b^{1/3})\exp\{-t\dot{\gamma}\} \quad \text{or} \tag{3.14}$$

$$t(V) = \frac{1}{\dot{\gamma}}\ln\frac{V_\infty^{1/3} - V_b^{1/3}}{V_\infty^{1/3} - V^{1/3}} \tag{3.15}$$

I will follow tradition and call this curve the von Bertalanffy growth curve despite its earlier origin and von Bertalanffy's contribution of introducing allometry, which I reject; see {12}. The von Bertalanffy growth rate equals

$$\dot{\gamma} \equiv (3/\dot{m} + 3fV_m^{1/3}/\dot{v})^{-1} \tag{3.16}$$

and the ultimate volumetric length

$$V_\infty^{1/3} \equiv fV_m^{1/3} - V_h^{1/3} \tag{3.17}$$

Time t in (3.14) is measured from hatching or birth. (Note that time and age are not the same.) The von Bertalanffy growth curve results for isomorphs at constant food density and temperature and has been fitted successfully to the data of some 270 species from many different phyla; see table 6.2 and [410]. The gain in insight since Pütter's original formulation in 1920 is in the interpretation of the parameters in terms of underlying processes. It appears that heating costs do not affect the von Bertalanffy growth rate $\dot{\gamma}$.

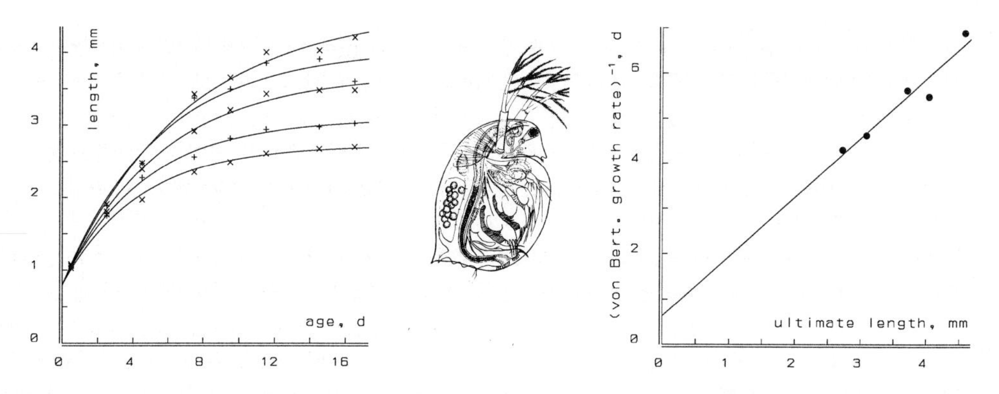

Figure 3.13: The left figure shows the length-at-age data of the waterflea *Daphnia magna* for various densities of the green alga *Chlorella* at 20 °C with von Bertalanffy growth curves. Data from [407]. The inverses of the estimated von Bertalanffy growth rates have been plotted against estimated ultimate lengths (right). The expected relationship is $\dot{\gamma}^{-1} = 3/\dot{m} + 3d_m L_\infty/\dot{v}$. The least squares fitted line gives estimates for $\dot{v}/d_m$ of 2.29 mm d^{-1} and for $\dot{m}$ of 4.78 d^{-1}, both of which seem to be too high in comparison with other species. Frequent moulting may contribute to the maintenance costs and so to the high estimate for the maintenance rate coefficient $\dot{m}$.

Being a rate, high temperature does elevate it, of course. Food density affects both the von Bertalanffy growth rate and the ultimate volume. The inverse of the von Bertalanffy growth rate is a linear function of the ultimate volumetric length; see figure 3.13. This is consistent with the Pütter's original formulation, which took this rate to be inversely proportional to ultimate length, as has been proposed again by Gallucci and Quinn [242].

The requirement that food density is constant for a von Bertalanffy curve can be relaxed if food is abundant. This is due to the hyperbolic functional response. As long as food density is higher than 4 times the saturation coefficient, food intake is higher than 80% of the maximum possible food intake, which makes it hardly distinguishable from maximum food intake. Since most birds and mammals have a number of behavioural traits aimed at guaranteed adequate food availability, they appear to have a fixed volume-age relationship. This explains the popularity of age-based models for growth in 'demand' systems. In the next chapter I will discuss deviations from the von Bertalanffy growth curve that can be understood in the context of the present theory.

In contrast, at low food densities, fluctuations in food density soon induce deviations from the von Bertalanffy curve. This phenomenon will be discussed further in the section on genetics and parameter variation, {112}.

Maximum volumetric length is reached at prolonged exposure to high food densities, where $f = 1$, which gives $V_\infty^{1/3} = V_m^{1/3} - V_h^{1/3}$. If the juvenile period ends upon exceeding volume V_p, the length of this period is $t(V_p)$ at constant food density, as given in (3.15).

In the discussion on population dynamics, it will become important to distinguish time, t, from age, a. The age at the end of the juvenile period, so at puberty, is thus $a_p = a_b + t(V_p)$, if a_b stands for the age at birth and fertilization initializes age.

Growth ceases, i.e. $\frac{d}{dt}V = 0$, if $[E] = (\{\dot{H}\} + [\dot{M}]V^{1/3})/\kappa\dot{v}$. If the energy density drops even further, some organisms, such as protozoa and coelenterates, shrink. Even animals with a skeleton, such as shrews of the genus *Sorex*, can exhibit a geographically varying winter size depression, known as the Dehnel phenomenon [250]. Molluscs seem be to able to reduce shell size [179]. Some animals only deviate from the κ-rule in situations of prolonged starvation, that is, they still follow first order dynamics for the use of energy reserves, pay maintenance (and heating) costs, the rest being spent on development and/or reproduction, whereas others deviate from first order dynamics for the utilization of energy, (3.7). These species only pay maintenance (and heating) costs, so

$$\frac{d}{dt}[E] = f\{\dot{A}_m\}V^{-1/3} - [\dot{M}] - \{\dot{H}\}V^{-1/3} \tag{3.18}$$

where V remains fixed. At constant food density, thus constant energy uptake rate, this dynamics implies that energy density either increases to the no-growth boundary or decreases to zero. Pond snails are a beautiful example of a species that follows both strategies for energy expenditure, depending on day length. When in a long day/short night cycle, they reproduce continuously, but they cease to do so in a short day/long night rhythm. This will be discussed further in the next chapter, {128}. Although food availability does not influence growth directly, it does so indirectly via reserve energy. Moreover, the maximum surface area-specific assimilation rate $\{\dot{A}_m\}$, and so energy conductance $\dot{v}$, relate to the food-energy conversion. Many herbivores, such as chickens, eat animal products in the early juvenile period to gain nitrogen, which they need for the synthesis of proteins. They experience a shift in diet during development. Mammals feed milk to their offspring, this needs little conversion and induces growth rates that cannot be reached with their later diet. Growth curves show a sharp kink at weaning.

Animals that have non-permanent exoskeletons (arthropods, insects) have to moult to grow. The rapid increase in size during the brief period between two moults, relates to uptake of water or air, not to synthesis of new structural biomass, which is a slow process occurring during the intermoult period. This minor deviation from the DEB model (see figure 3.11 and [409]) relates more to size measures than to model structure.

3.8.1 Embryonic growth

The DEB model takes the bold view that the only essential difference between embryos and juveniles is that the former do not feed. Although information on parameter values is still sparse, it indicates that no (drastic) changes of values occur at the transition from the embryonic to the juvenile state. I will first discuss eggs, which do not take up energy from the environment. (See [113] for an excellent introduction to eggs, with beautiful photographs.) Subsequently, I will deal with foetuses, which obtain energy reserves from the mother during development.

The idea is that the dynamics for growth, (3.12), and reserve density, (3.7), also apply to embryos in eggs in absence of food intake. The scaled functional response is thus taken to be $f = 0$. The dynamics for the reserve density then reduces to

$$\frac{d}{dt}[E] = -\dot{v}[E]V^{-1/3} \tag{3.19}$$

The initial volume is practically nil, so $V(0) = 0$. This makes the energy density infinitely large, so $[E](0) = \infty$. The (absolute) initial energy is a certain amount, $[E](0)V(0) = E_0$, which, however, is not considered to be a free parameter. Its value is determined from the condition of the energy reserves at hatching. Hatching occurs at age a_b, say, and initial energy density $[E_b]$, so $[E](a_b) = [E_b]$. The just-born juvenile still needs some energy reserves to cope with its metabolic needs. If all utilized energy is used for maintenance at hatching, a lower boundary for reserve energy density follows from $[\dot{M}]V_b = \dot{v}[E_b]V_b^{2/3}$, giving $[E_b] = [\dot{M}]V_b^{1/3}/\dot{v}$.

If food density is constant, the energy density will change from the one at hatching, called $[E_b]$, to $f[E_m]$ in juveniles. If energy density at hatching is about equal to $f[E_m]$, the growth curve will follow a von Bertalanffy curve. For initial energy densities less than $f[E_m]$, growth will be retarded compared to the von Bertalanffy growth curve; the opposite holds for initial densities larger than $f[E_m]$. Although the deviation from the von Bertalanffy growth curve will not last long, because the relaxation time for energy density is proportional to length, which is small at hatching, it is tempting to take the initial energy density to be equal to that of the mother at egg laying. This results in von Bertalanffy growth at constant food density even just after hatching, and it does not require additional parameters.

Tests on the realism of the initial condition that $[E_b]$ equals $[E]$ of the mother at egg laying are conflicting for daphnids. The triglycerides component of energy density is visible as a yellow colour and as droplets. I have observed that well-fed, yellow mothers of *Daphnia magna* give birth to yellow offspring, and poorly fed, glassy mothers give birth to glassy offspring. This is consistent with observations of Tessier *et al.* [710]. Later observations by Tessier as well as by Lisette Enserink, however, indicate an inverse relationship between food density and energy reserves at hatching. An increase of energy investment per offspring can also result in larger offspring rather than an increased reserve density at hatching. Large bodied offspring at low food availability has been described for the terrestrial isopod *Armadillium vulgare* [96]. Because of the relationship with energy costs for egg production, and so with reproduction rate, this response to resource depletion has implications for population dynamics. It can be viewed as a mechanism that aims to ensure adequate food supply for the existing individuals. The condition that energy density at hatching equals that of the mother at egg formation is made here for reasons of simplicity and theoretical elegance. No theoretical barriers exist for other formulations within the context of the DEB theory. Such formulations are likely to involve species-specific empirical or optimization arguments, however, which I have tried to avoid as much as possible.

Embryo development provides excellent opportunities to test the model for the dynamics of energy reserves, because of the huge change of energy density, which avoids the pathological conditions starving individuals face. As embryos do not feed, data on their development do not suffer from a major source of scatter.

The goodness of fit is remarkable, as illustrated in figure 3.14, where data on weight, yolk and respiration have been fitted simultaneously by Cor Zonneveld [790]. The total number of parameters is 5 excluding, or 7 including, respiration. As will be discussed later, {103}, respiration is taken to be proportional to the utilization rate. This makes up only 2.5 parameters per data set and thus approaches a straight line for simplicity when measured this way. The examples are representative of the data collected in table 3.1, which gives parameter estimates of some 40 species of snails, fish, amphibians, reptiles and birds. The model tends to underestimate embryo weight and respiration rate in the early phases of development. This is partly due to deviations in isomorphism, the contributions of extra-embryonic membranes (both in weight and in the mobilization of energy reserves), and the loss of water content during development. The estimates for the altricial birds such as the parrot *Agapornis* should be treated with some reservations, because neglected acceleration due to temperature increase during development substantially affects the estimates, as discussed on {135}.

The values for the energy conductance $\dot{v}$, as given in table 3.1, are in accordance with the average value for post-embryonic development, as given on {224}, which indicates that no major changes in energy parameters occur at birth. The maintenance rate constant $\dot{m}$ for reptiles and birds is about 0.08 d^{-1} at 30 °C, implying that the energy required to maintain tissue during 12 days at 30 °C is about equal to the energy necessary to synthesize the tissue from the reserves. The maintenance rate constant for fresh water species seems to be much higher, ranging from 0.3 to 2.3 d^{-1}. Data from Smith [670] on the rainbow trout *Salmo irideus*, now called *S. gairdnerii*, result in 1.8 d^{-1} and figure 3.13 gives over 10 d^{-1} for the waterflea *Daphnia magna* at 30 °C. The costs for osmosis might contribute to these high maintenance costs, as has been suggested on {78}.

Table 3.1 shows that about half of the reserves are used during embryonic development. The deviating values for altricial birds are artifacts, due to the mentioned acceleration of development by increasing temperatures. Congdon *et al.* [133] observed that the turtles *Chrysemus picta* and *Emydoidea blandingi* have 0.38 of the initial reserves at birth. Respiration measurements on sea birds by Pettit *et al.* [545] indicate values that are somewhat above the ones reported in the table. The extremely small value for the soft shelled turtle, see also figure 3.14, relates to the fact that these turtles wait for the right conditions to hatch, where they have to run the gauntlet as a cohort at night from the beach to the water, where a variety of predators are waiting for them.

The general pattern of embryo development in eggs is characterized by unrestricted fast development during the first part of the incubation period (once it has started the process) due to unlimited energy supply, at a rate that would be impossible to reach if the animal had to refill reserves by feeding. This period is followed by a retardation of development due to the increasing depletion of energy reserves. Due to the goodness of fit of the model in species that do not possess shells, retardation is unlikely to be due to limitation of gas diffusion across the shell, as has been frequently suggested for birds [579]. Such a limitation also fails to explain why respiration declines in some species after its peak value, here beautifully illustrated with the turtle data.

Large eggs, so large initial energy supplies, thus result in short incubation times if eggs of one species are compared. Crested penguins, *Eudyptes*, are known for egg dimorphism [749]; see figure 3.15. They first lay a small egg and, some days later a 1.5 times bigger one. As predicted by the DEB model, the bigger one hatches first, if fertile, in which case the parents cease incubating the smaller egg, because they are only able to raise one chick. They continue to incubate the small egg only if the big one fails to hatch. This is probably an adaptation to the high frequency of unfertilized eggs or other causes of loss of eggs

Figure 3.14: Yolk-free embryo weight (◇), yolk weight (×), and respiration rate (+) during embryo development, and fits on the basis of the DEB model. Data sources are indicated.

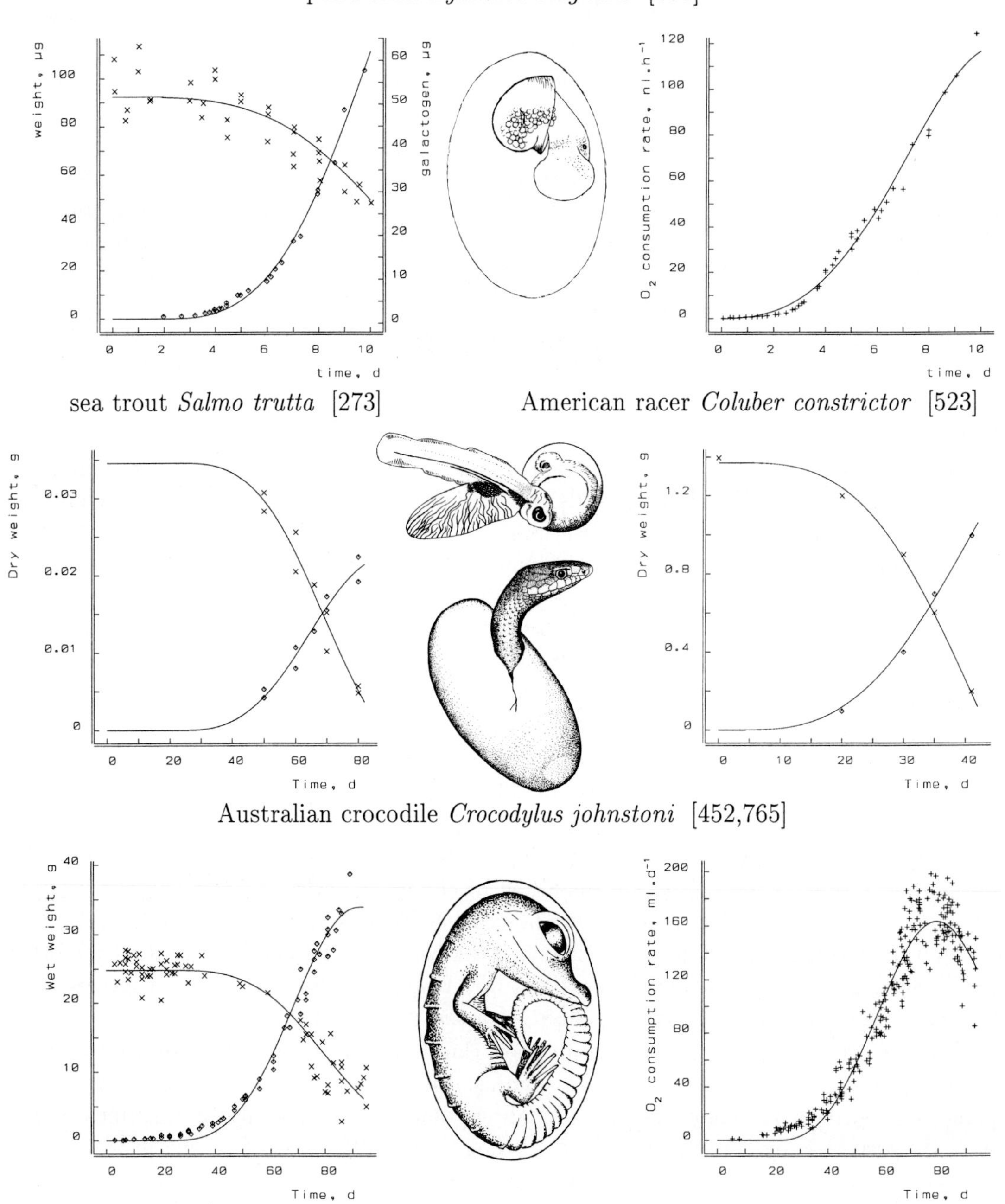

New Guinea soft-shelled turtle *Carettochelys insculpta* [755]

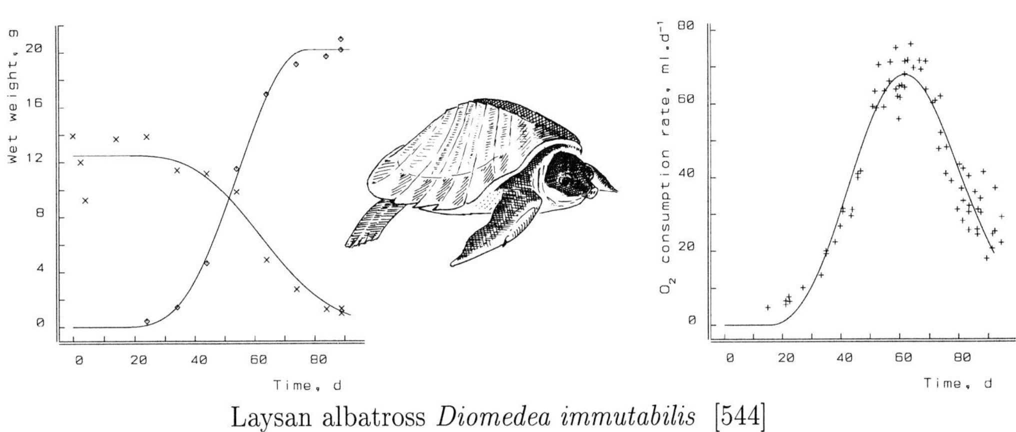

Laysan albatross *Diomedea immutabilis* [544]

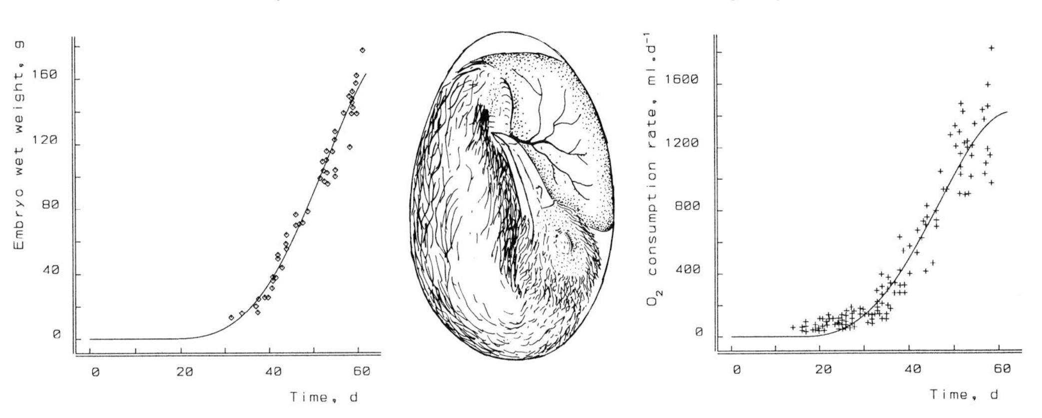

Table 3.1: Survey of re-analyzed egg data, and parameter values standardized to a temperature of 30 °C, taken from [790]. *1* Whitehead, pers. comm., 1989 ; *2* Thompson, pers. comm., 1989; 'galac.', stands for galactogen content.

species	temp. °C	type of data	$\dot{v}_{30}$ mm d^{-1}	$\dot{m}_{30}$ d^{-1}	E_b/E_0	reference
Lymnaea stagnalis	23	ED, galac, O	0.80	2.3	0.55	[335]
Salmo trutta	10	ED, YD	3.0	0.31	0.37	[273]
Rana pipiens	20	EW, O	2.5		0.87	[25]
Crocodylus johnstoni	30	EW, YW	1.9	0.060	0.31	[452]
	29, 31	O				[765]
Crocodylus porosus	30	EW, YW	2.7	0.024	0.19	[756]
	30	O				*1*
Alligator mississippiensis	30	EW, YW	2.7		0.34	[160]
	30	O				[716]
Chelydra serpentina	29	ED, YD	1.9		0.35	[523]
	29	O				[251]
Carettochelys insculpta	30	EW, YW, O	1.9	0.040	0.08	[755]
Emydura macquarii	30	EW, O	1.6	0.14	0.35	[716]
Caretta caretta	28-30	EW, O	3.0		0.65	[3,2]
Chelonia mydas	28-30	EW, O	3.0		0.57	[3,2]
Amphibolurus barbatus	29	ED, YD	0.92	0.061	0.47	[524]
Coluber constrictor	29	ED, YD	1.4		0.69	[525]
Sphenodon punctatus	20	HM, O	0.85	0.062	0.25	*2*
Gallus domesticus	39	EW, O, C	3.2	0.039	0.34	[610]
Gallus domesticus	38	EW, C	3.4		0.52	[77]
Leipoa ocellata	34	EE, YE, O	1.7	0.031	0.55	[743]
Pelicanus occidentalis	36.5	EW, O	3.2	0.10	0.77	[38]
Anous stolidus	35	EW, O	2.0	0.11	0.59	[546]
Anous tenuirostris	35	EW, O	1.8	0.20	0.59	[546]
Diomedea immutabilis	35	EW, O	2.5	0.069	0.57	[544]
Diomedea nigripes	35	EW, O	2.5	0.049	0.58	[544]
Puffinus pacificus	38	EW, O	0.92	0.084	0.61	[4]
Pterodroma hypoleuca	34	EW, O	1.9		0.20	[544]
Larus argentatus	38	EW, C	2.7	0.15	0.56	[182]
Gygis alba	35	EW, O	1.4		0.53	[543]
Anas platyrhynchos	37.5	EW	2.5	0.10	0.67	[569]
	37.5	O				[372]
Anser anser	37.5	EW	4.1	0.039	0.23	[609]
	37.5	O				[741]
Coturnix coturnix	37.5	EW, O	1.7		0.49	[741]
Agapornis personata	36	EW, O	0.8		0.79	[107]
Agapornis roseicollis	36	EW, O	0.84		0.81	[107]
Troglodytes aëdon	38	EW, O	1.4		0.82	[378]
Columba livia	38	EW	2.7		0.80	[373]
	37.5	O				[741]

EW: Embryo Wet weight
EE: Embryo Energy content
O: Oxygen consumption rate
YW: Yolk Wet weight
YE: Yolk Energy content
C: Carbon dioxide prod. rate
ED: Embryo Dry weight
YD: Yolk Dry weight
HW: Hatchling Wet weight

Figure 3.15: Egg dimorphism occurs standard in crested penguins (genus *Eudyptes*). The small egg is laid first, but it hatches later than the big one, which is 1.5 times as heavy. The DEB theory explains why the large egg requires a shorter incubation period. The illustration shows the Snares crested penguin *E. atratus*.

(aggression [749]), which occurs in this species.

Incubation periods only decrease for increasing egg size if the structural biomass of the hatchling is constant. The incubation period is found to increase with egg size in some beetle species, lizards and marine invertebrates [128,206,660]. In these cases, however, the structural biomass at hatching also increased with egg size. This is again consistent with the DEB theory, although it does not explanain the variation in egg sizes.

Foetal development differs from that in eggs in that energy reserves are supplied continuously via the placenta. The feeding and digestion processes are not involved. Otherwise, foetal development is taken to be identical to egg development, with initial reserves that can be taken to be infinitely large, for practical purposes. At birth, the neonate receives an amount of reserves from the mother, such that the reserve density of the neonate equals that of the mother. So the approximation $[E] = \infty$ for the foetus can be made for the whole gestation period and the dynamics of the reserve density (3.19) no longer applies, because the foetus lives on the reserves of the mother. In other words: unlike eggs, the development of foetuses is not restricted by energy reserves. Initially the egg and foetus develop in the same way, but the foetus keeps developing at an unrestricted rate till the end of the gestation time, while the development of the egg becomes retarded, due to depletion of the reserves. The approximation $[E] = \infty$ reduces the growth equation (3.12) to

$$\frac{d}{dt}V = \dot{v}V^{2/3} \quad \text{so} \tag{3.20}$$

$$V(t) = (\dot{v}t/3)^3 \tag{3.21}$$

This growth curve was proposed by Huggett and Widdas [341] in 1951. Payne and Wheeler [536] explained it by assuming that the growth rate is determined by the rate at which nutrients are supplied to the foetus across a surface that remains in proportion to the total

Figure 3.16: Foetal weight development in mammals.

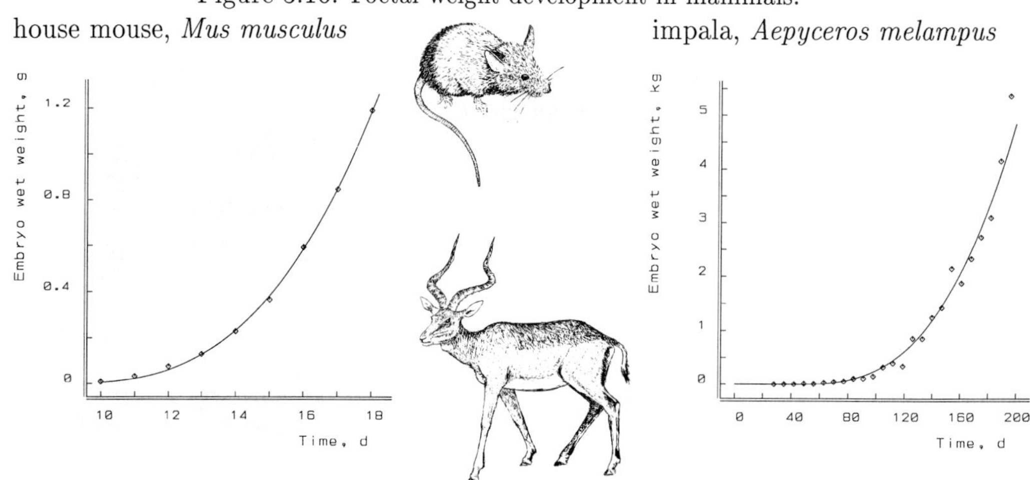

surface area of the foetus itself. This is consistent with the DEB model, which gives the energy interpretation of the single parameter.

The fit is again excellent; see figure 3.16. It is representative for the data collected in table 3.2 taken from [790]. A time lag for the start of foetal growth has to be incorporated, and this delay may be related to the development of the placenta, which possibly depends on body volume as well. The long delay for the grey seal *Halichoerus* probably relates to timing with the seasons to ensure adequate food supply for the developing juvenile. Variations in weight at birth are primarily due to variations in gestation period, not in foetal growth rate. For comparative purposes, energy conductance $\dot{v}$ is converted to 30 °C, on the assumption that the Arrhenius temperature is $T_A = 10200$ K and the body temperature is 37 °C for all mammals in the table. This is a rather crude conversion because the cat, for instance, has a body temperature of 38.6 °C. Weights were converted to volumes using a specific density of $[d_w] = 1 \text{ g cm}^{-3}$.

One might expect that precocial development is rapid, resulting in advanced development at birth and, therefore, comes with a high value for the energy conductance. The values collected in table 3.2, however, do not seem to have an obvious relationship with altricial-precocial rankings. The precocial guinea-pig and alpaca as well as the altricial humans have relatively low values for the energy conductance. The altricial-precocial ranking seems to relate only to the relative volume at birth V_b/V_m.

Egg costs

The embryo thus develops from state $(a, [E], V) = (0, \infty, 0)$ to state $(a_b, [E_b], V_b)$. The costs for growth and maintenance together with κ determine the energy costs for an egg, E_0. These costs and the incubation time thus follow from specifications at hatching. This back reasoning is necessary because the initial volume is taken to be infinitesimally small, which makes the initial reserve density infinitely large.

The derivation of the costs for an egg is a bit technical, I am afraid, due to the non-linearity of the dynamics. We will need the costs to go from an energy flux allocated

Table 3.2: The estimated energy conductance, $\dot{v}$, and its value corrected for a temperature of 30 °C, and the time lag for the start of development, t_l, for mammalian embryos.

Species (race)	$\dot{v}$ $mm\,d^{-1}$	(cv)	$\dot{v}_{30}$ $mm\,d^{-1}$	t_l d	(cv)	reference
Homo sapiens			0.84			
males	0.180	(0.3)		26.8	(2.0)	[758]
females	0.179	(0.4)		26.5	(2.9)	
Oryctolagus cuniculus	0.560	(0.9)	2.6	10.7	(1.5)	[432]
small litters	0.602	(1.5)		11.5	(2.4)	[34]
large litters	0.571	(1.5)		11.5	(2.4)	[34]
	0.504	(5.6)		10.4	(10)	[36]
Lepus americanus	0.573	(3.1)	2.7	13.1	(4.2)	[79]
Cavia porcellus	0.269	(3.3)	1.1	15.7	(8.3)	[180]
	0.239	(2.3)				[346]
Cricetus auratus	0.570	(2.1)	2.6	9.29	(1.3)	[573]
Mus musculus	0.333	(0.1)	1.5	8.45	(0.1)	[448]
Rattus norvegicus			2.5			
wistar	0.487	(0.5)		11.4	(0.3)	[217]
albino	0.531	(0.8)		12.2	(0.5)	[688]
	0.525	(0.2)		11.8	(0.2)	[341]
albino	0.568	(3.3)		12.7	(2.1)	[16]
albino	0.542	(3.1)		12.4	(2.0)	[222]
Clethrionomys glareolus	0.374	(9.3)	1.8	8.29	(11)	[140]
Aepyceros melampus	0.316	(1.2)	1.4	39.4	(3.8)	[210]
Odocoileus virginianus	0.296	(6.7)	1.3	34.9	(28)	[605]
	0.274	(1.6)		25.1	(8.5)	[733]
Dama dama	0.345	(6.4)	1.7	9.94	(46)	[21]
Cervus canadensis	0.336	(3.1)	1.5	24.9	(19)	[494]
Lama pacus	0.120	(7.6)	0.56	7.47	(83)	[218]
Ovis aries			1.9			
welsh	0.482	(5.6)		43.9	(12)	[341]
merino	0.341	(8.6)		14.9	(71)	[450]
	0.346	(4.6)		15.2	(32)	
	0.433	(4.4)		33.3	(13)	[130]
karakul	0.436	(3.7)		31.0	(13)	[193]
	0.403	(2.6)		27.5	(8.2)	[365]
hampshire ×	0.382	(1.5)		20.4	(7.9)	[775]
Capra hircus	0.339	(6.5)	1.7	24.3	(29)	[199]
	0.365	(4.5)		31.3	(14)	[35]
Bos taurus	0.475	(2.6)	2.3	59.5	(7.5)	[776]
Equus caballus	0.370	(11)	1.8	37.0	(81)	[481]
Sus scrofa	0.266	(0.6)		4.73	(12)	[750]
Yorkshire	0.283	(0.9)		5.49	(16)	[729]
Large white	0.383	(1.3)		23.6	(4.2)	[562]
Essex	0.321	(4.8)		14.1	(30)	
Felix catus	0.371	(1.2)	1.8	18.8	(2.3)	[139]
Pipistrellus pipistrellus			0.97			
1978	0.237	(1.9)		9.95	(2.9)	[575]
1979	0.181	(3.5)		13.7	(4.7)	
Halichoerus grypus	0.375	(10)	1.8	145	(9.2)	[314]

to reproduction to a reproductive rate. You will not miss a lot if you skip the rest of this section, if you are ready to accept the result that egg costs do not involve any new parameters. Costs for breeding by the parent are not included in this derivation.

The first step to derive the costs for an egg is to get rid of a number of parameters by turning to the dimensionless variables scaled energy density $e = [E]/[E_m]$, scaled volumetric length $l = (V/V_m)^{1/3}$ and scaled time $\tau = t\dot{m}$. Substitution into (3.19) and (3.12), reduces the coupled differential equations to

$$\frac{d}{d\tau}e = -g\frac{e}{l} \text{ and } \frac{d}{d\tau}l = \frac{g}{3}\frac{e-l}{e+g} \tag{3.22}$$

The ratio of these equations gives the Bernoulli equation

$$\frac{dl}{de} = -\frac{l}{3e}\frac{e-l}{e+g} \text{ or } \frac{dx}{de} = \frac{ex-1}{3e(e+g)} \tag{3.23}$$

where $x \equiv l^{-1}$ is only introduced because the resulting equation in x is of a solvable linear first order with variable coefficients. Its solution is

$$x(e) = v(e)\left(\int_{e_b}^{e} \frac{-de_1}{3(e_1+g)e_1 v(e_1)} + x(e_b)\right) \tag{3.24}$$

$$\text{with } v(e) = \exp\{\int_{e_b}^{e} \frac{de_1}{3(g+e_1)}\} = \left(\frac{g+e}{g+e_b}\right)^{1/3}$$

Substitution of $l = x^{-1}$ gives

$$\frac{1}{l} = \left(\frac{g+e}{g+e_b}\right)^{1/3}\left(\frac{1}{l_b} - \frac{(g+e_b)^{1/3}}{3g^{4/3}}\int_{\frac{e_b}{e_b+g}}^{\frac{e}{e+g}} s^{-1}(1-s)^{1/3}\,ds\right) \tag{3.25}$$

Assume that the condition at hatching is fixed at e_b and l_b and let $l \to 0$ and $e \to \infty$ such that $[E_m]V_m e l^3 = E_0$, say, which has the interpretation of the energy reserves in a freshly laid egg. Solving E_0 gives for $e_0 \equiv \frac{E_0}{[E_m]V_m}$

$$e_0 = \left(\frac{1}{l_b(g+e_b)^{1/3}} - \frac{B_{\frac{g}{e_b+g}}(\frac{4}{3},0)}{3g^{4/3}}\right)^{-3} \tag{3.26}$$

where $B_x(a,b) \equiv \int_0^x y^{a-1}(1-y)^{b-1}\,dy$ is the incomplete beta function. Its two term Taylor expansion in $[G]$ around the point $[G] = 0$ gives

$$e_0 \simeq \frac{2^6 e_b^4}{(4e_b/l_b - 1)^3} + \frac{4e_b/l_b - 16/7}{(4e_b/l_b - 1)^4} g 2^6 e_b^3 \tag{3.27}$$

Incubation time

The incubation time can be found by separating variables in (3.22) and substituting in (3.25). After some transformation, the result is

$$a_b = \frac{3}{\dot{m}}\int_0^{x_b} \frac{dx}{(1-x)x^{2/3}(\alpha - B_{x_b}(\frac{4}{3},0) + B_x(\frac{4}{3},0))} \tag{3.28}$$

where $x_b \equiv \frac{g}{e_b+g}$ and $\alpha \equiv 3gx_b^{1/3}/l_b$. Its two term-Taylor expansion in $[G]$ around the point $[G]=0$ gives after tedious calculation

$$a_b \simeq \frac{3\sqrt{2}}{\dot{m}}u^3\left(\frac{e_b}{g}+\frac{1}{4}-\frac{9}{28}u^4\right)\left(\frac{1}{2}\ln\frac{u^2+u\sqrt{2}+1}{u^2-u\sqrt{2}+1}+\arctan\frac{u\sqrt{2}}{1-u^2}\right)+\frac{9}{7\dot{m}}(u^4+\ln\{1+u^4\}) \tag{3.29}$$

where u stands for $(4e_b/l_b-1)^{-1/4}$. I owe you an apology for writing out such a threatful expression; the essence, however, is that no new parameters show up and that (3.29) can readily be implemented in computer code.

Foetal costs and gestation time

The energy costs for the production of a neonate is found by the addition of costs for development, growth and maintenance plus energy reserves at birth, i.e. $[E_b]V_b$. Expressed as a fraction of the maximum energy capacity of an adult, these costs are

$$e_0 = (\int_0^{a_b} \dot{C}(t)\,dt + [E_b]V_b)([E_m]V_m)^{-1}$$

Substitution of the κ-rule, $\kappa\dot{C} = [G]\frac{d}{dt}V + [\dot{M}]V$, and the growth curve (3.21) results in

$$e_0 = l_b^3(g+e_b+l_b3/4) \tag{3.30}$$

This expression does not include the costs for the placenta. These costs can easily be taken into account if they happen to be proportional to that of the rest of the foetuses; see {100}.

Gestation time (excluding any time lag) is

$$a_b = 3l_b/g\dot{m} = 3V_b^{1/3}/\dot{v} \tag{3.31}$$

3.8.2 Growth for non-isomorphs

The above derivation assumes isomorphism, but it can easily be extended to include changing shapes. The surface areas of organisms that change shape, such as filaments and rods, have to be corrected for this change by multiplying parameters containing surface area, $\{\dot{I}_m\}$ and $\{\dot{A}_m\}$ and thus $\dot{v}$ and V_m, by the shape correction function $\mathcal{M}(V)$. These organisms are ectothermic, so $\{\dot{H}\}=0$. For filaments, the shape correction function (2.4) transforms the change of energy density (3.7) and the growth rate (3.12) into

$$\frac{d}{dt}[E] = [\dot{A}_m](f-[E]/[E_m]) \tag{3.32}$$

$$\frac{d}{dt}V = \dot{\nu}\frac{[E]/[E_m]-(V_d/V_m)^{1/3}}{g+[E]/[E_m]}V \tag{3.33}$$

where V_d is the volume at division, and V_m is defined by $V_m^{1/3} = \frac{\dot{v}}{g\dot{m}}$. The length-specific energy conductance $\dot{\nu}$ is just an abbreviation for $\dot{\nu} \equiv \dot{v}V_d^{-1/3}$. It has dimension time^{-1}. Likewise, the notation $[\dot{A}_m] \equiv \{\dot{A}_m\}V_d^{-1/3}$ is introduced. If substrate density X and, therefore, the scaled functional response f are constant long enough, energy density tends to $[E] = f[E_m]$ and volume as a function of time since division becomes for $V(0) = V_d/2$

$$V(t) = \frac{1}{2} V_d \exp\{t\dot{\gamma}_f\} \quad \text{or} \tag{3.34}$$

$$t(V) = \dot{\gamma}_f^{-1} \ln\{2V/V_d\} \tag{3.35}$$

with $\dot{\gamma}_f \equiv \dot{\nu}\frac{f-(V_d/V_m)^{1/3}}{f+g}$. The time taken to grow from $V_d/2$ to V_d is thus $t(V_d) = \dot{\gamma}_f^{-1} \ln 2$.

Exponential growth can be expected if the surface area at which nutrients are taken up is proportional to volume. For filaments, this happens when the total surface area, or a fixed fraction of it, is involved. If uptake only takes place at tips, the number of tips should increase with total filament length to ensure exponential growth. This has been found for the fungi *Fusarium* [725], and *Penicillium* [506,558], which do not divide; see figure 3.17. The ascomycetous fungus *Neurospora* does not branch this way [201]; it has a mycelium that grows like a crust, see {145}.

Exponential growth of individuals should not be confused with that of populations. As will be discussed in the chapter on population dynamics, all populations grow exponentially at resource densities that are constant long enough, whatever the growth pattern of individuals. This is due to the simple fact that the progeny repeats the growth/reproduction behaviour of the parents. Only for filaments it is unnecessary to distinguish between the individual and the population level. This is a characteristic property of exponential growth of individuals and will be discussed on {162}.

The same derivation for growth can be made for rods on the basis of the shape correction function (2.7):

$$\frac{d}{dt}[E] = [\dot{A}_m]\left(\frac{\delta}{3}\frac{V_d}{V} + 1 - \frac{\delta}{3}\right)\left(f - \frac{[E]}{[E_m]}\right) \tag{3.36}$$

$$\frac{d}{dt}V = \dot{\nu}\frac{\delta V_d}{3V_\infty}\frac{[E]/[E_m]}{g + [E]/[E_m]}(V_\infty - V) \tag{3.37}$$

where $V_\infty \equiv V_d\frac{\delta}{3}(\frac{[E_m]}{[E]}(\frac{V_d}{V_m})^{1/3} - 1 + \frac{\delta}{3})^{-1}$ and, as before, $V_m^{1/3} \equiv \frac{\dot{\nu}}{g\dot{m}}$. If substrate density X and, therefore, the scaled functional response f are constant long enough, energy density tends to $[E] = f[E_m]$ and volume as a function of time since division becomes

$$V(t) = V_\infty - (V_\infty - V_d/2)\exp\{-t\dot{\gamma}_r\} \tag{3.38}$$

where $\dot{\gamma}_r \equiv \frac{V_d f \dot{\nu}\delta/3}{V_\infty(f+g)}$. The interpretation of V_∞ depends on its value.

- If $V_\infty = \infty$, i.e. if $f(1-\delta/3) = (V_d/V_m)^{1/3}$, the volume of rods grows linearly at rate $\frac{\dot{\nu}f}{f+g}V_d\frac{\delta}{3}$. This is frequently found empirically [29].

- If $0 < V_\infty < \infty$, V_∞ is the ultimate volume if the cell ceased to divide but continued to grow. For these values, $V(t)$ is a convex function and is of the same type as $V(t)^{1/3}$ for isomorphs, (3.14). Note that volume, and thus cubed length, grows skewly S-shaped for isomorphs. When V_∞ is positive, the cell will only be able to divide when $V_\infty > V_d$, thus when $f > (V_d/V_m)^{1/3}$.

Figure 3.17: DEB-based growth curves for cells of filaments and rods. The larger the aspect ratio, δ, the more the growth curve turns from the exponential to the satiation type, reflecting the different surface area to volume relationships.

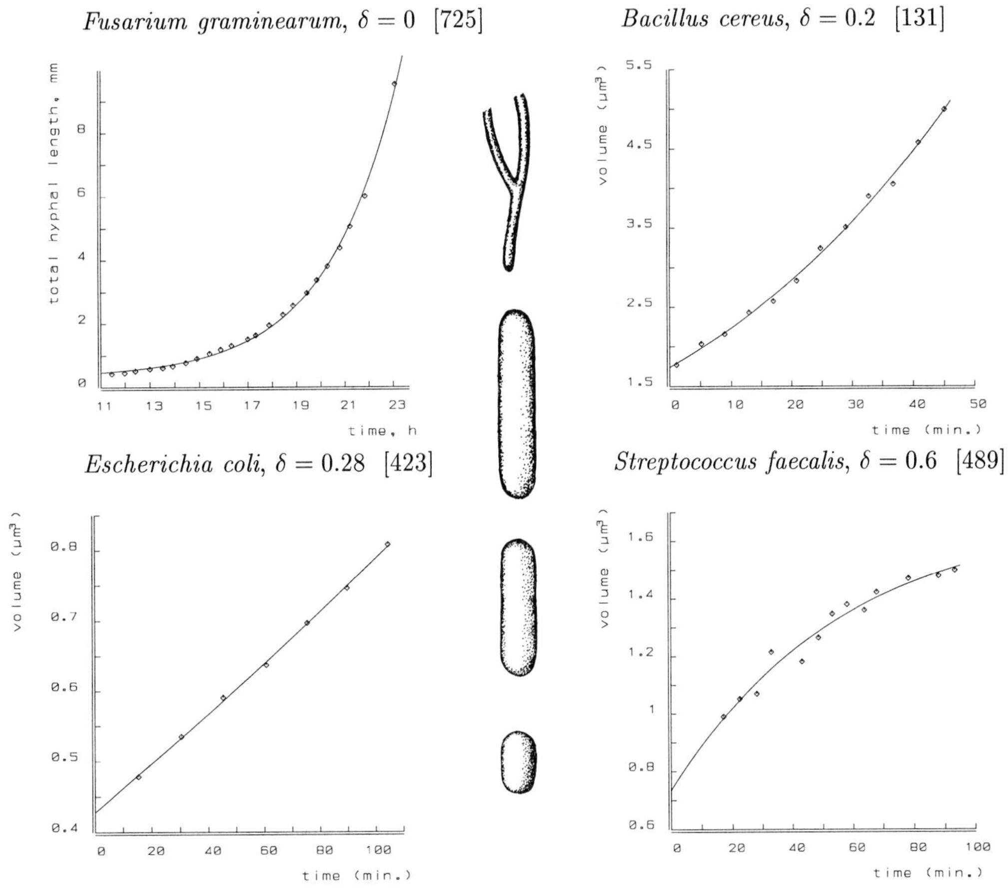

- If $\delta = 0$, $V_\infty = 0$ and the rod behaves as a filament, which grows exponentially.

- For $V_\infty < 0$, $V(t)$ is a concave function, tending to an exponential one. The cell no longer has an ultimate size if it ceased to divide. V_∞ is then no longer interpretated as ultimate size, but this does not invalidate the equations.

The shape of the growth curve, convex, linear or concave, thus depends on substrate density and the aspect ratio. Figure 3.17 illustrates the perfect fit of growth curves (3.38) with only three parameters: volume at 'birth', $V_d/2$, ultimate volume, V_∞, and growth rate, $\dot{\gamma}_r$. The figure beautifully reveals the effect of the aspect ratio; the larger the aspect ratio, the more important the effect of the caps, so a change from 1D-isomorphic behaviour to a 0D-isomorphic behaviour.

The time required to grow from $V_d/2$ to V at constant substrate density is found from

(3.38):

$$t(V) = \frac{(f+g)V_\infty}{f\dot{\nu}V_d\delta/3} \ln \frac{V_\infty - V_d/2}{V_\infty - V} \tag{3.39}$$

At the end of the cell cycle, the cell has to synthesize extra cell wall material. Since the cell grows in length only, the growth of surface material is directly tied to that of cytoplasm material. Straightforward geometry shows that the change in surface area A is given by $\frac{d}{dt}A = (16\pi\frac{1-\delta/3}{\delta V_d})^{1/3}\frac{d}{dt}V$. So the energy costs for growth can be partitioned as $[G] = [G_V] + \{G_A\}(16\pi\frac{1-\delta/3}{\delta V_d})^{1/3}$, where $\{G_A\}$ denotes the energy costs for the material in a unit surface area of cell wall and $[G_V]$ that for the material in a unit volume of cytoplasm. For reasons of symmetry, it is more elegant to work with $[G_A] \equiv \{G_A\}V_d^{-1/3}$ rather than $\{G_A\}$. The dimensions of $[G_V]$ and $[G_A]$ are then the same: energy per volume. At the end of the cell cycle, when cell volume is twice the initial volume, the surface material should still increase from $A(V_d)$ to $2A(V_d/2) = (1+\delta/3)A(V_d)$. This takes time, of course. If all incoming energy not spent on maintenance is used for the synthesis of this material, the change in surface area is given by $\frac{d}{dt}A = \frac{\dot{\nu}}{g_A}(fA - V_d/V_m^{1/3})$, where $g_A \equiv [G_A]/\kappa[E_m]$. So $A(t) = (A(0) - V_d/fV_m^{1/3})\exp\{tf\dot{\nu}/g_A\} + V_d/fV_m^{1/3}$. The time it takes for the surface area to reach $(1+\delta/3)V_d^{2/3}$, starting from $A(0) = V_d^{2/3}$, equals

$$t_A = \frac{g_A}{f\dot{\nu}}\left(\ln 2 + \ln\frac{V_\infty - V_d/2}{V_\infty - V_d}\right) \tag{3.40}$$

For the time interval between subsequent divisions, $t(V_d)$ must be added, giving

$$t_d = \frac{g_A}{f\dot{\nu}}\ln 2 + \left(\frac{g_A}{f\dot{\nu}} + \frac{(f+g)V_\infty}{f\dot{\nu}V_d\delta/3}\right)\ln\frac{V_\infty - V_d/2}{V_\infty - V_d} \tag{3.41}$$

The extra time for cell wall synthesis at the caps does not play a role for filaments, as their caps are comparatively small. It also does not play a significant role in unicellular eukaryotic isomorphs, because they do not have cell walls to begin with. The cell volume is full of membranes in these organisms, so the amount of membranes at the end of the cell cycle does not need to increase as abruptly as in bacteria, where the outer membrane and cell wall (if present) are the only surfaces. Comparable delays occur in ciliates for instance, where the cell mouth does not function during and around cell division.

Cooper [138] and Koch [398] argued that weight increase of bacterial cells is always of the exponential type, apart from minor contributions of cell wall, DNA, etc. If the activity of the carriers for substrate uptake is constant during the cell cycle, an implication of this model is that carriers should be produced at a rate proportional to the growth rate, and consequently to cell volume rather than to surface area. This would increase the number of carriers per unit of surface area of active membrane during the cell cycle. At the end of the cell cycle the number of carriers per unit of surface area should (instantaneously) drop by a factor of $(1+\delta/3)^{-1}$ due to the production of new membrane without carriers that separates the daughter cells. This factor amounts to $5/6 = 0.83$ for cocci and 1 for 1D-isomorphs. The factor stands for the ratio of the surface area of a body with volume V_d and two times the surface area of a body with volume $V_d/2$; so it is $2^{-1/3} = 0.79$ for

3D-isomorphs and $2^{-1/2} = 0.71$ for 2D-isomorphs. To my knowledge, such a reduction has never been demonstrated. The carrier density is assumed to be constant in the DEB theory. If the carrier density in the membrane is constant in case of exponential growth (in non-1D-isomorphs), the carrier activity should increase during the cell cycle. This requires the loss of homeostasis and/or complex regulation of carrier activity. In the DEB theory, the carrier activity is constant during the cell cycle. Although exponential growth of the cell seems an attractively simple model at first sight, theory to tie the growth rate to nutrient levels no longer comes naturally for such an extreme 'demand' type of system. Moreover, phenomena such as the small cell size in oligotrophic oceans, the growth of stalks in *Caulobacter*, the removal of disused DNA need other explanations than given in this book. Another point is of course, that if bacteria increase their weight exponentially, they would deviate from unicellular eukaryotes in this respect, where exponential growth is obviously untenable, cf. figure 1.1. The problem should then be adressed of what makes prokaryotes fundamentally different from eukaryotes in terms of energetics.

3.9 Development

Now that growth has been specified, the utilization rate for isomorphs can be evaluated from (3.8) and (3.12). It amounts to

$$\dot{C} = \frac{g[E]}{g + [E]/[E_m]}(\dot{v}V^{2/3} + \dot{m}V_h^{1/3}V^{2/3} + \dot{m}V) \tag{3.42}$$

Energy allocation to development is $(1-\kappa)\dot{C}$. Comparison of growth and reproduction at different food levels points to a problem: the volume at the first appearance of eggs in the broodpouch of daphnids seems to be independent of food density. It appears to be almost fixed; see figure 3.18. Let this volume be called V_p, where subscript p refers to puberty (transition juvenile/adult). The same holds for the volume at hatching, V_b, say, where subscript b refers to birth (transition embryo/juvenile). The problem is that the total energy investment in development depends on food density. Indeed, if feeding conditions are so poor that the ultimate volume is less than V_p, the cumulated energy investment into development becomes infinitely large, if the organism survives long enough. This seems to be highly unrealistic.

Horst Thieme [711] proposed a solution to this problem: split the energy allocated to development into two fluxes, the increase of the state of maturity and the maintenance of a certain degree of maturity. For a special choice of the maturity maintenance costs the total energy investment into the increase of the state of maturity does not depend on food density for ectotherms. This can be seen most easily from (3.11), when both sides are multiplied by $(1-\kappa)/\kappa$ to obtain the investment into development

$$(1-\kappa)\dot{C} = \frac{1-\kappa}{\kappa}\dot{M} + \frac{1-\kappa}{\kappa}[G]\frac{d}{dt}V \tag{3.43}$$

for juvenile ectotherms ($V < V_p$ and $\dot{H} = 0$). If the first term of the right hand side corresponds to maturity maintenance costs, the second one for the increase of the state of

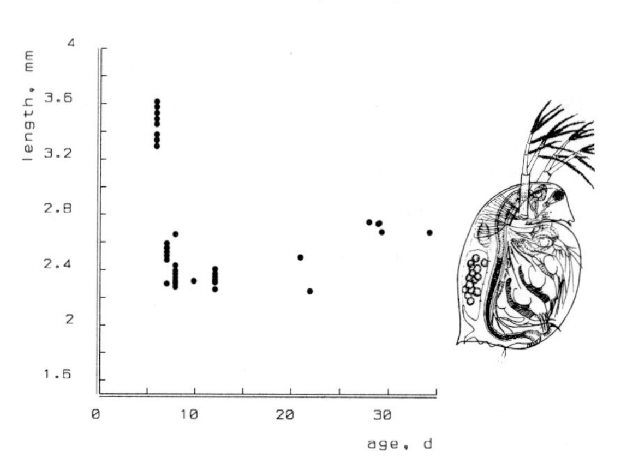

Figure 3.18: The carapace length of the daphnid *Daphnia magna* at 20 °C for 5 different food levels at the moment of egg deposition in the brood pouch. Data from Baltus [32]. The data points for short juvenile periods correspond with high food density and growth rate. They are difficult to interpret because length increase is only possible at moulting in daphnids.

maturity only depends on size, not on food density. Since the individual does not become more complex after attaining size V_p, the energy flow to maintain a certain degree of maturity must then be $\min\{V, V_p\}[\dot{M}]\frac{1-\kappa}{\kappa}$. It can be thought to relate to the maintenance of regulating mechanisms and of concentration gradients, such as those found in *Hydra*, that are responsible for the maintenance of the head/foot differentiation [258].

It took me quite a while to accept the existence of maturity maintenance as inevitable. Although the concept sounds a little esoteric, there are two hard observations in support of its existence. The first one concerns an experiment where food density is held constant at two levels, just below and above the food density that gives an ultimate size $V_\infty = V_p$. For ectotherms, such as daphnids, (3.17) implies that this food density is found from $f = (V_p/V_m)^{1/3} \equiv l_p$, so $X = Kl_p/(1 - l_p)$. If maturity maintenance did not exist, animals kept at the lower food density would never reproduce, while those at the higher food density would reproduce at a rate that might be substantial, depending on κ. Substitution of $V = V_p$ into (3.11) shows that the energy investment into development and/or reproduction tends to $\frac{1-\kappa}{\kappa}[\dot{M}]V_p$, which amounts to $4[\dot{M}]V_p$ for $\kappa = 0.2$, which is realistic for daphnids. This substantial difference in reproductive output as a result of a tiny difference in feeding rates has never been observed.

The second observation that points to the existence of maturity maintenance concerns pond snails, where the day/night cycle affects the fraction of utilized energy spent on maintenance plus growth [788], such that κ at equal day/equal night, κ_{md}, is larger than that at long day/short night, κ_{ld}. Apart from the apparent effects on growth and reproduction rates, volume at the transition to adulthood is also affected. If the cumulated energy investment into the increase of maturity does not depend on the value for κ and if the maturity maintenance costs are $\frac{1-\kappa}{\kappa}\dot{M}$, the expected effect is $\frac{V_{p,ld}}{V_{p,md}} = \frac{\kappa_{ld}(1-\kappa_{md})}{\kappa_{md}(1-\kappa_{ld})}$, which is consistent with the observations on the coupling of growth and reproduction investments to size at puberty [788]. Some species, such as birds, only reproduce well after the growth period. The giant petrel wanders seven years over Antarctic waters before it starts to breed for the first time. From a mathematical point of view, growth is asymptotic, so it is possible to choose V_p to be so close to V_∞ that the desired result is described adequately. This must be rejected, however, because it seems most unrealistic to have a model where decision rules depend on such small differences in volume in a world that is full of scatter.

The introduction of costs for maintaining a certain degree of maturity solves this problem, because the model is then energy-structured as well as size-structured. A transition from embryo to juvenile and from juvenile to adult occurs if the cumulative investment to increase the state of maturity exceeds specified amounts. If growth has almost ceased, this cumulative investment increases linearly, it therefore has no asymptote. The rate of increase of cumulated investment can be substantial, even if body size hardly increases, so this rule causes no problems for species that separate growth and reproduction in time.

There is, however, a problem connected with this introduction of maturity maintenance: it is hard to see why it should have just the value $\frac{1-\kappa}{\kappa}\dot{M}$; it is a fact that it produces the observed fixed-volume transition in daphnids and pond snails, but one would like to understand why. One solution might be to interpret $\frac{1-\kappa}{\kappa}$ as the basic parameter and try to explain why the relative allocation to development plus reproduction takes a value that relates to the costs of development. I still find this an unsatisfactory point in the theory. Of course, it is possible to introduce a free parameter for the maturity maintenance costs and use volume at first maturation for the estimation of its value, which then proves to be close to $\dot{M}_d = \frac{1-\kappa}{\kappa}\dot{M}$, because this value produces a volume that is independent of food density. If this free parameter has a different value, variations in volume at first maturation will result when food density varies, as has been observed for some species, according to the review by Bernardo [60]. Its introduction has the serious drawback that evaluation of the length of incubation and juvenile period become cumbersome, which causes problems especially at the population level. The fixed size transition should then be replaced by a fixed cumulative energy transition.

Note that growth and development are parallel processes in the DEB model, which links up beautifully with the concepts of acceleration and retardation of developmental phenomena such as sexual maturity [267]. These concepts are used to describe relative rates of development in species that are similar in other respects.

In embryos and juveniles, the energy spent on somatic maintenance and the maintenance of a certain degree of maturation can be combined, because both can be taken proportional to volume. The difference between the two only shows up in adults that still increase in size. Somatic maintenance remains proportional to size, while maturation maintenance stays constant at constant temperature. The same holds for the energy spent on growth and the increase of the degree of maturity. In embryos and juveniles, they can be combined, because both are taken proportional to volume increase. This means that for non-adults the κ-rule is not of quantitative relevance, and the model simplifies to the one for micro-organisms with respect to the use of energy.

Whether or not unicellulars and particularly prokaryotes invest in cell differentiation during the cell cycle is still open to debate. Dworkin [192] gives a review of development in prokaryotes and points to the striking similarities between myxobacteria and cellular slime molds and *Actinomyceta* and some fungi. A most useful aspect of the κ-rule is that this matter need not first be resolved, because this investment only shows up in the parameter values and not in the model structure. As stated in the introduction to this chapter, the energy invested in development according to the κ-rule can only be deduced from the transition to the adult state in metazoans. The utilization rate for rods can be obtained in the same way as for isomorphs: application of the shape correction function (2.7) to $\dot{v}$

in (3.42). This amounts to

$$\dot{C} = \frac{g[E]}{g + [E]/[E_m]} \left(\dot{\nu}\frac{\delta}{3}V_d + \left(\dot{m} + \dot{\nu}\left(1 - \frac{\delta}{3}\right)\right) V \right) \tag{3.44}$$

The utilization rate for filaments can be found by application of the shape correction function (2.4) to $\dot{\nu}$ in (3.42), which leads to

$$\dot{C} = \frac{g[E]}{g + [E]/[E_m]}(\dot{m} + \dot{\nu})V \tag{3.45}$$

It can also be obtained by letting $\delta \to 0$ in (3.44). Since both growth and maintenance are proportional to volume for filaments, the utilization rate is also proportional to volume.

3.10 Propagation

Organisms can achieve an increase in numbers in many ways. Sea anemones can split off foot tissue that can grow into a new individual. This is not unlike the strategy of budding yeasts. Colonial species usually have several ways of propagating. Fungi have intricate sexual reproduction patterns involving more than two sexes. Under harsh conditions some animals can switch from parthenogenic to sexual reproduction, others develop spores or other resting phases. It would not be difficult to fill a book with descriptions of all the possibilities. I will confine the discussion to the two most common modes of propagation: via egg and foetus or vegetatively, via division.

3.10.1 Reproduction

Energy allocation to reproduction equals the allocation to development plus reproduction minus the costs to maintain the state of maturity

$$(1 - \kappa)\dot{C} - \frac{1-\kappa}{\kappa}[\dot{M}]V_p \tag{3.46}$$

This is a continuous energy investment. The costs for egg (or foetus) development are fully determined, as has been discussed in the section on embryonic growth, {83}. The costs for the production of an egg can be written as E_0/q, where the dimensionless factor q between 0 and 1 relates to the overhead involved in the conversion from the reserve energy of the mother to the initial energy available for the embryo. Since these types of energy reserves are chemically related, the overhead is likely to be small in most cases so that q is close to 1. This might seem an odd way to introduce this overhead, but q can also be interpreted as an egg survival probability, which can be further modulated by predation and toxic compounds, as discussed in later chapters. This is practical because egg survival is frequently governed by different processes than survival of later stages. Substitution of utilization rate (3.42) into (3.46) leads to a mean reproduction rate for ectotherms of

$$\dot{R} = \frac{q}{e_0 V_m}(1-\kappa)\left(\frac{g[E]/[E_m]}{g + [E]/[E_m]}(\dot{\nu}V^{2/3} + \dot{m}V) - g\dot{m}V_p\right) \tag{3.47}$$

where the relative energy costs for embryo development e_0 are given in (3.26). Under no-growth conditions, i.e. when $\frac{[E]}{[E_m]} \leq (\frac{V}{V_m})^{1/3}$, individuals can no longer follow the κ-rule, because the allocation to maintenance would no longer be sufficient. Maintenance has priority over all other expenses. Individuals that still follow the storage dynamics (3.8) under no-growth conditions, must reproduce at mean rate $(\dot{C} - \dot{M} - \frac{1-\kappa}{\kappa}[\dot{M}]V_p)q/E_0$, so

$$\dot{R} = \frac{q}{e_0 V_m} g\dot{m} \left(\frac{[E]}{[E_m]} V_m^{1/3} V^{2/3} - \kappa V - (1-\kappa)V_p \right)_+ \tag{3.48}$$

At the border of the no-growth condition, i.e. when $\frac{[E]}{[E_m]} = (\frac{V}{V_m})^{1/3}$, both expressions for the reproduction rate are equal, so there is no discontinuity for changing energy reserves.

At constant food density where $[E] = f[E_m]$, the reproduction rate is according to (3.47) proportional to

$$\dot{R} \propto V^{2/3} + \frac{\dot{m}}{\dot{v}} V - \frac{g+f}{f} \frac{\dot{m}}{\dot{v}} V_p \tag{3.49}$$

where the third term is just a constant. Comparison of reproduction rates for different body sizes thus involves three compound parameters, i.e. the proportionality constant, the parameter $\dot{m}/\dot{v}$ and the third term, if all individuals experience the same food density for a long enough time. Figure 3.19 illustrates that this relationship is realistic, but that the notorius scatter for reproduction data is so large that access to the parameter $\dot{m}/\dot{v}$ is poor. The fits have been based on guestimates for the maintenance rate coefficient, $\dot{m} = 0.011$ d^{-1}, and the energy conductance, $\dot{v} = 0.433$ mm d^{-1} at 20 °C. Note that if the independent variable is a length measure rather than structural body volume, the shape coefficient $d_m = V^{1/3}L^{-1}$ has to be introduced since the guestimate for the energy conductance is expressed in volumetric length. For some length measure L, we have

$$\dot{R} \propto L^2 + \frac{\dot{m}}{\dot{v}} d_m L^3 - \frac{g+f}{f} \frac{\dot{m}}{\dot{v}} d_m L_p^3 \tag{3.50}$$

The practical significance of this remark is in the comparison between species, which will be discussed later, {217}. The main reason for the substantial scatter in reproduction data is that they are usually collected from the field, where food densities are not constant, and where spatial heterogeneities, social interactions, etc., are common.

The reproduction rate of spirorbid polychaetes has been found to be roughly proportional to body weight [312]. On the assumption by Strathmann and Strathmann [694] that reproduction rate is proportional to ovary size and that ovary size is proportial to body size (an argument that rests on isomorphy), the reproduction rate is also expected to be proportional to body weight. They observed that reproduction rate tends to scale with body weight to the power somewhat less than one for several other marine invertebrate species, and used their observation to identify a constraint on body size for brooding inside the body cavity. The DEB theory gives no direct support for this constraint; an allometric regression of reproduction rate against body weight would result in a scaling parameter between 2/3 and 1, probably close to 1, depending on parameter values.

The maximum (mean) reproduction rate for ectotherms of maximum volume $V_m = (\dot{v}/g\dot{m})^3$ amounts to

$$\dot{R}_m = \frac{q}{e_0}(1-\kappa)g\dot{m}(1 - V_p/V_m) \tag{3.51}$$

Figure 3.19: Reproduction rate as a function of body length for two randomly selected species. The data sources and DEB-based curves are indicated. The parameter that is multiplied by L^3 in both fits has been guestimated on the basis of common values for the maintenance rate coefficient and the energy conductance, with a shape coefficient of $d_m = 0.1$ for the goby and of $d_m = 0.5$ for the frog. Both the other parameters represent least squares estimates.

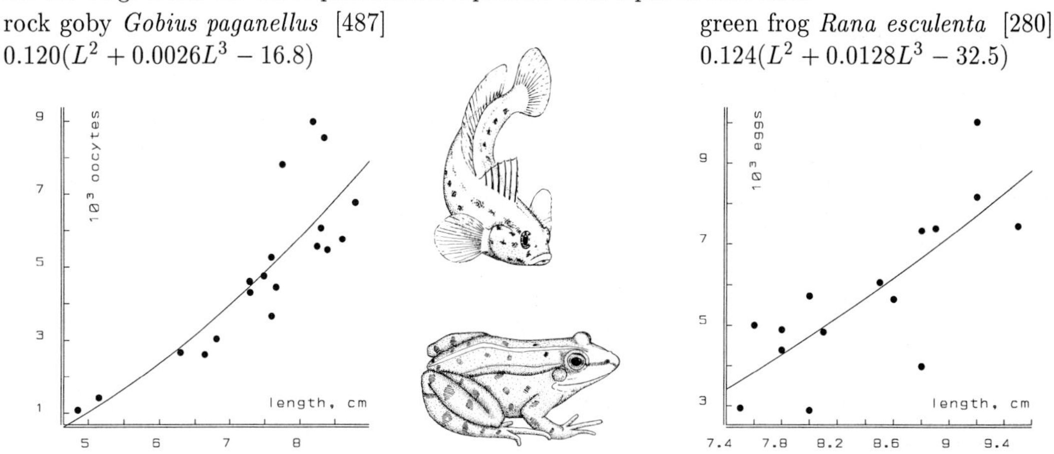

All these expressions only refer to mean reproduction rates. Individuals are discrete units, which implies the existence of a buffer, where the energy allocated to reproduction is collected and converted to eggs at the moment of reproduction. The translation of reproduction rate into number of eggs in figure 3.19 assumes that this accumulation is over a period of one year. The energy content of the buffer is denoted by $E_{\dot{R}}$.

Some species reproduce when enough energy for a single egg has been accumulated, others wait longer and produce a large clutch. There is considerable variation in the way the reproduction buffer is handled. If the reproduction buffer is used completely, the size of the clutch equals the ratio of the buffer content and the energy costs for one young, $qE_{\dot{R}}/E_0$, where E_0 is given in (3.26). This resets the buffer. So after reproduction $E_{\dot{R}} = 0$ and further accumulation continues from there. That is to say, if the bit of energy that was not sufficient to build the last egg will be lost. Fractional eggs do not exist. In the chapter on population dynamics, {171,207}, I will show that this uninteresting detail substantially affects dynamics at low population growth rates, which occur most frequently in nature. If food is abundant, the population will evolve rapidly to a situation in which food per individual is sparse and reproduction low if harvesting processes do not prevent this.

The strategies for handling this buffer are species-specific and are affected by environmental variables. Most species are able to synchronize the moment of reproduction with seasonal cycles such that food availability just matches the demand of the offspring. Clutch size in birds typically relates to food supply during a two-month period prior to egg laying and tends to decrease if breeding is postponed in the season [475]. The laying date is determined by a rapid increase in food supply. Since feeding conditions tend to improve during the season, internal factors must contribute to the regulation of clutch size. These conclusions result from an extensive study of the energetics of the kestrel *Falco tinnuncu-*

lus by Serge Daan and co-workers [171,458,474]. I see reproductive behaviour like this, for species that cease growth at an early moment in their life span, as variations on the general pattern that the DEB theory is aiming to grasp. Aspects of reproduction energetics for species that cease growth, are worked out under the heading 'imago' on {151}.

3.10.2 Division

If propagation is by division, the situation is comparable to the juvenile stage of species that propagate via eggs. The volume at division corresponds to the transition from juvenile to adult, so $V_d = V_p$. Donachie [175] pointed out that in fast growing bacteria the initiation of DNA duplication occurs at a certain volume V_p, but it requires a fixed and non-negligible amount of time t_D for completion. This makes the volume at division, V_d, dependent on the growth rate, so indirectly on substrate density, because growth proceeds during this period. The mechanism (in eukaryotic somatic cells) of division at a certain size is via the accumulation of cdc25 and cdc13 mitotic inducers, which are produced coupled to cell growth. (The name for the genes 'cdc' stands for cell devision cycle.) If these inducers exceed a threshold level, $p34^{cdc2}$ protein kinase is activated and mitosis starts [493,498]. During mitosis, $p34^{cdc2}$ is deactivated and the concentration of inducers resets to zero. This mechanism indicates that for shorter inter-division periods, the cell starts a new DNA duplication cycle when its volume exceeds $2V_p$, $4V_p$, $8V_p$ etc. The inter-division time for *Escherichia coli* can be as short as 20 minutes under optimal conditions, while it takes an hour to duplicate the DNA. In a dynamic environment, where (3.36) and (3.37) are supposed to apply, the implementation of this trigger is not simple. At constant substrate densities, the scaled cell length at division, $l_d \equiv (V_d/V_m)^{1/3}$, and the division interval, $t(l_d) \equiv t_d$, can be obtained directly. When i is an integer such that $2^{i-1} < V_d/V_p \leq 2^i$, V_d can be solved from

$$t_D = it(V_d) - t(2^{i-1}V_p) \qquad (3.52)$$

Figure 3.20 illustrates the derivation.

The volume at division V_d can be found numerically when (3.15), (3.35) or (3.39) is substituted for $t(V)$ into (3.52), for isomorphs, filaments or rods, respectively.

3.11 Respiration

Respiration, i.e. the use of oxygen or the production of carbon dioxide, can be taken to represent the total metabolic rate in an organism. Initially, eggs hardly use oxygen, but oxygen consumption rapidly increases during development; see figure 3.21. In juveniles and adults, oxygen consumption is usually measured in individuals that have been starved for some time, to avoid interpretation problems related to digestion. (For micro-organisms this is not possible without a substantial decrease of reserves.) As mentioned in the introductory chapter, the conceptual relationship between respiration and use of energy has already undergone some changes in history. Von Bertalanffy identified it with anabolic processes, while the Scope For Growth concept, {43}, relates it to catabolic processes. In the DEB model, the most natural identification is with the total use of energy from the reserves,

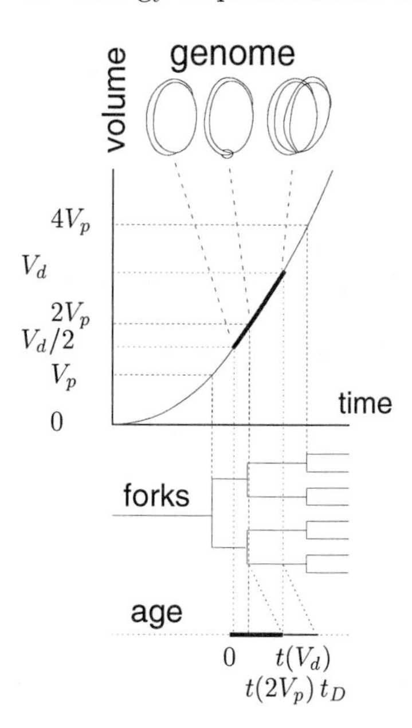

Figure 3.20: A schematic growth curve of a cell, where the fat part is used in steady state. This is the situation for $i = 2$, the number of forks switching between 1 and 3. If $V_d/V_p = 2^i$, equation (3.52) reduces to $t_D = it(2^iV_p) = it(V_d)$, with $t(2^{i-1}V_p) = 0$, which means that the time required to duplicate DNA is exactly i times the division interval. So, during each cell cycle, a fraction i^{-1} of the genome is duplicated, which implies that $2^i - 1$ DNA duplication forks must be visible during the cell cycle. At the moment that the number of forks jumps from $2^i - 1$ to $2^{i+1} - 1$, the cell divides and the number of forks resets to $2^i - 1$. This is obviously a somewhat simplified account, as cell division is not really instantaneous. If $V_d/V_p \neq 2^i$, the age of the cell at the appearance of the new set of duplication forks somewhere during the cell cycle is $t(2^{i-1}V_p)$, which thus has to be subtracted from $it(V_d)$ to arrive at the genome duplication time.

(3.42), with a fixed conversion factor from oxygen to energy use. This is consistent with the assumption of a constant chemical composition for the reserves. It is also consistent with the observation that respiration rate increases with reserve density [389], while reserves themselves do not use oxygen. Moreover, it explains the reduction of respiration during starvation; see {128}.

At constant food density, the proportionality between respiration and use of energy from the reserves implies that the respiration rate can be written as a weighted sum of a surface area and a volume. Figure 2.9 shows that it is indistinguishable from the standard allometric relationship. Apart from avoiding dimension problems, the surface area related costs for heating in endotherms, which have given rise to Rubner's surface law, also fit much more naturally. As mentioned above, this point of view solves the long standing problem of why the volume-specific respiration of ectotherms decreases with increasing size, when organisms of the same species are compared. This problem has been identified as one of the central problems of biology [777]. Many theories have been proposed, see e.g. [590] for a discussion, but all use too specific arguments to be really satisfactory; heating (but many species are ectothermic), muscle power (but movement costs are relatively unimportant), gravity (but aquatic species escape gravity). Peters [542] even argued to cease looking for a general explanation. The DEB theory, however, does offer a general explanation: the overhead for growth. A comparison of different species will be covered in a later chapter, {217}, where it will be shown that interspecies comparisons work out a bit differently.

The proportionality of respiration with the utilization rate is not, however, completely free from conceptual problems. This mapping gives double counts of energy flows at two places: energy that is fixed in structural biomass during growth and energy reserves deposited in eggs (by females). It is essential to realize that the costs for synthesis $[G]$ include overheads. So $[G]$ is larger than, and I think much larger than, the energy content of the structural biomass. On the other hand, the energy content of organisms, which is

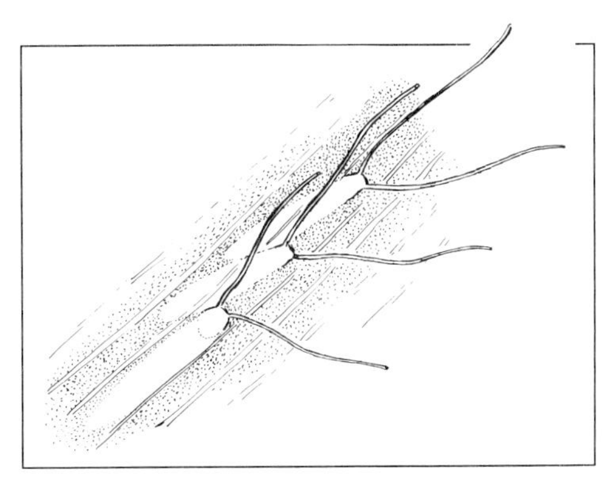

Figure 3.21: The water stick insect *Ranatra linearis* deposits its eggs in floating decaying plant material, where oxygen availability is usually poor. The eggs are easily spotted by the special respiratory organs that peek out of the plant. Just prior to hatching, eggs typically need a lot of oxygen, cf. figure 3.14.

frequently measured [147,527,540,512,572], includes energy reserves. These two problems complicate the interpretation of such measurements in terms of energy parameters.

The use of respiration measurements to estimate the parameters of the DEB model is limited. Respiration is taken proportional to utilization, so that it follows from (3.42) that the respiration rate is proportional to $V + V^{2/3}V_m^{1/3}(l_h + g)$ at constant food density. Respiration is thus a weighted sum of volume and surface area. If respiration data are available for different body sizes of a particular species of ectotherm, so that $l_h = 0$, the ratio of the weight coefficient for volume$^{2/3}$ and that for volume stands thus for $gV_m^{1/3} = \frac{\dot{v}}{\dot{m}} = \frac{\{\dot{A}_m\}[G]}{[E_m][M]}$, which has the dimension of length. Four original parameters are combined into this compound parameter. The two coefficients of the ratio are negatively correlated in a statistical sense, which implies that respiration data give poor access to the value of the compound parameter.

The maintenance rate coefficient $\dot{m}$ can be estimated easily if growth data together with respiration data are collected at a constant food density. The κ-rule implies that the respiration rate of ectotherms is proportional to energy allocation to growth plus maintenance, so according to (3.11) the respiration rate is proportional to $\frac{d}{dt}V + \dot{m}V$. The observation that respiration is proportional to a weighted sum of volume and change in volume goes back to the study of Smith [670] in 1957 on eggs of salmon. At constant food density, the change in volume is of the von Bertalanffy type, which makes respiration proportional to $3\dot{\gamma}(V_\infty^{1/3}V^{2/3} - V) + \dot{m}V$. This gives five parameters to be estimated from two data sets on respiration and growth : V_b, V_∞, $\dot{\gamma}$, a proportionality constant for respiration and the maintenance rate coefficient, $\dot{m}$. This gives 2.5 parameters per data set, which seems acceptable if the scatter is not too large.

In the section on mass-energy coupling, {192}, it will be shown how the respiration rates for micro-organisms can be tied to energy fluxes in a more rigorous way. I expect that the overheads for growth are much smaller for them, compared to animals, but I cannot substantiate this.

3.12 Aging

Since age is not a state variable, the steady shift in properties due to the poorly understood process of aging is only of secondary relevance to the DEB model. In a number of situations, however, one should consider life span which has well recognized roots in energetics. The frequently observed correlation between life span and the inverse volume-specific metabolic rate for different species (see e.g. [638]) has guided a lot of research. The impressive work of Finch [219] gives well over 3000 references. Animals tend to live longer at low food levels than at high ones. The experimental evidence, however, is rather conflicting on this point. For example, Ingle [349] found such a negative relationship, while McCauley [463] found a positive one for daphnids. This is doubtlessly due to the fundamental problem that death can occur for many reasons, such as food related poisoning, that are not directly related to aging. Some species such as salmon, octopus, *Oikopleura* die after (first) reproduction, cf. {149}. This cause, like many other causes of death, does not relate to aging. On approaching the end of the life span, the organism usually becomes very vulnerable, which complicates the interpretation of the life span of a particular individual in terms of aging. Experiments usually last a long time, which makes it hard to keep food densities at a fixed level and to prevent disturbances.

In a first naïve attempt to model the process of aging, it might seem attractive to conceive the senile state, followed by death as the next step in the sequence embryo, juvenile, adult, and then tie it to energy investment into development just as has been done for the transitions to the juvenile and adult stages. This is not an option for the DEB model, since at sufficiently low food densities, the adult state is never entered, even if the animal survives for nutritional reasons. This means that it would live for ever, as far as aging is concerned. Although species exist with very long life spans (excluding external causes of death [219]), this does not seem acceptable. Attempts to relate hazard rates directly to the accumulation of hazardous compounds formed as a spin off of respiration, such as oxidized lipids, have failed to produce realistic age-specific mortality curves: the hazard rate increases too rapidly for a given mean life span. See [388,518] for reviews on the role of secondary products from metabolism in aging. The same holds for the hazard tied to damage to membranes, if this damage accumulates at a rate proportional to volume-specific respiration. Accurate descriptions of survival data where aging can be assumed to be the major cause of death seem to call for an extra integration step, which points to DNA.

It has been suggested that free radicals, formed as a spin off of respiration, cause irreparable damage to the DNA in organisms [288,287,719]. The specific activity of antioxidants correlates with life span within the mammals [219]. The structure of the antioxidant enzyme manganese superoxide dismutase has recently been solved [621]. Although too unspecific to be of much help for molecular research, for energetics purposes the free radical hypothesis specifies just enough to relate the age-specific survival probability, and so life span, to energetics. The idea is that the hazard rate is proportional to damage density, which accumulates at a rate proportional to the concentration of changed DNA, while DNA changes at a rate proportional to utilization rate. Although it is not yet possible to draw firm conclusions on this point, this mechanism does provide the extra integration step that is required for an accurate description of data. It is further assumed that the cells with changed DNA do not grow and divide, while the density of affected cells is reduced owing to the propagation of the unchanged cells. This assumption is supported by the recent identification of gene chk1 [748], whose products are involved in the detection of DNA damage;

damaged DNA prevents entry into mitosis by controlling the activity of the protein that is produced by cdc2, cf. {103}. Because of the uncertainty in the coupling with molecular processes, I prefer to talk about damage and damage inducing compounds, rather than wrong proteins (or their products) and DNA. This idea can be worked out quantitatively as follows.

Let $[Q] \equiv Q/V$ denote the concentration of damage inducing compounds (changed DNA), which accumulate from value 0 in an embryo of age 0. Its dynamics can be obtained via the chain rule for differentiation: $\frac{d}{dt}[Q] = V^{-1}\frac{d}{dt}Q - [Q]\frac{d}{dt}\ln V$ and amounts to

$$\frac{d}{dt}[Q] = d_Q\frac{\dot{C}}{V} - [Q]\frac{d}{dt}\ln V \tag{3.53}$$

where d_Q is the contribution of the volume-specific utilization rate to the compounds per unit of energy. The second term stands for the dilution through growth, where cells with changed DNA become mixed with cells with unchanged DNA.

Substitution of (3.11) gives for ectotherms

$$\frac{d}{dt}[Q] = \frac{d_Q}{\kappa}[G]\frac{d}{dt}\ln V + \frac{d_Q}{\kappa}[\dot{M}] - [Q]\frac{d}{dt}\ln V \tag{3.54}$$

The concentration of damage inducing compounds as a function of time for ectotherms thus equals

$$[Q](t) = \frac{d_Q}{\kappa}[G]\left(1 - \frac{V(0)}{V(t)}\right) + \frac{d_Q}{\kappa}\frac{[\dot{M}]}{V(t)}\int_0^t V(t_1)\,dt_1 \tag{3.55}$$

As explained in the section on embryonic growth, {83}, the initial volume, $V(0)$, is infinitesimally small. The accumulated damage during the embryonic stage, however, is usually negligibly small. The high generation rate of damage inducing compounds is balanced by the high dilution rate through growth. The fact that the embryonic period is usually a very small fraction of the total life span ensures that one does not lose much information by starting from the moment of hatching.

Damage (wrong protein) accumulates at a rate proportional to the concentration of damage inducing compounds, so the damage density is proportional to $\int_0^t [Q](t_1)\,dt_1$. The hazard rate, $\dot{h}(t)$, is finally taken to be proportional to the damage density, which leads to:

$$\dot{h}(t) = \ddot{p}_a\int_0^t\left(1 - \frac{V(0)}{V(t_2)} + \frac{\dot{m}}{V(t_2)}\int_0^{t_2} V(t_1)\,dt_1\right)dt_2 \tag{3.56}$$

The proportionality constant $\ddot{p}_a$, here called the aging acceleration, absorbs both proportionality constants leading to this formulation of the age dependent hazard rate and is proportional to $d_Q[G]/\kappa$. This most useful property means that only a single parameter is necessary to describe the aging process.

The hazard rate relates to the survival probability according to the differential equation $\frac{d}{dt}\text{Prob}\{\underline{a}_\dagger > t\} = -\text{Prob}\{\underline{a}_\dagger > t\}\dot{h}(t)$ or $\dot{h}(t) = -\frac{d}{dt}\ln\,\text{Prob}\{\underline{a}_\dagger > t\}$. The survivor probability is thus

$$\text{Prob}\{\underline{a}_\dagger > t\} = \exp\{-\int_0^t \dot{h}(t_1)\,dt_1\} \tag{3.57}$$

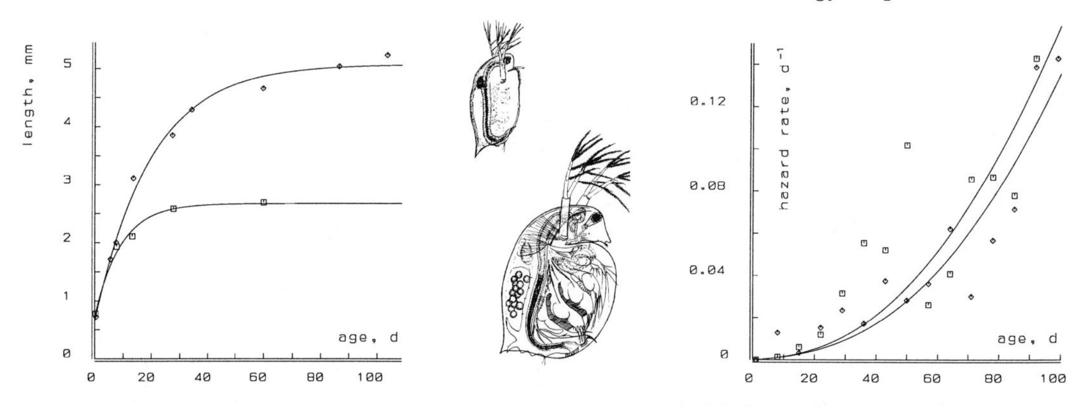

Figure 3.22: The growth curves of female (◇) and male daphnid (□)*Daphnia magna* at 18 °C and the observed hazard rates. Data from MacArthur and Baillie [445]. The growth curves are of the von Bertalanffy type with common length at birth. The hazard rates are fitted on the basis of the damage genesis discussed in the text, with a common aging acceleration of 2.587×10^{-5} d^{-2}. The difference in the hazard rates is due to the difference in ultimate lengths.

The mean life span equals $\mathcal{E}\underline{a}_\dagger = \int_0^\infty \text{Prob}\{\underline{a}_\dagger > t\}\, dt = \int_0^\infty \exp\{-\int_0^t \dot{h}(t_1)\, dt_1\}\, dt$. This hazard rate thus ties aging to energetics, which explains for instance why dormancy prolongs life span, cf. {131}.

Figure 3.22 shows that the fit with experimental data for male and female daphnids is quite acceptable, in view of the fact that the combined hazard curves have only one free parameter $\ddot{p}_a$ (so half a parameter per curve). The differences in survival probability of male and female daphnids can be traced back to difference in ultimate size (i.e. in the surface area-specific maximum assimilation rate $\{\dot{A}_m\}$).

It is instructive to compare this model with that of Weibull where $\text{Prob}\{\underline{a}_\dagger > t\} \equiv \exp\{-\int_0^t \dot{h}(t_1)\, dt_1\} = \exp\{-(\dot{p}_W t)^\beta\}$. The model was first proposed by Fisher and Tippitt [220] in 1928 as a limiting distribution of extreme values, and Weibull [757] has used it to model the failure of a mechanical device composed of several parts of varying strength, according to Elandt-Johnson and Johnson [198]. The (cumulative) hazard increases allometrically with time. Like many other allometrically based models for physiological quantities, it is attractively simple, but fails to explain, for instance, why the sexes of *Daphnia* have different shape coefficients β [415]. As long as both parameters of the Weibull model can be chosen freely, i.e. if only one data set is considered, it will be hard to distinguish it from the DEB-based model. See figure 3.23. The maintenance rate coefficient in the fit is here considered as a free parameter, so both curves then have two free parameters. This is done because the available estimate for the maintenance rate coefficient on the basis of egg development as reported in table 3.1 is rather far out of range. The resulting estimate of $\dot{m} = 0.073$ d^{-1} at 20 °C is much more realistic, which in itself lends strong support to my interpretation. It can be shown that the Weibull model with shape parameter 3 results if the growth period is short relative to the mean life span, {154}.

The Gompertz model for survival $\text{Prob}\{\underline{a}_\dagger > t\} = \exp\{\beta(1 - \exp\{\dot{p}_G t\})\}$ is also frequently used as a model for aging; see e.g. [779]. It can be mechanistically underpinned by a constant and independent failure rate for a fixed number of hypothetical critical ele-

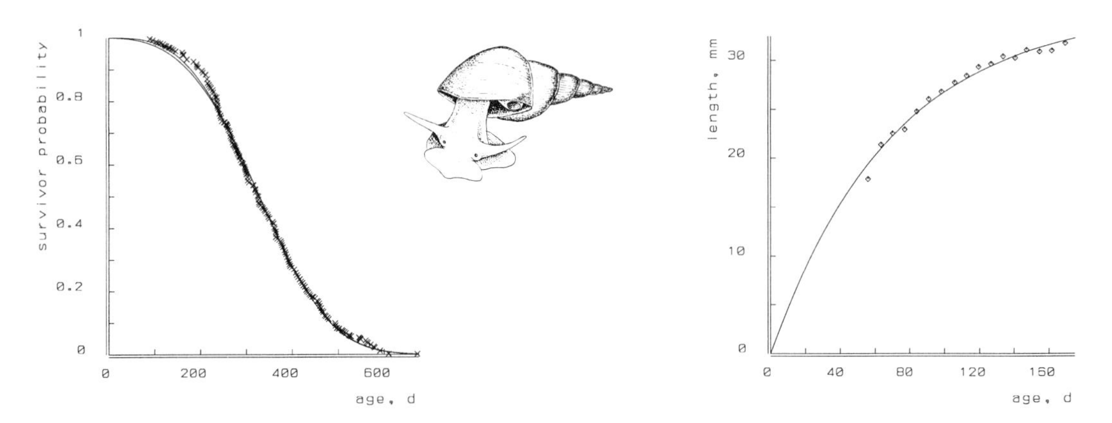

Figure 3.23: The survival probability and the growth curve of the pond snail *Lymnaea stagnalis* at 20 °C. Data from Slob and Janse [664] and Bohlken and Joosse [76,788]. The fitted growth curve is the von Bertalanffy one, giving an ultimate length of 35 mm and a von Bertalanffy growth rate of $\dot{\gamma} = 0.015$ d^{-1}. The survival curve was used to estimate both the maintenance rate constant, $\dot{m} = 0.073$ d^{-1} and the aging acceleration $\ddot{p}_a = 2.563 \times 10^{-6}$ d^{-2}. The Weibull curve with shape parameter 3.1 is plotted on top of the DEB model to show that both curves are hard to distinguish in practice.

ments. Death strikes if all critical elements cease functioning. The curvature of the survival probability then relates to the number of critical elements, which Witten [779] found to be somewhere between 5 and 15. Their nature still remains unknown. A property of this model is that the hazard rate does not approach zero for neonates (or embryos), which does not seem to be consistent with data [664]. Finch [219] favours the empirical description of aging rates given by the Gompertz model because its property of a constant mortality rate doubling time, $\dot{p}_G^{-1} \ln 2$, provides a simple basis for comparison of taxa.

The present formulation allows for a separation of the aging and energy based parameters. The estimation of the 'pure' aging parameter in different situations and for different species will hopefully reveal patterns that can guide the search for more detailed molecular mechanisms; however, many factors may be involved, cf. {154}. It has been suggested in the literature that the neural system may be involved in setting the aging rate. The fact that brain weight in mammals correlates very well with respiration rate [330], makes it difficult to identify factors that determine life span in more detail. The mechanism may be again via the neutralization of free radicals.

An indication for this pathway can be found in the age-specific survival probability for humans, see figure 3.24, which can be described well by a Weibull distribution with shape parameter 6.8. Compared with the data on ectotherms, we have here an extremely low hazard rate for the young ages, which increases rapidly after the age of 50 years. This pattern suggests that the system that is involved in the neutralization of free radicals is itself subjected to aging, while for ectotherms it is not necessary to build in this complication. As explained in the next chapter, {151}, a constant neutralization probability, combined with low mortality during growth, leads to survival curves which are close to the Weibull

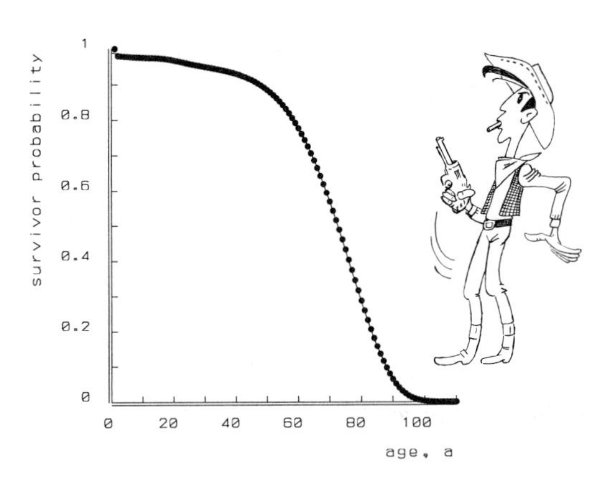

Figure 3.24: The survival curve for humans: white males in the USA in the period 1969-1971. Data from Elandt and Johnson [198]. The fitted empirical survival curve is $q\exp\{-\dot{p}t-(\dot{p}_W t)^{\beta}\}$, with $q = 0.988$, $\dot{p} = 0.0013$ a^{-1}, $\dot{p}_W = 0.01275$ a^{-1} and $\beta = 6.8$. The parameter q relates to neonate survival and $\dot{p}$ to death by accident.

curve with shape parameter 3. Aging as a result of free radicals is partially supported by the observation that the life spans of both ectotherms and endotherms correlate well with the specific activity of antioxidants [219]. It should be noted that if we compare an endotherm with a body temperature of 40 °C with that of an otherwise similar ectotherm at 20 °C, we should expect a 10 times shorter life span, on the basis of an Arrhenius temperature of 10000 K. Endotherms, therefore, have a problem to solve, which possibly involves additional mechanisms to remove free radicals.

One of the many questions that remain to be answered is how aging proceeds in animals that propagate by division rather than by eggs. Unlike eggs, they have to face the problem of initial damage. It might be that such animals have (relatively few) undifferentiated cells that can divide and replace the damaged (differentiated) ones. A consequence of this point of view is that the option to propagate by division is only open to organisms where the differentiation of specialized cells is not pushed to the extreme. If aging affects all cells at the same rate, it becomes hard to explain the existence of dividing organisms. This is perhaps the best support for the damage interpretation of the aging process. Theories that relate aging, for instance, to the accumulation of compounds as an intrinsic property of cellular metabolism, should address this problem. The same applies to unicellulars. If accumulated damage carries over to the daughter cells, it becomes hard to explain the existence of this life style. The assumption of the existence of cells with and without damage seems unavoidable. Organisms that live in anaerobic environments cannot escape aging, because other radicals will occur that have the same effect as oxygen. Note that if one follows the fate of each of the daughter cells, this theory predicts a limited number of divisions until death occurs, so that this event itself gives no support for aging theories built on cellular programming. Only the variation in this number can to some extent be used to choose between both approaches. The present theory can be worked out quantitatively for unicellulars as follows.

Since unicellulars cannot dilute changed DNA with unchanged DNA and cannot compensate for its effect, the hazard rate for unicellulars must equal $\dot{h}(t) = d_Q\dot{C}/V$, where d_Q is the net hit frequency per unit of energy density. (Note that the range of the cell volume is $(V_d/2, V_d)$, so that the volume-specific respiration rate is restricted, while for embryos, where V is assumed to be infinitesimally small initially, it does not have a boundary. Dilution by growth solves this problem for embryos.) From (3.45) it follows that the hazard rate for filaments is

$$\dot{h}(E) = d_Q(\dot{\nu} + \dot{m})\frac{g[E]}{g + [E]/[E_m]} \text{ , thus} \qquad (3.58)$$

$$\dot{h}(e) = \dot{p}_a e \frac{1+g}{e+g} \qquad (3.59)$$

where $\dot{p}_a \equiv d_Q[E_m]g\frac{\dot{\nu}+\dot{m}}{1+g}$ represents the maximum aging rate and $e \equiv [E]/[E_m]$ the reserve density as a fraction of the maximum capacity. At constant substrate densities, the scaled energy reserve density, e, equals the scaled functional response, f, so the hazard rate is constant and independent of the age of the filament. The hazard rate for rods is likewise found from (3.44):

$$\dot{h}(e,l) = \dot{p}_a e \frac{1+g}{e+g} \frac{((l_d/l)^3 - 1)\,\delta/3 + 1 + l_d/g}{1 + l_d/g} \qquad (3.60)$$

For the hazard rate of unicellular isomorphs we obtain from (3.42)

$$\dot{h}(e,l) = \dot{p}_a e \frac{1+g}{e+g} \frac{1+g/l}{1+g/l_d} \qquad (3.61)$$

In contrast to filaments, rods and isomorphs experience a reduction of the hazard rate during the cell cycle.

If DNA is changed, the cell will cease functioning. This gives a lower boundary for the (population) growth rate because the population will become extinct if the division interval becomes too long. To prevent extinction (in the long run) the survival probability to the next division should be at least 0.5, so the lower boundary for substrate density can be found from $\text{Prob}\{\underline{a}_\dagger > t_d\} = \exp\{-\int_0^{t_d} \dot{h}(t)\,dt\} = 0.5$. Substitution of (3.60) and (3.39) leads to the lower boundary for the substrate density for rods, which must be found numerically. It is tempting to relate this aging mechanism, which becomes apparent at low substrate densities only, to the occurrence of stringent responses in bacteria, as described by, for example, Cashel and Rudd [122]. This will be discussed further when populations are considered, {165}.

It is intriguing to realize that the present mechanism for aging implies that organisms use free radicals to change their DNA. Although most changes are lethal or adverse, some can be beneficial to the organism. Using a selection process, the species can exploit free radicals for adaptation to changing environments. By increasing the specific activity of antioxidants, a species can prolong the life span of individuals in non hostile environments, but it reduces its adaptation potential as a species if the environment changes. This trait defines an optimal specific activity for antioxidants that depends on the life history of the organism and the environment. Large body size, which goes with a long juvenile period, as will be discussed on {234}, requires efficient antioxidants to ensure survival to the adult state. It implies that large bodied species have little adaptation potential, which is further reduced by the long generation time; this makes them vulnerable from an evolutionary perspective. It is possibly one aspect of the extinction of the dinosaurs, although not all of them were large and they may have been endothermic. Endotherms appear to combine a high survival probability of the juvenile period with a high aging rate, thus having substantial adaptation potential during the reproductive phase; they reach this by a reduction of the efficiency of antioxydants during puberty.

The present formulation assumes that growth ceases as soon as DNA is changed. The background is that many genes are involved in the synthesis of one or more compounds that are essential to structural body mass and so to growth. A few genes are involved in suppressing unregulated growth of cells in multicellular organisms. If such genes are affected, tumors can develop. This theory can, therefore, also be used to work out the age-dependent occurrence rate of tumors as well as the growth rate of tumors, cf. {252}.

The energy parameters can be tied to the accumulated damage to account for the well known phenomenon that older individuals eat less and reproduce less than younger ones with the same body volume. Senescence can be modelled this way. This role of age in energetics is not worked out here to keep the model as simple as possible.

3.13 Genetics and parameter variation

The parameter values undoubtedly have a genetically determined component, which can to some extent be modulated phenotypically. As has hopefully been made clear, the processes of feeding, digestion, maintenance, growth, reproduction and aging are intimately related. They involve the complete cellular machinery. Although mechanisms for growth which involve just one gene, have been proposed [176], the DEB theory makes it likely that thousands are involved. This restricts the possibilities of population genetic theories to deal with auxiliary characters that do not have a direct link to energetics. (This is not meant to imply that such theories cannot be useful for other purposes.) In the context of quantitative genetics, some instructive points should be mentioned here. For this purpose a particular property of the DEB model, which I call the invariance property (just to have a name to refer to), should be discussed first. This property is at the basis of body size scaling relationships to be discussed later. These relationships express how species-specific characters depend on body size.

The invariance property of the DEB model is that two species with parameter sets that differ in a very special way behave identically with respect to energetics as long as food density is strictly constant. So they will have exactly the same energy dynamics, volume and reproduction ontogenies, and so on, for all life stages. The derivation of the relationship between both parameter sets is simple when two individuals are compared with the same body volume and with a maximum surface area-specific ingestion rate that differ by a factor z, so $\{\dot{I}_m\}_2 = z\{\dot{I}_m\}_1$. To behave identically, the ingestion rates must be equal: $\dot{I}_2 = \dot{I}_1$. Since their volumes are equal, $V_2 = V_1$, (3.2) implies that $f_2 = f_1/z$ or $K_2 = zK_1 + (z-1)X$. Since the assimilation rates must be the same, $\dot{A}_2 = \dot{A}_1$, it follows that $\{\dot{A}_m\}_2 = z\{\dot{A}_m\}_1$. They must have the same storage dynamics, so (3.7) implies $[E_m]_2 = z[E_m]_1$. Identical growth defined by (3.12) implies that the other parameters should be the same, so $V_{b,2} = V_{b,1}$, $V_{p,2} = V_{p,1}$, $V_{h,2} = V_{h,1}$, $\kappa_2 = \kappa_1$, $[\dot{M}]_2 = [\dot{M}]_1$, $[G]_2 = [G]_1$ and $\ddot{p}_{a,2} = \ddot{p}_{a,1}$.

If food density is not strictly constant, but fluctuates a little, both species behave in a different manner as far as energy is concerned. This is due to the non-linear relationship between the scaled functional response f and food density X. The change of f with respect to X is $\frac{d}{dX}f = K(K+X)^{-2} = (1-f)^2/K$. So if f approaches 1, the change in the ingestion rate, and so in the energy reserve density, becomes negligibly small. This overall homeostasis is probably selectively advantageous, because it implies that regulation systems have a much easier job to coordinate the various processes of energy allocation, which allows for optimization. The mechanism is not unlike the restriction of the tolerance range for temperature of enzymes of homeotherms relative to heterotherms. The invariance property has an interesting consequence with regard to selection processes. At a constant food density, the (constant) surface area-specific ingestion rate, surface area-specific assimilated energy, and reserve energy density can be regarded as achieved physiological characters. Small fluctuations in food density drive selection to a (genetic) fixation of these characters as the maximum possible ones: $\{\dot{I}_m\} \rightarrow \{\dot{I}\}$, $\{\dot{A}_m\} \rightarrow \{\dot{A}\}$ and $[E_m] \rightarrow [E]$. This phenomenon is known as 'dwarfing'.

The parameter values for different individuals are likely to differ somewhat. Differences in ultimate volume at constant food density testify to this basic fact. To what extent this has a genetic basis is not clear, but the heredity of size in different races of dogs and transgenic mice and turkeys reveals the genetic basis of growth and size. Since only a tiny fraction of available DNA in eukaryotic cells is in active use, one can easily imagine that changes in the pieces that are used, or in the intensity with which the active parts are used, can result in changes in energy parameters. These regulation processes can be subjected to phenotypic influence and to factors located in the cytoplasm, and so to maternal effects. An important statistical consequence of this point of view is that parameter estimates can in principle no longer be based on means: the mean of von Bertalanffy curves with different parameters is not a von Bertalanffy curve. This problem obviously grows worse with increasing scatter. The modelling of parameter variation can easily introduce a considerable number of new parameters. To select just one or two parameters to solve this problem seems arbitrary. An attractive choice might be to conceive the factor z, just introduced, to be a stochastic variable, which couples four energy parameters. This introduces stochasticity only at fluctuating food densities.

Chapter 4

Analysis of the DEB model

The purpose of this chapter is to summarize the core of the DEB model and to evaluate combinations of primary processes and their consequences. The last chapter treated them one by one, as far as possible. Now the models for these processes will gain colour as the processes change together in a variable environment. To avoid repetition and to reveal the strength of a model that treats dimensions well, this summary will be made in terms of dimensionless quantities. Although some tests of evaluations are presented on the basis of data collected from the literature, adequate tests require experimental programs specifically designed to test the theory.

4.1 Summary of the DEB model

The assumptions on which the DEB model in its simplest form is based are collected in table 4.1. The primary variables and parameters are listed table 4.2. The formulation focuses on volumes of isomorphs. Organisms that change in shape can be covered by multiplication of all surface area related parameters with an appropriate, dimensionless, shape correction function of volume, in the equations for ingestion and change in volume and reserves.

Although accumulated damage is the natural state variable (together with volume), it is mathematically equivalent to age (together with volume) in the present model for aging. Because age does not require an identification with molecular processes, it is preferred here. Let us focus on individuals that cut their energy expenses to somatic maintenance only, under non-growth conditions, and do not shrink. The equations for the scaled quantities that define the input/output relationships for ectotherms are according to (3.2), (3.7), (3.12), (3.56) and (3.47):

$$\text{ingestion } I(x,l) = (l > l_b)\imath_m f l^2 \text{ with } f = \frac{x}{1+x} \tag{4.1}$$

$$\text{reserve dyn. } \frac{d}{d\tau}e = (e \geq l)\frac{g}{l}((l > l_b)f - e) + (e < l)g((l > l_b)\frac{f}{l} - \kappa) \tag{4.2}$$

$$\text{length dyn. } \frac{d}{d\tau}l = \frac{g}{3}\frac{(e-l)_+}{e+g} \tag{4.3}$$

Table 4.1: The assumptions that lead to the DEB model as formulated for multicellular animals and modified for unicellulars.

---

1 Body volume, stored energy density and accumulated damage are the state variables.

2 Three life stages can exist: embryos, which do not feed, juveniles, which do not reproduce, and adults. The transition between stages depends on the cumulated energy invested in maturation.

3 The feeding rate is proportional to surface area and depends hyperbolically on food density.

4 Food is converted to energy at a fixed efficiency and added to the reserves.

5 The dynamics of energy density in reserve is a first order process, while maximum density is independent of the volume of the individual and homeostasis is observed.

6 A fixed fraction of energy, utilized from the reserves, is spent on somatic maintenance plus growth, the rest on maturity maintenance plus maturation or reproduction.

7 Somatic and maturity maintenance are proportional to body volume, but maturity maintenance does not increase after a given cumulated investment in maturation.

7a Heating costs for endotherms are proportional to surface area.

8 Costs for growth are proportional to volume increase.

9 The energy reserve density of the hatchling equals that of the mother at egg formation, the embryo beginning at an infinitesimally small size.

9a Foetuses develop in the same way as embryos in eggs, but at a rate unrestricted by energy reserves.

9b Unicellulars divide a fixed time after initiation of DNA duplication, which occurs upon exceeding a certain volume.

10 Under starvation conditions, individuals always give priority to somatic maintenance and follow one of two possible strategies:

10a They do not change the reserve dynamics (so continue to reproduce).

10b They cease energy investment in reproduction and maturity maintenance (thus changing reserves dynamics).

10c Most unicellulars and some animals shrink during starvation, but do not gain energy from this.

11 Aging related damage accumulates in proportion to the concentration of damage inducing compounds, which accumulate in proportion to the volume-specific metabolic rate. For unicellulars damage is lethal, therefore it does not accumulate.

12 Apart from 'accidents', the hazard rate is proportional to accumulated damage, but death occurs if somatic maintenance costs can no longer be paid.

Table 4.2: The primary variables and parameters of the DEB model, secondary compound parameters and dimensionless representations. The abbreviation 'spec' stands for surface area- or volume-specific.

variable	symbol	dim.	dim.less equivalent	variable	symbol	dim.	dim.less equivalent
food density	X	$L^3 l^{-2}$ or ^{3}	$x \equiv X/K$	volume	V	L^3	$l \equiv (V/V_m)^{1/3}$
storage density	$[E]$	eL^{-3}	$e \equiv [E]/[E_m]$	age	a	t	$\tau \equiv a\dot{m}$

parameter	symbol	dim	parameter	symbol	dim
saturation coeff.	K	$L^3 l^{-2}$ or ^{3}	spec.max.ingestion	$\{\dot{I}_m\}$	$L^3 L^{-2} t^{-1}$
spec.max.assimilation	$\{\dot{A}_m\}$	$eL^{-2}t^{-1}$	max.energy density	$[E_m]$	eL^{-3}
spec.maintenance	$[\dot{M}]$	$eL^{-3}t^{-1}$	spec.growth costs	$[G]$	eL^{-3}
partition fraction	κ	-	aging acceleration	$\ddot{p}_a$	t^{-2}
rel. reprod. overhead	q	-	volume at birth	V_b	L^3
volume at puberty	V_p	L^3	spec.heating costs	$\{\dot{H}\}$	eL^{-2}
aspect ratio	δ	-	spec.wall growth costs	$\{G_A\}$	eL^{-2}
DNA duplication time	t_D	t	aging rate	$\dot{p}_a$	t^{-1}
Arrhenius temp	T_A	T	volume at division	V_d	L^3

comp. parameter	symbol	dim	comp. parameter	symbol	dim
energy conductance	$\dot{v} \equiv \{\dot{A}_m\}/[E_m]$	Lt^{-1}	maintenance rate coef.	$\dot{m} \equiv [\dot{M}]/[G]$	t^{-1}
investment ratio	$g \equiv \frac{[G]}{\kappa[E_m]}$	-	max. volume	$V_m \equiv (\kappa\frac{\{\dot{A}_m\}}{[\dot{M}]})^3$	L^3
heating volume	$V_h \equiv (\{\dot{H}\}/[\dot{M}])^3$	L^3	scaled aging accel.	$p_a \equiv \ddot{p}_a/\dot{m}^2$	-
scaled birth length	$l_b \equiv (V_b/V_m)^{1/3}$	-	scaled puberty length	$l_p \equiv (V_p/V_m)^{1/3}$	-
scaled heating length	$l_h \equiv (V_h/V_m)^{1/3}$	-	scaled division length	$l_d \equiv (V_d/V_m)^{1/3}$	-
scaled ingestion rate	$\imath_m \equiv \{\dot{I}_m\}V_m^{2/3}/\dot{m}$	L^3	scaled reprod. rate	$q_R \equiv qg(1-\kappa)$	-
spec. energy cond.	$\dot{\nu} \equiv \dot{v}V_d^{-1/3}$	t^{-1}			

$$\text{hazard } h(\tau) = \frac{p_a}{(e>0)} \int_0^\tau \left(1 - \left(\frac{l(0)}{l(\tau_2)}\right)^3 + \int_0^{\tau_2} \left(\frac{l(\tau_1)}{l(\tau_2)}\right)^3 d\tau_1\right) d\tau_2 \quad (4.4)$$

$$\text{reproduction } R(e,l) = (e \geq l)(l > l_p)\frac{q_R}{e_0}\left(\frac{g+l}{g+e}el^2 - l_p^3\right) \quad (4.5)$$

where I, h, and R stand for $\dot{I}$, $\dot{h}$, and $\dot{R}$ with $\dot{m}^{-1}$ as the unit of time, while e_0 is given in (3.26). The 7 parameters of this set of equations, which fully determine feeding, growth, survival and reproductive behaviour, are: l_b, l_p, $\imath_m$, g, κ, p_a and q_R. The conversion to (unscaled) time, food and body volume involves 3 additional parameters. Since all rates are thought to depend on temperature in the same way, as a first approximation, the choice of $\dot{m}^{-1}$ as the unit of time makes the dimensionless system independent of temperature. The respiration and reproduction rates are given in figure 4.1 as functions of scaled reserves and length.

At constant food density, the scaled length of an individual (including endotherms) as a function of scaled age is

$$l(\tau_b + \tau) = f - l_h - (f - l_h - l_b)\exp\left\{\frac{-\tau/3}{1+f/g}\right\} \quad (4.6)$$

Figure 4.1: Stereo view of respiration (upper) and reproduction (lower) rates in ectotherms that cease reproduction during starvation, as fractions of their maxima. The rates (z-axis) are given as functions of scaled length (x-axis) and scaled functional response (y-axis). Chosen parameters: $l_b = 0.1$, $l_p = 0.2$, $g = 0.1$ and $\kappa = 0.666$.

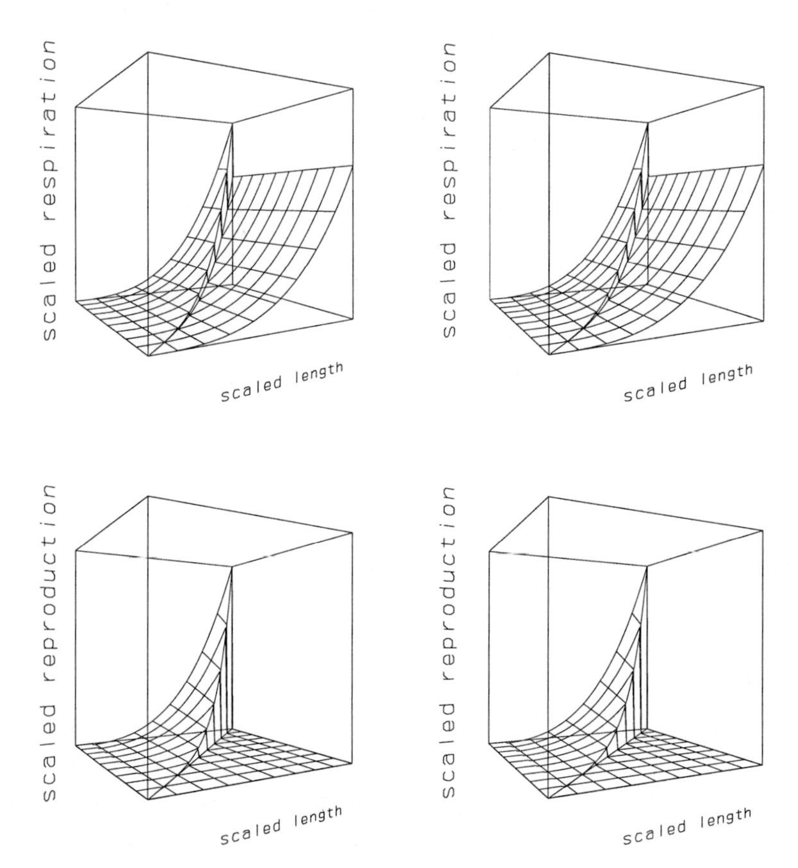

The symbol l_h is defined as $l_h \equiv (V_h/V_m)^{1/3}$ and stands for the reduction in ultimate scaled length, due to the energy drain used for heating. Note that age is initialized at the start of embryonic development, not at birth. The length of the juvenile period at constant food density on the basis of (4.6) is

$$\tau_p - \tau_b = 3\left(1 + \frac{f}{g}\right) \ln\left\{\frac{f - l_h - l_b}{f - l_h - l_p}\right\} \tag{4.7}$$

This expression can be of importance in life history studies. The ability of a bird to fly and to take part in migration, for instance, relates to a cumulative energy investment into development in a way similar to the ability to reproduce. The expression, therefore, gives the time the animal is bound to the breeding site.

Growth in volume is at a maximum if $\frac{d^2}{d\tau^2} l^3 = 0$. This occurs if $f = 1$ and $l = (1 - l_h)2/3$

and amounts to

$$\max \frac{d}{d\tau} l^3 = \frac{4}{27} \frac{g}{1+g} (1 - l_h)^3 \tag{4.8}$$

While growth in length only decreases after birth, growth in volume first increases and then decreases, which results in a sigmoid volume-time curve. Later in this chapter, I will discuss deviations from this pattern.

The adjustments that an individual must make during growth in the rate constants for the first order dynamics of the energy reserves (not as density) to maintain homeostasis are small if $\frac{d}{dt} \ln V \ll V^{-1/3}\{\dot{A}_m\}[E_m]^{-1}$; this condition becomes $0 \ll l + l_h + g$ in the scaled symbols.

For filaments, the scaled dynamics for length and energy density amount to

$$\frac{d}{d\tau} l = \frac{g}{3} \frac{l}{l_d} \frac{e - l_d}{e + g} \tag{4.9}$$

$$\frac{d}{d\tau} e = g(f - e)/l_d \tag{4.10}$$

Similarly we have for rods:

$$\frac{d}{d\tau} l = \frac{g}{3} \frac{l}{l_d} \frac{e}{e+g} \left(\frac{\delta}{3} \frac{l_d^3}{l^3} - \frac{l_d}{e} + 1 - \frac{\delta}{3} \right) \tag{4.11}$$

$$\frac{d}{d\tau} e = \frac{g}{l_d} \left(\frac{\delta}{3} \frac{l_d^3}{l^3} + 1 - \frac{\delta}{3} \right) (f - e) \tag{4.12}$$

4.1.1 Equivalent assumptions

Assumptions about storage dynamics and the κ-rule can only be tested directly if energy flows in individuals are measured, which is a difficult task. This problem can be solved by replacing assumptions 5 and 6 by three other assumptions that are mathematically equivalent for isomorphs.

At constant food density:

- growth is of the von Bertalanffy type after birth
- the ultimate length is proportional to the (scaled) functional response
- the inverse of the von Bertalanffy growth rate is linear in the (scaled) functional response

The demonstration of the equivalence is somewhat technical; this subsection can be skipped without loss of continuity. The significance of the exercise is to show that some assumptions and consequences are interchangeable, which is important for judging the testability of the DEB model. The strategy to prove the equivalence is first to assume that food density is constant (from birth) and then use the property of dynamic systems that the behaviour of a system only depends on the value of the state variables, that is on the position in the state space and not on the trajectory through the state space.

Suppose that κ can be any function of the state variables $[E]$ and V and that food density is constant. The task is to show that it is constant with respect to these variables.

It is always possible to write the sum of the energy spent on growth plus maintenance as $[G]\frac{d}{dt}V + [\dot{M}]V = \kappa([E], V)\dot{C}$ with $\dot{C} = \dot{A} - \frac{d}{dt}E$. This formulation acknowledges that at each time increment the individual decides how much of the utilized energy it allocates to growth plus maintenance.

Von Bertalanffy growth can be written as $\frac{d}{dt}V = \dot{\rho}V^{2/3} - 3\dot{\gamma}V$, where $\dot{\rho}$ as well as $\dot{\gamma}$ are independent of V. At constant food density X, the reserves after birth can be written as a function E of volume V and the scaled input $f = X/(K+X)$, so $\frac{d}{dt}E = \frac{d}{dt}V\,\frac{\partial}{\partial V}E$. This allows for the solution of $\frac{\partial}{\partial V}E$: $\frac{\partial}{\partial V}E = \frac{\alpha_1+\beta_1 V^{1/3}}{\alpha_2+\beta_2 V^{1/3}}$ with $\alpha_1 = \{\dot{A}_m\}f - \dot{\rho}(f)[G]/\kappa$, $\alpha_2 = \dot{\rho}(f)$, $\beta_1 = -[\dot{M}]/\kappa + 3\dot{\gamma}(f)[G]/\kappa$ and $\beta_2 = -3\dot{\gamma}(f)$.

The concept 'state variable' implies that the change in the state variables only depends on the values of the state variables; the dynamics of the reserves, therefore, do not depend directly on allocation, so $\frac{d}{d\kappa}\frac{\partial}{\partial V}E = 0$ for all values of V. It follows that

$$\frac{d}{d\kappa}\frac{\partial}{\partial V}E = \frac{c_0 + c_1 V^{1/3} + c_2 V^{2/3}}{(\alpha_2 + \beta_2 V^{1/3})^2} \tag{4.13}$$

with $c_0 = \alpha_2^2\frac{d}{d\kappa}(\alpha_1/\alpha_2)$, $c_1 = \beta_2^2\frac{d}{d\kappa}(\alpha_1/\beta_2) + \beta_1^2\frac{d}{d\kappa}(\alpha_2/\beta_1)$ and $c_2 = \beta_2^2\frac{d}{d\kappa}(\beta_1/\beta_2)$, has to vanish for all values of V, so that $\frac{d}{d\kappa}(\alpha_1/\alpha_2) = 0$ and $\frac{d}{d\kappa}(\beta_1/\beta_2) = 0$. This gives $\frac{d}{d\kappa}\dot{\rho}(f) = \dot{\rho}^2(f)[G]/(f\{\dot{A}_m\}\kappa^2)$ and $3\frac{d}{d\kappa}\dot{\gamma}(f) = 9\dot{\gamma}^2(f)[G]/(\kappa[\dot{M}]) - 3\dot{\gamma}(f)/\kappa$. Solution of these differential equations gives $\dot{\rho}(f) = \frac{\{\dot{A}_m\}f}{[G]/\kappa + \frac{\partial}{\partial V}E}$ and $3\dot{\gamma}(f) = \frac{[\dot{M}]/\kappa}{[G]/\kappa+\frac{\partial}{\partial V}E}$. Since $\dot{\rho}$ and $\dot{\gamma}$ are independent of V, $\frac{\partial}{\partial V}E$ is also independent of V, so E has the form $E = \chi(f) + \psi(f)V$.

Growth does not depend on assimilation energy directly, which implies that $\frac{\partial}{\partial V}E$, which is equal to $\psi(f)$, in $\dot{\rho}$ and $\dot{\gamma}$ has to be replaced by $E/V - \chi(f)/V$; this is only independent of f and V for $\chi(f) = 0$. So we have $E = \psi(f)V$ or $[E] = \psi(f)$. Similarly, f in $\dot{\rho}$ has to be written as a function of E, so f is replaced by $\psi^{-1}([E])$, where ψ^{-1} is the inverse function of ψ, i.e. $\psi^{-1}(\psi(f)) \equiv f$. We now obtain $\frac{d}{dt}V = \dot{\rho}V^{2/3} - 3\dot{\gamma}V = \frac{V^{2/3}\{\dot{A}_m\}\psi^{-1}([E]) - V[\dot{M}]/\kappa}{[G]/\kappa+[E]}$.

Since volume and storage are the only state variables on which growth depends, the values in the past should be irrelevant. We can, therefore, apply the equation to fluctuating food densities as long as the values of the state variables are within the domain that can be reached at constant food densities. The dynamics of the volume-specific storage $\frac{d}{dt}[E] = V^{-1}\frac{d}{dt}E - EV^{-2}\frac{d}{dt}V$ can now be inferred from the balance equation $\frac{d}{dt}E = \dot{A} - \dot{C}$ and the κ-rule. It is $\frac{d}{dt}[E] = \{\dot{A}_m\}V^{-1/3}(f - \psi^{-1}([E]))$.

Since $1/\dot{\gamma}$ is linear in f, ψ is linear in f. The conservation law for energy also dictates that $\dot{\gamma} = 0$ if $f = 0$, which makes ψ proportional to f, say $\psi(f) = [E_m]f$. This implies that ψ^{-1} is proportional to $[E]$ and vice versa. In other words the volume-specific storage obeys a simple first order process if and only if $\dot{\gamma}^{-1}$ is linear in f.

4.1.2 State space

Like all system theoretical models, it is possible to represent the individual as a point in the three dimensional state space spanned up by scaled length, scaled energy reserve density and accumulated damage. As time passes, the point moves through the state space. Individuals appear at birth in the plane through $l = l_b$, and disappear at death. In a population there are many individuals around, so many points are moving simultaneously through the state space for individuals. The next chapter discusses the possibility of following them, {206}. This section serves as an introduction for the population level.

Since damage only serves to remove individuals in the DEB model, the discussion can be restricted to scaled length and scaled reserve density. Figure 4.2 summarizes the movements through the state space for isomorphs, rods and filaments, for three levels of functional response. This representation is known as a direction field, where the length of the line segments represents the rate of movement.

4.1.3 Scatter structure of weight data

For simplicity's sake, the processes of feeding and growth have been modelled deterministically, so far. This is not very realistic, as (feeding) behaviour especially is notoriously erratic. This subsection discusses growth if feeding follows a special type of random process, known as an alternating Poisson process or a random telegraph process. Because of the resulting complexity, I rely here on computer simulation studies. Some analytical properties of this input process are studied on {264}, in relation to the behaviour of one compartment models.

Suppose that feeding occurs in meals that last an exponentially distributed time interval $\underline{t}_1$ with parameter $\dot{\lambda}_1$, so $\text{Prob}\{\underline{t}_1 > t\} = \exp\{-t\dot{\lambda}_1\}$. The mean length of a meal is then $\dot{\lambda}_1^{-1}$. The time intervals of fasting between the meals is also exponentially distributed, but with parameter $\dot{\lambda}_0$. Food intake during a meal is copious so $f = 1$, while $f = 0$ during fasting. The mean value for f is $\mathcal{E}\underline{f} = \dot{\lambda}_0(\dot{\lambda}_0 + \dot{\lambda}_1)^{-1}$. This on/off process is usually smoothed out by the digestive system, but let us here assume that this is of minor importance. According to (4.2) and (4.3), growth in scaled length of juveniles (that are able to shrink) is given by

$$\frac{d}{d\tau}e = \frac{g}{l}(f - e) \quad \text{and} \quad \frac{d}{d\tau}l = \frac{g}{3}\frac{e - l}{e + g}$$

where τ is the scaled time, $t\dot{m}$, as before. Only one parameter, g, is involved in this growth process, and two others, λ_0 and λ_1, occur in the description of the on/off process of f. (Note that the λ's do not have dots, because we now work in scaled time, which is dimensionless.) The process is initiated with $l(0) = l_b$ and $e(0)$ equals the scaled energy density of a randomly chosen adult.

Figure 4.3 shows the results of a computer simulation study, where scaled weight relates to scaled length and scaled energy density, according to (2.9) as

$$W_w([d_{wv}]V_m)^{-1} = (1 + e[d_{we}]/[d_{wv}])l^3$$

At this moment, I do not understand mathematically why the deterministic growth curve is a bit above the mean of (500) random growth curves. The resemblance of the scatter structure with experimental data is striking, see for instance figure 2.7. This does not imply, however, that the feeding process is the only source of scatter. Differences of parameter values between individuals are usually important as well. The results do suggest a mechanism behind the generally observed phenomenon that scatter in weights increases with the mean.

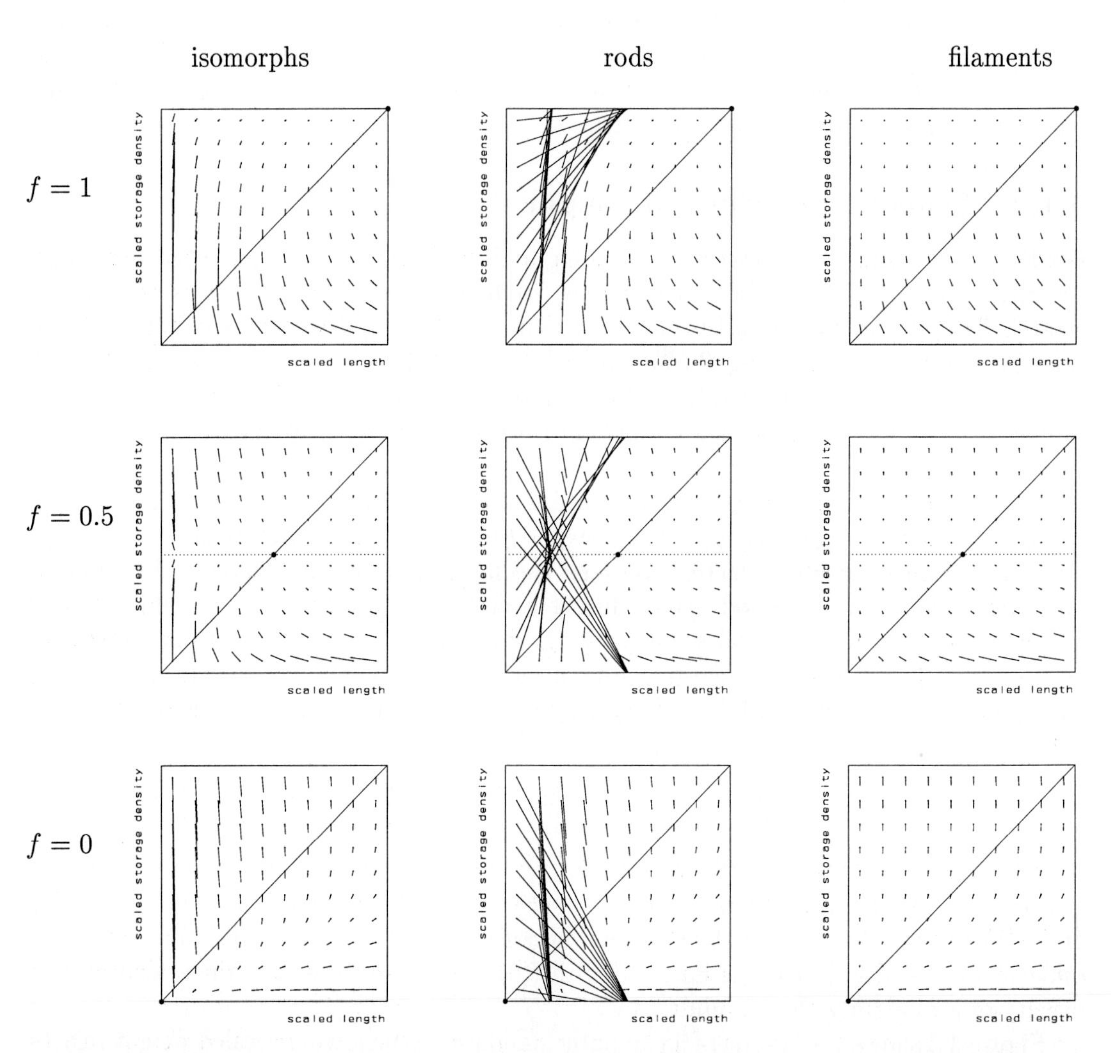

Figure 4.2: The direction field for the DEB model for isomorphs, rods and filaments and three values for the functional response f. The length of the line segments equals the change of the state variables during a period $\dot{m}^{-1}$. The diagonal line represents the isocline $\frac{d}{dt}l = 0$; the fat dot is an absorbing state. The parameters are $g = 0.03$, $l_d = 0.8$ (rods and filaments) and $\delta = 0.6$ (rods only). Shrinkage during starvation is allowed in all cases, to facilitate comparison.

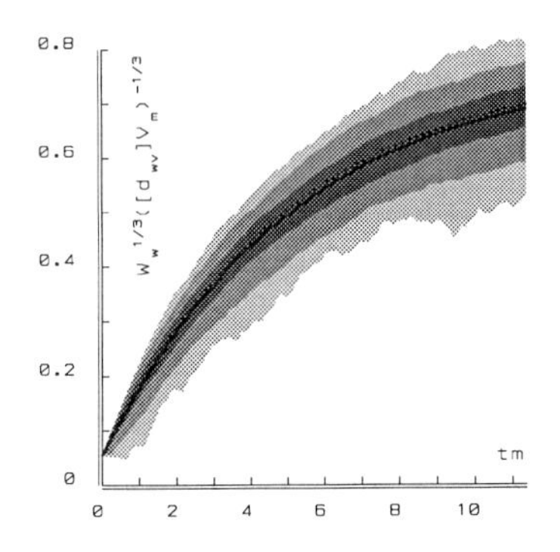

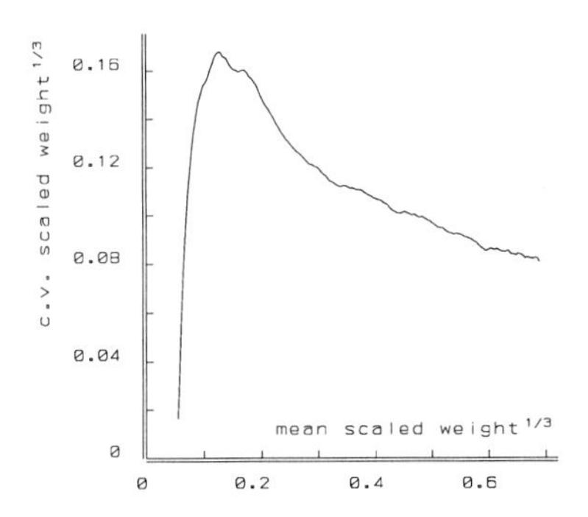

Figure 4.3: Computer simulated scaled weight$^{1/3}$ is plotted against scaled time in the left figure, if feeding follows an alternating Poisson process. The shade areas give frequency intervals of 99, 90 and 50%, the drawn curve gives the mean and the dotted one gives the deterministic growth curve, if feeding is constant at the same mean level. The coefficient of variation is given in the right figure. The parameters are $\lambda_0 = 11.666$, $\lambda_1 = 5$, $g = 1$, $l_b = 0.05$ and $[d_{we}]/[d_{wv}] = 0.5$.

4.1.4 Yield

The conversion of one form of biomass into another is basic to the population and ecosystem levels. It is, therefore, introduced here at the level of individuals, as conversion efficiency can be most easily studied at this level. Food is converted to faeces, to the biovolume of the individual and, for females, to the biovolume of offspring.

A yield factor measuring (gross) conversion efficiency is defined as the ratio of the produced biovolume and the food ingested. (The term 'yield' is frequently used in the microbiological literature; in eco-physiology it is known as the gross production efficiency.) This measure does not account for energy reserves. In relation to conversions in food chains, the energy reserves of prey obviously contribute to its nutritional value so that exclusion of reserves restricts the usefulness of yield coefficients. A problem hampering the inclusion of reserves in efficiency measures is the artificial conversion of reserve energy to biovolume or vice versa. This conversion is artificial because growth takes time and, therefore, maintenance is intrinsically involved; thus this conversion depends on the energy uptake capacity of the organism. This is why large bodied species are more efficient than small species: their absolute growth rate is higher (as will be discussed in the chapter on comparison of species, {217}), so that less energy is lost in maintenance processes. (Large endotherms are, additionally, more efficient than small ones because of the relatively smaller amount of heat loss due to a more favourable volume/surface area ratio. This efficiency is widely recognized.) Biovolume of offspring should not be added at egg production, because the embryo is assumed to have a negligibly small volume. The mother has already paid for development to the juvenile stage, so it seems natural to add biovolume at birth to the volume increase of the mother. The time delay caused by incubation is neglected in the yield factor. The only satisfactory way to include energy reserves is on the basis of free energy. This is feasible for micro-organisms, see {201}, but difficult for animals where food and faeces are hard to define thermodynamically; this hampers access to the free energy of structural biomass and reserves.

The yield factor, with dimension volume biomass per volume food, is thus defined by

$$Y_i = (\frac{d}{dt}V + \dot{R}V_b)/\dot{I} \tag{4.14}$$

This measure only makes sense in situations of constant food density, where the reserve density does not change. At fluctuating food density it is possible that growth and reproduction allocation occurs without ingestion, which makes the measure meaningless. Considered as a function of food density and volume, this yield factor only depends on the partition coefficient κ, the energy investment ratio g, and a proportionality constant, $\dot{v}/\{\dot{I}_m\}$, standing for the ratio of energy yield of a unit volume of ingested food and the maximum stored energy density. The latter proportionality constant, which has dimension volume biomass per volume food, converts a dimensionless yield factor y_i, into Y_i. For ectotherms that also reproduce under no-growth conditions, this scaled yield factor is for $y_i \equiv Y_i\{\dot{I}_m\}/\dot{v}$ given by

$$y_i(l,f) = \frac{f - l + (l > l_p)(1-\kappa)q\left(f(g+l) - (f+g)l_p^3/l^2\right)l_b^3/e_0}{f(f+g)} \quad \text{for } l \leq f \tag{4.15}$$

$$= (l > l_p)q(f - \kappa l - (1-\kappa)l_p^3/l^2)l_b^3/fe_0 \quad \text{for } l > f \tag{4.16}$$

where e_0, given in (3.26), is the energy invested in an egg as a fraction of the maximum stored energy in an individual of maximum volume. This scaled yield factor is illustrated in figure 4.4.

The maximum scaled yield is

$$y_{i,\max} = l_b^{-1}\left(1 + \sqrt{1 + g/l_b}\right)^{-2} \tag{4.17}$$

which is reached for $l = l_b$ and $f = l_b + \sqrt{l_b^2 + gl_b}$. For $l_b \to 0$ and $f \to 0$, $y_{i,\max} \to g^{-1}$, which means that an animal of zero volume would spend no energy on development or energy storage; it will just convert all energy it can obtain from food into biomass. It also means that all real world animals, for which $l > 0$ holds, are much less efficient converters.

Although (4.15) and (4.16) look rather massive, it is surprising that they do not contain parameters such as the maintenance cost, except in a hidden form in the scaling of length as a fraction of maximum length. The yield has a very weak local maximum for (female) adults. The volume and functional response at this local maximum must be obtained numerically from (4.15).

The yield factor (4.15) and (4.16), from now on called instantaneous yield, has a limited use for studies at a longer time scale, because of its instantaneous nature. It will (rapidly) change in time, because the animal changes its volume. The non-instantaneous yield factor, Y_n, defined by the ratio of the cumulated biomass production from birth onwards and the cumulated amount of ingested food is more informative. For $y_n \equiv Y_n\{\dot{I}_m\}/\dot{v}$, we have

$$y_n(l,f) = \frac{l^3 - l_b^3 + l_b^3\int_0^{t(l)} \dot{R}(t)\,dt}{g\dot{m}f\int_0^{t(l)} l^2\,dt} \tag{4.18}$$

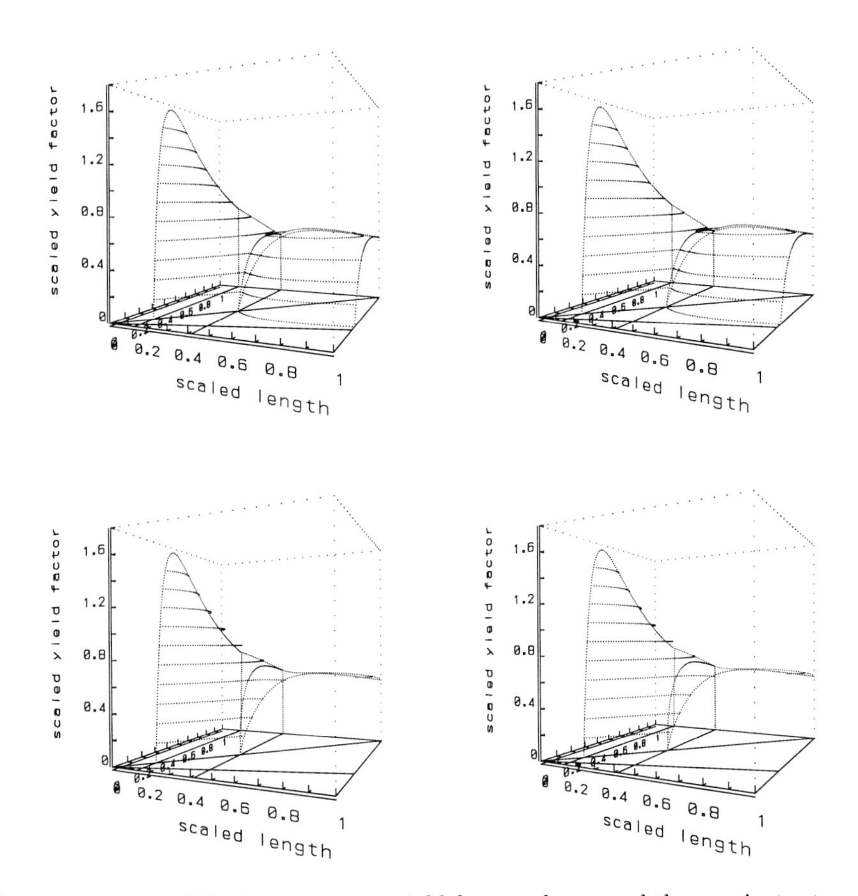

Figure 4.4: Stereo view of the instantaneous yield factor, above, and the non-instantaneous one, below, for a female ectotherm as a function of length and, in depth, functional response (or stored energy density). Parameters: $l_b = 0.133$, $l_p = 0.417$, $g = 0.033$ and $\kappa = 0.3$. The figures show that the conversion efficiency from food into biomass rapidly decreases for increasing body size till some plateau where the decrease in growth is compensated by the increase in reproduction.

where $l(t) = f - (f - l_b)\exp\{\frac{-tg\dot{m}/3}{f+g}\}$ and $t(l) = 3\frac{f+g}{g\dot{m}}\ln\{\frac{f-l_b}{f-l}\}$. Substitution of these functions and of (4.5) for the reproduction rate $\dot{R}$ into (4.18) reveals that the integration can readily be carried out and made explicit. The result is not a nice small formula, but a line filling one, illustrated in figure 4.4.

The animal, whose functional response remains constant from birth onwards, will obey $l < f$. This restriction was not necessary for y_i, which allows that an animal first grows to a large volume at abundant food and then stays at a low food density for a sufficiently long time for the amount of stored energy to be adapted. At $l = l_b$, the non-instantaneous yield equals the instantaneous one, conceptually. For $l = f$, they are also equal, because the animal here does not change volume for an infinitely long period. Fig.4.4 points to the counter intuitive result that yield at high food densities is a bit lower than at moderate ones. One would think that growth is fastest at high food densities, so that relatively little energy is lost through maintenance. The result can be explained by increasing energy investment in storage. Is it just coincidence that laboratory cultures of many species of animals do better at 70% of the *ad libitum* amount of food?

The non-instantaneous yield factor will prove to be identical to conversion efficiency at the population level, if we harvest at a fixed age. See next chapter, {181}.

4.2 Changing and poor feeding conditions

4.2.1 Step up/down

The difference between age-based and size-based models becomes apparent in situations of changing food densities. As long as food density remains constant, size-based models can always be converted into age-based ones, which makes it impossible to tell the difference.

Figure 4.5 shows the result of an experiment with *Daphnia magna* at 20 °C, exposed to constant high food densities with a single instantaneous switch to a lower food density at 1, 2 or 3 weeks. The reverse experiment with a single switch from low to high food densities has also been done, together with continuous exposure to both food densities. Figure 3.13 has already shown that the maintenance rate coefficient $\dot{m}$ and energy conductance $\dot{v}$ can be obtained by comparison of growth at different constant food densities. These compound parameters, together with ultimate and maximum lengths and the common length at birth have been obtained from the present experiment without a switch. These five parameters completely determine growth with a switch, both up and down, leaving no free parameters to fit in this situation. The excellent fit gives strong support to the DEB theory.

4.2.2 Mild starvation

If a growing individual is starved for some time, it will (like the embryo) continue to grow (at a decreasing rate) till it hits the non-growth boundary of the state space ($e = l$). Equation (3.25) describes the e, l-path. Depending on the amount of reserves, the change in volume will be small for animals not far from maximum size. Strömgren and Cary [696] found that mussels in the range of 12–22 mm grew 0.75 mm. If the change in size is

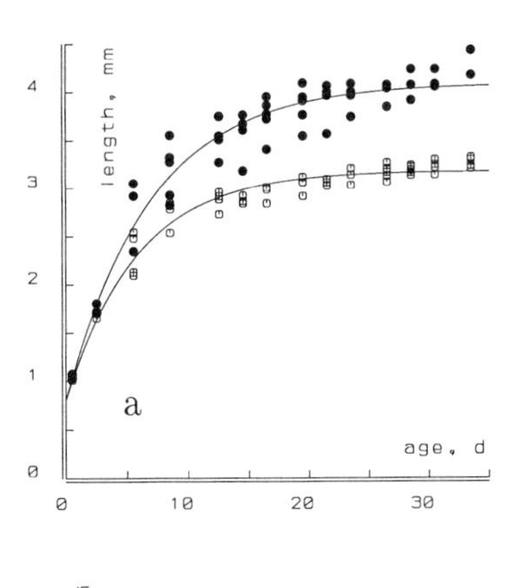

Figure 4.5: Length-at-age for the water-flea *Daphnia magna* at 20 °C feeding at a high (•) and a low (○) constant density of the green alga *Chlorella pyrenoidosa* (a), and with a single interchange of these two densities at 1 (b,e), 2 (c,f) or 3 (d,g) weeks. The curves b to g describe the slow adaptation to the new feeding regime. They are completely based on the 5 parameters obtained from a, so no additional parameters were estimated. From [410].

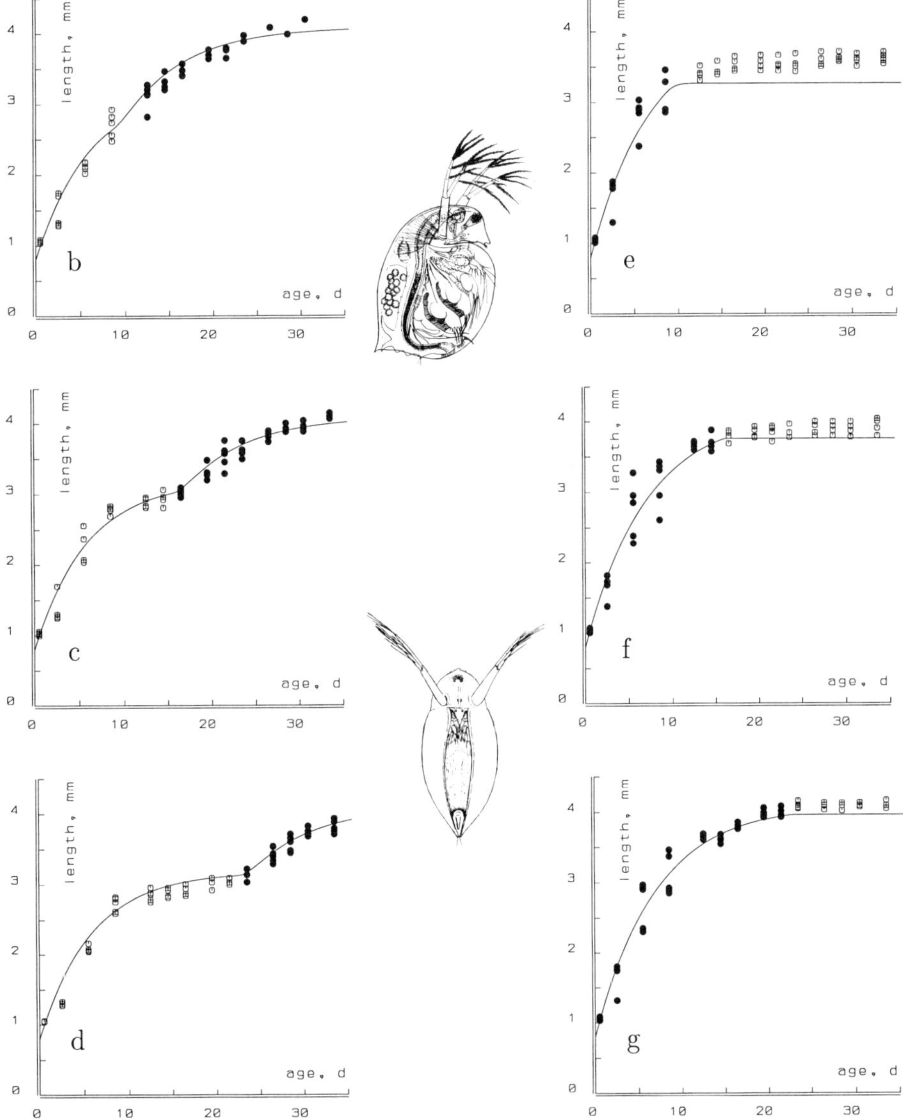

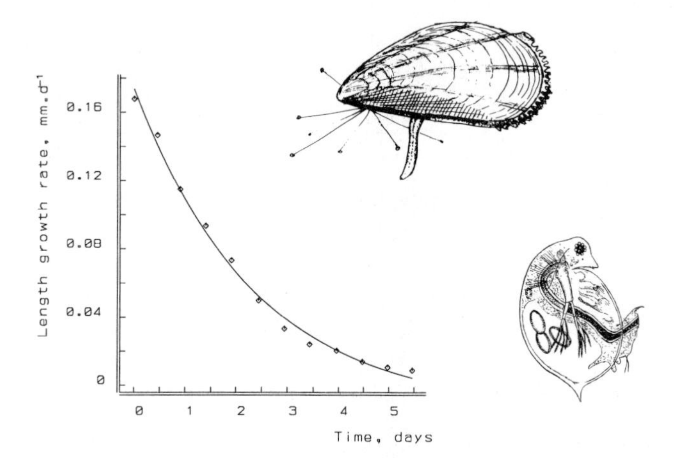

Figure 4.6: Growth rate in the starved mussel *Mytilus edulis* at 21.8 °C. Data from Strömgren and Cary [696]. The parameter estimates (and standard deviations) are $\frac{g}{e(0)} = 12.59(1.21)$, $\dot{m} = 2.36(0.99)\ 10^{-3}$ d^{-1} and $\dot{v} = 2.52(0.183)$ mm d^{-1}.

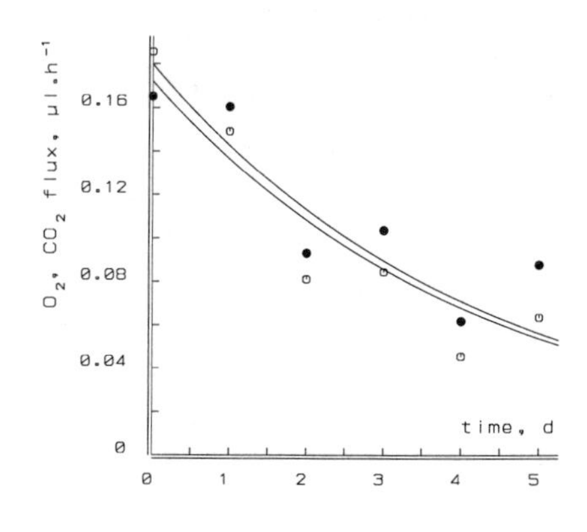

Figure 4.7: The oxygen consumption rate (•) and the carbon dioxide production rate (○) in starved *Daphnia pulex* of 1.62 mm at 20 °C. Data from Richman [595]. The exponential decay rate is 0.23 (0.032) d^{-1}.

neglected, the scaled reserve density changes as $e(\tau) = e(0)\exp\{-g\tau/l\}$ and the growth of scaled length is $\frac{d}{d\tau}l = \frac{g}{3}\frac{\exp\{-g\tau/l\}-l/e(0)}{\exp\{-g\tau/l\}+g/e(0)}$. Figure 4.6 confirms this prediction.

Respiration during starvation is proportional to the use of reserves; see {103}. It should, therefore, decrease exponentially in time at a rate of $\dot{v}V^{-1/3}$ if size changes can be neglected. See (3.7). Figure 4.7 confirms this prediction for a daphnid. If a shape coefficient of $[d_m] = 0.6$ is used to transform the length of the daphnid into a volumetric one, the energy conductance becomes $\dot{v} = 0.6\times1.62\times0.23 = 0.22$ mm d^{-1}. This value seems to be somewhat small in comparison with the mean energy conductance of many species, cf. {224}. The next section suggests an explanation in terms of changes in allocation rules to reproduction during starvation.

4.2.3 Prolonged starvation

If the reserve density drops below the non-growth barrier $e = l$, a variety of possible physiological behaviours seems to occur, depending on the species and environmental factors. Deviation from the κ-rule is necessary, because the standard allocation to growth plus maintenance is no longer sufficient for maintenance, even if growth ceases. Pond snails seem to continue energy allocation to reproduction during prolonged starvation under a light/dark 16:8 cycle (summer conditions, denoted by LD), but they cease reproduction under a 12:12 cycle (spring/autumn conditions, denoted by MD) [76,788]. This makes sense because under summer conditions, an individual can expect high primary production rates, so if it has consumed a plant, it will probably find another one in the direct neighbourhood. Under spring/autumn conditions, however, it can expect a long starvation period. By ceasing allocation to reproduction, it can increase its survival period by a factor of two; see figure 4.9. Another aspect is that offspring have a remote survival probability if there is no food around. They are more vulnerable than the parent, as follows from energy reserve dynamics. These dynamics can be followed on the basis of the assumption that LD snails do not change the rule for utilization of energy from the reserves, and both MD and LD snails do not cut somatic maintenance.

If starvation is complete and volume does not change, i.e. $f = 0$ and l is constant, the energy reserves will be $e(t) = e(0)\exp\{-g\dot{m}t/l\}$; see (3.22). Dry weight is a weighted sum of volume and energy reserves, so according to (2.10) for LD snails we must have

$$W_d(l,t) = V_m l^3([d_{dv}] + [d_{de}]e(0)\exp\{-g\dot{m}t/l\}) \tag{4.19}$$

if the buffer of energy allocated to reproduction is emptied frequently enough ($E_{\dot{R}}$ small). For MD snails, where $e(t) = e(0) - ([\dot{M}]/[E_m])t$, dry weight becomes

$$W_d(l,t) = V_m l^3([d_{dv}] + [d_{de}]e(0) - [d_{de}]([\dot{M}]/[E_m])t) \tag{4.20}$$

So dry weight of LD snails decreases exponentially and that of MD snails linearly. Figure 4.8 confirms this. It also supports the length dependence of the exponent.

When storage levels become too low for maintenance, some species can decompose their structural biomass to some extent. If feeding conditions then become less adverse, recovery may be only partial. The distinction between structural biomass and energy reserves fades at extreme starvation. The priority of storage materials over structural biomass is perhaps even less strict in species that shrink during starvation. Species with (permanent or non-permanent) exoskeletons usually do not shrink in physical dimensions, but the volume-specific energy content nonetheless decreases during starvation.

If we exclude the possibility of prolonging life through decomposition of structural body mass and if death strikes when the utilization rate drops below maintenance level, the time till death by starvation can be evaluated.

In animals such as LD snails, that do not change storage dynamics, the utilization rate, $-\frac{d}{dt}[E]$, equals the maintenance rate, $[\dot{M}]$ for $[E]/[E_m] = V^{1/3}[\dot{M}]/\{\dot{A}_m\}$ or $e = \kappa l$. Since $e(t) = e(0)\exp\{-\dot{m}tg/l\}$, death strikes at $t_\dagger = \frac{l}{\dot{m}g}\ln\frac{e(0)}{\kappa l}$. This only holds if length increase is negligibly small.

In animals such as MD snails, which change storage dynamics to $\frac{d}{dt}e = [\dot{M}]/[E_m]$ or $e(t) = e(0) - t[\dot{M}]/[E_m]$, death strikes when $e = 0$, that is at $t_\dagger = e(0)[E_m]/[\dot{M}] = \frac{e(0)}{\kappa\dot{m}g}$. This only holds as long as there is no growth, so $e(0) < l$. In practice, this is a more stringent condition than the previous one. The first part of the starvation period usually includes a period where growth continues, because $e > l$. This complicates the analysis of starvation data, as illustrated in the following example. In a starvation experiment with MD snails, individuals were taken from a standardized culture and initially fed *ad libitum* for 4 days prior to complete starvation. If we assume that food density in the culture has been constant, so $e(0) = f_c$, say, with f_c being about 0.7, and $f = 1$ during the 4 days prior to the starvation experiment, the change in length is negligibly small. The initial storage density is $e(0) = 1 - (1 - f_c)\exp\{-4\dot{m}g/l\}$, according to (3.7). The time till growth ceases is found again from (3.7) and the boundary condition $l = e(0)\exp\{-t\dot{m}g/l\}$. (Although

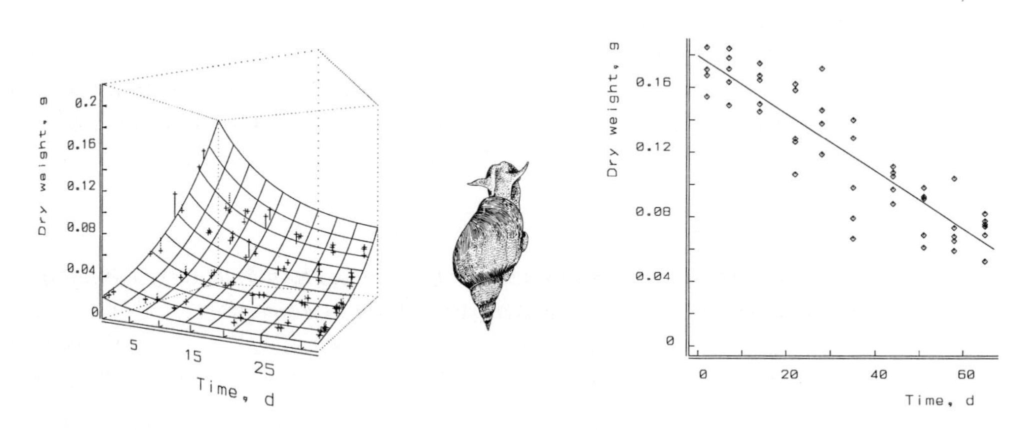

Figure 4.8: Dry weight during starvation of LD (left) and MD (right) pond snails *Lymnaea stagnalis* at 20 °C. The left figure gives dry weights (z-axis) as a function of starvation time (x-axis) and length (y-axis: 1.6–3.3 cm). In the right figure, the length of the MD pond snails was 3 cm. From [788]. The surface and curve are fitted DEB-based expectations.

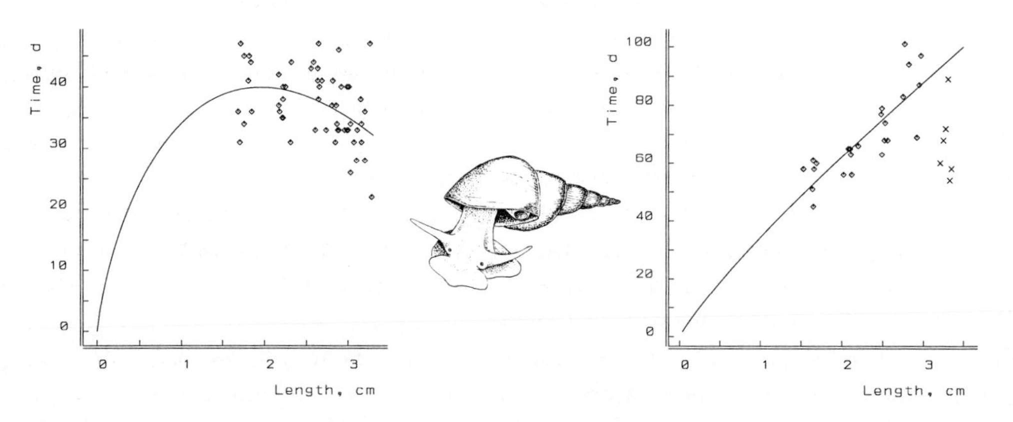

Figure 4.9: Survival time during starvation of LD (left) and MD (right) pond snails as a function of length. From [788]. The data points × in the right figure are not included in the DEB-based fit. These large individuals had deformations of the shell.

length increase is negligibly small, energy allocation to growth can be substantial.) After a period $l(\kappa \dot{m} g)^{-1}$ death will strike, so

$$t_\dagger = \frac{l}{\dot{m} g} \left(\frac{1}{\kappa} + \ln \left\{ l^{-1} (1 - (1 - f_c) \exp\{-4\dot{m} g / l\}) \right\} \right) \tag{4.21}$$

Figure 4.9 confirms model predictions for the way survival time depends on length in LD and MD snails, and shows that MD snails can prolong life by a factor of two by not reproducing during starvation. In contrast to the situation concerning embryonic growth, this confirmation gives little support to the theory, because the shape of the survival time-length curve is very flexible for the LD case, although there are only two free parameters. The upper size class of the MD snails has been left out of the model fit, because the shape of their shell suggested a high age, which probably affected energy dynamics.

4.2.4 Dormancy

Some species manage to escape adverse feeding conditions (and/or extreme temperature or drought) by switching to a torpor state where growth and reproduction are ceased and maintenance (and heating) costs strongly reduced. The finding that metabolic rate in homeotherms is proportional to body weight during hibernation [375], suggests that maintenance costs are reduced by a fixed proportion.

As heating is costly, a reduction in the elevation of body temperature saves a lot of energy. Bats and kolibries reduce their body temperature in a daily cycle. This probably relates to the relatively long life span of bats (for their size) [219]. Although most bird embryos have a narrow temperature tolerance range, swifts survive significant cooling. This relates to the food gathering behaviour of the parents. Dutch swifts are known to collect mosquitoes above Paris at a distance of 500 km, if necessary. During hibernation, not only is the body temperature reduced, but other maintenance costs as well.

Hochachka and Guppy [323] found that the African lungfish *Protopterus* and the South American lungfish *Lepidosiren* reduce maintenance costs during torpor in the dry season, by removing ion channels from the membranes. This saves energy expenses for maintaining concentration gradients over membranes, which proves to be a significant part of the routine metabolic costs. This metabolic arrest also halts aging. The life span of lungfish living permanently submerged, so always active, equals the cumulative submerged periods of lungfish that were regularly subjected to desiccation. This is consistent with the DEB interpretation of aging.

If maintenance cannot be reduced completely in a torpor state, it is essential that some reserves are present, {41}. This probably explains how individuals frequently survive adverse conditions as freshly laid eggs because the infinitesimally small embryo requires little maintenance. It only has to delay development. The start of the pupal stage in holometabolic insects is also very suitable for inserting a diapause in order to survive adverse conditions, {151}.

4.2.5 Determination of sex

The determination of sex in some species is coupled to dormancy in a way that can be understood in the context of the DEB model. Daphnids use special winter eggs, packed in an ephippium. The diploid female daphnids usually develop diploid eggs that hatch into new diploid females. If food densities rapidly switch from a high level to a low one and the energy reserves are initially high, the eggs hatch into diploid males, which fertilize females that now produce haploid eggs [665]. The fertilized eggs, the 'winter eggs', develop into new diploid females. The energy reserves of a well-fed starving female are just sufficient to produce males, to wait for their maturity, and to produce winter eggs. The trigger for male/winter egg development is not food density itself, but change of food density. If food density drops gradually, females do not switch to the sexual cycle [405], cf. figure 5.12. Sex determination in species such as daphnids is controlled by environmental factors, so that both sexes are genetically identical [109,299]. Technicians from the TNO laboratories informed me that a randomly assembled cohort of neonates from a batch moved to one room proved to consist almost exclusively of males after some days of growth, while in another cohort from the same batch moved to a different room all individuals developed into females as usual. This implies that sex determination in *Daphnia magna*, and probably in all other daphnids and most rotifers as well, can be affected even after hatching. More observations are needed. Male production does not seem to be a strict prerequisite for winter egg production [391]. Kleiven, Larsson and Hobæk [391] found that crowding and shortening of day length also affect male production in combination with a decrease in food availability at low food densities.

The switch to sexual reproduction as a reaction to adverse feeding conditions frequently occurs in unrelated species, such as slime molds, myxobacteria, oligochaetes (*Nais*) and plants.

4.2.6 Geographical size variations

The energy constraints on distribution, apart from barriers to migration, consist primarily in the availability of food in sufficient quantity and quality. The second determinant is temperature which should be in the tolerance range for the species for a long enough period. If it drops below the lower limit, the species must possess adequate avoidance behaviour (migration, dormancy) to survive.

The minimum food density for survival relates to metabolic costs. If an individual is able to get rid of all other expenses, mean energy intake should not drop below $[\dot{M}]V + \{\dot{H}\}V^{2/3}$ for an individual of volume V, so the minimum ingestion rate, known as the maintenance ration, should be $\frac{\{\dot{I}_m\}}{\{\dot{A}_m\}}([\dot{M}]V + \{\dot{H}\}V^{2/3})$. For a 3 mm daphnid at 20 °C this minimum ingestion rate is about 6 cells of *Chlorella* (diameter 4 μm) per second [404]. The minimum scaled food density X/K is $x_s = \frac{l_h + l}{1/\kappa - l_h - l}$.

This minimum applies to mere survival for an individual. For prolonged existence, reproduction is essential to compensate at least for losses due to aging. The ultimate volumetric length, $fV_m^{1/3} - V_h^{1/3}$, should exceed that at puberty, $V_p^{1/3}$, which leads to the minimum scaled food density $x_R = \frac{l_h + l_p}{1/\kappa - l_h - l_p}$.

Several factors determine food density. It is one of the key issues of population dynamics, which is discussed in the next chapter, {159}. The fact that von Bertalanffy growth curves frequently fit data from animals in the field indicates that they live at relatively constant (mean) food densities. In the tropics, where climatic oscillations are at a minimum, many populations are close to their 'carrying capacity', i.e. the individuals produce a small number of offspring, just enough to compensate for losses. It also means that the amount of food per individual is small, which reduces them in ultimate size. Towards the poles, seasonal oscillations divide the year into good and bad seasons. In bad seasons, populations are thinned, so in the good seasons a lot of food is available per surviving individual. Breeding periods are synchronized with the good seasons, which means that the growth period coincides with food abundance. So food availability in the growth season generally increases with latitude [424]. The effect is stronger towards the poles, which means that body size tends to increase towards the poles for individuals of one species. Figure 4.10 gives two examples. Other examples are known from, for instance, New Zealand including extinct species such as the moa *Dinornis* [110]. Note that size increase towards the poles also comes with a better ability to survive starvation and a higher reproduction rate, traits that will doubtlessly be of help in coping with harsh conditions.

Geographical trends in body sizes can easily be distorted by regional differences in soils, rainfall or other environmental qualities affecting (primary) production. Many species or races differ sufficiently in diets to hamper a geographically based body size comparison. For example, the smallest stoats are found in the north and east of Eurasia, but in the south and west of North America [384]. The closely related weasels are largest in the south, both in Eurasia and in North America. Patterns like these can only be understood after a careful analysis of the food relationships. Simpson and Boutin [659] observed that muskrats *Ondatra zibethicus* of the northern population in Yukon Territory were smaller and have a lower reproduction rate, than the southern population in Ontario. They could relate these differences to feeding conditions, which were better for the southern population, this time.

Bergmann [57] observed the increase in body size towards the poles in 1847, but he explained it as an effect of temperature. Large body size goes with small surface area-volume ratios, which makes endotherms more efficient per unit body volume. This explanation has been criticized [471,641,646]. It is indeed hard to see how this argument applies in detail. Animals do not live on a unit of body volume basis, but as a whole individual [471]. It is also hard to see why the argument applies within a species only and why mice, foxes and bears can live together in the Arctic. The tendency to increase body size towards the poles also seems to occur in ectotherms, which needs another explanation. The DEB theory offers an alternative explanation for the phenomenon because of the relationship between food availability and ultimate body volume. Temperature alone works in the opposite direction within this context. If body temperature has to be maintained at some fixed level, individuals in the Arctic are expected to be smaller while living at the same food density, because they have to spend more energy on heating, which reduces their growth potential. The effect will, however, be small since insulation tends to be better towards the poles.

It is interesting to note that species with distribution areas large enough to cover climatic gradients generally tend to split up in isolated races or even subspecies. The dif-

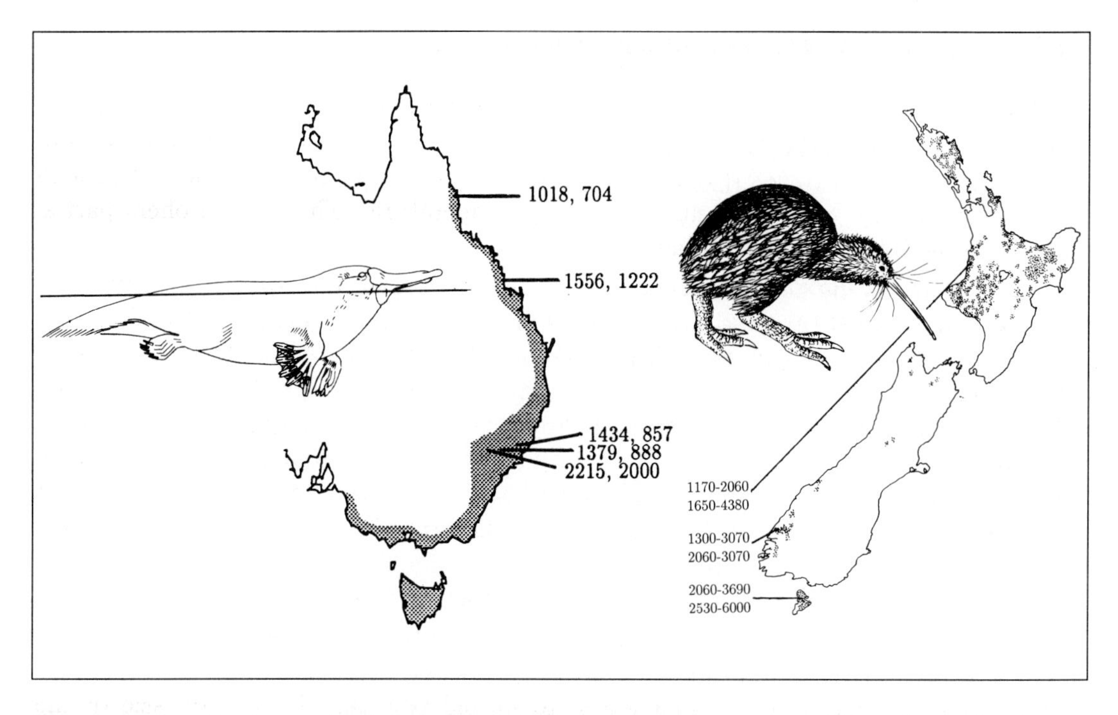

Figure 4.10: The brown kiwi *Apteryx australis* in subtropical north of New Zealand is heavier than in the temperate south. The numbers give ranges of weights of male and female in grams, calculated from the length of the tarsus using a shape coefficient of d_m= 1.817 $g^{1/3}\,cm^{-1}$. Data from Fuller [239]. A similar gradient applies to the platypus *Ornithorhynchus anatinus* in Australia. The numbers give the mean weights of male and female in grams as given by Strahan [693]. The DEB theory relates adult weights to food availability and so to the effect of seasons. This interpretation is supported by the observation that platypus weights increase with seasonal differences at the same latitude in New South Wales. The seasons at the three indicated sites are affected by the Great Dividing Range in combination with the easterly winds.

ferences in ultimate size have usually become genetically fixed. This is typical for 'demand' systems where regulation mechanisms set fluxes at predefined values which are obtained through adaptation. Within the DEB theory this means that the parameter values are under genetic control and that the minimum food level at which survival is possible is well above the level required for maintenance. The matter will be taken up again on {238}.

4.3 Reconstruction problems

The DEB model is simple enough to allow reconstructions of body temperature and/or food intake from growth observations and knowledge of some energy parameters. This subsection and the following one illustrate how this can be done. The spin off is a warning against jumping to conclusions in cases where essential information for the interpretation of data is lacking.

4.3.1 Temperature reconstruction

The body temperature of endotherms can be well above the environment temperature. Cooling can complicate model testing and/or parameter estimation. Altricial birds provide an excellent case to illustrate the problem of the energy interpretation of growth measurements in the case of an unknown body temperature. This section offers partial solutions to the interpretation problem.

Birds usually become endothermic around hatching; precocial species usually make the transition just before hatching, and altricial ones some days after. The ability to keep the body temperature at some fixed level is far from perfect at the start, so the body temperature depends on that of the environment and the behaviour of the parent(s) during that period. Unless insulation is perfect, the parents cannot heat the egg to their own body temperature. There will be a few degrees difference, but this is still a high temperature, which means that the metabolic rate of the embryo is high. So it produces an increasing amount of heat as a byproduct of its general metabolism before the start of endothermic heating.

This process of pre-endothermic heating can be described by: $\frac{d}{dt}T_b = d_T\dot{C} - \dot{k}_{be}(T_b - T_e)$, where T_b is the body temperature of the embryo, T_e the temperature of the environment, d_T the heat generated per unit of utilized energy and $\dot{k}_{be}$ the heat flux from the egg to the environment. The latter is here taken to be independent of the body size of the embryo, because the contents of the egg are assumed to be homogeneous with respect to the temperature. (The Brunnich's guillemot seems to need a 40 °C temperature difference between one side of the egg and the other to develop [591].) Energy conductance and the maintenance rate coefficient depend on temperature according to the Arrhenius relationship with an Arrhenius temperature of 10000 K. Figure 4.11 illustrates the development of the lovebird *Agapornis*, with changing body temperature. The curves are hardly different compared with a constant temperature, but the parameter values differ substantially. The magnitude of the predicted temperature rise depends strongly on the parameter values chosen. The information contained in the data of figure 4.11 did not allow a reliable estimation of all parameters; the temperature difference of 4 °C is arbitrary, but not unrealistic.

It is interesting that the red-headed lovebird, *A. pullaria* from Africa, and at least 11 other parrot species in South America, Australia and New Guinea breed in termite nests, where they profit from the heat generated by the termites. Breeding Golden-shouldered parrots, *Psephotus chrysopterygius*, in captivity failed frequently, until it became known that one has to heat the nest to 33 °C for some days before hatching and for two weeks after.

The significance of this exercise is the following: the least squares fitted curves remain almost exactly the same for a constant temperature and a changing one, but the parameter estimates for e.g. energy conductance differ considerably. It follows that these data are not suitable for estimating energy parameters unless the temperature is known as a function of time. This holds specially for altricial birds because they hatch too early to show the reduction in respiration rate that gives valuable information about parameter values. The few studies on bird development that include temperature measurements indicate that

Figure 4.11: Embryo weight and respiration ontogeny in the parrot *Agapornis personata*. Data from Bucher [107]. The curves are DEB model predictions accounting for a temperature increase of 4 °C during development; see text. The temporary respiration increase at day 23 relates to hatching. It is not part of the model.

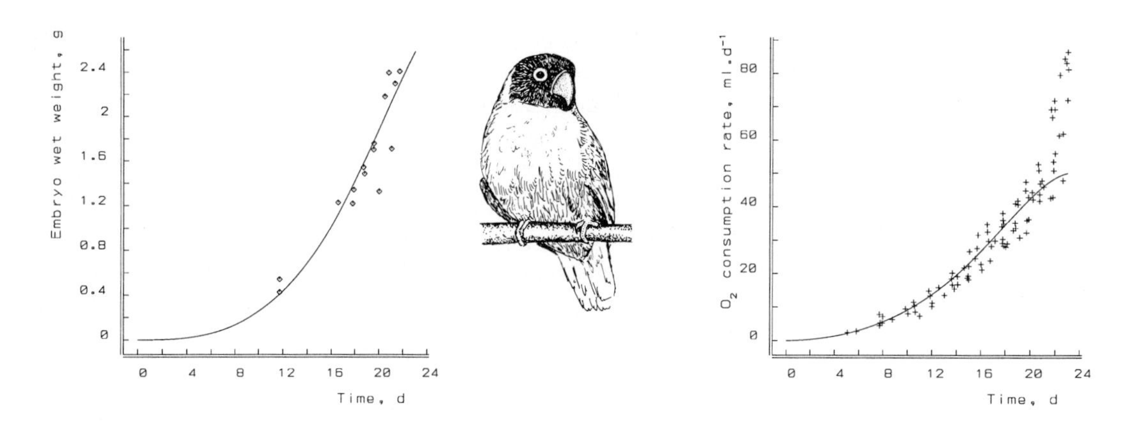

the temperature change during incubation is not negligibly small. Drent [182] found an increase from 37.6 to 39 °C in the precocial herring gull *Larus argentatus*.

The change in body temperature also causes deviations from the simplest formulation of the DEB model after hatching in some species. My conjecture is that they are the main cause of the (empirical) logistic growth curve fitting better than the von Bertalanffy curve for birds living at food abundance; as body temperature is measured in only a few exceptional studies, it makes sense to study the inverse argument. Given the observed growth pattern and the DEB model, can the body temperature ontogeny be recovered at abundant food?

At abundant food, (3.13) reduces to $\frac{d}{dt}l = \dot{\gamma}(1-l)$, where the von Bertalanffy growth rate $\dot{\gamma} = \frac{\dot{m}}{3}\frac{g}{1+g}$ is now considered not as a constant but as a function of time, since the temperature and thus the maintenance rate coefficient $\dot{m}$ change. Integration gives

$$l(t) = 1 - (1 - l(0)) \exp\left\{-\int_0^t \dot{\gamma}(t_1)\, dt_1\right\} \text{ with} \tag{4.22}$$

$$\dot{\gamma}(t) = \dot{\gamma}_\infty \exp\{T_A(T_\infty^{-1} - T_b(t)^{-1})\} \tag{4.23}$$

where $\dot{\gamma}_\infty$ is the ultimate growth rate when the body temperature is kept constant at some target temperature in the range 39 – 41 °C, or $T_\infty = 312$ (non-passerines) or 314 K (passerine birds). Body temperature is thus given by

$$T_b(t) = \left(\frac{1}{T_\infty} - \frac{1}{T_A}\ln\frac{\frac{d}{dt}l}{\dot{\gamma}_\infty(1-l)}\right)^{-1} \tag{4.24}$$

Given an observed growth and size pattern, this equation tells us how to reconstruct the temperature. The reconstruction of body temperature, therefore, rests on the assumption of (time inhomogeneous) von Bertalanffy growth (4.22) and an empirical description of the observed growth pattern. It is a problem, however, that both the growth rate and the length difference with its asymptote 1 vanish, which means that their ratio becomes undetermined if inevitable scatter is present. General purpose functions such as polynomials or splines to describe size-at-age are not suitable in this case.

A useful choice for an empirical description of growth is

$$\frac{d}{dt}l = \frac{\dot{\gamma}_\infty}{k}(l^{-k} - 1)l \quad \text{or} \tag{4.25}$$

$$l(t) = (1 - (1 - l(0)^k)\exp\{-\dot{\gamma}_\infty t\})^{1/k} \tag{4.26}$$

because it covers both von Bertalanffy growth (shape parameter $k = 1$), and the frequently applied logistic growth ($k = -3$) and all shapes in between. For the shape parameter $k = 0$, the well known Gompertz curve arises: $l(t) = l(0)^{\exp\{-\dot{\gamma}t\}}$. Nelder [504] called this model the generalized logistic equation. It was originally proposed by Richards [592] to describe plant growth. The graph of volume as a function of age is skewly sigmoid with an inflection point at $V/V_\infty = (1 - k/3)^{3/k}$ for $k \leq 3$. Substitution of (4.25) into (4.24) gives

$$T_b(t) = \left(\frac{1}{T_\infty} - \frac{1}{T_A}\ln\frac{1}{k}\frac{1 - l^{-k}}{1 - l^{-1}}\right)^{-1} \tag{4.27}$$

Note that if growth is of the von Bertalanffy type, so $k = 1$, this reconstruction amounts to $T_b(t) = T_\infty$, which does not come as a surprise. This interpretation of growth data implies that the growth parameters of the logistic, Gompertz and von Bertalanffy growth curves are comparable in their interpretation and refer to the target body temperature. The DEB theory gives the physiological backgrounds. Figure 4.12 gives examples of reconstructions, which indicate that the body temperature at hatching can be some 10 °C below the target and it increases almost as long as growth lasts. The reconstruction method has been tested on several data sets where the body temperature has been measured during growth [789]. It has been found to be quite accurate given the scatter in the temperature data. Figure 4.12 gives one example. Although the Arrhenius temperature can be estimated from combined weight/temperature data, its value proved to be poorly defined.

An important conclusion from this exercise is that deviations from von Bertalanffy growth at food abundance in birds can be explained by changes in body temperature.

4.3.2 Food intake reconstruction

Many data sets on growth in the literature do not provide adequate information about food intake. Sometimes it is really difficult to gain access to this type of information experimentally. The blue mussel *Mytilus edulis* filters what is called 'particulate organic matter' (POM). Apart from the problem of monitoring the POM concentration relevant to a particular individual, its characterization in terms of nutritional value is problematic. The relative abundances of inert matter, bacteria and algae change continuously. In the search for useful characterizations, it can be helpful to invert the argument: given an observed size and temperature pattern can the assimilation energy be reconstructed in order to relate it to measurements of POM? The practical gain of such a reconstruction is

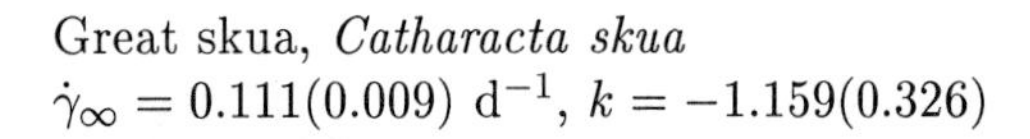

Great skua, *Catharacta skua*
$\dot{\gamma}_\infty = 0.111(0.009)$ d^{-1}, $k = -1.159(0.326)$

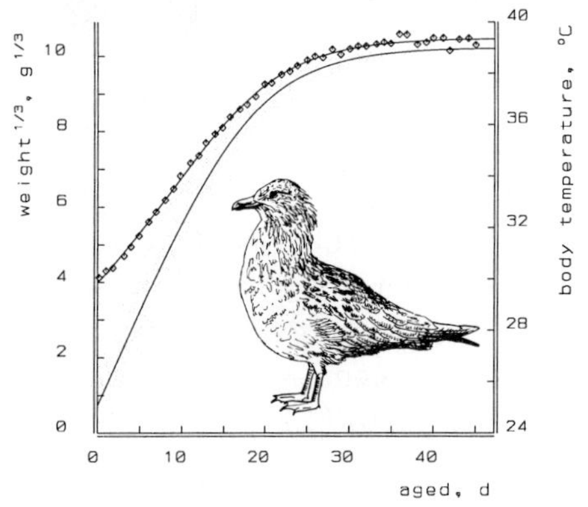

Long-tailed skua, *Stercorarius longicaudus*
$\dot{\gamma}_\infty = 0.267(0.035)$ d^{-1}, $k = -2.538(0.804)$

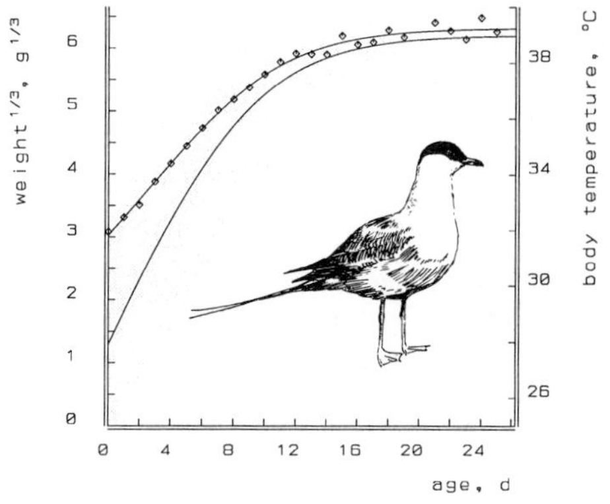

Manx shearwater, *Puffinus puffinus*
$\dot{\gamma}_\infty = 0.114(0.008)$ d^{-1}, $k = -2.483(0.467)$

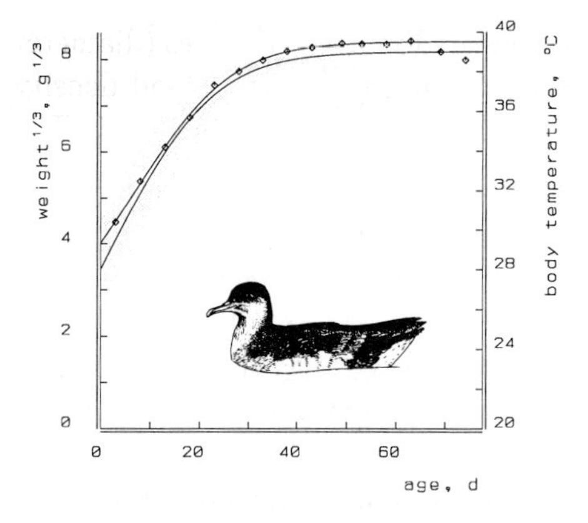

Guillemot, *Uria aalge*
$\dot{\gamma}_\infty = 0.125(0.037)$ d^{-1}, $k = -0.883(1.707)$

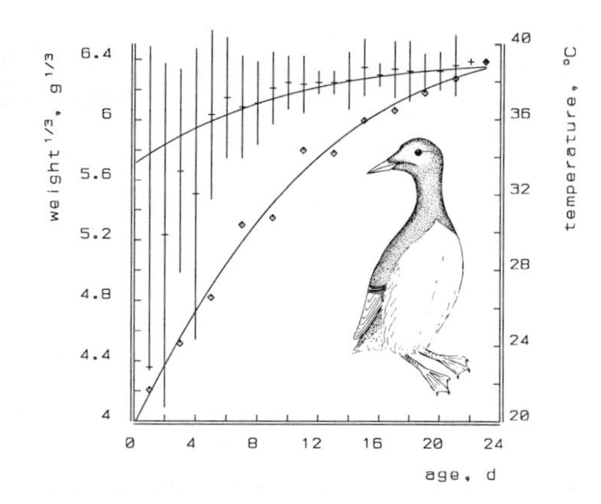

Figure 4.12: The empirical, generalized, logistic growth curves have been fitted to measured data for some birds. The von Bertalanffy growth rate $\dot{\gamma}_\infty$ at the ultimate body temperature and shape parameter k are given. On the basis of these fits the body temperature was reconstructed, on the assumption that $T_\infty = 312$ K and $T_A = 10000$ K. The shaded areas around the body temperature curves indicate the 95% confidence interval based on the marginal distribution for k. The reconstruction method is tested on the guillemot data (lower right figure) where measured body temperatures were available. The bars indicate the standard deviation. Both temperature parameters, $T_\infty = 312.3(2.32)$ K and $T_A = 8225(16300)$ K, have been estimated from the combined weight/temperature data. Data from Furness, de Korte in [241], Thompson in [99] and [449] respectively.

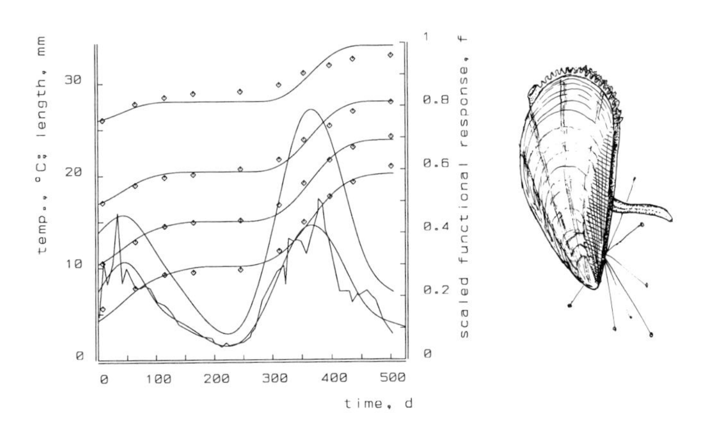

Figure 4.13: The reconstruction of the scaled functional response since the first of August from mean length-time data for four length classes of the mussel *Mytilus edulis* as reported by Kautsky [374] (upper four curves). The reconstruction (the curve in the middle with two peaks) is based on a cubic spline description of the measured temperature (lower curve and capricious line) and the parameter values $L_m = 100$ mm, $g = 0.13$, $\dot{m}_{15} = 0.03$ d^{-1} and $T_A = 7600$ K.

in the use of correlation measures to determine the nutrition value of bacteria, alga, etc. Since the correlation coefficient is a linear measure, a direct correlation between bacteria numbers and mussel growth, for instance, only has limited value because assimilation and growth are related in a complex way, whereas bacteria numbers and assimilation are related linearly.

Kautsky [374] measured mussels from 4 size classes kept individually in cages (diameter 10 cm) at a depth of 15 m in the Baltic at 7 °/$_{oo}$. Suppose that (the mean) food density changes slowly enough to allow an approximation of the energy reserves with $e = f$. The growth equation (3.13) then reduces to

$$\frac{d}{dt}l = \frac{(f(t) - l)_+}{3(f(t) + g)} g\dot{m}_{15}(T(t) > T_0) \exp\left\{T_A\left(\frac{1}{288} - \frac{1}{T(t)}\right)\right\} \tag{4.28}$$

where $\dot{m}_{15}$ is the maintenance rate coefficient at the chosen reference temperature of 15 °C = 288 K and T_0 is at the lower end of the tolerance range. The next step is to choose cubic spline functions to describe the observed temperature pattern $T(t)$ and the unobserved scaled functional response $f(t)$. The reconstruction of $f(t)$ from length-time data then amounts to the estimation of the knot values of the spline at chosen time points, given realistic choices for the growth parameters. Figure 4.13 shows that the simultaneous least squares fit of the numerically integrated growth description (4.28) is acceptable in view of the scatter in the length data (not shown), which increases in time in the upper size class in the original data. The scaled functional response (i.e. the hyperbolically transformed food abundance in terms of its nutritional value) appears to follow the temperature cycle during the year. Such a reconstructed food abundance can be correlated with POM and chlorophyll measurements to evaluate their significance for the mussel.

If food intake changes too fast to approximate the reserve density with its equilibrium value, the reserve density should be reconstructed as well. Such a reconstruction will be illustrated with the penguin as an example. Figure 4.14 shows that von Bertalanffy growth makes sense for penguins, which indicates that body temperature is constant and food is abundant. The deviation at the end of the growth period probably relates to the refusal

Figure 4.14: Weight ontogeny of the small adelie penguin *Pygoscelis adeliae* (left) and the large emperor penguin *Aptenodytes forsteri* (right). Data from Taylor [705] and Stonehouse [685]. The adelie data follow the fitted von Bertalanffy growth curve, which suggests food abundance during the nursery period. The cubic spline through the emperor data is used to reconstruct food intake $fV^{2/3} = \dot{I}/\{\dot{I}_m\}$. $[d_{wv}] = 0.3 \text{ g cm}^{-3}$, $[d_{we}] = 0.7 \text{ g cm}^{-3}$, $g = 0.1$, $\dot{v} = 0.6 \text{ cm d}^{-1}$, $l_h = 0.01$, $V_m = 6000 \text{ cm}^3$, $e_0 = 0.6$.

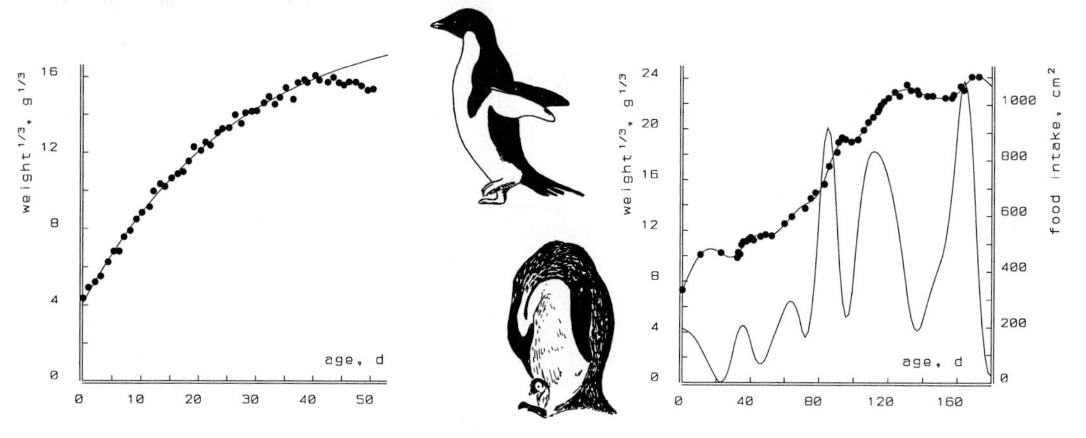

of the parents to feed the chicks in order to motivate them to enter the sea. The small bodied adelie penguin manages to synchronize its breeding cycle with the local peaks in plankton density in such a way that it is able to offer the chicks abundant food. Typically there are two such peaks a year in northern and southern cold and temperate seas. The plankton density drops sharply when the chicks are just ready to migrate to better places. This means that a larger species, such as the king penguin, is not able to offer its chicks this continuous wealth of food, because its chicks require a longer growth period (see the chapter on comparison of species for an explanation, {217}). So they have to face the bleak period between plankton peaks. (Food for king penguins, squid and fish, follows plankton in abundance.) The parents do not synchronize their breeding season with the calendar; they follow a 14 to 17 month breeding cycle [658]. The largest living penguin, the emperor penguin, has also to use both plankton peaks for one brood, which implies a structural deviation from a simple von Bertalanffy growth curve.

Given weight-time data, food intake can be reconstructed on the basis of the DEB theory. The relationship between (wet) weights, volumes and energy reserves according to (2.9) for juveniles (where $E_{\dot{R}} = 0$) is $[W_w] = [d_{wv}] + [d_{we}]e$ and is thus not considered to be a constant. Growth according to (3.12) and (3.7) is given by

$$\frac{d}{dt}W_w = [W_w]\frac{d}{dt}V + [d_{we}]V\frac{d}{dt}e \quad (4.29)$$

$$= \dot{v}V^{2/3}\left(\frac{[W_w]}{e+g}\left(e - l_h - (V/V_m)^{1/3}\right) + [d_{we}](f-e)\right) \quad (4.30)$$

Solution of f and substitution of (3.7) gives

$$f = e + \frac{[W_w]^{2/3}}{\dot{v}[d_{we}]W_w^{2/3}} \frac{d}{dt} W_w - \frac{[W_w]/[d_{we}]}{g+e} \left(e - l_h - \left(\frac{W_w}{V_m[W_w]} \right)^{1/3} \right) \quad (4.31)$$

$$\frac{d}{dt} e = \frac{[W_w]/[d_{we}]}{W_w} \frac{d}{dt} W_w - \dot{v} \frac{[W_w]/[d_{we}]}{g+e} \left((e - l_h) \left(\frac{[W_w]}{W_w} \right)^{1/3} - V_m^{-1/3} \right) \quad (4.32)$$

The steps to reconstruct feeding are as follows: first fit a cubic spline through the weight data, which gives $W_w(t)$ and so $\frac{d}{dt}W_w(t)$. Use realistic values for $e(0)$, $[d_{wv}]$, $[d_{we}]$, g, V_m, l_h and $\dot{v}$ and recover $e(t)$ through numerical integration of (4.32) and then $f(t)$ by substitution. Figure 4.14 gives an example. The peaks in the reconstruction will probably be much sharper if stomach contents of the chick are taken into account. This reconstruction can be useful in cases where feeding behaviour that is hard to observe directly is studied and knowledge concerning energetics from captive specimens is available. The significance of this example is to show that the DEB theory hardly poses constraints for growth curves in general. The simple von Bertalanffy growth curve only emerges under the conditions of constant food density and temperature.

4.4 Special case studies

The purpose of this section is to show that some biological details, that seem to falsify the DEB theory at first sight, can still be understood within the context of the theory; However, they require careful analysis of data. Blind application will soon lead to inconsistencies between theory and data. This section focuses on the interpretation of data.

4.4.1 Diffusion limitation

The purpose of this subsection is to show why small deviations from the hyperbolic functional response can be expected under certain circumstances, and how the functional response should be corrected.

Any submerged body in free suspension has a stagnant water mantle of a thickness that depends on the roughness of its surface, its electrical properties and on the turbulence in the water. The uptake of nutrients by cells that are as small as that of a bacterium can be limited by the diffusion process through this mantle [396]. Logan [436,437] related this limitation to the flocculation behaviour of bacteria at low food densities. The existence of a diffusion limited boundary layer is structural in Gram-negative bacteria such as *Escherichia* [394], which have a periplasmic space between an inner and outer membrane. The rate of photosynthesis of aquatic plants [668] and algae [600] can also be limited by diffusion of CO_2 and HCO_3^- through the stagnant water mantle that surrounds them. Since diffusion limitation affects the functional response, it is illustrative to analyze the deviations in a bit more detail. For this purpose I will reformulate some results that originate from Best [68] and Hill and Whittingham [319] in 1955.

Suppose that the substrate density in the environment is constant and that it can be considered as well mixed beyond a distance r_1 from the centre of gravity of a spherical cell of radius r_0. Let X_1 denote the substrate density in the well mixed environment and X_0 that at the cell surface. The aim is now to evaluate uptake in terms of substrate density in the environment, given a model for substrate uptake at the cell surface.

The build-up of the concentration gradient from the cell surface is fast compared with other processes, such as growth; the gradient is here assumed to be stationary. The mass flux over a sphere with radius r according to Fick's diffusion law is proportional to the substrate density difference in the adjacent inner and outer imaginary tunics (i.e. 3D-annulus). Together with the conservation law for mass this leads to what is known as the Laplace's equation $\frac{d}{dr}\left(r^2\frac{d}{dr}X\right) = 0$. The boundary condition $X(r_1) = X_1$, determines the solution $X(r) = X_1 - (X_1 - X_0)\frac{1-r_1/r}{1-r_1/r_0}$. The mass flux at r_0 is according to Fick's law $4\pi r_0^2 \dot{D}\frac{d}{dr}X(r_0)$, where $\dot{D}$ is the diffusivity. It must be equal to the uptake rate $\dot{I} = \dot{I}_m X_0/(K + X_0)$. This gives the relationship between the density at the cell surface and the density in the environment as a function of the thickness of the mantle

$$X_0 = \frac{1}{2}X_c + \frac{1}{2}\sqrt{X_c^2 + 4X_1K} \tag{4.33}$$

with $X_c \equiv X_1 - K - \frac{\dot{I}_m}{4\pi\dot{D}r_0}\left(1 - \frac{r_0}{r_1}\right)$. Since the cell can only 'observe' the substrate density in its immediate surroundings, X_0 must be taken as the argument for the hyperbolic functional response and not X_1. If $X_1 \ll K$, X_0 is about proportional to X_1, but for large X_1, (small) deviations from hyperbolic responses are to be expected. Note that for large diffusivities $X_0 \to X_1$, as might be expected. For relatively thick water mantles, especially, it is not important that the cell is spherical. The approximate relationship $V \simeq r_0^3\pi 4/3$ will be appropriate for most rods.

For Gram-negative bacteria, which have an inactive outer membrane with a limited permeability for substrate transport, the relationship between the substrate density at the active inner membrane and that in the well-mixed environment is a bit more complicated. On the assumption that the substrate flux through the outer membrane is proportional to the difference of substrate densities on either side of the outer membrane, the permeability affects the last factor in the expression for X_c. The substrate density X_c in (4.33) is given by $X_c \equiv X_1 - K - \frac{\dot{I}_m}{4\pi\dot{D}r_0}\left(1 - \frac{r_0}{r_1} + \frac{r_0\dot{D}}{r_2^2\dot{P}}\right)$, where r_2 is the radius at which the outer membrane occurs and $\dot{P}$ is the permeability of that membrane (dimension length.time^{-1}). The periplasmic space is typically some 20–40% of the cell volume [503], so that $r_0/r_2 \simeq 0.89$. If $r_2\dot{P} \gg \dot{D}$, the resistance of the outer membrane for substrate transport is negligible. Figure 4.15 illustrates how substrate density decreases towards the inner membrane.

Increasing water turbulence and active motion by flagellas will reduce the thickness of the water mantle. Its effect on mass transfer is usually expressed by the Sherwood number, which is defined as the ratio of mass fluxes with and without turbulence. If $X_1 \ll K$, the Sherwood number is independent of substrate density, and amounts to $\left(1 + \frac{\dot{I}_m}{4\pi\dot{D}Kr_0}\right)\left(1 + \frac{\dot{I}_m(1-r_0/r_1)}{4\pi\dot{D}Kr_0}\right)^{-1}$. For larger values of X_1, the Sherwood number becomes dependent on substrate density and increasing turbulence will less easily increase mass

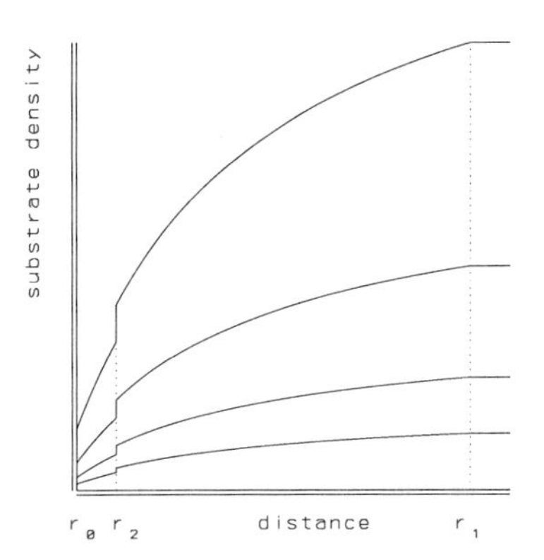

Figure 4.15: Substrate density as a function of the distance from the cell centre for a Gram-negative bacterium. The inner membrane is at distance r_0, the outer membrane at distance r_2 and beyond distance r_1, the medium is completely mixed. Four different choices for substrate densities X_1 in the medium have been made, to illustrate that the higher X_1, the more the substrate density at the inner membrane X_0 is reduced.

Figure 4.16: Stereo view of the substrate uptake rate of a cell in suspension relative to that in completely stagnant water, as a function of the substrate density in the medium (x-axis) and the thickness of the water mantle (y-axis). Parameter choice: $\dot{I}_m = 4\pi \dot{D} K r_0$

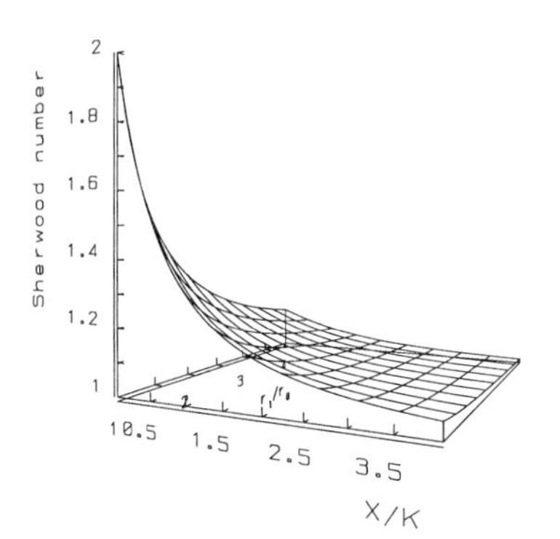

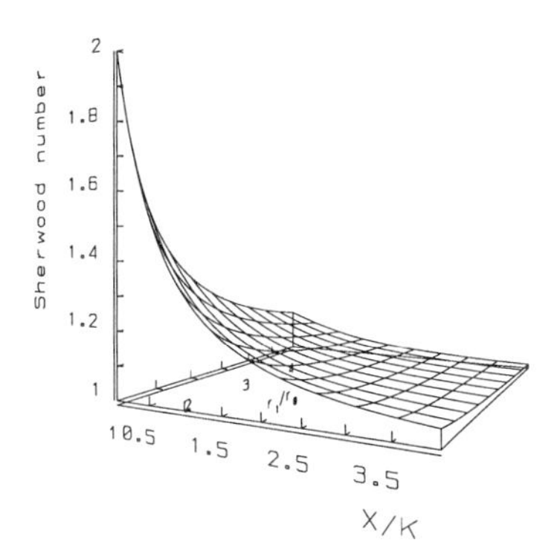

transfer, because uptake will be rate limiting; see figure 4.16. This probably defines the conditions for producing sticky polysaccharides which result in the development of films of bacteria on hard substrates or of flocs. If a cell attaches itself, it looses potentially useful surface area for uptake, but increases mass transfer via convection. Although the quantitative details for the optimization of uptake can be rather complex, the qualitative implication that cells usually occur in free suspension when substrate densities are high, and in flocs when they are low can be understood from Sherwood numbers.

Since diffusivity is proportional to (absolute) temperature, see e.g. [24], and uptake rates tend to follow the Arrhenius relationship, {44}, the temperature dependence of diffusion limited uptake is likely to depend on temperature in a more complex way.

It is conceivable that slowly moving or sessile animals exhaust their immediate surroundings in a similar way to that described here for bacteria in suspension, if the transport of food in the environment is sufficiently slow. Trapping devices suffer from this problem too [355]. Patterson [534] showed by changing the flow rate that the physical state of the boundary layer surrounding the symbiosis of coral and algae directly affects nutrient transfer. The shape, size and polyp-wall thickness of scleractinian corals could be related to diffusion limitation of nutrients. Some processes of transport can be described accurately by diffusion equations, although the physical mechanism may be different [517,663]. For $K \ll \dot{I}_m(4\pi \dot{D} r_0)^{-1}$, the functional response approaches Holling's type I [332], also known as Blackman's response [72], where ingestion rate is just proportional to food density up to some maximum. This response is at the root of the concept of limiting factors, which still plays an important role in physiology. This exercise thus shows that the two types of Holling's functional response are related and mixtures are likely to be encountered.

4.4.2 Growth of 0D- and 2D-isomorphs

As mentioned in the subsection on changing shapes on {31}, 2D-isomorphs grow in diameter, not in length, which leads to the shape correction function $\mathcal{M}(V) = (V_d/V)^{1/6}$. The growth equation for 2D-isomorphs can be obtained from (3.12) by multiplying $\dot{v}$ and V_m by $\mathcal{M}(V)$, because they contain the surface area dependent parameter $\{\dot{A}_m\}$, and putting $V_h = 0$, assuming that they are ectothermic. This leads to

$$\frac{d}{dt}V = \frac{\dot{v}}{e+g}\left((V_d/V)^{1/6} e V^{2/3} - V_m^{-1/3} V\right) \tag{4.34}$$

$$\frac{d}{dt}l = \dot{\gamma}(e\sqrt{l_d/l} - l) \tag{4.35}$$

where $\dot{\gamma} = \frac{\dot{m}g}{3(e+g)}$ represents the von Bertalanffy growth rate and $l = (V/V_m)^{1/3}$ the scaled (volumetric) length, as before. If substrate density is constant for a long enough time, so $e = f$, scaled length is given by

$$l(t) = \left(f\sqrt{l_d} - \left(f\sqrt{l_d} - l_b^{3/2}\right)\exp\{-t\dot{\gamma}3/2\}\right)^{2/3} \tag{4.36}$$

$$t(l) = \frac{2}{3\dot{\gamma}} \ln \frac{f\sqrt{l_d} - l_b^{3/2}}{f\sqrt{l_d} - l^{3/2}} \tag{4.37}$$

where l_b is the scaled length at birth, i.e. just after division at $t = 0$. Division into two equal daughters implies that $l_d = 2^{1/3} l_b$. Figure 4.17 shows that the growth curves are more convex than the von Bertalanffy one, which makes it easier to test whether uptake is coupled to surface area in unicellulars as well. The expected growth curves for rods are so close to exponential that this test is hardly feasible for rods. Experimental data to test this idea are not available, but I hope that this account inspires someone to have a look.

The shape correction function for 0D-isomorphs, such as biofilms, is $\mathcal{M}(V) = (V_d/V)^{2/3}$, so that the growth equation becomes

$$\frac{d}{dt}V = \frac{\dot{v}}{e+g}(eV_d^{2/3} - V_m^{-1/3} V) \tag{4.38}$$

$$\frac{d}{dt}l^3 = \dot{\gamma}(el_d^2 - l^3) \tag{4.39}$$

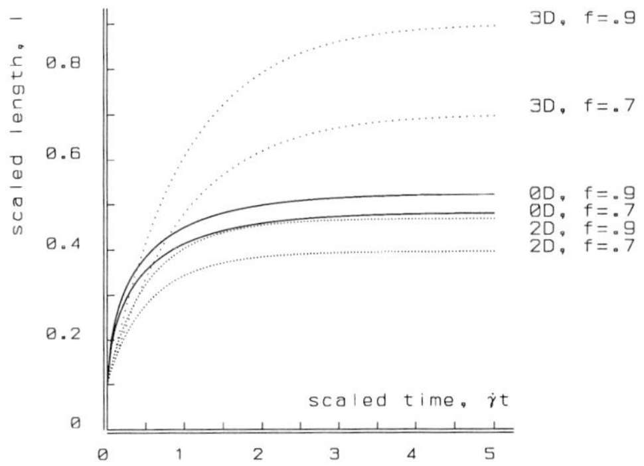

Figure 4.17: Expected growth curves for 0D- and 2D-isomorphs compared with those for 3D-isomorphs at constant substrate densities. Parameters: $l_b = 0.1$, $f = 0.7$ and 0.9 and $l_d = l_b 2^{1/3}$

At constant substrate or food density, where $e = f$, the growth curve is

$$l(t) = \left(f l_d^2 - (f l_d^2 - l_b^3) \exp\{-t\dot{\gamma}\}\right)^{1/3} \quad (4.40)$$

$$t(l) = \dot{\gamma}^{-1} \ln \frac{f l_d^2 - l_b^3}{f l_d^2 - l^3} \quad (4.41)$$

Crusts

Crusts, i.e. biofilms of limited extent that grow on hard surfaces, are mixtures of 0D-isomorphs in the center and 1D-isomorphs in the periphery where the new surface is covered. Bacterial colonies on an agar plate, conceived as super-organisms, are among crusts. When crusts grow, an increasing proportion of the biomass behaves as a 0D-isomorph. With an extra assumption about the transfer of biomass from one mode of growth to the other, the growth of the crust on a plate is determined and can be worked out as follows for constant substrate density.

Suppose that biomass in the outer annulus of diameter L_ϵ of the circular crust is growing exponentially in an outward direction, while it is building up a layer of thickness L_ϵ. This biomass thus behaves as a 1D-isomorph. All other biomass is growing as a 0D-isomorph. When the crust has diameter $L_r(t)$, the volume of the 1D-isomorph is

$$\frac{\pi}{4} L_\epsilon (L_r^2(t) - (L_r(t) - L_\epsilon)^2) = \frac{\pi}{4} L_\epsilon^2 (2L_r(t) - L_\epsilon)$$

In a period dt, the 1D-isomorph increases by $\dot{\gamma}_f \, dt$ times this volume, see (refeqn:Vf), which must be equal to the volume $\frac{\pi}{4} L_\epsilon (L_r^2(t + dt) - L_r^2(t))$, if the layer has thickness L_ϵ. Since $\dot{\nu} = \dot{m}g/l_d$, the volume-specific growth rate of a 1D-isomorph relates to the von Bertalanffy growth rate as $\dot{\gamma}_f = 3\dot{\gamma}(f/l_d - 1)$. This leads to $\frac{d}{dt} L_r = \dot{\mu} L_\epsilon (1 - L_\epsilon / 2L_r)$, from which it follows that the diameter of the crust is growing linearly in time for $L_\epsilon \ll L_r$. This linear growth in diameter has been observed experimentally by Fawcett [212], and the linear growth model originates from Emerson [201] in 1950 according to Fredricson *et al.* [232]. Figure 4.18 shows that this linear growth applies to lichen growth on moraines. Richardson [594] discusses the value of gravestones for the study of lichen growth, because of the reliable dates. Lichen growth rates are characteristic for the species, so that the

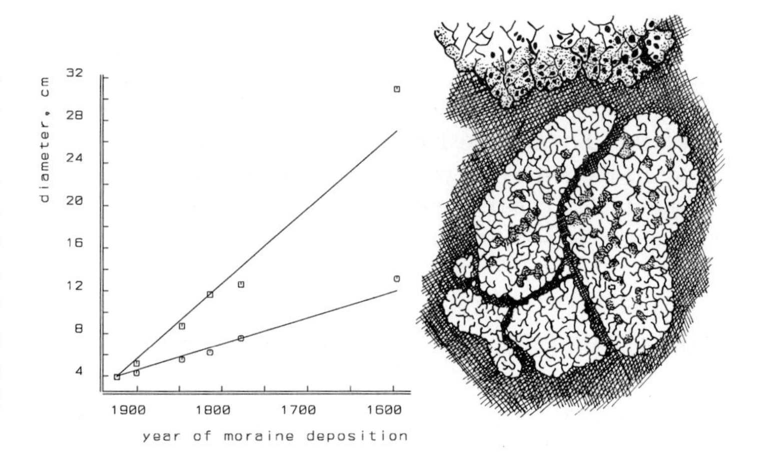

Figure 4.18: The lichens *Aspicilia cinerea* (above) and *Rhizocarpon geographicum* (below) grow almost linearly in a period of more than three centuries on moraine detritus of known age in the European Alps. Data from Richarson [594]. Linear growth is to be expected from the DEB model, when such lichens are conceived as dynamic mixtures of 0D- and 1D-isomorphs.

diameter distribution of the circular patches can be translated into arrival times, which can then be linked to environmental factors, for instance.

If substrate transport in the vertical direction on the plate is sufficient to cover all maintenance costs, and transport in the horizontal direction is small, the growth rate of the 0D-isomorph on top of an annulus of surface area dA is

$$\frac{d}{dt}V = \frac{f\{\dot{A}_m\}\,dA - [\dot{M}]V}{[G] + [E_m]f}$$

The denominator stands for the volume-specific costs for structural biomass and reserves. Division by the surface area of the annulus gives the change in height L_h

$$\frac{d}{dt}L_h = \frac{f\{\dot{A}_m\} - [\dot{M}]L_h}{[G] + [E_m]f} = 3\dot{\gamma}(fV_m^{1/3} - L_h) \text{ or } \frac{d}{dt}l_h = 3\dot{\gamma}(f - l_h)$$

with the scaled height $l_h \equiv L_h V_m^{-1/3}$. The initial growth rate in scaled height is $3\dot{\gamma}(f - l_\epsilon)$. The parameter $l_\epsilon \equiv L_\epsilon V_m^{-1/3}$ can be eliminated, on the assumption that the growth rate in the outward direction equals the initial growth rate in the vertical direction, which gives $l_\epsilon = 2l_d$ for $l_d \ll f$. For $l_d \ll l_r$ with $l_r \equiv L_r V_m^{-1/3}$, the end result amounts to

$$l_h(t, l_r) = f - (f - 2l_d)\exp\left\{\frac{l_r/2}{f - l_d} - 3\dot{\gamma}t\right\} \tag{4.42}$$

The scaled height of the crust is thus growing asymptotically to f, so that the total volume increases as $\pi f(l_d\dot{\mu}t)^2$. Different crust shapes can be obtained by accounting for horizontal transport of biomass and diffusion limitation of food transport to the crust.

Growth in thickness of a biofilm on a plane, which behaves as a 0D-isomorph, is thus similar to that of a spherical biofilm on a small core in suspension, which behaves as a 3D-isomorph. Films are growing in a von Bertalanffy way in both situations, if growth

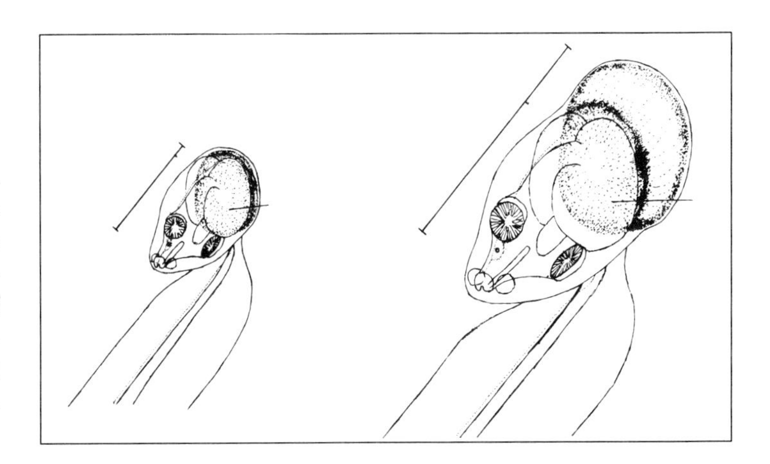

Figure 4.19: The larvacean *Oikopleura* grows isomorphically; during its short life it accumulates reproductive material at the posterior end of the trunk. The energy interpretation of data on total trunk lengths should take account of this.

via settling of suspended cells on the film is not important. Note that if maintenance is small, so that asymptotic depth of the film is large, increase in diameter is linear with time, so that volume increases as time^3, as has been found for foetuses in (3.21) by a different reason. This mode of growth is called the 'cube root' phase by Emerson [201], who found it applicable to submerged mycelia of the fungus *Neurospora*.

The spatial expansion of geographical distribution areas of species, such as the musk rat in Europe, and of infectious diseases, cf. [82,83,300], closely resembles that of crusts. Although these population phenomena differ in many respects from the growth of crusts as (super) individuals, the reason why the expansion proceeds at a constant rate is basically the same from an abstract point of view: material in the border area grows exponentially and the inner area hardly contributes to the expansion.

4.4.3 Reproduction measurement from length data

Larvaceans of the genus *Oikopleura* are an important component of the zooplankton of all seas and oceans and have an impact as algal grazers comparable with that of copepods. *Oikopleura* sports an heroic way of reproduction which leads to instant death. During its week-long life at 20 °C and abundant food, it accumulates energy for reproduction which is deposited at the posterior end of the trunk; see figure 4.19. Except for this accumulation of material for reproduction, the animal remains isomorphic. The total length of the trunk, L_t, including the gonads, can be partitioned into the true trunk length, L, and the length of the gonads, L_R. Since the reproduction material is deposited on a surface area of the trunk, the length of the gonads is about proportional to the accumulated investment of energy into reproduction divided by the squared true trunk length. Fenaux and Gorsky [215] provided data where both the true and the total trunk length have been measured under laboratory conditions. This gives the possibility of testing the consequences of the DEB theory for reproduction.

Let $e_{\dot{R}}(t_1, t_2)$ denote the cumulative investment of energy into reproduction between t_1

Figure 4.20: The total trunk length, L_t (□ and upper curve, left), the true trunk length, L (◇ and lower curve, left) and the dry weight (right) for *Oikopleura longicauda* at 20 °C. Data from Fenaux and Gorsky [215]. The DEB-based curves account for the contribution of the cumulated energy, allocated to reproduction, to total trunk length and to dry weight. The parameter estimates (and standard deviations) are L_m= 822 (37) μm, $l_b = l_p$= 0.157 (0.006), $\dot{m}$= 1.64 (0.14) d^{-1}, g= 0.4, $V_R q_R$= 0.0379 (0.0083) mm^3, excluding the last L_t data point. Given these parameters, the weight data give d_{dl}= 0.0543 (0.0131) g cm^{-3}, d_{dr}= 15.2 (4.20) μg, the last data point is excluded.

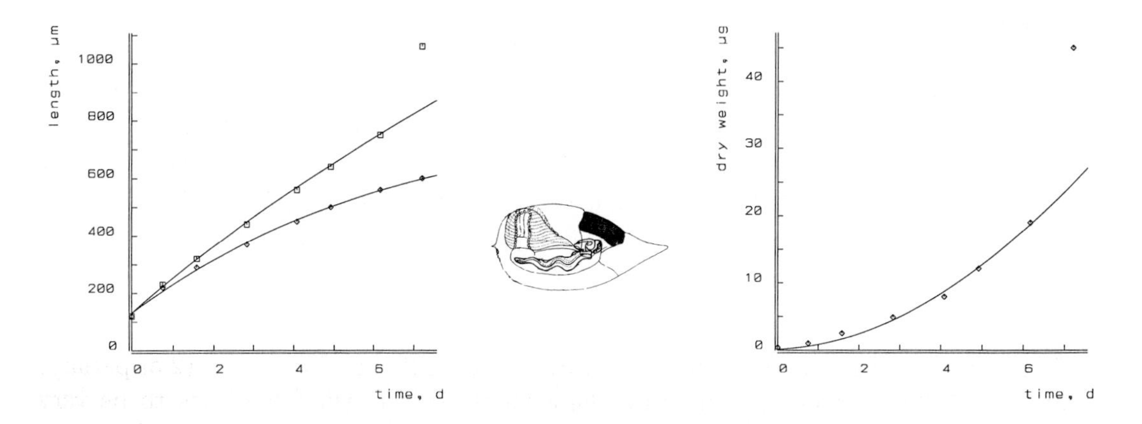

and t_2, as a fraction of the maximum energy reserves $[E_m]V_m$, then (4.5) gives for adults

$$e_{\dot{R}}(t_1, t_2) = \dot{m} q_R \int_{t_1}^{t_2} \left(\frac{g + l(t)}{g + e(t)} e(t) l^2(t) - l_p^3 \right) dt \tag{4.43}$$

Oikopleura has a non-feeding larval stage and starts investing in reproduction as soon as it starts feeding, so $L_b = L_p$. It thus lacks a juvenile stage in the present definition, and the larva should be classified as an embryo. The total trunk length then amounts to $L_t(t) = L(t) + V_R e_{\dot{R}}(0,t)/L^2(t)$. The volume V_R is a constant that converts the scaled cumulative reproductive energy per squared trunk length into the contribution to the total length. At abundant food, the true trunk length follows the von Bertalanffy growth curve $L(t) = L_m - (L_m - L_b)\exp\{-\frac{t g \dot{m}}{3(1+g)}\}$ and $e(t) = 1$, where L_m denotes the maximum length, i.e. $L_m = V_m^{1/3}/d_m$. If the data set $\{t_i, L(t_i), L_t(t_i)\}_{i=1}^n$ is available, the 5 parameters L_b, L_m, $\dot{m}$, g and $V_R q_R$ can be estimated in principle. Dry weight relates to trunk length and reproductive energy as $W_d(t) = [d_{dl}]L^3(t) + d_{dr} e_{\dot{R}}(0,t)$, where the two coefficients give the contribution of cubed trunk length and cumulative scaled reproductive energy to dry weight. If dry weight data are available as well, there are 7 parameters to be estimated from three curves.

Figure 4.20 gives an example. The data appear to contain too little information to determine both $\dot{m}$ and g, so either $\dot{m}$ or g has to be fixed. The more or less arbitrary choice $g = 0.4$ is made here. The estimates are tied by the relationship that $\frac{\dot{m}g}{1+g}$ is almost constant. The high value for the maintenance rate coefficient $\dot{m}$ probably relates to the investment of energy into the frequent synthesis of new filtering houses.

4.4.4 Suicide reproduction

Like *Oikopleura*, salmon, eel and most cephalopods die soon after reproduction. The distribution of this type of behaviour follows an odd pattern in the animal kingdom. Tarantula males die after first reproduction, but the females reproduce frequently and can survive for 20 years. Death does not follow the Weibull-type aging pattern and probably has a different mechanism. Because the (theoretical) asymptotic size is not approached in cephalopods, they also seem to follow a different growth pattern. I believe, however, that early death, not the energetics, makes them different. The arguments are as follows.

The surface area in von Bertalanffy growth is almost linear in time across a fairly broad range of surface areas not close to zero or the asymptote. This has led Berg and Ljunggren [56] to propose an exactly linear growth of the surface area for yeast until a certain threshold is reached; see figure 1.1. Starting from an infinitesimally small size, which is realistic for most cephalopods, the volume increases with cubed time: $V(t) = (\frac{\dot{v}ft}{3(f+g)})^3$. Over a small trajectory of time, this closely resembles exponential growth, as has been fitted by Wells [761], for instance.

Squids show a slight decrease in growth rate towards the end of their lives (2 or perhaps 3 years [697]), just enough to indicate the asymptotic size, which happens to be very different for female and male in *Loligo pealei*. It will be explained in the section of primary scaling relationships, {218}, that the costs for growth $[G]$ in the von Bertalanffy growth rate $\dot{\gamma} = \frac{[\dot{M}]/3}{[G]+\kappa[E_m]f}$ hardly contributes in large bodied species because it is independent of asymptotic length, while maximum energy density is linear therein. So $\dot{\gamma} \simeq \frac{[\dot{M}]}{3\kappa[E_m]f}$. The product $\dot{\gamma}V_\infty^{1/3} \simeq \dot{v}/3$ should then be independent of ultimate size. On the basis of data provided by Summers [697], the product of ultimate length and the von Bertalanffy growth rate was estimated to be 0.76 and 0.77 dm a^{-1} for females and males respectively. The equality of these products supports the interpretation in terms of the DEB model. The fact that the squids die well before approaching the asymptotic size only complicates parameter estimation.

A large (theoretical) ultimate volume goes with a large maximum growth rate. If the maximum growth rates of different species are compared on the basis of size at death, the octopus *Octopus cyana* grows incredibly quickly, as argued by Wells [761]. Assuming that the maximum growth rate is normal, however, a (theoretical) ultimate volume can be inferred by equating $\dot{\gamma}V_\infty^{1/3}$ for the octopus to that for the squid, after correction of temperature differences. Summers did not indicate the temperature appropriate for the squid data, but on the assumption that it has oscillated between 4 and 17 °C and that $T_A = 12500$ K, the growth rate has to be multiplied by 9.3 to arrive at the temperature that Wells used, i.e. 25.6 °C. The data of Wells indicate a maximum growth rate of $\frac{4}{9}\dot{\gamma}V_m = 25.5$ $\text{dm}^3\,\text{a}^{-1}$. The ultimate volume is thus $\left(\frac{9\times25}{4\times9.3\times0.77}\right)^{3/2} = 22$ dm^3 for the octopus. This is three times the volume at death.

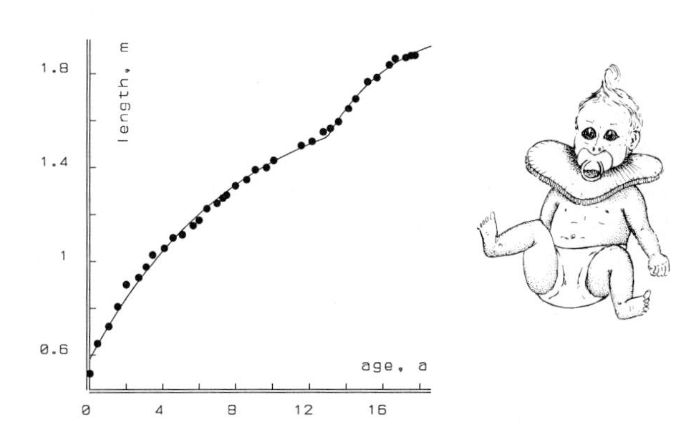

Figure 4.21: Length-at-age of man, de Montbeillard's son, in the years 1759-1777. Data from Cameron [120]. The curve is the von Bertalanffy one with an instantaneous change of the ultimate length from 177(4.6) cm to 201(8.2) cm and of the von Bertalanffy growth rate from 0.123(0.0093) a^{-1} to 0.285(0.094) a^{-1} at the age of 13(0.215) a.

4.4.5 Changing parameter values

As discussed in the sections on diet and starvation, {60,128}, some species can change energy allocation through κ. Two further examples are given here.

Changes at puberty

Growth curves suggest that some species, e.g. humans, change the partition coefficient κ and the maximum surface area-specific assimilation rate $\{\dot{A}_m\}$ at puberty in situations of food abundance; see figure 4.21. These changes amount to changes in the ultimate length and the von Bertalanffy growth rates via $L_m = (\kappa\{\dot{A}_m\}/[\dot{M}] - V_h^{1/3})/d_m$ and $\dot{\gamma} = [\dot{M}](3[G]+\kappa[E_m])^{-1}$. Suppose that the volume-specific maintenance costs $[\dot{M}]$, the volume-specific growth costs $[G]$, and the heating volume V_h do not change at puberty. Table 3.1 suggests that $\dot{m} \equiv [\dot{M}]/[G]$ will be about 0.1 d^{-1} at 37 °C. If a man of 180 cm weighs 75 kg and if half this weight is structural biomass, the shape coefficient is approximately $d_m = 0.19$. For $V_h^{1/3}$ is 10 cm, the observed changes in ultimate length and the von Bertalanffy growth rate correspond with a change by a factor 2.8 for $\{\dot{A}_m\}$ and by a factor 0.426 for κ. This analysis can only be provisional. Deviations from strict isomorphism may affect estimates.

Changes in response to the photoperiod

The allocation of energy to reproduction in the pond snail *Lymnaea stagnalis* depends on the photoperiod, as has been discussed under 'prolonged starvation', {128}. The photoperiod also effects the allocation under non-starvation conditions. This is obvious from the ultimate length. Snails kept under a 12:12 cycle (MD conditions) have a larger ultimate length than under a 16:8 cycle (LD conditions) [788]. MD snails also have a smaller von Bertalanffy growth rate and a smaller volume at puberty, cf. {98}, but MD and LD snails are found to have the same energy conductance of $\dot{v} \simeq 1.55$ mm d^{-1} at 20 °C. This is a strong indication that the photoperiod only affects the partition coefficient κ.

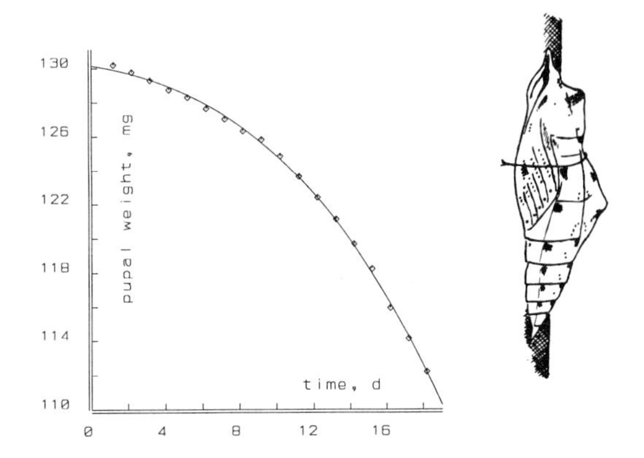

Figure 4.22: The wet weight development of the male pupa of the green-veined white butterfly *Pieris napi* at 17 °C until eclosion, after having spent 4 months at 4 °C. Data from Forsberg and Wiklund [225]. The fitted curve is $W_w(t) = 130.56 - (\frac{7.16+t}{0.104})^3$, with weight in mg and time in days, as is expected from the DEB theory.

4.4.6 Pupa and imago

Insects that develop pupae do not grow in the adult stage, called the imago. They are thus much less flexible in their allocation of energy. The use of energy in the pupal stage strongly suggests the embryo development pattern, or, more specifically, the foetal development pattern since the energy reserves at eclosion are usually quite substantial so that there is hardly any growth retardation due to reserve depletion. This resemblance to a development pattern is not a coincidence because the adult tissue develops from a few tiny imaginal disks, the structural biomass of the caterpillar being first converted to reserves for the pupa. So the initial structural volume of the pupa is very small indeed. Since no energy input from the environment occurs until development is completed, pupal weight decreases, reflecting the use of energy. This can be worked out quantitatively as follows.

As discussed under foetal development, growth is given by $\frac{d}{dt}V = \dot{v}V^{2/3}$, so that, if temperature is constant, $V^{1/3}(t) = V_0^{1/3} + \dot{v}t$, where V_0 represents the structural volume of the imaginal disks. The energy in the reserves decreases due to growth, maintenance and development, so that

$$E(t) = E_0 - \frac{[G]}{\kappa}V(t) - \frac{[\dot{M}]}{\kappa}\int_0^t V(t_1)\,dt_1 \tag{4.44}$$

$$= E_0 - \frac{[G]}{\kappa}(V_0^{1/3} + \dot{v}t)^3 - \frac{[\dot{M}]}{4\kappa\dot{v}}(V_0^{1/3} + \dot{v}t)^4 + \frac{[\dot{M}]}{4\kappa\dot{v}}V_0^{4/3} \tag{4.45}$$

Together with the contribution of the structural volume, this translates via (2.9) into the wet weight development

$$W_w(t) = \frac{[d_{we}]E_0}{[E_m]} - (g[d_{we}] - [d_{wv}])(V_0^{1/3} + \dot{v}t)^3 - \frac{[d_{we}]}{4V_m^{1/3}}\left(\left(V_0^{1/3} + \dot{v}t\right)^4 - V_0^{4/3}\right) \tag{4.46}$$

Tests against experimental data quickly show that the contribution of the third term, which relates to maintenance losses, is too small to be noticed. So the weight-at-time curve reduces to a three parameter one. It fits the data excellently; see figure 4.22. Just as in foetuses, the start of the development of the pupa can be delayed, in a period known as the diapause. The precise triggers that start development are largely unknown.

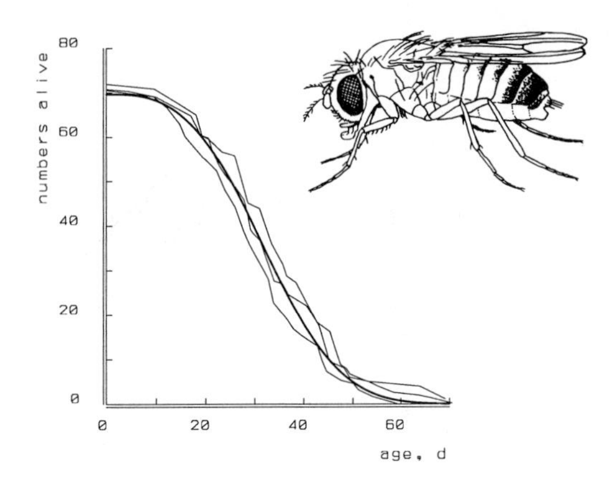

Figure 4.23: The survival curves of the female fruitfly *Drosophila melanogaster* at 25 °C and unlimited food. Data from Rose [612]. The fitted survival curve is $\exp\{-(\dot{p}_i t)^3\}$ with $\dot{p}_i = 0.0276$ (0.00026) d^{-1}.

Imagos do not grow, so if the reserve dynamics (3.7) still applies, the catabolic rate reduces to $\dot{C} = \{\dot{A}_m\}V^{2/3}e$. If food density is constant or high, and aging during the pupate state is negligible, the change in damage inducing compounds is

$$\frac{d}{dt}[Q] = d_Q[\dot{C}] = d_Q\{\dot{A}_m\}V^{-1/3}f = \frac{d_Q}{\kappa}[G]\dot{m}f/l$$

where $l \equiv (V/V_m)^{1/3}$ and $V_m^{1/3} \equiv \kappa\{\dot{A}_m\}/[\dot{M}]$ as before. Note that, in this case, V_m cannot be interpreted as the maximum body volume and κ cannot be interpreted as a partition coefficient. Energy derived from food is spent on (somatic plus maturity) maintenance at a constant rate (at constant temperature) in imagos; the rest is spent on reproduction. The loss of the interpretations for V_m and κ is not a problem; the term $[\dot{M}]/\kappa$ represents the sum of the somatic and maturity volume-specific maintenance costs, so $V_m^{1/3}$ represents the ratio of the maximum surface area-specific assimilation rate and the volume-specific total maintenance costs.

The hazard rate and the survival probability simplify to

$$\dot{h}(t) = \frac{t^2}{2}\ddot{p}_a\dot{m}f/l \quad (4.47)$$

$$\text{Prob}\{\underline{a}_\dagger > a_p + t|\underline{a}_\dagger > a_p\} = \exp\left\{-\frac{1}{6}t^3\ddot{p}_a\dot{m}f/l\right\} = \exp\left\{-(\dot{p}_i t)^3\right\} \quad (4.48)$$

for $\dot{p}_i \equiv (\frac{1}{6}\ddot{p}_a\dot{m}f/l)^{1/3}$ having the interpretation of an aging rate. This is thus the Weibull model with a fixed shape parameter of 3. The mean age at death as an imago then equals $\Gamma(\frac{1}{3})(3\dot{p}_i)^{-1} \simeq 0.54(\ddot{p}_a\dot{m}f/l)^{-1/3}$, where Γ stands for the gamma function $\Gamma(x) \equiv \int_0^\infty t^{x-1}\exp\{-t\}\,dt$.

Experimental results of Rose, figure 4.23, suggest that this is realistic. He showed that longevity can be prolonged in female fruitflies by selecting offspring from increasingly older females for continued culture [612]. It cannot be ruled out, however, that this effect has a simple nutrient/energy basis with little support for evolutionary theory. Selection for digestive deficiency also results in a longer life span. Reproduction, feeding, respiration and, therefore, aging rates must be coupled due to the conservation law for energy. This is

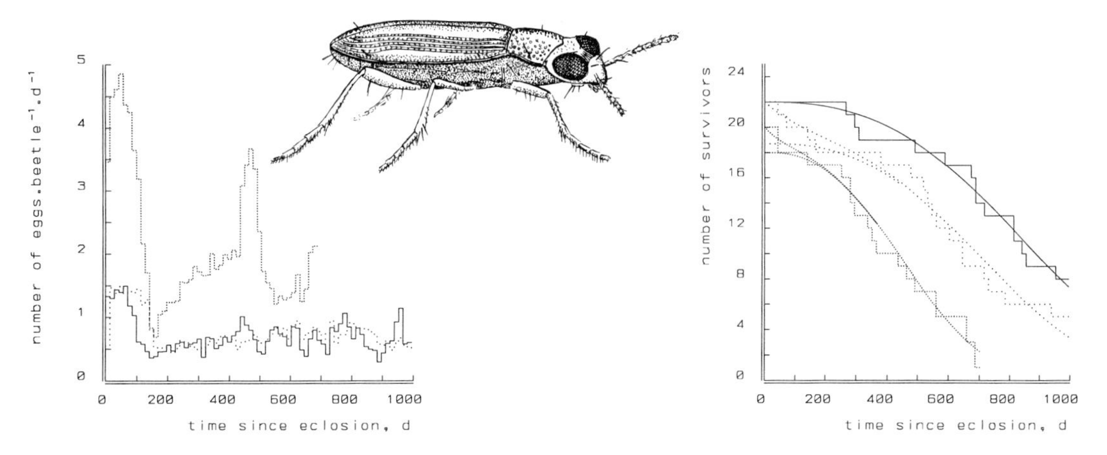

Figure 4.24: The reproduction rate (left figure) of the carabid beetle *Notiophilus biguttatus* feeding on a high density (stippled graphs) of springtails at 20/10 °C (densely stippled) and at 10 °C (sparsely stippled) and a lower density (drawn graph) at 20/10 °C. The survival probability of these cohorts since eclosion is given in the right figure. Data from Ger Ernsting, personal communication and [205]. The survival probability functions (right) are based on the observed reproduction rates with estimated parameter $\ddot{p}_a e_0 (qgl^3)^{-1} = 0.63(0.02)$ a^{-2} for high food level and high temperature, 0.374(0.007) a^{-2} for low food level and high temperature, 0.547(0.02) a^{-2} for high food level and low temperature. The contribution of maintenance costs to aging is determined from respiration data. A small fraction of the individuals at the high food levels died randomly at the start of the experiments.

beautifully illustrated with experimental results from Ernsting [205], who collected carabid beetles *Notiophilus biguttatus* from the field shortly after eclosion, kept them at a high and a low level of food (springtail *Orchesella cincta*) at 16 h 20 °C:8 h 10 °C, and measured survival and egg production. A third cohort was kept at 10 °C at a high feeding level. He showed that the respiration rate of this 4–7 mg beetle is linear in the reproduction rate: $0.84 + 0.041\dot{R}$ in J d^{-1} at 20/10 °C and $0.57 + 0.051\dot{R}$ at 10 °C. This linear relationship is to be expected for imagos on the basis of the above mentioned interpretations. It allows a reconstruction of the respiration rate during the experiment from reproduction data and a detailed description of the aging process. This is more complex than the Weibull model, because metabolic activity was not constant, despite standardized experimental conditions. The quantitative details are as follows.

The catabolic rate is partitioned into the maintenance and reproduction costs as

$$\dot{C} = \dot{M}/\kappa + \dot{R}E_0/q = (\dot{m} + \dot{R}e_0(qgl^3)^{-1})[G]V/\kappa$$

where the scaled egg cost e_0 is given in (3.26). This gives the hazard rate and survival probability

$$\begin{aligned} \dot{h}(t) &= \frac{1}{2}\ddot{p}_a \dot{m} t^2 + \frac{\ddot{p}_a e_0}{qgl^3} \int_0^t \int_0^{t_1} \dot{R}(t_2)\, dt_2\, dt_1 \\ \text{Prob}\{\underline{a}_\dagger > a_p + t | \underline{a}_\dagger > a_p\} &= \exp\left\{ -\frac{1}{6}\ddot{p}_a \dot{m} t^3 - \frac{\ddot{p}_a e_0}{qgl^3} \int_0^t \int_0^{t_1} \int_0^{t_2} \dot{R}(t_3)\, dt_3\, dt_2\, dt_1 \right\} \end{aligned}$$

Although e_0 depends on the reserve energy density of the beetle, and so on feeding behaviour, variations will be negligibly small for the present purpose since food dependent differences in egg weights have not been found. The low temperature cohort produced slightly heavier eggs, which is consistent with the higher respiration increment per egg. The estimation procedure is now to integrate the observed $\dot{R}(t)$ three times and to use the result in the estimation of the two compound parameters $\frac{1}{6}\ddot{p}_a\dot{m}$ and $\ddot{p}_a e_0 (qgl^3)^{-1}$ of the survivor function from observations.

Figure 4.24 confirms this relationship between reproduction, and thus respiration, and aging. The contribution of maintenance in respiration is very small and could not be estimated from the survival data. The mentioned linear regressions of respiration data against the reproduction rate indicates, however, that $qg\dot{m}l^3/e_0 = 0.84/0.041 = 20.4$ d^{-1} at 20/10 °C or $0.57/0.051 = 11$ d^{-1} at 10 °C. This leaves just one parameter $\ddot{p}_a e_0 (qgl^3)^{-1}$ to be estimated from each survival curve. The beetles appear to age a bit faster per produced egg at high than at low food density. This might be due to eggs being more costly at high food density, because of the higher reserves at hatching. Another aspect is that at high food density, the springtails induced higher activity, and so higher respiration, by physical contact. Moreover, the substantial variation in reproduction rate at high food density suggests that the beetles had problems with the conversion of energy, that is allocated to reproduction, to eggs, which led to an increase in q and a higher respiration per realized egg. Note that these variations in reproduction rate are hardly visible in the survival curve, which is due to triple integration. The transfer from the field to the laboratory seemed to induce early death for a few individuals at the high food levels. This is not related to the aging process but, possibly, to the difference from field conditions.

The Weibull model for aging with a fixed shape parameter of 3 should not only apply to holometabolic insects, but to all ectotherms with a short growth period relative to the life span. Gatto [245] found, for instance, a perfect fit for the bdelloid rotifer *Philodina roseola* where the growth period is about 1/7-th of the life span. Notice that constant temperature and food density are still necessary conditions for obtaining the Weibull model.

The presented tests on pupal growth and survival of the imago support the applicability of the DEB theory to holometabolic insects, if some elementary facts concerning their life history are taken into account. This suggests new interpretations for experimental results.

4.4.7 Food induced aging acceleration

Some data sets, such as that of Robertson and Salt on the rotifer *Asplanchna girodi* feeding on the ciliate *Paramecium tetraurelia* [606] and figure 4.25, indicate that the hazard rate increases sharply with food density. Although the shapes of the hazard curves are well described by (3.56), this equation does not predict the extreme sensitivity to food density. This particular data set shows that aging acceleration is linear in the food density, which suggets that something that is proportional to food density affects the build-up of damage inducing compounds or the transformation of these compounds into damage. One possibility is nitrite derived from the lettuce used to culture the ciliates; nitrite is known for its mutagenic capacity [328], cf. {284}. This argument can only be speculative, but it might not be unrealistic because if the medium in which the ciliates have been cultured

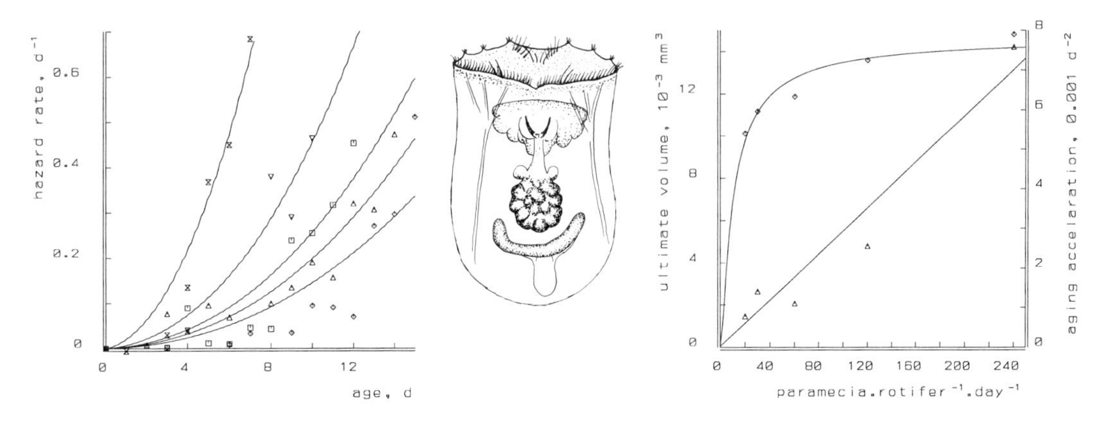

Figure 4.25: The hazard rates for the rotifer *Asplanchna girodi* for different food levels: 20 (◇) 30 (△) 60 (□) 120 (▽) and 240 (⋈) paramecia rotifer^{-1} d^{-1} at 20 °C. Data from Robertson and Salt [606]. The one-parameter hazard curves were based on the scaled food densities as estimated from the ultimate volumes (◇, right), which gave f= 0.877, 0.915, 0.955, 0.977, 0.988. The resulting five aging accelerations are plotted in the right figure (△). They proved to depend linearly on food density, with an intercept that is consistent with the aging acceleration found for daphnids.

contained such a factor, it is likely that it builds up in the rotifer proportionally to the concentration in the environment. The hazard rate at zero food density should reveal the 'pure' aging acceleration $\ddot{p}_a$. For this reason the intercept of the linear relationship was chosen at 3.27×10^{-5} d^{-2}. This is the temperature corrected value that has been found for daphnids in figure 3.22. The data are consistent with this interpretation but it will be unsatisfactory as long as the aging modifying factor has not been identified.

4.4.8 Segmented individuals

Figure 4.26 shows that von Bertalanffy growth also applies to isomorphic annelids living in rich culture media; however, some thread-like annelids grow in length only, so their change in shape corresponds to that of filaments. The water nymph *Nais elinguis* is an example of an aquatic oligochaete with such a small diameter (0.19 mm, aspect ratio δ = 0.02) that no advanced circulation system is necessary to ensure adequate gas exchange. Its segmentation restricts mass exchange in the longitudinal direction to the extent that the assumptions for the digestion process can be tested on the basis of growth performance in an intriguing way. The digestion process will be discussed on {247} more detail. It will be assumed here that the assimilation energy input decreases linearly with length to zero at the posterior end of the body, so the reserve density in the anterior part is higher than in the posterior part. Water nymphs usually propagate asexually by division, which implies that the freshly separated anterior part has a relatively high initial reserve density, and thus high initial growth, while the opposite holds for the posterior part. The DEB theory provides the tools to evaluate this difference in growth quantitatively [583].

Let l° denote a dimensionless length measure of the worm, which takes value 0 at the posterior end and value 1 at the anterior end. The (total) assimilation energy that comes in equals $\int_0^1 \dot{A}(l^\circ)\, dl^\circ$, which for filaments is given by $[\dot{A}_m]fV$; see (3.3). Since $\dot{A}(l^\circ)$ according to the digestion model, {247}, is proportional to l°, the proportionality constant can be found from the equation $[\dot{A}_m]fV \propto \int_0^1 l^\circ\, dl^\circ$, which results in $\dot{A}(l^\circ)dl^\circ = 2[\dot{A}_m]fVl^\circ dl^\circ$. The reserve energy dynamics at l° are thus found by substitution of $2fl^\circ$ for f into (3.32), which gives $\frac{d}{dt}e(l^\circ) = \dot{\nu}(2fl^\circ - e(l^\circ))$. Growth according to (3.33) is now: $\frac{d}{dt}V = \dot{\nu}V \int_{l^\circ=0}^1 \frac{e(l^\circ)-(V_d/V_m)^{1/3}}{e(l^\circ)+g}\, dl^\circ$.

Worms such as *Nais* typically live in sludge from sewage treatment plants, so food is always abundantly available. The main variable is not food abundance, but food quality, which relates to the value of $\{\dot{A}_m\}$, while $f = 1$. If food quality changes slowly, the equations for the energy reserve density and body volume can be solved explicitly. The result is

$$e(l^\circ, t) = 2fl^\circ - (2fl^\circ - e(l^\circ, 0)) \exp\{-\dot{\nu}t\} \tag{4.49}$$

$$V(t) = V(0) \exp\left\{\dot{\nu} \int_0^t \int_0^1 \frac{e(l^\circ, t_1) - (V_d/V_m)^{1/3}}{e(l^\circ, t_1) + g}\, dl^\circ\, dt_1\right\} \tag{4.50}$$

A peculiar consequence is that now the initial storage density for the anterior and the posterior part of the just divided worm differ. If the worm divides when its energy density is already at its equilibrium value $e(l^\circ) = 2fl^\circ$, the initial energy density of the anterior part is $e(l^\circ, 0) = 2f(l_d^\circ + l^\circ(1 - l_d^\circ))$, while that of the posterior part is $e(l^\circ, 0) = 2fl_d^\circ l^\circ$, where l_d° is the value for l° where the division occurs. *Nais* appears not to divide into two equally long parts; the anterior part is usually somewhat longer. More in particular, the initial volume is given by $V(0) = (1 - l_d^\circ)V_d$ for the anterior part and $V(0) = l_d^\circ V_d$ for the posterior one. Note that since ingestion $\dot{I} = [\dot{I}_m]fV$ is linear in the body volume for filaments, see (3.3), it drops sharply at division. This relates to the gut residence time for food particles, which is constant (so independent of length or volume). The fit with measured data is quite acceptable; see figure 4.27. The population dynamical consequences are evaluated on {173}.

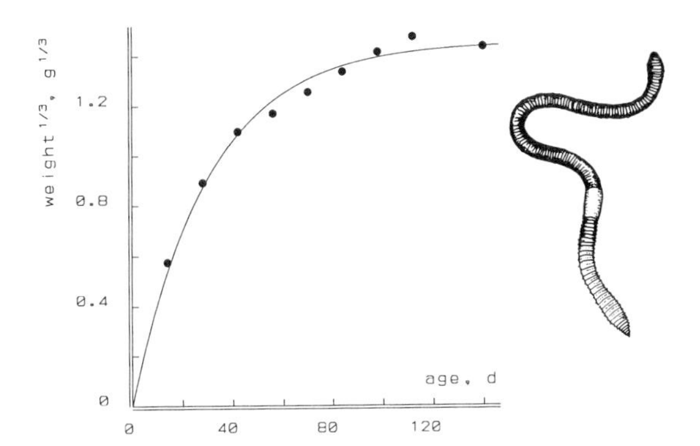

Figure 4.26: Weight-at-age curve for the isomorphic earthworm *Dendrobeana veneta* at 20 °C and abundant food. The adults produce 1 cocoon a week in this medium. Data provided by Hans Bos (pers. comm.). The data points represent means of 100 individuals. The ultimate wet weight$^{1/3}$ is 1.453 (0.024) g$^{1/3}$ and the von Bertalanffy growth rate $\dot{\gamma} = 0.033$ (0.002) d^{-1}.

Figure 4.27: The sludge eating filamentous water nymph *Nais elinguis* divides into an anterior part (◇) which is somewhat longer than the posterior one (□) and has a higher initial storage density due to incomplete mixing along the longitudinal axis. This gives the anterior part an advantage in terms of growth. Data by Christa Ratsak [583].

Chapter 5

Living together

The primary purpose of this chapter is to evaluate the consequences of the DEB model for individuals at the population level if extremely simple rules are defined for the interaction between individuals and the energy balance of the whole system.

Most models of population dynamics treat individuals as identical objects, so that a population is fully specified by total number or total biomass. Such populations are called non-structured populations. This obviously leads to attractive simplicity. I will discuss some doubts about their realism on {160}, doubts that can be removed by turning to structured populations. Structured populations are populations where the individuals differ from each other by one or more characteristics, such as age, which affect feeding, survival and/or reproduction. The DEB model provides an attractive, albeit somewhat complicated, structure. I will show the connection between non-structured and DEB-structured populations step by step. The introduction of a structure does not necessarily lead to realistic population models due to the effects of many environmental factors that typically operate at population level: spatial heterogeneity, seasonality, erratic weather, climatic changes, processes of adaptation and selection, subtle species interactions and so on. The occurrence of infectious diseases is perhaps one of the most common causes of decline and extinction of species, which typically operates in a density-dependent way. This means that population dynamics, as discussed in this chapter, still has to be embedded in a wider framework to arrive at realistic descriptions of population dynamics. The subject of this book is restricted to phenomena that have a direct bearing on the set of simple mechanisms for the uptake and use of energy listed in table 4.1.

The interaction between individuals of the same species is here restricted to feeding on the same resource. This point of view might seem a caricature in the eyes of a behavioural ecologist. The general idea, however, is not to produce population models that are as realistic as possible, but to study the consequences of feeding on the same resource. A comparison is then made with non-structured populations dynamics and with real world populations to determine the pay-off between realism and model complexity. If DEB-structured population dynamics predictions are not realistic, while the DEB model is at the individual level, this will give a key to factors that will be of importance in this situation. The basic energetics must be right before the significance of the more subtle factors can be understood. My fear is that most of the factors shown to be relevant will be specific for a particular species, a particular site and a particular period in time. This casts doubts on the extent to which general theory is applicable and on the feasibility of systems ecology. The main application of population dynamics theory here concerns a mental exercise pertaining to evolutionary theory, with less emphasis on direct testing in real world populations. The theory should, however, be able to predict population behaviour in simplified environments, such as those found in laboratory set-ups, in bio-reactors and the like, so that it has potential practical applications.

The significance of the population level for biological insight at all organization levels is manifold. It not only sets food availability and predation pressure for each individual, but it also defines the effect of all changes in life history which is pertinent to evolutionary theory. All other individuals belong to the environment of the particular individual, whose fitness is being judged. Fitness, whatever its detailed meaning, relates to the production of offspring, thus it changes the environment of the individual. This is one of the reasons why fitness arguments, which are central to evolutionary theories, should always involve the population level.

5.1 Non-structured populations

The chemostat, a popular device in microbiological research, will be used to make the transition from the intensively studied non-structured populations to DEB-structured populations. In a chemostat, food (substrate) is supplied at a constant rate to a population. Food density in the inflowing medium is denoted by X_r and the medium is flowing through the chemostat at throughput rate $\dot{p}$ times the volume of the chemostat V_c. Together with the initial conditions (food and biovolume density) these controls determine the behaviour of the system, in particular the food (substrate) density X_0 and the biovolume density X_1 as functions of time. The index 0 in the notation for food density is added for reasons of symmetry with X_1: the biovolume density of predators, i.e. the ratio of the sum of the individual volumes and the volume of the chemostat, V_c. So $X_1 = \sum_{i=1}^N V_i/V_c$, if there are N individuals in the population.

Batch cultures, which do not have a supply of food other than that initially present, are a special case of chemostat cultures, where $\dot{p} = 0$. I start with the Lotka–Volterra model, which was and probably still is the standard predator/prey model in ecology. In a sequence of related models, the effect of the stepwise introduction of biological detail that leads to DEB-structured populations will be studied.

The chemostat as a model can also be realistic for particular situations outdoors. An important difference between chemostat models and many population dynamical models is that food (substrate) does not propagate in the formulation here, while exponential or logistic growth is the standard assumption in most literature. The reason is that I want to stick to mass and energy balance equations in a strict way. The growth rate of food should, therefore, depend on its resource levels, which should be modelled as well. In the section on food chains, {212}, higher trophic levels, X_2, $X_3 \cdots$ will be introduced, not lower ones.

5.1.1 Lotka–Volterra

The Lotka–Volterra model assumes that the predation frequency is proportional to the encounter rate with prey (here substrate), on the basis of what is known as the law of mass action, i.e. the product of the densities of prey and predator. It can be thought of as a linear Taylor approximation of the hyperbolic functional response around food density 0: $f = \frac{X_0}{K+X_0} \simeq X_0/K$ for $X_0 \ll K$. The ingestion rate is taken to be proportional to body volume, as is appropriate for filaments, so that the sum of all ingestion rates by individuals in the population is found by adding the volumes of all individuals and applying the same proportionality constant.

The Lotka–Volterra model for chemostats with throughput rate $\dot{p}$ is

$$\frac{d}{dt}X_0 = \dot{p}X_r - [\dot{I}_m]\frac{X_0}{K}X_1 - \dot{p}X_0 \quad (5.1)$$

$$\frac{d}{dt}X_1 = Y[\dot{I}_m]\frac{X_0}{K}X_1 - \dot{p}X_1 \quad (5.2)$$

where Y stands for the yield factor, i.e. the conversion efficiency from prey to predator biomass; this is taken to be constant here. This model does not account for maintenance or energy reserves, so that in the context of the DEB model we have $Y = \frac{\kappa\{\dot{A}_m\}}{[G]\{\dot{I}_m\}}$, with $[\dot{M}] = 0$ and $[E_m] = 0$. At the individual level, this model implicitly assumes that the feeding rate is proportional to the volume of the individual. This aspect corresponds with the filament case of the DEB model; see (3.3). The analysis of the population dynamics can best be done with the dimensionless quantities $\tau \equiv t\dot{p}$, $[I_m] \equiv [\dot{I}_m]/\dot{p}$, $x_r \equiv X_r/K$, $x_0 \equiv X_0/K$, $x_1 \equiv X_1/K$. These substitutions turn (5.1) and (5.2) into

$$\frac{d}{d\tau}x_0 = x_r - [I_m]x_0x_1 - x_0 \quad (5.3)$$

$$\frac{d}{d\tau}x_1 = Y[I_m]x_0x_1 - x_1 \quad (5.4)$$

The equilibrium is found by solving x_0 and x_1 from $\frac{d}{d\tau}x_0 = \frac{d}{d\tau}x_1 = 0$. The positive solutions are $x_0^* = (Y[I_m])^{-1}$ and $x_1^* = Yx_r - [I_m]^{-1}$. The yield factor in this model, has a double interpretation. It stands for the conversion efficiency from food into biomass at both the individual level (this is how it was introduced in the previous chapter) and the population level. To see this, one has to realize that food influx is at rate $\dot{p}Kx_r$ and food output is at rate $\dot{p}Kx_0^* = \frac{\dot{p}K}{Y[I_m]}$ at equilibrium. So total food consumption is $\dot{p}K(x_r - \frac{1}{Y[I_m]})$. Biomass output is $\dot{p}Kx_1^* = \dot{p}K(Yx_r - [I_m]^{-1})$. The conversion efficiency at the population level thus amounts to $\frac{\dot{p}K(Yx_r - [I_m]^{-1})}{\dot{p}K(x_r - (Y[I_m])^{-1})} = Y$. This is so simple that it seems trivial. That this impression is false soon becomes obvious when we introduce more elements of the DEB machinery; the conversion efficiency at the population level then behaves differently from that at the individual level for non-filaments.

The linear Taylor approximation around the equilibrium of the coupled system (5.3) and (5.4) equals for $\mathbf{x}^T \equiv (x_0, x_1)$ and $\mathbf{x}^{*T} \equiv (x_0^*, x_1^*)$

$$\frac{d}{d\tau}\mathbf{x} \simeq \begin{pmatrix} -[I_m]x_1 - 1 & -[I_m]x_0 \\ Y[I_m]x_1 & Y[I_m]x_0 - 1 \end{pmatrix}_{\mathbf{x}=\mathbf{x}^*} (\mathbf{x} - \mathbf{x}^*) \quad (5.5)$$

$$\simeq \begin{pmatrix} -Y[I_m]x_r & -Y^{-1} \\ Y^2[I_m]x_r - Y & 0 \end{pmatrix} \begin{pmatrix} x_0 - \frac{1}{Y[I_m]} \\ x_1 - Y(x_r - \frac{1}{Y[I_m]}) \end{pmatrix} \tag{5.6}$$

The eigenvalues of the matrix with coefficients, the Jacobian, are -1 and $-Y[I_m]x_r + 1$, so that this system does not oscillate. See Edelstein-Keshet [195], and Yodzis [786] for valuable introductions to this subject, and Hirsch and Smale [321], Ruelle [618], and Arrowsmith and Place [22,23] for more advanced texts. Mathematical texts on nonlinear dynamics systems are now appearing at an overwhelming rate [52,181,278,352,714], especially with a focus on 'chaos', but simple biological problems still seem too complex to analyze analytically. Figure 5.1 compares the dynamics of the Lotka–Volterra model with other simplifications of the DEB model.

Although this model cannot produce oscillations, with a minor change it can, by feeding the outflowing food (substrate) back into the bio-reactor. This is technically a simple operation. Most microbiologists even neglect the small outflow in open systems in their mass balances. The situation is covered by deletion of the third term in (5.3), i.e. $-x_0$. The eigenvalues of the Jacobian then become $-\frac{1}{2}Y[I_m]x_r \pm \frac{1}{2}\sqrt{(Y[I_m]x_r)^2 - 4Y[I_m]x_r}$. For $Y[I_m]x_r < 4$, the eigenvalues are complex, thus the system is oscillatory.

5.1.2 Monod, Marr–Pirt and Droop

If the hyperbolic functional response is used in the Lotka–Volterra model, rather than the linear Taylor approximation, we arrive with some reconstructions of the original formulations at the well-known model of Monod. Marr [455] and Pirt [556] extended this model to account for maintenance, while Droop [183,184] extended it in another way to account for (nutrient) reserves. Maintenance or reserves have been introduced directly at the population level, however, which presents the problem of reconstructing the implicit assumptions at the individual level. This problem can most easily be solved with the DEB model for filaments, (3.3), (3.32) and (3.33).

The energy reserve density follows the functional response according to a first order process; see (3.32). So, if e_1 and e_2 denote the scaled energy density of two particular individuals, the difference decays exponentially with a relaxation time of $\dot{\nu}^{-1}$, because $\frac{d}{dt}(e_1 - e_2) = -\dot{\nu}(e_1 - e_2)$. Even if substrate density changes so rapidly that the energy reserve density is not at its equilibrium, and even if the initial energy densities of the individuals differ, the energy reserve densities of all individuals soon follow the same time-curve. It follows that $\frac{d}{dt}X_1 \propto \sum_i \frac{d}{dt}V_i \propto \sum_i V_i$. So the change of the sum of the volumes equals the sum of the changes of each volume, which are simple functions of volumes in the DEB model for filaments; see (3.33). The structured population of filaments collapses to a non-structured one. In order to compare its dynamics with classic models, I now assume that the specific energy conductance is large enough with respect to changes of food density, $\frac{d}{dt}\ln X_0 \ll \dot{\nu}$, meaning that the energy reserves are close to their equilibrium value $e = f$. This condition will be removed in the subsection on DEB filaments on {166}. The result is now that reconstructions of the models of Marr–Pirt, Droop and Monod are

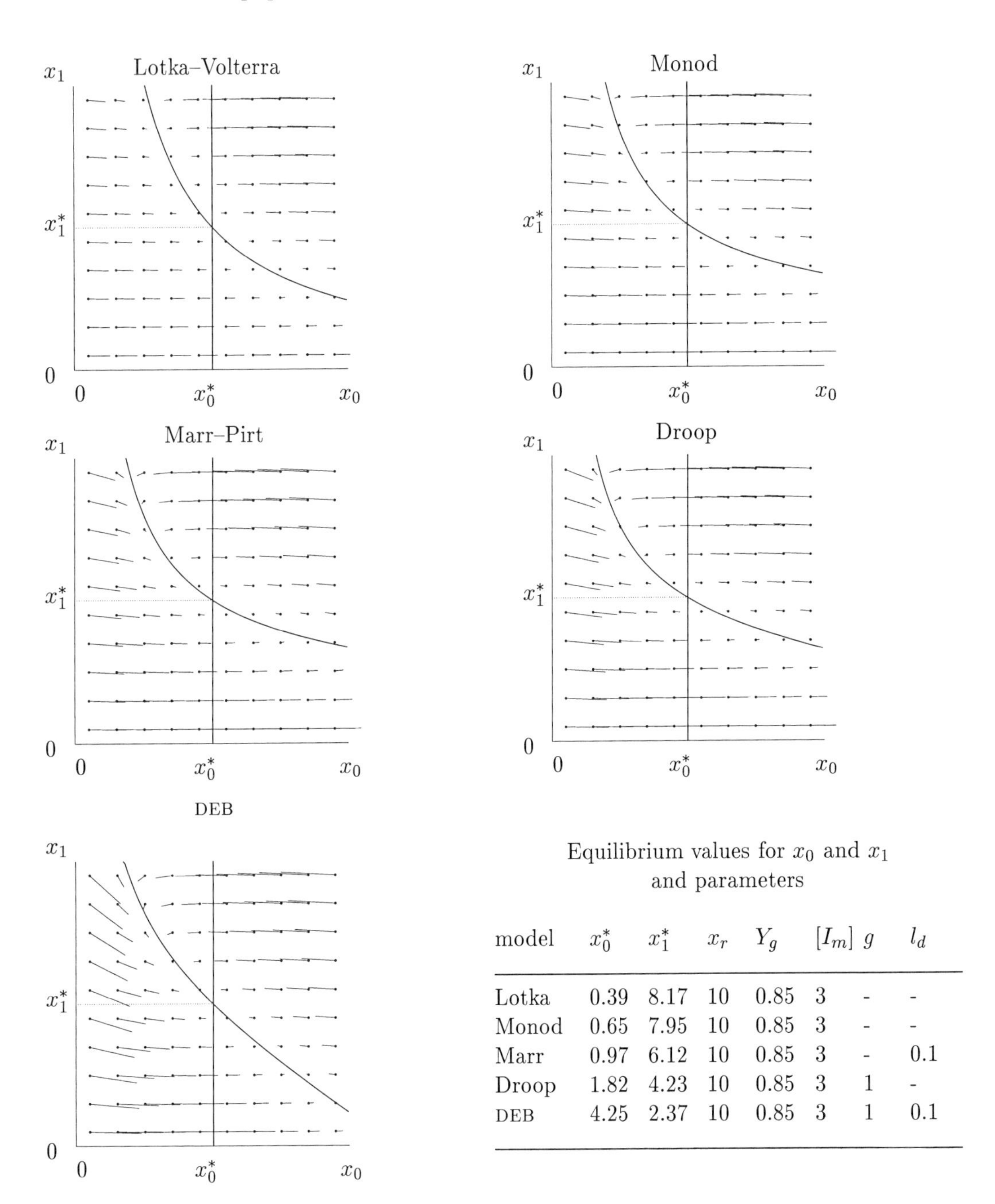

Equilibrium values for x_0 and x_1 and parameters

model	x_0^*	x_1^*	x_r	Y_g	$[I_m]$	g	l_d
Lotka	0.39	8.17	10	0.85	3	-	-
Monod	0.65	7.95	10	0.85	3	-	-
Marr	0.97	6.12	10	0.85	3	-	0.1
Droop	1.82	4.23	10	0.85	3	1	-
DEB	4.25	2.37	10	0.85	3	1	0.1

Figure 5.1: The direction fields and isoclines for the DEB model for filaments in a chemostat with reserves at equilibrium, and the various simplifications of this model. The lengths and directions of the line segments indicate the change in scaled food density x_0 and scaled biovolume x_1. The isoclines represent x_0, x_1-values where $\frac{d}{d\tau}x_0 = 0$ or $\frac{d}{d\tau}x_1 = 0$. All parameters and variables are made dimensionless, as indicated in the text. Figure 5.3 gives the direction field when the reserves are not in equilibrium.

special cases of the DEB model for filaments. It reads

$$\frac{d}{d\tau}x_0 = x_r - [I_m]fx_1 - x_0 \tag{5.7}$$

$$\frac{d}{d\tau}x_1 = Y[I_m]fx_1 - x_1 \tag{5.8}$$

with $f = \frac{x_0}{1+x_0}$. The yield factor Y is only constant in the Monod model. The growth dynamics for filaments, (3.33), can be used to show that the conversion efficiency equals

$[\dot{M}]$ / $[E_m]$	0	$\neq 0$	0	$\neq 0$
0	Monod $\frac{\kappa[\dot{A}_m]}{[G][\dot{I}_m]}$	Marr–Pirt $\frac{\kappa[\dot{A}_m]}{[G][\dot{I}_m]}\frac{f-l_d}{f}$	Monod Y_g	Marr–Pirt $Y_g\frac{f-l_d}{f}$
$\neq 0$	Droop $\frac{\kappa[\dot{A}_m]}{[G][\dot{I}_m]}\frac{g}{f+g}$	DEB for fil. $\frac{\kappa[\dot{A}_m]}{[G][\dot{I}_m]}\frac{g}{f}\frac{f-l_d}{f+g}$	Droop $Y_g\frac{g}{f+g}$	DEB for fil. $Y_g\frac{g}{f}\frac{f-l_d}{f+g}$

In the microbiological literature, Y_g is known as the 'true' yield, i.e. the yield excluding maintenance losses. In the Lotka–Volterra and the Monod model, the (actual) yield equals the 'true' yield, $Y = Y_g$, but in the Marr–Pirt, Droop and DEB models we find that $Y < Y_g$ and that Y is a function of food density, while Y_g is a constant. The conversion from food into biomass cannot be constant for models allowing for maintenance; this is obvious if one realizes that maintenance has priority over growth. So if feeding conditions are poor, a larger fraction of the available energy is spent on maintenance, compared with good feeding conditions.

The biologically interesting equilibrium values x_0^* and x_1^* can easily be obtained from (5.7) and (5.8), but the result is line filling. The linear Taylor approximation in the equilibrium for the Monod case is:

$$\frac{d}{d\tau}\mathbf{x} \simeq \begin{pmatrix} -\frac{x_r+x_0^{*2}}{x_0^*+x_0^{*2}} & -\frac{1}{Y_g} \\ \frac{x_r-x_0^*}{[I_m]x_0^{*2}} & 0 \end{pmatrix} (\mathbf{x}-\mathbf{x}^*) \tag{5.9}$$

The eigenvalues of the Jacobian are -1 and $-\frac{1}{Y_g[I_m]}(x_r - \frac{1}{Y_g[I_m]-1})(Y_g[I_m]-1)^2$, so that the system does not oscillate. The linear Taylor approximation of the functional response is accurate for small equilibrium values of food density, and thus a high value for $Y_g[I_m]$, which means that the Monod and the Lotka–Volterra models for the chemostat are very similar. The Monod model has less tendency to oscillate than the Lotka–Volterra model. This becomes visible if the substrate is fed back to the bio-reactor. (Thus we omit the term $-x_0$ in (5.7).) Contrary to the Lotka–Volterra model, the eigenvalues of the Jacobian cannot become complex, so that the system cannot oscillate.

the direction field of the DEB model for filaments in which energy reserves are allowed to deviate from their equilibrium values. The functional response in the equilibrium of the Monod model is only 0.4, for the chosen parameter values, which results in a close similarity with the Lotka–Volterra model. The direction fields of the Marr–Pirt and Droop models are rather similar, so that the effect of the introduction of maintenance and reserves are more or less the same. When introduced simultaneously, as in the DEB model, the effect is enhanced. Note that the isocline $\frac{d}{d\tau}x_0 = 0$ hits the axis $x_1 = 0$ at $x_0 = x_r$, which is just outside the frame of the picture for the DEB model, but far outside for the Lotka–Volterra model. For very small initial values for x_0 and x_1, the direction fields show that x_0 will first increase very rapidly to x_r, without a significant increase of x_1, then the $\frac{d}{dt}x_0 = 0$-isocline is crossed and the equilibrium value x_0^*, x_1^* is approached with strongly decreasing speed. This means that x_0 falls back to a very small value for Lotka's model, but much less so for the DEB model. The most obvious difference between the models is in the equilibrium values, where $x_1^* \gg x_0^*$ in Lotka's model, but the reverse holds in the DEB model. The other models take an intermediate position. The approach of x_0, x_1 to the equilibrium value closely follows the $\frac{d}{dt}x_0 = 0$-isocline if $x_1 > x_1^*$ in all models. The speed in the neighbourhood of the isocline is much less than further away from the isocline, and the differences in speed are larger for Lotka's model than for the DEB model. These extreme differences in speed mean that the numerical integration of this type of differential equations needs special attention.

5.1.3 Death

The usefulness of the chemostat in microbiological research lies mainly in the continuous production of cells that are in a particular physiological state. This state depends on the dilution rate. In equilibrium situations, this rate is usually equated to the population growth rate. The implicit assumption being made is that cell death plays a minor role. As long as the dilution rate is high, this assumption is probably realistic, but if the dilution rate is low, its realism is doubtful. Low dilution rates go with low substrate densities and long interdivision intervals. In the section on aging, the hazard rate for filaments has been tied to the respiration rate and so, indirectly, to substrate densities in (3.59). The law of large numbers tells that the hazard rate can be interpreted as a mean (deterministic) death rate for large populations. The dynamics for the dead biovolume, $x_\dagger$ reads

$$\frac{d}{dt}x_\dagger = \dot{h}x_1 - \dot{p}x_\dagger \tag{5.10}$$

with $\dot{h}$ denoting the hazard rate. It can easily be seen that in the equilibrium, we must have that $\dot{p}x_\dagger^* = \dot{h}x_1^*$, so the fraction of dead biovolume equals $\frac{x_\dagger^*}{x_1^*+x_\dagger^*} = \frac{\dot{h}}{\dot{p}+\dot{h}}$. The dynamics of the biomass should account for this loss, thus

$$\frac{d}{dt}x_1 = Y[\dot{I}_m]fx_1 - (\dot{p}+\dot{h})x_1 \tag{5.11}$$

Substitution of the expression for the hazard rate and the yield and the condition $\frac{d}{dt}x_1 = 0$ leads to the equilibrium value for f: $\frac{g(\dot{m}+\dot{p})}{\dot{\nu}-\dot{p}-\dot{p}_a(1+g)}$. Back-substitution into the hazard rate and the yield finally results in

$$\frac{x_\dagger^*}{x_1^*+x_\dagger^*} = \frac{\dot{m}+\dot{p}}{\dot{m}+(\dot{m}+\dot{\mu}_m^\circ)\dot{p}/\dot{p}_a} \tag{5.12}$$

Figure 5.2: The fraction of dead cells depends hyperbolically on the population growth rate, and increases sharply for decreasing population growth rates. The three curves correspond with $\dot{m}/\dot{\mu}_m = 0.05$, $\dot{p}_a/\dot{\mu}_m = 0.01$ (lower), $\dot{m}/\dot{\mu}_m = 0.1$, $\dot{p}_a/\dot{\mu}_m = 0.01$ (middle) and $\dot{m}/\dot{\mu}_m = 0.05$, $\dot{p}_a/\dot{\mu}_m = 0.1$ (upper curve). For high growth rates, the dead fraction is close to $\dot{p}_a/\dot{\mu}_m$, which will be very small in practice. The curves make clear that experimental conditions are extremely hard to standardize at low growth rates.

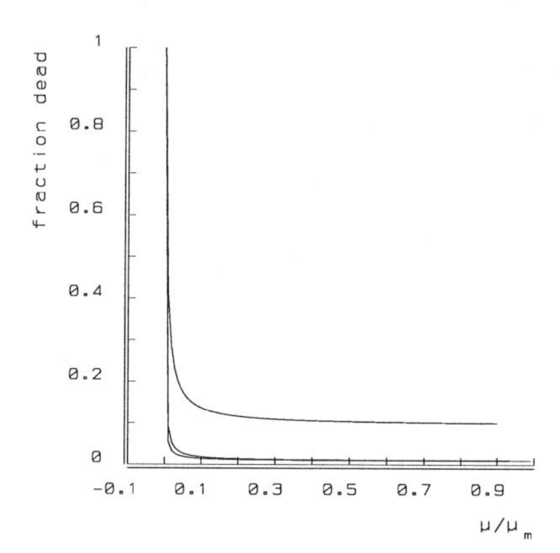

where $\dot{\mu}_m^\circ = \frac{\dot{\nu}-\dot{m}g}{1+g}$ is the gross maximum population growth rate. (The net maximum population growth rate is $\dot{\mu}_m = \dot{\mu}_m^\circ - \dot{p}_a$ and $\dot{p} \leq \dot{p}_m \leq \dot{\mu}_m \leq \dot{\mu}_m^\circ$, cf. (5.54). Since most microbiological literature does not account for death, and saturation coefficients are usually small, these different maximum rates are usually not distinguished. The concept 'population growth rate' is introduced on {169}.) Figure 5.2 illustrates how the dead fraction depends on the population growth rate.

The significance of the fraction of dead cells is not only of academic interest. Since it is practically impossible to distinguish the living from the dead, it can be used to 'correct' the measured biomass for the dead fraction to obtain the living biomass, {189}.

In the section on aging, {105}, I speculated that prokaryotes might not die instantaneously, but first switch to a physiological state called 'stringent response'. The fraction (5.12) can then be interpreted as the fraction of individuals that is in stringent response. A typical difference between both types of cells is the intracellular concentration of ppGpp, which is usually expressed per gram of total biomass. This quantification implicitly assumes that all cells in the population behave in the same way physiologically, rather than that the population can be partitioned into cells that are in stringent response and cells that are not. Which presentation is the more realistic remains to be studied.

5.1.4 DEB filaments

Figure 5.1 gives the direction fields of the various simplifications of the DEB model and figure 5.3 gives The full DEB model for filaments in chemostats is in need of an auxiliary equation for energy reserves, which amounts to the following three coupled equations:

$$\frac{d}{d\tau}x_0 = x_r - [I_m]fx_1 - x_0 \tag{5.13}$$

$$\frac{d}{d\tau}e = Y_g[I_m]g(f - e) \tag{5.14}$$

$$\frac{d}{d\tau}x_1 = Y_g[I_m]g\frac{e - l_d}{e + g}x_1 - x_1 - p_a\frac{1+g}{e+g}ex_1 \tag{5.15}$$

These coupled equations can be reduced to one integro-differential equation by integration of $\frac{d}{d\tau}e$ and $\frac{d}{d\tau}x_1$, and substitution of the results into the differential equation for x_0. This

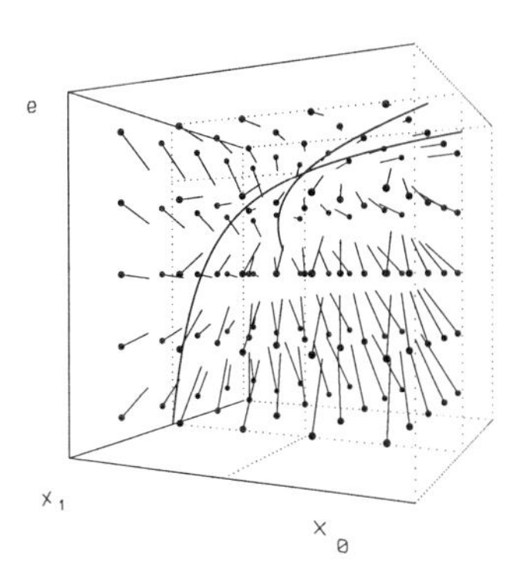

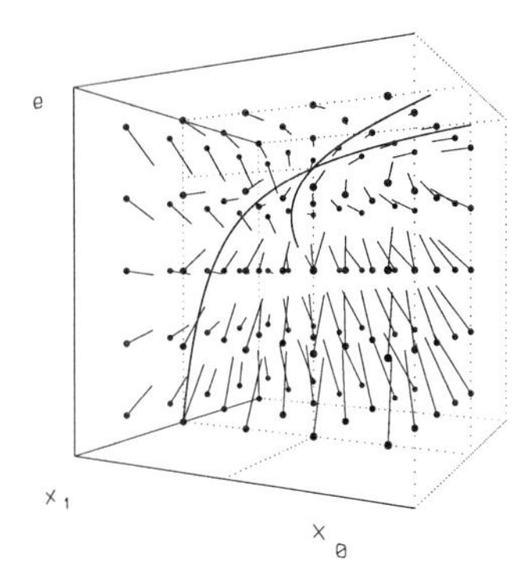

Figure 5.3: Stereo view of the direction field and isoclines for the DEB model for filaments in a chemostat. The parameter values are the same as in figure 5.1 and the projection of this direction field on the x, y-plane reduces to the direction field given in figure 5.1, where the reserves have been set at equilibrium.

is of little help, however, because this equation also has to be solved numerically.

The direction field of this model is given in figure 5.3. Mortality is excluded, $p_a = 0$, to facilitate comparison with the situation where reserves are in equilibrium; see figure 5.1.

A special case of conceptual interest can be solved analytically. This case relates to batch cultures, where no input or output (of substrate or biomass) exists, the biomass just developing on the substrate that is present at the start of the experiment. If the saturation coefficient, the maintenance costs and aging rate are small, DEB filaments will show a growth pattern which might be called expo-logistic. Initially they will grow exponentially and after a certain time (which corresponds to the depletion of the substrate) they switch to logistic growth, depleting their reserves. The biomass-time curve is smooth, even at the transition from one mode of growth to the other.

Worked out quantitatively, we get the following results. The functional response f is initially 1, since K is small with respect to X_0. If the inoculum is from a culture that did not suffer from substrate depletion, we have $e = 1$ and $X_1(t) = X_1(0)\exp\{\dot{\mu}_m t\}$, so the population growth rate is maximal, i.e. $\dot{\mu}_m = (\dot{\nu} - \dot{m}g)(1+g)^{-1}$. The substrate concentration develops as $X_0(t) = X_0(0) - \int_0^t [\dot{I}_m] X_1(t_1)\, dt_1$. It becomes depleted at t_0, say, where $X_0(t_0) = 0$. Substitution gives

$$X_0(t) = X_0(0)(\exp\{\dot{\mu}_m t_0\} - \exp\{\dot{\mu}_m t\})(\exp\{\dot{\mu}_m t_0\} - 1)^{-1}$$

where depletion occurs at time $t_0 = \frac{1}{\dot{\mu}_m} \ln\left\{1 + \frac{X_0(0)}{X_1(0)} \frac{\dot{\mu}_m}{[\dot{I}_m]}\right\}$. The reserves then decrease exponentially, i.e. $e(t_0 + t) = \exp\{-\dot{\nu}t\}$. The biovolume thus behaves as $X_1(t_0 + t) = X_1(t_0)\exp\left\{\int_0^t \frac{\dot{\nu}e(t_0+t_1) - \dot{m}g}{e(t_0+t_1)+g}\, dt_1\right\}$. For small maintenance costs, $\dot{m} \to 0$, this reduces to $X_1(t_0+t) = X_1(t_0)\frac{1+g}{\exp\{-\dot{\nu}t\}+g}$. This is the solution of the equation $\frac{d}{dt}X_1 = \dot{\nu}\left(1 - \frac{X_1(t)}{X_1(0)}\frac{g}{1+g}\right)X_1$, the well known logistic growth equation. If the maintenance costs are not negligibly small, the integral for $X_1(t)$ has to be evaluated numerically. Biovolume will first rise to a maximum and then collapse at a rate that depends on the maintenance costs. This behaviour offers the possibility to determine these costs experimentally. The quantitative evaluation can easily be extended to include fed batch cultures for instance, which have food (substrate) input and no output of food or biomass, but this does involve numerical work.

Similar biovolume-time curves can also arise if the reserve capacity rather than the saturation coefficient is small. If maintenance and aging are negligible as before, the batch culture can be described by $\frac{d}{dt}X_0 = -[\dot{I}_m]fX_1$ and $\frac{d}{dt}X_1 = Y_g[\dot{I}_m]fX_1$. We must also have $X_1(t) = X_1(0) + Y_g(X_0(0) - X_0(t))$. Substitution and separation of variables gives

$$[\dot{I}_m]Y_g t = \frac{KY_g}{X_1(\infty)} \ln \frac{X_1(t)(X_1(\infty) - X_1(0))}{X_1(0)(X_1(\infty) - X_1(t))} + \frac{1}{2} \ln \frac{X_1(t)}{X_1(0)}$$

Although this expression looks very different from the corresponding one for small saturation coefficients, the numerical values are practically indistinguishable, as shown in figure 5.4, where both population growth curves have been fitted to data on *Salmonella*. The only way to distinguish a difference is in the simultaneous fit for both biomass and substrate. This illustrates the rather fundamental problem of model identification for populations, even in such a simple case as this with only 4 free parameters. (To reduce the number of free parameters, maintenance and aging were taken to be negligible for both special cases.) Although *Salmonella* is a rod shaped bacterium, it is treated here as a filament because of its small aspect ratio; full treatment of rods is much more complicated, as shown later in this chapter. The conclusion to be drawn is that these data are not very informative and models for individual dynamics are soon too complicated to be of much help with the interpretation.

If other information is available to allow a choice between various possibilities, such as in the case of very efficient histidine uptake by deficient *Salmonella* strains, cf. {284}, the growth of batch cultures can be used to estimate the reserve capacity. This has been done in figure 5.5 to illustrate that under particular circumstances, the DEB model implies mass fluxes, as discussed in more detail on {192}.

5.1.5 Realism

Filaments have the unique property that they grow proportionally to their volume as individuals, which makes them an ideal paradigm for the connection between non-structured populations and structured ones. The definition of an individual is hard to make for filaments and, in the DEB model, of no importance; it indeed makes no difference if the population consists of one single large filament or many small ones. For isomorphs this the situation is different, of course. The simplicity of non-structured population dynamics comes with several unrealistic phenomena that have the potential to devalue any conclusion about real world populations. I will discuss some of them on {171,171,174}.

5.2 Structured populations

It is not my intention to review the rapidly growing literature on structured population dynamics, but for those who are unfamiliar with the topic, some basic notions are introduced

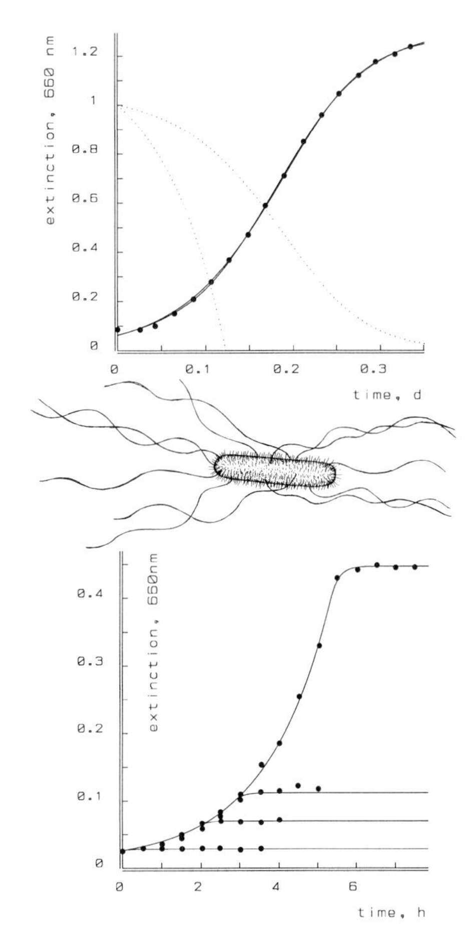

Figure 5.4: A batch culture of *Salmonella typhimurium* strain TA98 at 37 °C in Vogel and Bonner medium with glucose, (excess) histidine and biotin added. Two models have been fitted and plotted: one assumes that the saturation coefficient is negligibly small, but the reserves capacity is substantial, while the other does the opposite. Only the substrate density will tell the difference (stippled curves), but this is not measured. Parameters (s.d.): $\dot{\nu} = 18.6$ (0.37) d^{-1}, $g = 0.355$ (0.063), $X_0(0)/[\dot{I}_m] = 0.020$ (0.0036) d or $X_1(\infty) = 1.28$ (0.02), $YK = 1.31$ (0.86), $Y[\dot{I}_m] = 23.6$ (11.5).

Figure 5.5: Batch cultures of a histidine deficient strain of *S. typhimurium*, with initially only 0, 0.5, 1 or 5 μg histidine ml^{-1} in the medium, cease growth due to histidine depletion. The fit is based on the assumption of negligible maintenance requirements for histidine, which implies that the extinction plateau is a linear function of the added amount of histidine. The parameters (s.d.) are $[\dot{I}_m] = 8$ (0.44) μg his $\mathrm{ml}^{-1}\,\mathrm{h}^{-1}$, $\dot{\nu} = 5.3$ (1.2) h^{-1} and $g = 7.958$ (0.00205). One extinction unit corresponds with 7.56×10^8 cells ml^{-1}, so that the yield is $Y_g = \frac{\dot{\nu}}{[\dot{I}_m]g} = 0.0834$ ml μg his^{-1}. This corresponds with 1.1×10^{-10}g his $\mathrm{cell}^{-1} = 3.15 \times 10^5$ molecules his cell^{-1} with a maximum of 4×10^4 molecules histidine in the reserve pool.

below to help develop intuition. See DeAngelis and Gross [159], Ebenman and Persson [194], Heijmans [303], Łomnicki [439] and Metz and Diekmann [478] for reviews.

5.2.1 Stable age distributions

If food density is constant or high (with respect to the saturation coefficient), the distribution of individual states in the population, such as age and volume, stabilizes, while the numbers grow exponentially. This distribution can be evaluated in a relatively simple way, which makes it possible to evaluate statistics such as the mean volume and its variance, mean life span, etc. Situations may occur where the individual states change cyclically, so that such a stable distribution does not exist. The distribution of individual states has a limited practical value, because it only holds at prolonged constant food densities. How long food density must remain constant for state distributions to stabilize is hard to tell in specific cases and impossible in general. The main value of stable distributions lies in finding practical approximations for the behaviour of population models based on individuals. The derivation of stable state distributions is most easy via the stable age distribution, which I will explain briefly. A more extensive treatment is given by Frauenthal [230].

Let $n(a, t)\, da$ denote the number of females at time t with an age somewhere in the interval $(a, a + da)$, where da is an infinitesimally small time increment. The total number of individuals is thus $N(t) = \int_0^\infty n(a, t)\, da$. Individuals that have age a at t must have been born at $t - a$ and must be still alive to be counted in n, so we have the recursive relationship $n(a, t) = n(0, t - a)\text{Prob}\{\underline{a}_\dagger > a\}$, where $n(0, t)da$ denotes the number of births in $(t, t + da)$. The birth rate relates to the reproduction rate as $n(0, t) = \int_0^\infty n(a, t)\dot{R}(a)\, da$, where $\dot{R}(a)$ is the reproduction rate of an individual of age a. If we substitute the birth rate into the recursive relationship, we arrive at the integral equation

$$n(0, t) = \int_0^\infty n(0, t - a)\text{Prob}\{\underline{a}_\dagger > a\}\dot{R}(a)\, da \tag{5.16}$$

Rather than specifying the number of births before the start of the observations at $t = 0$, we specify the founder population $n(a, 0) = n_0(a)$ and write

$$n(0, t - a) = n_0(a - t)/\text{Prob}\{\underline{a}_\dagger > a - t\} \quad \text{for } a > t$$

The integral in (5.16) can now be partitioned and gives what is known as the renewal equation

$$n(0, t) = \int_0^t n(0, t - a)\text{Prob}\{\underline{a}_\dagger > a\}\dot{R}(a)\, da + \int_t^\infty \frac{\text{Prob}\{\underline{a}_\dagger > a\}}{\text{Prob}\{\underline{a}_\dagger > a - t\}} n_0(a - t)\dot{R}(a)\, da \tag{5.17}$$

The second term thus relates to the contribution of the individuals that were present in the founder population. Depending on the survival probability and age-dependent reproduction rate, its importance decreases with time. Suppose that it is negligibly small at some time t_1 and that the solution of (5.17) is of the form $n(0, t) = n(0, 0)\exp\{\dot{\mu}t\}$, for some value of $\dot{\mu}$ and $n(0, 0)$. Substitution into (5.17) gives for $t > t_1$:

$$n(0, 0)\exp\{\dot{\mu}t\} = \int_0^{t_1} n(0, 0)\exp\{\dot{\mu}(t - a)\}\text{Prob}\{\underline{a}_\dagger > a\}\dot{R}(a)\, da \quad \text{or} \tag{5.18}$$

$$1 = \int_0^{t_1} \exp\{-\dot{\mu}a\}\text{Prob}\{\underline{a}_\dagger > a\}\dot{R}(a)\, da \tag{5.19}$$

The latter equation is known as the characteristic equation. It is possible to show that, under some smoothness restrictions on reproduction as a function of age, this equation has exactly one real root for the population growth rate $\dot{\mu}_1$. The other roots are complex and have a real part smaller than $|\dot{\mu}_1|$. The general solution for $n(0, t)$ is a linear combination $\sum_i n_i(0, 0)\exp\{\dot{\mu}_i t\}$. For large t, the exponential $\exp\{\dot{\mu}_1 t\}$ will be dominant, so the asymptotic solution will be $n_1(0, 0)\exp\{\dot{\mu}_1 t\}$; because the other roots are of little practical interest, the index will be dropped and $\dot{\mu}$ is thus taken to be the dominant root. The smoothness restrictions on $\dot{R}(a)$ are violated if, for instance, reproduction is only possible at certain particular ages. In this case, the information about the age distribution of the founder population is not lost.

The stable age distribution, i.e. the distribution of the ages of a randomly taken individual, $\underline{a}$, is defined by $\phi_{\underline{a}}(a)\, da \equiv n(a, t)da/N(t)$ for $t \to \infty$. As before, we have for large t

$$n(a, t) = n(0, t - a)\text{Prob}\{\underline{a}_\dagger > a\} = n(0, 0)\exp\{\dot{\mu}(t - a)\}\text{Prob}\{\underline{a}_\dagger > a\}$$

As $N(t) \equiv \int_0^\infty n(a,t)\, da$ serves only to normalize the distribution, we get the simple relationship between the age distribution and the survivor probability of the individuals

$$\phi_{\underline{a}}(a) = \frac{\exp\{-\dot{\mu}a\}\text{Prob}\{\underline{a}_\dagger > a\}}{\int_0^\infty \exp\{-\dot{\mu}a_1\}\text{Prob}\{\underline{a}_\dagger > a_1\}\, da_1} \tag{5.20}$$

Note that $\underline{a}$ is defined for the population level, while $\underline{a}_\dagger$ is the age at which a particular individual dies, so it is defined for the individual level. For a stable age distribution, the adage 'older and older, rarer and rarer' always holds. The mean age in the population is thus

$$\mathcal{E}\underline{a} = \int_0^\infty a\phi_{\underline{a}}(a)\, da = \frac{\int_0^\infty a \exp\{-\dot{\mu}a\}\text{Prob}\{\underline{a}_\dagger > a\}\, da}{\int_0^\infty \exp\{-\dot{\mu}a\}\text{Prob}\{\underline{a}_\dagger > a\}\, da} \tag{5.21}$$

5.2.2 Reproducing neonates

There is no way to prevent neonates from giving rise to new neonates in unstructured populations. This artifact of the formulation can dominate population dynamics at lower growth rates. Comparison with a simple age-structured population, where individuals reproduce at a constant rate after an certain age a_p, can illustrate this.

In a constant environment, any population grows exponentially given time, structured as well as non-structured. (Real populations will not do so because the environment will soon change due to food depletion.) Let $N(t)$ denote the number of individuals at time t. The numbers follow $N(t) = N(0)\exp\{\dot{\mu}t\}$, where the population growth rate $\dot{\mu}$ is found from the characteristic equation

$$1 = \int_0^\infty \text{Prob}\{\underline{a}_\dagger > a\}\dot{R}(a)\exp\{-\dot{\mu}a\}\, da \tag{5.22}$$

Suppose that death plays a minor role, so $\text{Prob}\{\underline{a}_\dagger > a\} \simeq 1$, and that reproduction is constant after age a_p, so $\dot{R}(a) = (a > a_p)\dot{R}$, where, with some abuse of notation, $\dot{R}$ in the right argument is taken to be a constant. Substitution into (5.22) gives

$$\exp\{-\dot{\mu}a_p\} = \dot{\mu}/\dot{R} \tag{5.23}$$

This equation ties the population growth rate $\dot{\mu}$ to the length of the juvenile period and the reproduction rate. It has to be evaluated numerically. For unstructured populations, where $a_p = 0$ must hold, the population growth rate equals the reproduction rate, $\dot{\mu} = \dot{R}$. For increasing a_p, $\dot{\mu}$ falls sharply; see figure 5.6. This means that neonates giving birth to new neonates contribute significantly in unstructured populations.

5.2.3 Discrete individuals

The formulation of the reproduction rate such as $\dot{R}(a) = (a > a_p)\dot{R}$ treats the number of individuals as a continuous variable. Obviously, this is unrealistic, because individuals are discrete units. It would be more appropriate to gradually fill a buffer with energy allocated to reproduction and convert it to a new individual as soon as enough energy has

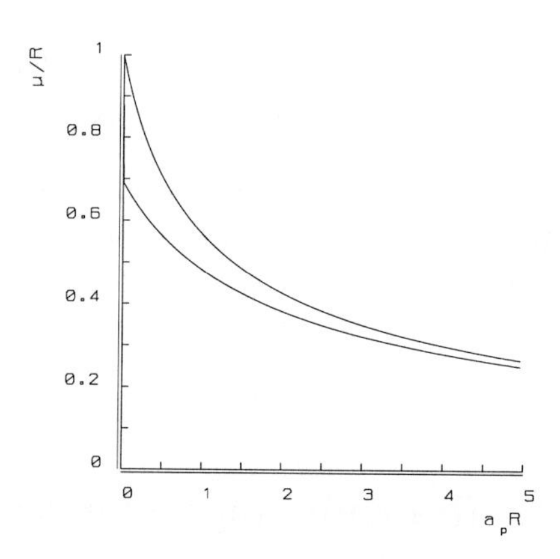

Figure 5.6: For a constant reproduction rate $\dot{R}$ in the adult state, the population growth rate depends sensitively on the length of the juvenile period, as shown in the upper curve. The unit of time is $\dot{R}^{-1}$ and mortality is assumed to be negligible. The lower curve also accounts for the fact that individuals are discrete units of biomass. The required accumulation of reproductive effort to produce such discrete units reduces the population growth rate even further, especially for short juvenile periods. Note that the effect of food availability is not shown in this figure, because it only affects the chosen unit of time.

been accumulated. In that case, the reproduction rate becomes $\dot{R}(a) = (a = a_p + i/\dot{R})/da$, for $i = 1, 2, \cdots$. It is zero almost everywhere, but at regular time intervals it switches to ∞ over an infinitesimally small time interval da, such that the mean reproduction rate as an adult over a long period is $\dot{R}$ as before. Giving death a minor role, the characteristic equation becomes

$$1 = \sum_{i=1}^{\infty} \exp\{-\dot{\mu}(a_p + i/\dot{R})\} = \exp\{-\dot{\mu}/\dot{R} - \dot{\mu}a_p\} \left(1 - \exp\{-\dot{\mu}/\dot{R}\}\right)^{-1} \tag{5.24}$$

In analogy with (5.23) this can be rewritten as

$$\exp\{-\dot{\mu}a_p\} = \exp\{\dot{\mu}/\dot{R}\} - 1 \tag{5.25}$$

to reveal the effect of individuals being discrete units rather than continuous flows of biomass; see figure 5.6. The effect is most extreme for $a_p = 0$, where $\dot{\mu} = \dot{R}\ln 2$, which is a fraction of some 0.7 of the continuous biomass case. If young are not produced one by one, but in a litter, which requires longer accumulation times of energy, the discreteness effect is much larger. For a litter size n and a reproduction rate of $\dot{R}(a) = (a = a_p + in/\dot{R})n/da$, the population growth rate is $n^{-1}\ln\{1+n\}$ times the one for continuous biomass with the same mean reproduction rate and negligibly short juvenile period.

The effect of the discrete character of individuals is felt most strongly at low reproduction rates. Since populations tend to grow rapidly into a situation where reproduction drops sharply due to food limitation, this problem is rather fundamental. Reproduction, i.e. the conversion of the energy buffer to offspring, is usually triggered by independent factors (a two-day moulting cycle in daphnids, seasonal cycles in many other animals). If reproduction is low, details of buffer handling become dominant for population dynamics. Energy that is not sufficient for conversion into the last young dominates population dynamics. Whether it gets lost or remains available for the next litter makes quite a difference and, unfortunately, we know little about what exactly does happen.

5.2.4 Differing daughters

When a naidid divides into two new individuals, an anterior piece and a posterior one, the difference between both individuals is visible until the time they divide again. If the anterior pieces reach a certain length at division L_d in a time t_{da} and the posterior ones in a time t_{dp} with $t_{da} < t_{dp}$, then there will be less anterior pieces than posterior ones in a growing population. The section on segmented individuals, {155}, explains why the latter is to be expected and relates the interdivision time to the process of food uptake and digestion. If N_a denotes the number of anterior pieces, N_p the number of posterior ones and N_+ the total number of individuals after some time, the relationship

$$1 = (N_a/N_+)^{t_{da}/t_{dp}} + N_a/N_+ \tag{5.26}$$

exists between the fraction of anterior pieces and the relative interdivision times. It can be used to estimate t_{da}/t_{dp} for instance, from an observation on N_a/N_+. This is quite a help, because a direct observation of a complete cycle is difficult in practice.

This relationship can most easily be derived by focusing on the special case where $t_{dp} = kt_{da}$ for $k = 1, 2, \cdots$. We choose the unit of time such that $t_{da} = 1$. The number of anterior pieces after the i-th division , $N_{a,i}$, then follows the generalized Fibonacci series

$$N_{a,i} = N_{a,i-1} + N_{a,i-k} \quad \text{for } i \geq k \tag{5.27}$$

with $N_{a,0} = N_{a,1} = \cdot = N_{a,k} = 1$. The number of posterior pieces equals $N_{p,i} = N_{a,i+k-1}$ and the total number of individuals $N_{+,i} = N_{a,i+k}$. Fibonacci (Leonardo de Pisa, 1175–1250) noted that this series is asymptotically a geometrical one, $N_{a,i} = \alpha N_{a,i-1}$. The multiplier α is found from the equation $\alpha^k = \alpha^{k-1} + 1$. Fibonacci studied the series for $k = 2$ and it is believed that Kepler was, in 1611, the first to recognize the connection between the series and the golden mean $\alpha = (1 + \sqrt{5})/2$ for $k = 2$; he used it to study 'phyllotaxis' [195].

The multiplier α relates to the population growth rate as $\dot{\mu} = t_{da}^{-1} \ln \alpha$, so that the population growth rate can be found from

$$1 = \exp\{-\dot{\mu}t_{da}\} + \exp\{-\dot{\mu}t_{dp}\} \tag{5.28}$$

This result can be obtained for arbitrary choices of t_{da} and t_{dp}, via a formulation in terms of partial differential equations, worked out on {180}.

The division intervals for anterior and posterior naidids obviously depend on their initial lengths. Figure 5.7 shows that for the model described under 'segmented individuals', {155}, the population growth rate is maximized if the new anterior part is somewhat larger than the posterior one, which is actually what has been observed in most individuals [583]. For extremely short anterior parts, the population growth rate rises again, but this is probably unrealistic because the complete physiological machinery has to be contained and because the morphological changes to produce a new head must then occur in a very short period.

These generalizations to differing daughters obviously also apply to unicellulars such as budding yeasts, for instance, where the differences are large enough to call the daughters still mother and baby. The buds leave scars on the mother cell, which contribute to the difference.

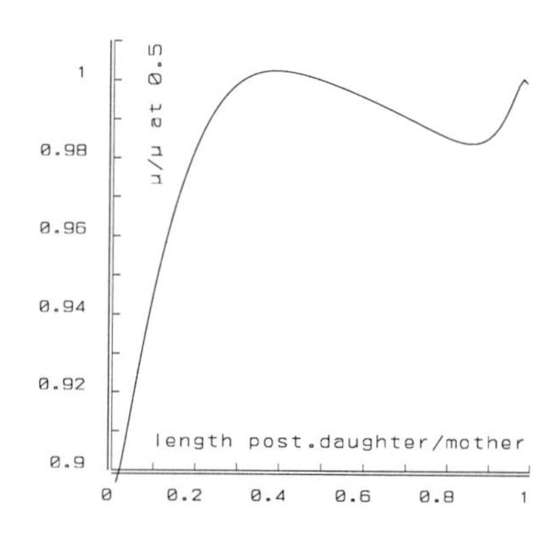

Figure 5.7: The population growth rate of a segmented DEB filament as a function of the ratio of the length of the posterior daughter and the mother. The parameters are realistic for *Nais elinguis* at 20 °C: $\dot{m} = 0.05$ h^{-1}, $\dot{\nu} = 0.126$ h^{-1}, $g = 1.76$ and $L_d = 7.19$ mm. The observed length ratio is about 0.4. For poorer food qualities, so smaller values of $\dot{\nu}$, the effect of the length ratio on the population growth rate becomes larger, but the optimal ratio does not change much.

5.2.5 Maintenance

It is difficult to incorporate the notion of maintenance into a non-structured population model in a satisfactory way because maintenance is a property of individuals that depends on their size, while non-structured population models do not specify sizes. It is built into the popular family of logistic population growth models in an implicit way. The logistic model originates from Pearl [537] in 1927; Emlen [202] has evaluated evolutionary aspects of population dynamics based on the logistic family. The core is the differential equation

$$\frac{d}{dt}N \propto N(1 - N/N_\infty) \tag{5.29}$$

where the ultimate number of individuals in the population, N_∞, is treated as a parameter. It is known as the 'carrying capacity', where food supply is just sufficient to maintain the population. Reproduction is then just enough to compensate for the (usually minor) losses and almost all food is used for maintenance by the individuals. As food is not modelled explicitly, no use of balance equations for energy can be made and the relationship with actual populations must for this reason remain somewhat vague. To follow food consumption more closely, it is necessary to introduce food density explicitly. Logistic population growth should be treated as an empirical description. In the discussion on filaments, {167}, it has been shown that batch cultures can grow in an almost logistic manner for different reasons (energy reserves, saturation coefficients); this illustrates that different individual traits can work out similarly at the population level.

5.3 DEB-structured populations

5.3.1 Population growth rates

The relationship between population growth rate and division interval can be obtained as follows. When the substrate density is constant for a sufficiently long period and death has little effect, the population of dividing individuals will grow exponentially at rate $\dot{\mu} = a_d^{-1} \ln 2$, where the division interval a_d tends to some fixed value at constant food densities. This relationship, which is well known in microbiology, is obvious if one realizes that starting from a single, just divided, individual in an environment that has not changed over a long period, the development of the population in terms of cell numbers is given by $N(t) = 2^{t/a_d} = \exp\{\dot{\mu}t\}$ if the observations are done at $t = 0, a_d, 2a_d, \cdots$. From a strict point of view, the development of cell numbers in continuous time is a step function. If we start from a large population rather than a single individual, the cell numbers will be close to $N(t) = N(0)2^{t/a_d} = N(0)\exp\{\dot{\mu}t\}$, but not exactly so, due to the deterministic nature of the growth and division process. This preserves information about the age distribution of the founder population, as explained on {170}. In practice more than enough scatter is found in almost all aspects of the growth and division process. We can, therefore, assume for practical purposes that information about the founder population rapidly fades, even without formulating these stochastic processes explicitly.

The relationship between population growth rate and the division interval can also be obtained from a formulation that allows for the production of neonates by letting the mother cell disappear at the moment of division, where two baby cells appear. Thus we write $\text{Prob}\{\underline{a}_\dagger > a\} = (a \leq a_d)$ and $\dot{R}(a)\,da = 2(a = a_d)$. Substitution into the characteristic equation (5.22) gives $1 = 2\exp\{-\dot{\mu}a_d\}$.

The division interval a_d is given in (3.15), (3.39) or (3.35). Substitution gives the expressions for the population growth rates at constant substrate densities and for their relative values with respect to the maximum population growth rate that are collected in figure 5.8. The scaled length at division, l_d, is a function of f, due to the fixed period required to duplicate DNA. It has to be solved numerically from (3.52), but for most practical purposes, it can probably be treated as a constant. For small aspects ratios, δ, the expressions for rods reduce to that for filaments, while for an aspect ratios of $\delta = 0.6$ rods resemble isomorphs. The table in figure 5.8 therefore illustrates how the population growth rate of dividing DEB isomorphs reduces stepwise to well known classic models. It also illustrates why many microbiologists do not like models that explicitly deal with substrate density; the saturation coefficient for uptake is usually very small for most combinations of micro-organisms and substrate types, and the saturation coefficient for population growth is even smaller, so that problems arise in measuring such low densities. Natural populations of micro-organisms tend either to grow at the maximum rate, or not to grow at all. This on/off behaviour is a major obstacle in the analysis of population dynamics.

The population growth rate is plotted against the substrate concentration for the rods *Escherichia coli* and *Klebsiella aerogenes* in figure 3.6 and 5.9, and for the isomorph *Colpidium* also in figure 5.9. The curves closely resemble simple hyperbolic functions, which indicates that these curves contain little information about some of the parameter values of the individual-based DEB model, particularly the energy investment ratio g. Since the goodness of fit is quite acceptable, the modest conclusion can only be that these population responses give little reason to change assumptions about the energy behaviour of individuals. Figure 5.9 also illustrates that the scatter in population responses tends to increase dramatically with body size. There has been substantial debate about the goodness of fit of hyperbolic functions to this type of data.

It should be noted that if the saturation coefficient is small, small systematic additive errors in the determination of low substrate levels have a substantial effect on the goodness of fit, as illustrated in these figures. Measurement errors as small as 12 and 80 ng glucose

Figure 5.8: The population growth rate $\dot\mu$ for dividing organisms as it simplifies when expressed as a fraction of its maximum $\dot\mu_m$ and small maintenance costs $[\dot M]$ and/or storage capacity $[E_m]$. The last three rows in the 'filaments'-column correspond to the models by Marr–Pirt, Droop and Monod. These models are graphically compared with the DEB model for filaments in the figure below. The symbols l_1 and V_1 stand for l_d and V_d for $f = 1$.

	isomorphs	rods	filaments
$\dot\mu$	$\frac{g\dot m}{f+g}\frac{\frac{1}{3}\ln 2}{\ln\frac{f-l_d 2^{-1/3}}{f-l_d}}$	$\frac{(1-\delta/3)f/l_d-1}{(f+g)/g\dot m}\frac{\ln 2}{\ln\frac{2(1-l_d/f)}{1-l_d/f+\delta/3}}$	$\frac{f/l_d-1}{(f+g)/g\dot m}$
$\frac{\dot\mu}{\dot\mu_m}$	$\frac{1+g}{f+g}\frac{\ln\frac{1-l_1 2^{-1/3}}{1-l_1}}{\ln\frac{f-l_d 2^{-1/3}}{f-l_d}}$	$\frac{1+g}{f+g}\frac{(1-\delta/3)f/l_d-1}{(1-\delta/3)/l_1-1}\frac{\ln\frac{2(1-l_1)}{1-l_1+\delta/3}}{\ln\frac{2(1-l_d/f)}{1-l_d/f+\delta/3}}$	$\frac{1+g}{f+g}\frac{f/l_d-1}{1/l_1-1}$
$\frac{\dot\mu}{\dot\mu_m}\Big\vert\, [E_m]\to 0$	$\frac{\ln\frac{1-l_1 2^{-1/3}}{1-l_1}}{\ln\frac{f-l_d 2^{-1/3}}{f-l_d}}$	$\frac{(1-\delta/3)f/l_d-1}{(1-\delta/3)/l_1-1}\frac{\ln\frac{2(1-l_1)}{1-l_1+\delta/3}}{\ln\frac{2(1-l_d/f)}{1-l_d/f+\delta/3}}$	$\frac{f/l_d-1}{1/l_1-1}$
$\frac{\dot\mu}{\dot\mu_m}\Big\vert\, [\dot M]\to 0$	$f\frac{1+g}{f+g}\left(\frac{V_1}{V_d}\right)^{1/3}$	$f\frac{1+g}{f+g}\left(\frac{V_1}{V_d}\right)^{1/3}$	$f\frac{1+g}{f+g}$
$\frac{\dot\mu}{\dot\mu_m}\Big\vert\, [E_m],[\dot M]\to 0$	$f\left(\frac{V_1}{V_d}\right)^{1/3}$	$f\left(\frac{V_1}{V_d}\right)^{1/3}$	f

In all models, Monod, Marr–Pirt, Droop and DEB, uptake as a fraction of its maximum depends hyperbolically on substrate density X as a fraction of the saturation coefficient K, as indicated by the thick curve. In the Monod model $\dot\mu \propto \frac{x}{x+1}$, this curve coincides with the population growth rate $\dot\mu$ as a fraction of its maximum $\dot\mu_m$. The Marr–Pirt model $\dot\mu \propto \frac{x-l_d/(1-l_d)}{x+1}$, which includes maintenance, has a translation to the right. The Droop model $\dot\mu \propto \frac{x}{x+g/(1+g)}$, which includes storage, has a smaller saturation coefficient, whereas the DEB model for filaments $\dot\mu \propto \frac{x-l_d/(1-l_d)}{x+g/(1+g)}$ has both. All four curves are hyperboles with horizontal asymptote 1.

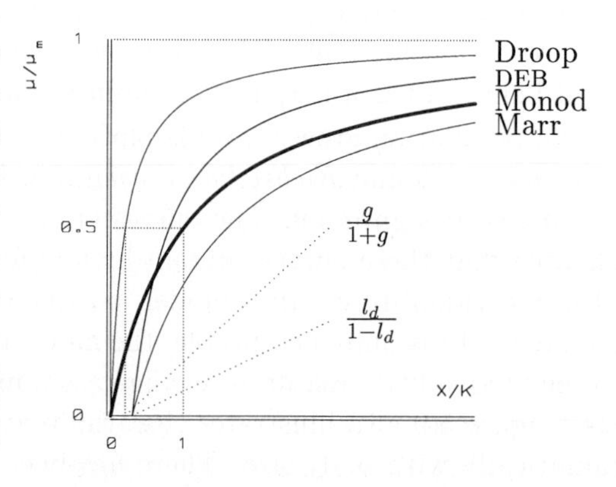

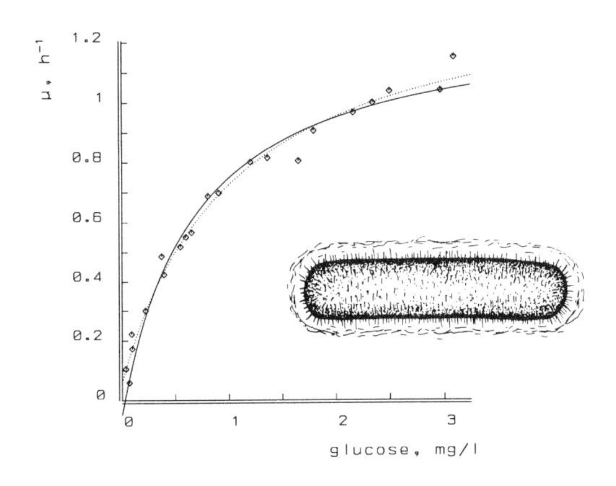

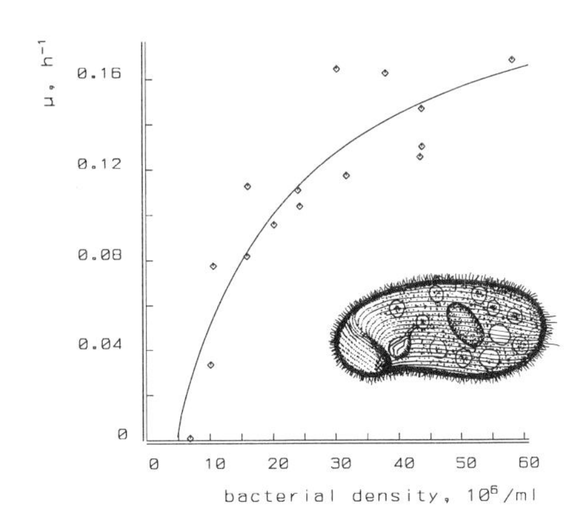

Figure 5.9: The population growth rate as a function of the concentration of substrate or food. The left figure concerns the rod *Klebsiella aerogenes* feeding on glucose at 35 °C. Data from Rutgers *et al.* [623]. The right figure concerns the isomorphic ciliate *Colpidium campylum* feeding on suspensions of *Enterobacter aerogenes* at 20 °C. Data from Taylor [706]. The dotted curve in the left figure accounts for a small additive error in the measurement of the concentration of glucose.

per litre convert the drawn curves into the better fitting dotted ones. In view of the technical difficulties in obtaining such measurements under these experimental conditions and the possible existence of absorbed glucose, I am inclined to consider this goodness of fit problem of academic interest only. Moreover, it may be recalled that diffusion limitations cause deviations from the hyperbolic functional response; see {141}.

If propagation is via eggs, the population growth rate in constant environments has to be evaluated numerically from the characteristic equation (5.22).

5.3.2 Stable age and size distributions

Volume distribution has an intimate relationship with the growth of dividing individuals, as has been widely recognized [131,170,293,456,745]. It can most easily be expressed in terms of its survivor function. If death plays a minor role, (5.20) gives the stable age distribution for $\text{Prob}\{\underline{a}_\dagger > a\} = (a < a_d)$ with $a_d = \dot{\mu}^{-1} \ln 2$. The survivor function of the stable age distribution is thus: $\text{Prob}\{\underline{a} > a\} \equiv \int_a^{a_d} \phi_{\underline{a}}(a_1)\, da_1 = (a < \dot{\mu}^{-1} \ln 2)(2 \exp\{-\dot{\mu} a\} - 1)$. The stable age distribution only exists at constant food densities, where volume increases if age increases. It was first derived by Euler in the 18-th century [398]. The remarks on the need for scatter for stability of age distributions also apply to size distributions. See Diekmann *et al.* [168,169] or a more technical discussion.

If growth is deterministic and division occurs at a fixed size and the baby cells are of equal size, no stable age distribution exists. If there is some scatter in size at division, a stable age distribution exists, unless growth is exponential [53], because the information of the age distribution of the founder populations never gets lost. If sisters do not have exactly the same size, a stable age distribution exists, even if growth is exponential. The age distribution has a weaker status, that of an eigenfunction: if the founder population has this particular age distribution, the age distribution will not change, while all other age distributions for the founder population will change cyclically with period a_d. In practice, however, scatter in growth rate and size of baby cells will be more than sufficient for a rapid convergence to the stable age distribution.

The survivor function of the stable volume distribution is

$$\text{Prob}\{\underline{V} > V\} = \text{Prob}\{\underline{a} > t(V)\} = 2\exp\{-\dot{\mu}t(V)\} - 1$$

where $t(V)$ is the age at which volume V is reached. The probability density is thus

$$\phi_{\underline{V}}(V)\,dV = (V \geq V_d/2)(V \leq V_d)2\dot{\mu}\exp\{-\dot{\mu}t(V)\}\,dt \tag{5.30}$$

For isomorphs, $t(V)$ is given in (3.15). Since scaled length, l, has a monotonous relationship with volume; we have $\text{Prob}\{\underline{l} > l\} = \text{Prob}\{\underline{V} > V\}$. The survivor function of the stable length distribution for isomorphs that divide at scaled length l_d becomes:

$$\text{Prob}\{\underline{l} > l\} = 2^{1+\ln\frac{f-l}{f-l_d 2^{-1/3}}/\ln\frac{f-l_d 2^{-1/3}}{f-l_d}} - 1 \tag{5.31}$$

The same can be done for rods, which leads to

$$\text{Prob}\{\underline{l} > l\} = 2^{\ln\frac{1-l_d/f}{(1-l_d/f-\delta/3)(l/l_d)^3+\delta/3}/\ln\frac{2(1-l_d/f)}{1-l_d/f+\delta/3}} - 1 \tag{5.32}$$

and for filaments

$$\text{Prob}\{\underline{l} > l\} = (l_1/l)^3 - 1 \tag{5.33}$$

These relationships can be important for testing assumptions about the growth process by means of the stable length distribution. Actual stable length distributions reveal that the scaled length at division, l_d, is not identical for all individuals, but has some scatter, which is close to a normal distribution [399]. Unlike what has been derived previously for *Nias*, it is now assumed that the size-age curve does not depend on the size of the baby cell. As soon as a small baby cell has grown to the size of a larger baby cell, the rest of their growth curves are indistinguishable. Let $\phi_{\underline{V}_b}$ denote the probability density of the number of baby cells of volume V, i.e. cells of an age less than an arbitrarily small period Δt, and $\phi_{\underline{V}_d}$ the probability density of the number of mother cells of volume V, i.e. cells which will divide within the period Δt. A practical way to determine $\phi_{\underline{V}_b}(V)\,dV$ and $\phi_{\underline{V}_d}(V)\,dV$ empirically is to make photographs at t and $t + \Delta t$ of the same group of cells and select cells that are divided at $t + \Delta t$, but not at t. The photograph at t can be used to obtain $\phi_{\underline{V}_d}(V)\,dV$ and that at $t + \Delta t$ to obtain $\phi_{\underline{V}_b}(V)\,dV$. When N denotes the total number of cells in the population, the number of cells with a volume in the interval $(V, V + dV)$ is $N\phi_{\underline{V}}(V)\,dV$. Painter and Marr [528] argued that the change in this number is given by

$$\frac{d}{dt}N\phi_{\underline{V}} = 2\frac{d}{dt}N\phi_{\underline{V}_b} - \frac{d}{dt}N\phi_{\underline{V}_d} - N\frac{\partial}{\partial V}\left(\phi_{\underline{V}}\frac{dV}{dt}\right) \tag{5.34}$$

The first term stands for the increase due to birth, the second one for loss due to division and the third term for loss due to growth. Since the stable volume distributions do not depend on time and $\frac{d}{dt}N = \dot{\mu}N$, some rearrangement of terms gives

$$\frac{\partial}{\partial V}\left(\phi_{\underline{V}}\frac{dV}{dt}\right) = \dot{\mu}\left(2\phi_{\underline{V}_b} - \phi_{\underline{V}_d} - \phi_{\underline{V}}\right)$$

This is a linear inhomogeous differential equation in $\phi_{\underline{V}}(V)$, with solution

$$\phi_{\underline{V}}(V) = \frac{dt}{dV}\dot{\mu}\exp\{-\dot{\mu}t(V)\}\int_{V_{\min}}^{V}\exp\{\dot{\mu}t(V_1)\}(2\phi_{\underline{V}_b}(V_1) - \phi_{\underline{V}_d}(V_1))\,dV_1 \qquad (5.35)$$

where $V_{\min}$ is the smallest possible cell volume and, since $\phi_{\underline{V}}(V_{\max}) = 0$, $\dot{\mu}$ satisfies [745]

$$\int_{V_{\min}}^{V_{\max}}\exp\{\dot{\mu}t(V_1)\}(2\phi_{\underline{V}_b}(V_1) - \phi_{\underline{V}_d}(V_1))\,dV_1 = 0 \qquad (5.36)$$

The connection with the previous deterministic rules for division can be made as follows. When mother cells divide into two equally sized baby cells, we have $\phi_{\underline{V}_b}(V) = 2\phi_{\underline{V}_d}(2V)$. So, $\phi_{\underline{V}_b}(V)\,dV = (V = V_d/2)$ and $\phi_{\underline{V}_d}(V)\,dV = (V = V_d)$ when division always occurs at V_d. Substitution into (5.35) gives (5.30) and into (5.36) gives $t_d^{-1}\ln 2$, as before. When division always occurs at V_d, so $\phi_{\underline{V}_d}(V)\,dV = (V = V_d)$, but the sizes of the baby cells are V_a and V_p, this gives $\phi_{\underline{V}_b}(V)\,dV = (V = V_a)/2 + (V = V_p)/2$ with $V_a + V_p = V_d$ and $V_a < V_p$ cf. {173}. Substitution into (5.35) gives

$$\phi_{\underline{V}}(V)\,dV = (V \geq V_a)(V \leq V_d)\dot{\mu}\exp\{\dot{\mu}(t(V_a) + t(V_p)(V \geq V_p) - t(V))\}\,dt$$

and substitution into (5.36) gives (5.28).

If $\phi_{\underline{V}_d}$ is log-normal and $\underline{V}_b = \underline{V}_d/2$, we do not have any problems with tails when $V_\infty < 0$, i.e. when all individuals will eventually divide. Although $\dot{\mu}$ has to be obtained numerically from (5.36), a very good approximation is still $\dot{\mu}t(V_d) = \ln 2$. Substitution into (5.35), together with (3.39) gives for rods:

$$\phi_{\underline{V}}(V) = \int_0^V H(V_1)\left(2\exp\left\{-\frac{(\ln 2V_1/V_d)^2}{2\sigma^2}\right\} - \exp\left\{-\frac{(\ln V_1/V_d)^2}{2\sigma^2}\right\}\right)d\ln V_1 \qquad (5.37)$$

$$H(V_1) \equiv \frac{\ln 2}{\sqrt{2\pi\sigma^2}}\frac{\exp\left\{(\ln 2)\left(\ln\frac{V_\infty - V}{V_\infty - V_1}\right)\left(\ln\frac{V_\infty - V_d/2}{V_\infty - V_d}\right)^{-1}\right\}}{(V_\infty - V)\ln\frac{V_\infty - V_d/2}{V_\infty - V_d}} \qquad (5.38)$$

Although this expression looks massive, it has only three parameters: V_d, V_∞ and σ^2. Additional knowledge of δ can be used to determine $\frac{g\dot{m}}{\dot{v}f}$ from V_∞ and knowledge of $\dot{\mu}$ can be used to determine $\frac{f+g}{f\dot{v}}$.

Figure 5.10 gives the stable length distribution for *Escherichia coli*, together with the model fit with a log-normal distribution for the length at division, as in (5.37). Since the curves approach the x-axis very closely for large cell lengths, the approximation $\dot{\mu} = t_d^{-1}\ln 2$ has been appropriate. Although the goodness of fit is quite acceptable and only three parameters occur, the one relating to the growth process, V_∞, is not well fixed by the data. Again, the conclusion must be that this population response is consistent with what can be deduced from the individual level, but population behaviour gives poor access to individual behaviour.

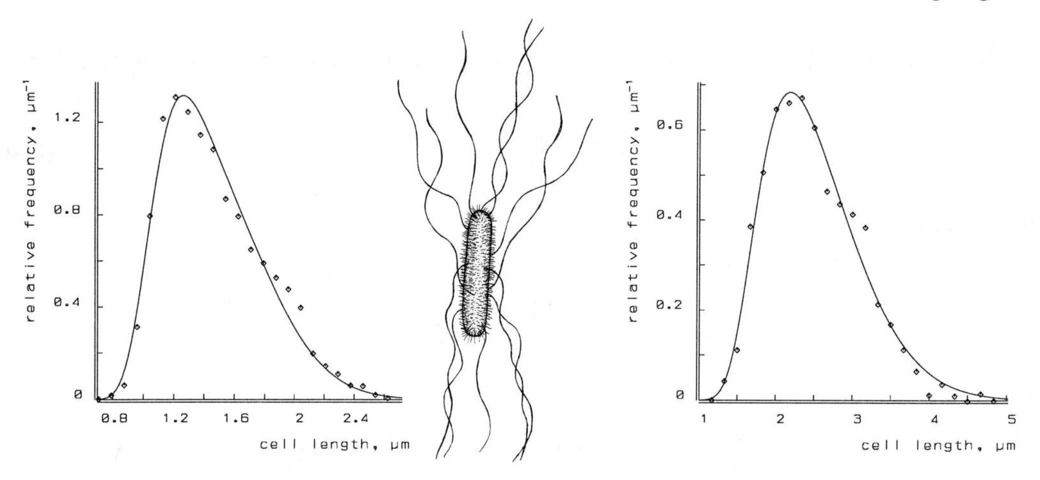

Figure 5.10: The probability density of the length of *E. coli* B/r A (left) and K (right) at a population growth rate of 0.38 and 0.42 h^{-1} respectively at 37 °C. Data from Koppes *et al.* [420]. For an aspect ratio of $\delta = 0.3$, the three parameters are $V_d = 0.506\ \mu m^3$, $V_\infty = -0.001\ \mu m^3$ and $\sigma^2 = 0.026$ and $V_d = 2.324\ \mu m^3$, $V_\infty = -1\ \mu m^3$ and $\sigma^2 = 0.044$. Because of the relatively large variance of the volume at division, these frequency distributions give poor access to the single parameter that relates to the growth process V_∞.

Differing daughters

The derivation of the stable size distribution is much simpler if an individual of volume V_d divides into V_a and V_p, with $V_p < V_a$ and $V_a + V_p = V_d$, as has been discussed for *Nais* on {173}. Unlike the details of the *Nais*-case, the anterior and posterior parts are assumed to follow the same growth curve in this derivation, which serves as an exercise for techniques that will be used later, {206}. See Heijmans [302] for a more extensive study of this subject.

Suppose that we have a chemostat with throughput rate $\dot{p}$ and let $n(t, V)\, dV$ denote the number of individuals at time t with a volume somewhere in the interval $(V, V + dV)$. This number decreases by growth and by wash out, so

$$\frac{\partial}{\partial t} n(t, V) = -\frac{\partial}{\partial V}\left(n(t, V)\frac{dV}{dt}\right) - \dot{p}n(t, V) \tag{5.39}$$

The conservation law for mass leads to the boundary conditions

$$\begin{aligned} n(t, V_p)\left.\frac{dV}{dt}\right|_{V_p} &= n(t, V_d)\left.\frac{dV}{dt}\right|_{V_d} \quad \text{and} \\ n(t, V_a^+)\left.\frac{dV}{dt}\right|_{V_a^+} &= n(t, V_a^-)\left.\frac{dV}{dt}\right|_{V_a^-} + n(t, V_d)\left.\frac{dV}{dt}\right|_{V_d} \end{aligned}$$

where V_a^- denotes a number infinitesimally smaller than V_a and V_a^+ a number infinitesimally larger than V_a. The stable volume distribution is found for $t \to \infty$ from $\phi_{\underline{V}}(V) \equiv n(\infty, V)/\int_V n(\infty, V)\, dV$, where $\dot{\mu} = \dot{p}$. The population growth rate is given by $\dot{\mu} =$

$\phi_{\underline{V}} \left.\frac{dV}{dt}\right|_{V_d}$. When (5.39) is devided by $\int_V n(\infty, V)\, dV$ and for $t \to \infty$, so that $\frac{\partial}{\partial t} n(t, V) = 0$, we get

$$\frac{\partial}{\partial V}\left(\phi_{\underline{V}} \frac{dV}{dt}\right) = -\dot{\mu}\phi_{\underline{V}} \qquad (5.40)$$

with boundary conditions

$$\phi_{\underline{V}} \left.\frac{dV}{dt}\right|_{V_p} = \dot{\mu} \quad \text{and} \quad \phi_{\underline{V}} \left.\frac{dV}{dt}\right|_{V_a^+} = \phi_{\underline{V}} \left.\frac{dV}{dt}\right|_{V_a^-} + \dot{\mu}$$

The solution of the stable volume distribution is

$$\begin{aligned} \phi_{\underline{V}}(V) &= \dot{\mu}\frac{dt}{dV} \exp\{-\dot{\mu}t(V)\} \quad \text{for } V_p \leq V < V_a \\ &= \dot{\mu}(1 + \exp\{\dot{\mu}t(V_a)\})\frac{dt}{dV} \exp\{-\dot{\mu}t(V)\} \quad \text{for } V_a \leq V < V_d \end{aligned}$$

The requirement that $\int_V \phi_{\underline{V}}(V)\, dV = 1$ leads to the value for $\dot{\mu}$ as given by (5.28), after substitution of $t(V_p) = 0$, $t(V_d) = t_{dp}$ and $t(V_d) - t(V_a) = t_{da}$.

5.3.3 Mean size of individuals

For the purpose of testing the DEB theory and estimating the parameters, it is easier in practice to measure mean length, rather than length at division. Mean volume and mean surface area are also necessary for theoretical purposes, for instance to relate the number of individuals to total biovolume, or to study the impact of the predator on prey. In the steady state of exponential population growth, these statistics are easy to obtain via the stable age distribution, particularly if death plays a minor role.

For dividing individuals with an age between 0 and a_d, the stable age distribution is given by $\phi_{\underline{a}}(a)\, da = 2\dot{\mu} \exp\{-\dot{\mu}a\}\, da = \frac{2 \ln 2}{a_d} 2^{-a/a_d}\, da$. For reproducing immortal individuals, the stable age distribution is $\phi_{\underline{a}}(a)\, da = \dot{\mu} \exp\{-\dot{\mu}a\}\, da$. The expected value of scaled length to the power i amounts to $\mathcal{E}\underline{l}^i = \int \phi_{\underline{a}}(a) l(a)^i\, da$. Substitution of (4.6) or (3.38) gives the results collected in table 5.1.

Note that the mean length increases less steeply with increasing substrate density or $\dot{\mu}$ than length at division, because the mean age reduces. Figure 5.11 shows that the mean volume of rods depends on population growth rate in the predicted way.

5.4 Yield at the population level

The conversion efficiency of food into biomass is almost always taken as a constant in non-structured population models, as well as in structured ones that do not take maintenance into account. I expect that this will prove to be the most crucial difference between these models and structured population models that do take account of maintenance. The Marr–Pirt model and its descendants are the only non-structured population models I know of that account for maintenance. I have shown that it is a special case of the DEB model

Table 5.1: Size statistics for individuals in populations growing at rate $\dot{\mu}$ with a scaled functional response f.

rods

$$\mathcal{E}\underline{V} = V_\infty \left(1 - \left(1 + \ln \frac{V_\infty - V_d/2}{V_\infty - V_d} / \ln 2\right)^{-1}\right)$$

dividing isomorphs, with $\alpha \equiv \frac{f - l_d/\sqrt[3]{2}}{f - l_d}$

$$\begin{aligned}
\mathcal{E}\underline{l} &= f - \frac{f - (2^{2/3} - 1)l_d}{1 + \ln\alpha/\ln 2} \\
\mathcal{E}\underline{l}^2 &= f^2 - 4f(f - l_b)\frac{1 - \frac{1}{2\alpha}}{1 + \ln\alpha/\ln 2} + 2(f - l_b)^2\frac{1 - \frac{1}{2\alpha^2}}{1 + 2\ln\alpha/\ln 2} \\
\mathcal{E}\underline{l}^3 &= f^3 - 6f^2(f - l_b)\frac{1 - \frac{1}{2\alpha}}{1 + \ln\alpha/\ln 2} + 6f(f - l_b)^2\frac{1 - \frac{1}{2\alpha^2}}{1 + 2\ln\alpha/\ln 2} - 2(f - l_b)^3\frac{1 - \frac{1}{2\alpha^3}}{1 + 3\ln\alpha/\ln 2}
\end{aligned}$$

reproducing isomorphs, with $\dot{\gamma} \equiv (3/\dot{m} + 3fV_m^{1/3}/\dot{v})^{-1}$

$$\begin{aligned}
\mathcal{E}\underline{l}^2 &= f^2 - 2f\frac{f - l_b}{1 + \dot{\gamma}/\dot{\mu}} + \frac{(f - l_b)^2}{1 + 2\dot{\gamma}/\dot{\mu}} \\
\mathcal{E}\underline{l}^3 &= f^3 - 3f^2\frac{f - l_b}{1 + \dot{\gamma}/\dot{\mu}} + 3f\frac{(f - l_b)^2}{1 + 2\dot{\gamma}/\dot{\mu}} - \frac{(f - l_b)^3}{1 + 3\dot{\gamma}/\dot{\mu}}
\end{aligned}$$

for filaments, where phenomena at individual level can be directly related to those at the population level. Generally, however, the conversion efficiencies of food into biomass are different for these levels. This fact is not at all specific for the DEB model; all individual-based models are likely to imply this difference.

The conversion efficiency has been evaluated for the individual level in the previous chapter, {123}. This section treats conversion efficiency at the population level for parthenogenetically reproducing female ectotherms in spatially homogeneous equilibrium situations, cf. [411]. The discussion can, however, be easily extended to cover a fixed sex ratio. This will give more insight when the conversion of food into biomass is evaluated in combination with food and biomass density in a chemostat environment. In this situation food density is only indirectly under experimental control via the concentration in the inflowing medium X_r, and via the throughput rate $\dot{p}$. Outflowing food is fed back into the bio-reactor. Let us take the substrate influx, $\dot{p}X_r$, to be constant, as it is in a situation of constant primary production in a given environment, or in a sewage treatment plant with a constant influx of sewage water (i.e. the substrate for the microbial community). We take individuals out of the bio-reactor at probability rate $\dot{p}$ per individual and study what happens if we increase $\dot{p}$, if we compare steady state situations only. For the purpose of

Figure 5.11: The mean volume of *E. coli* as a function of population growth rate at 37 °C. Data from Trueba [726]. For a chosen aspect ratio $\delta = 0.28$, a maintenance rate coefficient $\dot{m} = 0.05\ \mathrm{h}^{-1}$ and an investment ratio $g = 1$, the least squares estimates (with s.d.) of the volume at the start of the DNA replication is V_p=0.454(0.069) $\mu\mathrm{m}^3$, the time required for division is $t_D = 1.03(0.081)$ h and the energy conductance $\dot{v} = 31.3(32)\ \mu\mathrm{m\,h}^{-1}$.

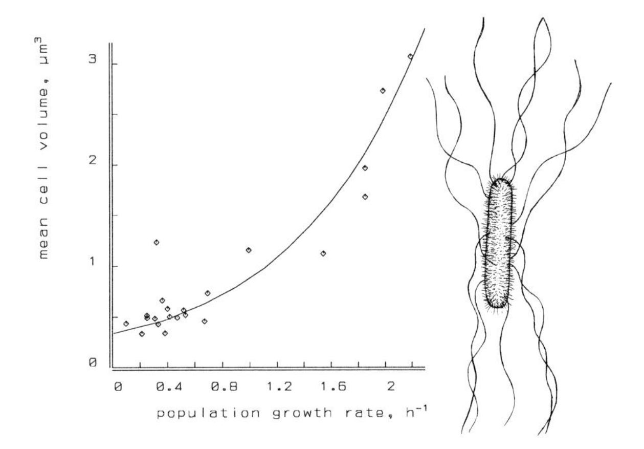

conceptual clarity, death is once again assumed to play a minor role, so that the conversion process of food into biomass is subjected to the harvesting process only.

Before analysing the conversion process via the yield factor in more detail, it is helpful to point out the fundamental difference between the population and the individual level. If we do not harvest at all and only supply food to the bio-reactor, the population will eventually grow to a size where food input just matches the maintenance needs of the individuals. In this situation no individual is able to grow or reproduce (otherwise we would not have a steady state). The conversion efficiency is then zero. Figure 5.12 illustrates this situation for experimental *Daphnia* populations. If we increase the harvesting rate, we increase conversion efficiency, at least initially. This illustrates that the conversion process is controlled by the way the population is sandwiched between food input and harvesting. Individual energetics only set the constraints.

In many field situations, the harvesting rate will not be set intentionally. The process of aging can be considered, for instance, as one of the ways of harvesting resulting from intrinsic causes, but this does not affect the principle. The present aim is to study the behaviour of the yield factor in steady state situations, so $\dot{\mu} = 0$, and compare the different life styles: filaments, rods and isomorphs, propagating via division and eggs. For this purpose, we strip the population of as many details as possible and think of it in terms of the diagram given in figure 5.13

The yield factor can be obtained in a two-step procedure. First, the equilibrium substrate density is obtained from the characteristic equation with $\dot{\mu} = 0$:

$$1 = \int_0^\infty \mathrm{Prob}\{\underline{a}_\dagger > a\}\dot{R}(a)\,da \tag{5.41}$$

which thus gives the food density at which each individual can, on average, just replace itself before being harvested. When the age at division is a_d and $\dot{R}(a)\,da = 2(a = a_d)$ we get $\mathrm{Prob}\{\underline{a}_\dagger > a\} = (a < a_d)\exp\{-\dot{p}a\}$ for dividing individuals. For these organisms, (5.41) reduces to $\exp\{-\dot{p}a_d\} = 0.5$ or $a_d = \dot{p}^{-1}\ln 2$ and the stable age distribution is $\phi_{\underline{a}}(a) = (a < \dot{p}^{-1}\ln 2)2\dot{p}\exp\{-\dot{p}a\}$. For reproducing individuals, the survivor function is $\mathrm{Prob}\{\underline{a}_\dagger > a\} = \exp\{-\dot{p}a\}$ and the stable age distribution is simply $\phi_{\underline{a}}(a) = \dot{p}\exp\{-\dot{p}a\}$. The reproduction rate is given in (3.47). Note that if we harvest randomly, which results in an exponential survivor function, we may as well assume that the individuals stay in the

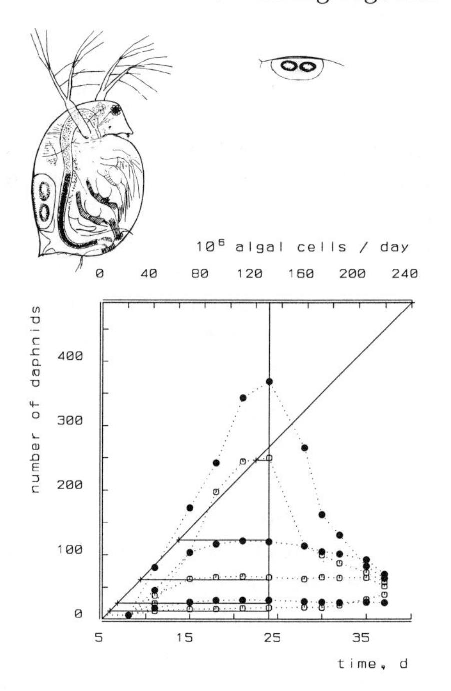

Figure 5.12: Populations of daphnids, *Daphnia magna* fed a constant supply of food, the green alga *Chlorella pyrenoidosa* at 20 °C, grow to a maximum number of individuals that is directly proportional to food input [405]. From this experiment, it can be concluded that each individual requires 6 algal cells per second just for maintenance. No deaths occurred before day 24. A reduction of food input to 30×10^6 cells day^{-1} after day 24, resulted in almost instant death if the populations were at carrying capacity. The 240×10^6 cells day^{-1} population was still growing when the food supply was suddenly reduced, so the energy reserves were high, and it produced many winter eggs. The daily food supply related to the cumulated number of winter eggs as

240	120	60	30	12	6	10^6 cells d−6
38	1	3	1	0	0	wintereggs

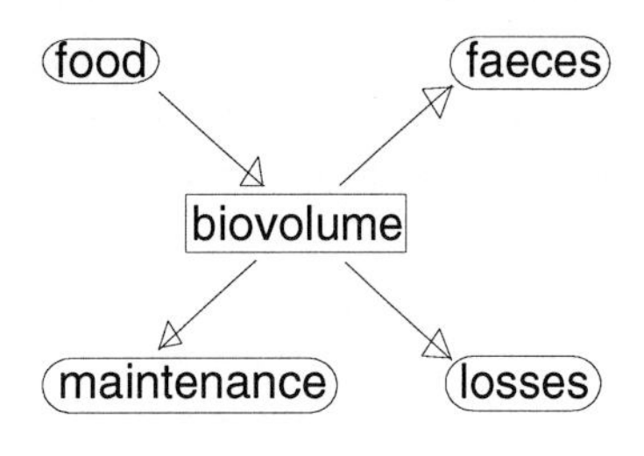

Figure 5.13: The population, quantified as the sum of the volumes of the individuals, converts food into faeces, while extracting energy. Part of this energy becomes lost in maintenance processes and part of it is deposited in losses, i.e. the cumulated harvest. The harvesting effort determines the allocation rules and sets the population size and so its impact on resources.

population and that the population is growing exponentially at rate $\dot{p}$. This symmetry of the role of $\dot{\mu}$ and $\dot{p}$ in the characteristic equation becomes lost if we harvest in a non-random way.

The second step is to find the equilibrium number of individuals from the balance equation for food:

$$\dot{p}X_r = \sum_{i=1}^{N} \dot{I}_i = \{\dot{I}_m\} f N \mathcal{E} \underline{V}^{2/3} \tag{5.42}$$

which simply states that all supplied food is eaten by the population, because we do not harvest food. The ingestion rates for filaments, rods and isomorphs are given in (3.3) and (3.2). Substitution results in the balance equations collected in table 5.2. The mean volume and surface area that occur in these equations are given in table 5.1.

The yield factor, i.e. the ratio of the harvested biomass to the supplied food, is

$$Y = \frac{\int_a \dot{h}(a)\phi_{\underline{a}}(a)V(a)\,da}{\{\dot{I}_m\} f \int_a \phi_{\underline{a}}(a)V^{2/3}(a)\,da} \tag{5.43}$$

Table 5.2: The food balance equation, the throughput rate $\dot{p}$, and the scaled yield, $y \equiv Y\{\dot{I}_m\}/\dot{v}$, as functions of the scaled functional response, $f \equiv X_0/(X_0+K)$.

filaments:

$$\begin{aligned}
\dot{p}X_r &= N[\dot{I}_m] f \mathcal{E} \underline{V} \\
\frac{\dot{p}}{\dot{\nu}} &= \frac{f - l_d}{f+g} \\
y &= \frac{1}{f}\frac{f-l_d}{f+g}
\end{aligned}$$

rods:

$$\begin{aligned}
\dot{p}X_r &= N[\dot{I}_m] f \left(\frac{\delta}{3}V_d + (1-\frac{\delta}{3})\mathcal{E}\underline{V}\right) \\
\frac{\dot{p}}{\dot{\nu}} &= f\frac{l_d/f - 1 + \delta/3}{f+g}\ln 2 \left(\ln \frac{l_d/f - 1 - \delta/3}{2l_d/f - 2}\right)^{-1} \\
y &= \frac{1}{f+g}\left(1 + \frac{l_d}{f}\frac{\ln\left\{1+\frac{\delta/3}{1-l_d/f}\right\}/\ln 2 - 1}{l_d/f - 1 + \delta/3}\right)^{-1}
\end{aligned}$$

dividing isomorphs, $l_b = l_d 2^{-1/3}$, $l_p = l_d$, $\alpha \equiv \frac{f-l_b}{f-l_p}$:

$$\begin{aligned}
\dot{p}X_r &= N\{\dot{I}_m\} f \mathcal{E} \underline{V}^{2/3} \\
\frac{\dot{p}}{\dot{\nu}} &= \frac{\ln 2}{\ln \alpha}\frac{l_p/3}{f+g} \\
y &= \frac{\ln 2}{\ln \alpha}\frac{1}{3(f+g)}\frac{\mathcal{E}\underline{l}^3}{f\mathcal{E}\underline{l}^2}
\end{aligned}$$

reproducing isomorphs, $\alpha \equiv \frac{f-l_b}{f-l_p}$:

$$\begin{aligned}
\dot{p}X_r &= N\{\dot{I}_m\} f \mathcal{E} \underline{V}^{2/3} \\
\frac{\dot{p}}{\dot{\nu}} &= \frac{\dot{p}}{\dot{\gamma}}\frac{l_p/3}{f+g} \quad \text{with } \dot{p}/\dot{\gamma} \text{ implicit from :} \\
\frac{e_0/3}{(1-\kappa)q} &= \frac{f+g}{\alpha^{\dot{p}/\dot{\gamma}}}\frac{f^3 - l_p^3}{\dot{p}/\dot{\gamma}} - f^2\frac{f-l_b}{\alpha^{1+\dot{p}/\dot{\gamma}}}\frac{2g+3f}{1+\dot{p}/\dot{\gamma}} + f\frac{(f-l_b)^2}{\alpha^{2+\dot{p}/\dot{\gamma}}}\frac{g+3f}{2+\dot{p}/\dot{\gamma}} - \frac{(f-l_b)^3}{\alpha^{3+\dot{p}/\dot{\gamma}}}\frac{f}{3+\dot{p}/\dot{\gamma}} \\
y &= \frac{\dot{p}/\dot{\gamma}}{3(f+g)}\frac{\mathcal{E}\underline{l}^3}{f\mathcal{E}\underline{l}^2}
\end{aligned}$$

For random harvesting, this reduces to X_1/X_r. The total biomass in the bio-reactor is $X_1 = N\mathcal{E}\underline{V}$. Since we assumed that $\dot{p}X_r$ is constant, X_r increases if $\dot{p}$ decreases and visa versa. If $\dot{p} \to 0$, the biomass approaches $X_1 = \dot{p}X_r([\dot{I}_m]l_d)^{-1} = \dot{p}X_r V_m^{1/3}/\{\dot{I}_m\}$. For increasing harvesting rates, the biomass in the bio-reactor will decrease. The biomass density as a fraction of its maximum equals

$$x = \frac{1}{f}\frac{\int_a \text{Prob}\{\underline{a}_\dagger > a\} l^3(a)\,da}{\int_a \text{Prob}\{\underline{a}_\dagger > a\} l^2(a)\,da} \tag{5.44}$$

$$= \frac{\{\dot{I}_m\}}{\dot{p}X_r} N\mathcal{E}\underline{V} \tag{5.45}$$

For random harvesting, this reduces to $x = X_1\{\dot{I}_m\}g\dot{m}(\dot{v}\dot{p}X_r)^{-1} = yg\dot{m}/\dot{p} = yl_d\dot{\nu}/\dot{p}$, where y stands for the dimensionless scaled yield $y \equiv Y\{\dot{I}_m\}/\dot{v}$. For isomorphs, this amounts to $x = \mathcal{E}\underline{l}^3(f\mathcal{E}\underline{l}^2)^{-1}$.

Table (5.2) gives the formula for filaments, rods and dividing and reproducing ectothermic isomorphs. For the purpose of producing a figure like 5.14, it is easier to choose values for the scaled functional response and solve the corresponding harvesting rate, rather than the other way around. It serves as an entry in figure 5.14 and its range from l_b to 1 is known beforehand, while the maximum harvesting rate must be obtained numerically. To aid the comparison of the different life styles, energy investment in wall material as well as the time for DNA duplication have been neglected for the rods, and incubation time has been neglected for reproducing isomorphs.

Figure 5.14 shows that all four yield factors, total biovolumes and food densities are quite comparable, despite the differences at the individual level. The three types of dividing organisms are especially hard to distinguish. The difference between filaments and rods has been maximized in the figure, by taking the aspect ratio to be equal to its maximum value $\delta = 0.6$, the value for cocci. Most rods will resemble filaments even more closely. If the DNA duplication time for rods is not negligibly small, the yield factor has less tendency to decrease at high harvesting rates. The comparison of reproducing and dividing organisms is hampered by the problem of selecting appropriate parameter values. The role of κ remains hidden in the compound parameters l_d and g for dividing organisms, while it becomes more explicit in the reproduction rate. The problem is how realistic is it to take the costs of development as equal in both cases. The scaled length at birth is taken to be equal to the scaled length just after division, in figure 5.15; again, it is not obvious how the comparison can best be done.

Figure 5.15 does make clear, however, that the yield factor is far from constant, due to the priority of the maintenance rate, which is incorporated in the scaled length l_d. The yield factor appears to have a maximum value at moderate harvesting efforts. The main reason for the lower yield factors at the very high harvesting rates lies in the quality of the individuals. The yield factor measures volume only and neglects the fact that energy reserve density increases with increasing food availability and so with increasing harvesting efforts. It should be noted that the maximum sustainable harvesting effort occurs at a substantial total biovolume. When applied in fisheries for example, the massage is that it is very difficult to judge the maximum 'safe' harvesting quota on the basis of standing crops.

Figure 5.14: Stereo view of the scaled yield factor y as a function of the scaled total biovolume x of the population and the scaled functional response f for populations subjected to constant food input and different harvesting efforts at steady state. The four curves relate from low to high maxima to dividing filaments, rods and isomorphs, and reproducing isomorphs (right curve). The parameter values are $l_d = l_p = 0.133\ 2^{1/3}$, $g = 0.033$, $\delta = 0.6$ (rods only), $\kappa = 0.3$ and $l_b = 0.133$ (reproducing isomorphs only).

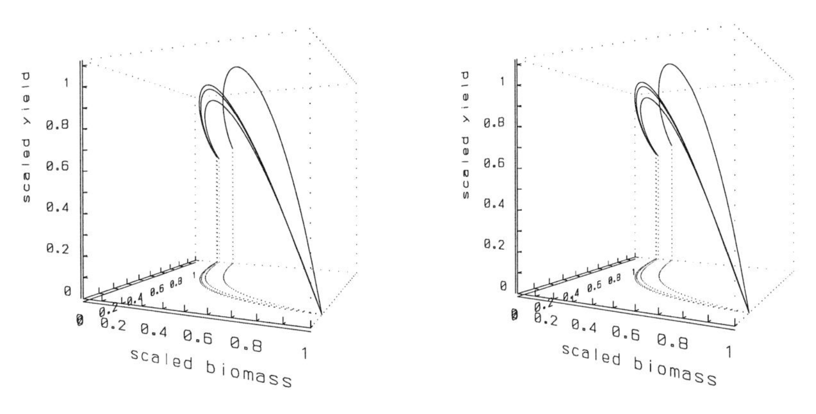

Figure 5.15: Stereo view of the yield factor for reproducing isomorphs subjected to random and age-specific harvesting. Parameters: $l_b = 0.133$, $l_p = 0.417$, $g = 0.033$ and $\kappa = 0.3$.

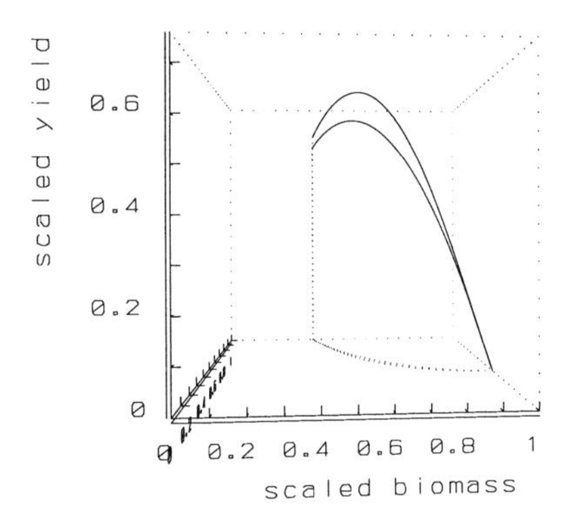

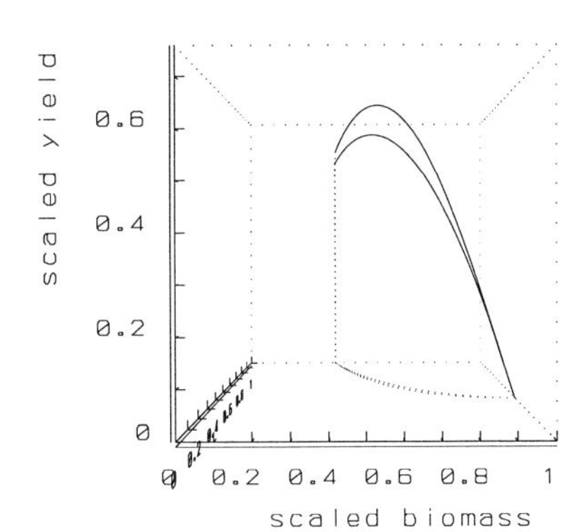

Figure 5.15 also shows that low harvesting rates go with high biomass densities; this partly explains the global distribution of standing crops. Primary production depends mainly on irradiation, which has a limited variation from the poles to the equator. This means that food input is more or less constant. Harvesting rate depends on temperature, and this means that the rate is reduced in the polar areas. The high biomass densities found in the polar regions are a direct consequence of this.

Random harvesting is one particular way to select individuals. In some situations, it makes sense to study the effect of harvesting only individuals that are larger than a particular size (i.e. older than a particular age in equilibrium situations). The aging process, for example, harvests older individuals at a higher rate. Random harvesting is compared with age-specific harvesting in figure 5.15. If the individuals are harvested at age $a_\dagger$, the stable age distribution of the population becomes $\phi_{\underline{a}}(a) = (a < a_\dagger)/a_\dagger$. The scaled functional response in the equilibrium situation of harvesting at age $a_\dagger$ must again be found from the characteristic equation (5.41), with the survivor function for the individuals: $\text{Prob}\{\underline{a}_\dagger > a\} = (a < a_\dagger)$. The scaled biomass again equals $x = \mathcal{E}\underline{l}^3(f\mathcal{E}\underline{l}^2)^{-1}$. The scaled yield factor equals the non-instantaneous yield factor as introduced for individuals, (4.18), because each individual only replaces itself in equilibrium situations. So $\int_0^{t(l)} \dot{R}(t)\, dt = 1$, which reduces (4.18) to $y_n = \frac{l(a_\dagger)^3}{g\dot{m}f\int_0^{t(l)} l^2\, dt} = \frac{l(a_\dagger)^3/3}{(g+f)f\dot{\gamma}a_\dagger\mathcal{E}\underline{l}^2}$. The difference between the present yield and that at the individual level is that in the latter instance we had direct experimental control over food density, while in the former it was indirect. Age-specific harvesting proves to be almost the same as random harvesting (as a continued process). This remarkable result thus indicates that it matters little how we select individuals for harvesting; a more realistic incorporation of the aging process will have the same result.

If we allow three ways of harvesting simultaneously, i.e. random, partial numbers of neonates and complete numbers of a cohort of a specific age, the maximum yield is obtained by harvesting neonates as well as individuals of a certain age, while not harvesting randomly. The age at maximum yield with additional harvesting of neonates is the same as without harvesting neonates. Since the way of selecting individuals hardly affects the yield factor, this result is of little practical value. Its significance lies in the result obtained by Beddington and Taylor [51], that maximum sustainable yield in a population of constant size, on the basis of a discrete time Leslie model, is obtained by complete removal of one age class, and the partial removal of a second one. Although I did not study all possible choices for the survival function, it seems that this result also applies to the present continuous time model with interaction between individuals.

It should be noted that the discussion in this section is a bit too simple. A stable age distribution does not exist for at least part of the appropriate range of parameter values for reproducing isomorphs [415]. The DEB model appears to generate cycles in fed batch situations, which could be partly reduced by allowing a small scatter among parameter values between different individuals as will be discussed later in this chapter, {210}. When stable age distributions do not exist, the analytical analysis of the yield factor becomes extremely complicated. Computer simulations indicated that the mean conversion factor closely follows the expected value based on the above mentioned analysis.

A few remarks must be made on behalf of the comparison of these results with the extensive literature on conversion efficiencies. See e.g. Getz and Haight [252], who use age-structured discrete-time models in the context of forestry and fisheries, and van Straalen [692], who surveyed concepts relating to turnover rates in age-structured continuous time models. Studies that involve real populations usually consider effects of experimental or man-induced harvesting efforts only. So they exclude 'naturally' occurring losses, which usually remain unknown. The practical significance of such an approach is obvious, because these natural loss rates are hard to control experimentally and the primary interest is usually to evaluate the effect of (commercial) harvesting programs. I did not make this distinction because the effects of these programs depend on the naturally occurring loss rates within the context of the DEB model.

Another basic difference from most of the literature is that the present evaluation of conversion efficiency allows for the interaction between individuals with regard to competition for food. By this I mean that if individuals are harvested at an increasing rate, their numbers are likely to reduce, which means higher food availability and so a higher growth rate for the remaining individuals. This feedback is rarely taken into account, because most models are age-structured, implying a fixed relationship between size and age, independent of food supply. The realism of this feedback is illustrated by the fact that most species of whales now tend to produce young at an earlier age than in the past when they were much more abundant.

5.4.1 Product and weight yield for DEB filaments

The product formation rate, such as that for penicillin, is of considerable practical and commercial interest. Microbial product formation is also the basis of many intriguing symbiontic relationships between organisms. The bacterium *Vibrio alginolyticus* excretes tetrodotoxin, which is used by chaetognats and the blue ringed octopus *Hapalochlaena maculata* to capture prey via sodium channel blocking. The tetraodontid fish *Fugu vermicularis* and the starfish *Astropecten polyacanthus* receive some protection against predators from this toxin [718]. This subsection deals with the conversion of substrate into biomass and products of DEB filaments in chemostats on the basis of C-moles. It serves as an introduction to the subsection on mass-energy coupling.

Chemostats have two basic control variables: throughput rate, $\dot{p}$, and substrate density in the supply, X_r. (Note that for reasons of consistency of notation, substrate density is expressed as volume of substrate per volume of environment; the coefficient d_{mx} converts it to moles.) To show how production rates depend on these control variables, we start from (5.13) – (5.15) in unscaled form and append product formation in the DEB model for filaments.

Micro-organisms can produce products via several routes. If the DEB model still applies in the strict sense, the mere fact that product formation costs energy implies that product formation must be a weighted sum of the energy fluxes assimilation, maintenance and growth investment. The energy drain to product formation can then be considered as overhead costs in these three processes. The necessity to tie product formation to all the three energy fluxes in general becomes obvious in a closer analysis of fermentation; see {195}. If product formation is independent of one or more energy fluxes, mass balance equations dictate that more than one product must be made under anaerobic conditions and that the relative amounts of these products must depend on the population growth rate in a very special way. Note that in the Monod case, maintenance and reserves are absent, so that assimilation is proportional to growth investment, which leaves just a single energy flux available to couple to product formation. In the Marr–Pirt case, reserves are absent, so that assimilation is proportional to maintenance plus growth investment, which leaves two energy fluxes available to couple to product formation. Maintenance and reserves together allow for a three dimensional base for product formation.

The change in product density becomes

$$\frac{d}{dt}P = \frac{[d_{PA}]}{[E_m]}[\dot{A}_m]fX_1 + \frac{[d_{PM}]}{[E_m]}\frac{[\dot{M}]}{\kappa}X_1 + \frac{[d_{PG}]}{[E_m]}\frac{[G]}{\kappa}\frac{d}{dt}X_1 \tag{5.46}$$

The maximum reserve capacity $[E_m]$ is introduced just to convert energy fluxes to volume fluxes, so that the parameters $[d_{P*}]$ convert volumes to moles. The last term refers to product formation coupled to the rate of synthesis of structural biomass. Since both assimilation and maintenance are proportional to biomass, product formation is a weighted sum of biomass X_1 and growth $\frac{d}{dt}X_1$ in steady state situations. This rate of product formation was proposed by Leudeking and Piret [433] in 1959. They studied lactic acid fermentation by *Lactobacillus delbruekii.* Leudeking–Piret kinetics has proved extremely useful and versatile in fitting product formation data for many different fermentations [29].

In a chemostat with a throughput rate $\dot{p}$ and a density X_r of substrate in the supply flow, the dynamics of the substrate, reserves, biomass and product density are

$$\frac{d}{dt}X_0 = \dot{p}X_r - [\dot{I}_m]fX_1 - \dot{p}X_0 \quad \text{with } f \equiv X_0/(K + X_0) \tag{5.47}$$

$$\frac{d}{dt}e = \dot{\nu}(f - e) \tag{5.48}$$

$$\frac{d}{dt}X_1 = \left(\frac{\dot{\nu}e - g\dot{m} - \dot{p}_a e(1+g)}{e+g} - \dot{p}\right)X_1 \tag{5.49}$$

$$\frac{d}{dt}P = [d_{PA}]\dot{\nu}fX_1 + [d_{PM}]\dot{m}gX_1 + [d_{PG}]g\frac{\dot{\nu}e - \dot{m}g}{e+g}X_1 - \dot{p}P \tag{5.50}$$

where $\dot{\nu}$ relates to the 'true' yield and the maximum gross population growth rate as $\dot{\nu} = g[\dot{I}_m]Y_g = g\dot{m} + \dot{\mu}_m^\circ(1+g)$. In equilibrium we have

$$e = f = \frac{g(\dot{m} + \dot{p})}{\dot{\nu} - \dot{p} - \dot{p}_a(1+g)} \tag{5.51}$$

$$X_1 = \frac{X_r\dot{p}}{[\dot{I}_m]f}\left(1 - \frac{fK/X_r}{1-f}\right) \tag{5.52}$$

$$\frac{P}{X_1} = [d_{PA}]\frac{\dot{\nu}f}{\dot{p}} + [d_{PM}]\frac{\dot{m}g}{\dot{p}} + [d_{PG}]\frac{g}{\dot{p}}\frac{\dot{\nu}f - \dot{m}g}{f+g} \tag{5.53}$$

This equilibrium only exists if the throughput rate $\dot{p}$ is less than the maximum one

$$\dot{p}_m = \frac{\dot{\nu} - \dot{p}_a(1+g) - g\dot{m}(1 + K/X_r)}{1 + g(1 + K/X_r)} \tag{5.54}$$

otherwise biomass washes out completely. If the concentration of substrate in the feed is large in comparison with the saturation coefficient, the maximum throughput rate approaches the maximum net population growth rate. If the death rate is small at the same time, the maximum throughput rate approaches the maximum gross population growth rate $\dot{\mu}_m^\circ = \frac{\dot{\nu}-\dot{m}g}{1+g}$.

Since structural biovolume cannot be measured directly, total biomass expressed as C-moles has a practical value; see {37}. Analogous to (2.10), the living biomass density is $W_1 = ([d_{mv}] + [d_{me}]e)X_1$. (It is a density because X_1 is a density.) In equilibrium, where $e = f$, (5.12) can be applied to find the total biomass in a chemostat, including the dead filaments. We have to multiply X_1 by $\frac{\dot{p}_a\dot{m}+\dot{p}(\dot{\mu}_m^\circ+\dot{m})}{\dot{p}(\dot{m}+\dot{\mu}_m-\dot{p}_a)}$ to arrive at the total biomass.

The molar yield of a quantity is the ratio of its production rate in moles and the molar uptake, which equals $d_{mx}[\dot{I}_m]fX_1 = d_{mx}\dot{p}(X_r - X_0)$. The index m in ${}_mY$ is used to distinguish yields on the basis of C-moles from that on the basis of volume. The yield of dry weight, ${}_mY_{W_1} \equiv \frac{\dot{p}W_1}{d_{mx}\dot{p}(X_r-X_0)}$, and product ${}_mY_P \equiv \frac{\dot{p}P}{d_{mx}\dot{p}(X_r-X_0)}$ per mole of substrate that has been taken up then amounts to

$$
\begin{aligned}
{}_mY_{W_1} &= \frac{[d_{mv}] + [d_{me}]f}{d_{mx}[\dot{I}_m]f}\,\frac{\dot{p}_a\dot{m} + \dot{p}(\dot{\mu}_m^\circ + \dot{m})}{\dot{\mu}_m^\circ - \dot{p}_a + \dot{m}} && (5.55)\\
{}_mY_P &= \frac{[d_{PA}]}{d_{mx}[\dot{I}_m]}\dot{\nu} + \frac{[d_{PM}]}{d_{mx}[\dot{I}_m]}\frac{\dot{m}g}{f} + \frac{[d_{PG}]}{d_{mx}[\dot{I}_m]}\frac{g}{f}\,\frac{\dot{\nu}f - \dot{m}g}{f+g} && (5.56)
\end{aligned}
$$

These equations can be used to optimize reactor performance, if the financial costs and benefits for substrate, biomass and products are known.

If death is negligible, so $\dot{p}_a = 0$ and $\dot{\mu}_m^\circ = \dot{\mu}_m = \frac{\dot{\nu}-\dot{m}g}{1+g}$ and $\dot{p} = \dot{\mu}$ at equilibrium, the molar yields of reserves, structural biomass, total biomass and product reduce to

$$
\begin{aligned}
{}_mY_E &= t_E\dot{\mu} & {}_mY_{X_1} &= t_{X_1}\tfrac{\dot{\mu}}{g}\tfrac{\dot{\nu}-\dot{\mu}}{\dot{m}+\dot{\mu}}\\
{}_mY_{W_1} &= {}_mY_E + {}_mY_{X_1} & {}_mY_P &= t_{PA}\dot{\nu} + (t_{PM}\dot{m} + t_{PG}\dot{\mu})\tfrac{\dot{\nu}-\dot{\mu}}{\dot{m}+\dot{\mu}}
\end{aligned} \qquad (5.57)
$$

where time parameters

$$(\, t_{X_1} \;\; t_E \;\; t_{PA} \;\; t_{PM} \;\; t_{PG} \,) \equiv (d_{mx}[\dot{I}_m])^{-1}(\, [d_{mv}] \;\; [d_{me}] \;\; [d_{PA}] \;\; [d_{PM}] \;\; [d_{PG}] \,)$$

are introduced to simplify the notation; they can readily be estimated from measurements, as will be shown in the next subsection. The practical application is to get rid of the parameter $[\dot{I}_m]$, which is usually difficult to obtain because of its association with conversion coefficients.

Use of the time parameters t_{X_1} and t_E allows for a simple expression for the biomass density in terms of the molecular weights of structural biomass and energy reserves at constant substrate density:

$$W_1 = ([d_{mv}] + [d_{me}]f)X_1 = d_{mx}(w_{X_1}t_{X_1}/f + w_Et_E)\dot{\mu}(X_r - X_0) \qquad (5.58)$$

where biomass density W_1 is expressed in C-mole per volume.

Parameter values can depend on properties of the substrate, such as the chemical potential $\tilde{\mu}_{X_0}$. Since $[\dot{A}_m]/d_{mx}[\dot{I}_m]$ stands for the assimilation energy per C-mole of substrate,

Figure 5.16: The molar yield of biomass corrected for a fixed population growth rate of $\dot{\mu} = 0.2$ h^{-1} is proportional to the chemical potential of substrate, expressed per C-mole in combustion reference. Data from Rutgers [622] for *Pseudomonas oxalaticus* (•) and from van Verseveld, Stouthamer and others [473,735,736,737,738] for *Paracoccus denitrificans* (∘) under aerobic conditions with NH_4^+ as nitrogen source, corrected for a temperature of 30 °C. No product, or a negligible amount, is formed during these experiments [735].

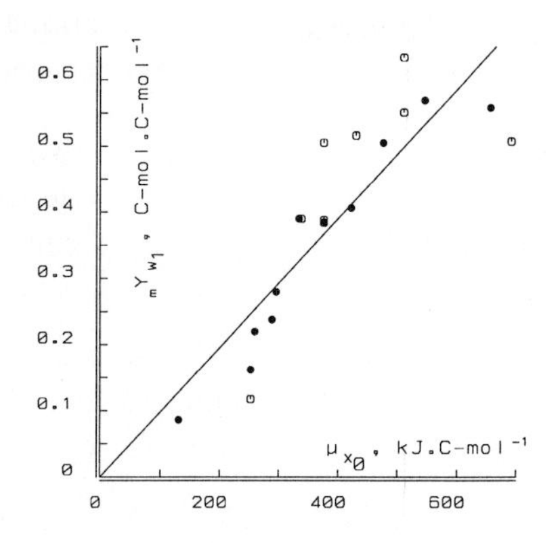

it seems reasonable to assume that $[\dot{A}_m]$ is proportional to the chemical potential of substrate. If we tie the maximum reserve capacity $[E_m]$ to $[\dot{A}_m]$, see {218} for arguments, the specific energy conductance $\dot{\nu}$ would become independent of the chemical potential and the molar yield of biomass $_mY_{W_1}$ would become proportional to the chemical potential for a fixed value of $\dot{\mu}$ via $[E_m]$ in g and in $[d_{me}]$. This is confirmed in figure 5.16.

5.4.2 Mass-energy coupling

Some microbiologists find a description of input-output relationships of individuals hard to accept on the basis of energy fluxes that cannot be measured directly. They prefer a description in terms of mass fluxes, which are of interest in their own right. A rigorous derivation of the respiration rate can be made on the basis of mass balance equations. In the first chapter, {4}, I discussed the problem that detailed descriptions of mass fluxes pose. A crude description, however, is feasible and instructive. The DEB model allows for such a crude description, because it assumes homeostasis for both structural biomass and (energy) reserves. The significance of the discussion in this subsection is that the partitioning of biomass into a structural and a reserve component poses no fundamental problems in the analysis of mass fluxes. On the contrary, mass flux analysis can be used to decompose the biomass into both components in a unique way and to partition mass fluxes over uptake, maintenance and growth. One of the major advantages of the present approach is that it is no longer necessary to assume that the composition of biomass is fixed. It can depend on the nutritional state (in a special way).

Macro-chemical reaction equation

I will illustrate the construct for filaments in steady state and follow the microbiological tradition of expressing mass in C-moles; see {37}. The death rate is assumed to be small, just to simplify the expressions. The discussion will be restricted to growth on a single energy substrate, or on a mixture of substrates with fixed relative uptake preferences. So we are seeking a specification of the macro-chemical reaction equation

$$\overset{\text{E-substrate},X_0}{\mathrm{CH}_{n_{HX_0}}\mathrm{O}_{n_{OX_0}}\mathrm{N}^{n+X_0}_{n_{NX_0}}} + \overset{\text{N-substrate},N}{{}_mY_N\mathrm{H}_{n_{HN}}\mathrm{O}_{n_{ON}}\mathrm{N}^{n+N}_{n_{NN}}} + \overset{\text{oxygen},O}{{}_mY_O\mathrm{O}_2} \overset{\dot{k}}{\rightarrow} \overset{\text{biomass},X_1}{{}_mY_{X_1}\mathrm{CH}_{n_{HX_1}}\mathrm{O}_{n_{OX_1}}\mathrm{N}_{n_{NX_1}}} +$$

$$+ \underset{\text{reserves},E}{{}_mY_E \mathrm{CH}_{n_{HE}}\mathrm{O}_{n_{OE}}\mathrm{N}_{n_{NE}}} + \underset{\text{carbonate},C}{{}_mY_C \mathrm{CHO}_3^{-1}} + \underset{\text{water},A}{{}_mY_H \mathrm{H}_2\mathrm{O}} + \underset{\text{protons},+}{{}_mY_+ \mathrm{H}^{+1}} + \underset{\text{product},P}{{}_mY_P \mathrm{CH}_{n_{HP}}\mathrm{O}_{n_{OP}}\mathrm{N}_{n_{NP}}^{n_{+P}}}$$

The symbols N, O, C, A and $+$ will be used as names for the different compounds, as indicated. For the moment I assume that the n's are known, but on {196} I will solve the problem to obtain n_{*E} and n_{*X_1} from observations on n_{*W_1}, where $*$ stands for H, O, or N and W_1 for total biomass, including the structural and reserve components. The yield coefficients for supplied compounds are taken to be negative, while yield coefficients for produced compounds are taken positive. Some of the yield coefficients, such as for water, ${}_mY_A$, can for this reason be both positive and negative, depending on substrate type and organism, but most of the yield coefficients usually have one sign only. The five yields that relate to 'minerals' and electric charge will be solved from the mass balance equations, while the DEB model specifies the ones corresponding to biomass, reserves and products, see (5.57), as well as the transformation rate. The method of determining the coefficients on the basis of the DEB model is simple if one realizes that the coefficients represent molar yields. The description will include more than one product.

From energy to mass

The balance equations for the four elements and the electric charge can be expressed conveniently as

$$\begin{pmatrix} 1 & 0 & 0 & 0 & 0 \\ 1 & 2 & 0 & n_{HN} & 1 \\ 3 & 1 & 2 & n_{ON} & 0 \\ 0 & 0 & 0 & n_{NN} & 0 \\ -1 & 0 & 0 & n_{+N} & 1 \end{pmatrix} \begin{pmatrix} {}_mY_C \\ {}_mY_A \\ {}_mY_O \\ {}_mY_N \\ {}_mY_+ \end{pmatrix} = - \begin{pmatrix} 1 & 1 & 1 & 1 & 1 & \cdots \\ n_{HX_0} & n_{HX_1} & n_{HE} & n_{HP_1} & n_{HP_2} & \cdots \\ n_{OX_0} & n_{OX_1} & n_{OE} & n_{OP_1} & n_{OP_2} & \cdots \\ n_{NX_0} & n_{NX_1} & n_{NE} & n_{NP_1} & n_{NP_2} & \cdots \\ n_{+X_0} & 0 & 0 & n_{+P_1} & n_{+P_2} & \cdots \end{pmatrix} \begin{pmatrix} {}_mY_{X_0} \\ {}_mY_{X_1} \\ {}_mY_E \\ {}_mY_{P_1} \\ {}_mY_{P_2} \\ \vdots \end{pmatrix} \tag{5.59}$$

Each column of both matrices gives the elemental composition of a compound in the macrochemical reaction equation. Inclusion of more elements, such as phosphorus and sulfur, only appends extra rows. By definition we have ${}_mY_{X_0} \equiv -1$. The following notation is introduced to prevent losing one's way in the jungle of coefficients

$$\mathbf{Y}_M^T \equiv \begin{pmatrix} {}_mY_C & {}_mY_A & {}_mY_O & {}_mY_N & {}_mY_+ \end{pmatrix} \tag{5.60}$$

$$\mathbf{Y}_D^T \equiv \begin{pmatrix} -1 & {}_mY_{X_1} & {}_mY_E & {}_mY_{P_1} & {}_mY_{P_2} & \cdots \end{pmatrix} \tag{5.61}$$

$$\mathbf{u} \equiv \begin{pmatrix} 1 & 0 & 0 & 0 & 0 \\ 1 & 2 & 0 & n_{HN} & 1 \\ 3 & 1 & 2 & n_{ON} & 0 \\ 0 & 0 & 0 & n_{NN} & 0 \\ -1 & 0 & 0 & n_{+N} & 1 \end{pmatrix}^{-1} \quad \mathbf{n} \equiv \begin{pmatrix} 1 & 1 & 1 & 1 & 1 & \cdots \\ n_{HX_0} & n_{HX_1} & n_{HE} & n_{HP_1} & n_{HP_2} & \cdots \\ n_{OX_0} & n_{OX_1} & n_{OE} & n_{OP_1} & n_{OP_2} & \cdots \\ n_{NX_0} & n_{NX_1} & n_{NE} & n_{NP_1} & n_{NP_2} & \cdots \\ n_{+X_0} & 0 & 0 & n_{+P_1} & n_{+P_2} & \cdots \end{pmatrix} \tag{5.62}$$

The DEB model specifies $\mathbf{Y}_D$, see (5.57), which is indicated by the index D. The five remaining yield coefficients $\mathbf{Y}_M$ can be solved from the mass balance equations (5.59).

The solution of the yield coefficients for the 'minerals' reads

$$\mathbf{Y}_M = -\mathbf{unY}_D \tag{5.63}$$

For NH_4^+ as N-substrate, the matrix $\mathbf{u}$ becomes

$$\mathbf{u} = \begin{pmatrix} 1 & 0 & 0 & 0 & 0 \\ 1 & 2 & 0 & 4 & 1 \\ 3 & 1 & 2 & 0 & 0 \\ 0 & 0 & 0 & 1 & 0 \\ -1 & 0 & 0 & 1 & 1 \end{pmatrix}^{-1} = \begin{pmatrix} 1 & 0 & 0 & 0 & 0 \\ -1 & \frac{1}{2} & 0 & -\frac{3}{2} & -\frac{1}{2} \\ -1 & -\frac{1}{4} & \frac{1}{2} & \frac{3}{4} & \frac{1}{4} \\ 0 & 0 & 0 & 1 & 0 \\ 1 & 0 & 0 & -1 & 1 \end{pmatrix} \tag{5.64}$$

The matrix $\mathbf{u}$ of coefficients has an odd interpretation in terms of reduction degrees. The third row, i.e. the one that relates to oxygen yield, represents the ratio of the reduction degree of the elements C, H, O, N and the electric charge and that of O_2, which is -4. That is to say, N-atoms count for -3 in these reduction degrees, whatever their real values in the rich mixture of components that are present. The third row of $\mathbf{un}$ thus represents the ratio of the reduction degrees of X_0, X_1, E and P and that of O. Sandler and Orbey [626] discuss the concept of generalized degree of reduction.

The rate of the macro-chemical reaction is the total ingestion rate, expressed as number of 'molecules' per time, so $\dot{k} = d_{mx}[\dot{I}_m]fX_1$ or $d_{mx}\dot{\mu}(X_r - X_0)$ in a chemostat at equilibrium. Via (5.51) this amounts to

$$\dot{k} = d_{mx}\dot{\mu}\left(X_r - K\frac{g(\dot{m}+\dot{\mu})}{\dot{\nu} - \dot{\mu} - g(\dot{m}+\dot{\mu})}\right)$$

where $\dot{\mu}$ and X_r are the control variables of the chemostat.

These expressions show that the DEB model fully specifies the macro-chemical equation.

Respiration

The mass flux consideration now allows a more rigorous interpretation of the respiration rate in terms of energy fluxes. More specifically, it is possible to associate oxygen, carbon dioxide and other mass fluxes with the three energy fluxes, assimilation, maintenance and growth investment, as has been done for products in (5.57):

$$_mY_* = t_{*A}\dot{\nu} + (t_{*M}\dot{m} + t_{*G}\dot{\mu})\frac{\dot{\nu} - \dot{\mu}}{\dot{m} + \dot{\mu}} \tag{5.65}$$

where $*$ stands for C, A, O, N, $+$, X_0, X_1 or E. An important difference between 'mineral' products and the other masses is that the time parameters for 'mineral' products are not free parameters; their values are fully determined by the mass balance equations. The time parameters can be found by equating (5.65) to (5.63). The result is

$$\mathbf{t}_M \equiv \begin{pmatrix} t_{CA} & t_{CM} & t_{CG} \\ t_{AA} & t_{AM} & t_{AG} \\ t_{OA} & t_{OM} & t_{OG} \\ t_{NA} & t_{NM} & t_{NG} \\ t_{+A} & t_{+M} & t_{+G} \end{pmatrix} = -\mathbf{un} \begin{pmatrix} t_{X_0A} & t_{X_0M} & t_{X_0G} \\ t_{X_1A} & t_{X_1M} & t_{X_1G} \\ t_{EA} & t_{EM} & t_{EG} \\ t_{P_1A} & t_{P_1M} & t_{P_1G} \\ t_{P_2A} & t_{P_2M} & t_{P_2G} \\ \vdots & \vdots & \vdots \end{pmatrix} \equiv -\mathbf{un}\,\mathbf{t}_D \tag{5.66}$$

The first three rows of $\mathbf{t}_D$ are:

$$\begin{pmatrix} t_{X_0A} & t_{X_0M} & t_{X_0G} \\ t_{X_1A} & t_{X_1M} & t_{X_1G} \\ t_{EA} & t_{EM} & t_{EG} \end{pmatrix} = \begin{pmatrix} -\dot{\nu}^{-1} & 0 & 0 \\ 0 & 0 & t_{X_1}/g \\ t_E & -t_E & -t_E \end{pmatrix} \quad (5.67)$$

The derivation is simple after converting (5.65) and (5.63) to second degree polynomials in $\dot{\mu}$, via multiplication with $(\dot{m}+\dot{\mu})$, and equating the three polynomial coefficients separately. The matrix notation of the time parameters can be used to rewrite (5.65) as

$$\mathbf{t}_M \begin{pmatrix} \dot{\nu} \\ \dot{m}\frac{\dot{\nu}-\dot{\mu}}{\dot{m}+\dot{\mu}} \\ \dot{\mu}\frac{\dot{\nu}-\dot{\mu}}{\dot{m}+\dot{\mu}} \end{pmatrix} = \mathbf{Y}_M \quad \text{and} \quad \mathbf{t}_D \begin{pmatrix} \dot{\nu} \\ \dot{m}\frac{\dot{\nu}-\dot{\mu}}{\dot{m}+\dot{\mu}} \\ \dot{\mu}\frac{\dot{\nu}-\dot{\mu}}{\dot{m}+\dot{\mu}} \end{pmatrix} = \mathbf{Y}_D$$

The interpretation of the time parameters can, for instance, be illustrated for reserves. Reserves are supplied via assimilation and used via maintenance and growth investment, with equal weights for all the three energy fluxes. The absolute values of t_{EA}, t_{EM} and t_{EG} are therefore the same, they only differ in sign. Assimilation and maintenance do not contribute to growth investment, thus $t_{X_1A} = t_{X_1M} = 0$, while t_{X_1G} is positive. Substrate is used at a rate that depends on assimilation, but maintenance and growth do not affect substrate directly, so $t_{X_0M} = t_{X_0G} = 0$.

The significance of the decomposition of the mass fluxes into three components that relate to energy fluxes is in mass conversions in transient, i.e. non-equilibrium, situations. Although yield coefficients change in value when death is introduced, the time parameters are unaffected. The dynamics of product density (5.46) also describes uptake of oxygen and production of carbonate and protons by substituting $t_{*A}d_{mx}[\dot{I}_m]$ for $[d_{PA}]$, $t_{*M}d_{mx}[\dot{I}_m]$ for $[d_{PM}]$ and $t_{*G}d_{mx}[\dot{I}_m]$ for d_{PG}, with $*$ standing for O, C or $+$. The result for oxygen consumption, for instance, is

$$(d_{mx}[\dot{I}_m]X_1)^{-1}\frac{d}{dt}O = t_{OA}\dot{\nu}f + t_{OM}\dot{m}g + t_{OG}g\frac{\dot{\nu}e - \dot{m}g}{e+g}$$

Fermentation

Fermentation is an anaerobic process in which organic compounds act as electron donor as well as electron acceptor. Usually several products are made rather than just one. These products can be valuable substrates under aerobic conditions, but under anaerobic conditions mass balances force organisms to leave them untouched. Cellulose is fermented to products such as acetate, propionate, butyrate and valerate in cows [632], which micro-organisms cannot use as substrates under anaerobic conditions (much to the benefit of the cow!).

Under anaerobic conditions we have ${}_mY_O = 0$. Substitution of this value in (5.65) shows that this translates into the three constraints

$$\mathbf{0}^T = (\ t_{OA} \quad t_{OM} \quad t_{OG}\ )$$

where the time parameters are given in the third row of (5.66). The practical implementation of these constraints in non-linear regressions is via Lagrange multipliers, which can be found in standard texts on calculus. An interesting consequence of these constraints is that there are no free parameters for product formation if just one product is made. Mass balances then fully determine how this product is tied to the three energy fluxes. If $t_{P_iM} = t_{P_iG}$ applies, (5.65) shows that product formation is linear in the population growth rate, which reduces the number of constraints to two. This special case corresponds with production associated with assimilation and the κ-rule for energy allocation to production processes.

From mass to energy

Let us now consider the inverse problem: what information about the parameter values of the DEB model can be obtained from observations about the coefficients of the macrochemical reaction? A practical point is that elements associated with the energy reserves are not separable from those associated with structural biomass. This makes the relative abundances of the elements dependent on the population growth rate. If we assume that these relative abundances have been measured, it is possible to substitute

$$ {}_mY_{W_1}\mathrm{CH}_{n_{HW_1}}\mathrm{O}_{n_{OW_1}}\mathrm{N}_{n_{NW_1}} = {}_mY_{X_1}\mathrm{CH}_{n_{HX_1}}\mathrm{O}_{n_{OX_1}}\mathrm{N}_{n_{NX_1}} + {}_mY_E\mathrm{CH}_{n_{HE}}\mathrm{O}_{n_{OE}}\mathrm{N}_{n_{NE}} \qquad (5.68) $$

with ${}_mY_{W_1} = {}_mY_{X_1} + {}_mY_E$ and $n_{*W_1} = {}_mY_{W_1}^{-1}({}_mY_{X_1}n_{*X_1} + {}_mY_E n_{*E})$, where $*$ stands for H, O, N or $+$.

For the special case that $n_{*X_1} = n_{*E} = n_{*W_1}$, which is in fact the standard assumption in the microbiological literature, and for NH_4^+ as N-substrate, (5.63) together with (5.64) give a simple relationship between the molar yields of biomass and oxygen if no products are formed:

$$ {}_mY_O = -\begin{pmatrix} -1 & -\frac{1}{4} & \frac{1}{2} & \frac{3}{4} & \frac{1}{4} \end{pmatrix} \begin{pmatrix} 1 & 1 \\ n_{HX_0} & n_{HW_1} \\ n_{OX_0} & n_{OW_1} \\ n_{NX_0} & n_{NW_1} \\ n_{+X_0} & n_{+W_1} \end{pmatrix} \begin{pmatrix} -1 \\ {}_mY_{W_1} \end{pmatrix} $$

This result follows directly from the mass balance equation and does not use any model assumption on resource uptake and use. The only assumption is that of constant biomass composition. In view of the interpretation of the third row of $\mathbf{u}$, the result can be rewritten as $\gamma_{X_0}\,{}_mY_{X_0} = \gamma_O\,{}_mY_O + \gamma_{W_1}\,{}_mY_{W_1}$, where γ_* represents the reduction degree of X_0, O or W_1. This result is well known from the microbiological literature [306,608]. Figure 5.17 compares the measured oxygen yield with the yield that has to be expected on the basis of this relationship and measured values of biomass yields ${}_mY_{W_1}$, for a wide variety of organisms and 15 substrates that differ in n_{HX_0} and n_{OX_0}, but all have $n_{NX_0} = 0$. The substantial scatter shows that the error of measurement is large and/or that the biomass composition is not equal for all organisms and is not independent of the growth rate. It is thus assumed here that, in general, $n_{*X_1} \neq n_{*E}$ and that n_{*W_1} depends on the population growth rate $\dot{\mu}$.

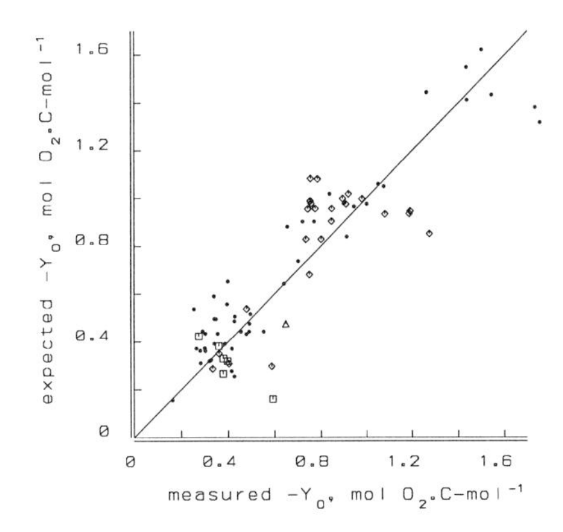

Figure 5.17: The expected molar yield for oxygen as a function of the measured value based on the assumption of a constant and common biomass composition of $n_{HW_1} = 1.8$, $n_{OW_1} = 0.5$ and $n_{NW_1} = 0.2$ for a wide variety of bacteria (•), yeasts (◇), fungi (□) and the green alga *Chlorella* (△). The expectation is based on measured yields for biomass. Data gathered from the literature by Heijnen and Roels [306] on aerobic growth on a wide variety of substrates without product formation and NH_4^+ as nitrogen substrate.

Just as for products and 'minerals', biomass and reserve yields can be decomposed into components associated with assimilation, maintenance and growth such as

$$\begin{aligned} {}_mY_{W_1} &= t_{W_1A}\dot{\nu} + (t_{W_1M}\dot{m} + t_{W_1G}\dot{\mu})\frac{\dot{\nu} - \dot{\mu}}{\dot{m} + \dot{\mu}} \\ \text{with } \begin{pmatrix} t_{W_1A} & t_{W_1M} & t_{W_1G} \end{pmatrix} &= \begin{pmatrix} t_{X_1A} & t_{X_1M} & t_{X_1G} \end{pmatrix} + \begin{pmatrix} t_{EA} & t_{EM} & t_{EG} \end{pmatrix} \end{aligned}$$

So instead of measured values for ${}_mY_{X_1}$ and ${}_mY_E$ and known values for n_{HX_1}, n_{OX_1}, n_{NX_1}, n_{HE}, n_{OE} and n_{NE}, we now have measured values for ${}_mY_{W_1}$, n_{HW_1}, n_{OW_1} and n_{NW_1}. The parameters that should be estimated now include n_{HX_1}, n_{OX_1}, n_{NX_1}, n_{HE}, n_{OE} and n_{NE}. The identification of the parameters that can be estimated is a bit more complicated than usual, particularly if no products are produced. The nature of the problem can best be illustrated with a simple model that relates an observed variable y to different values of a known variable x. If this model is $y = abx$, where a and b are parameters, they cannot both be estimated. Only the product ab can be estimated. The same problem occurs here in a more complicated situation.

Parameter identification

The identification of estimable (compound) parameters is rather straightforward, if different values of $\dot{\mu}$ have been considered. The expression for ${}_mY_{W_1}$ can be written out as ${}_mY_{W_1} = \frac{\dot{\mu}}{\dot{m}+\dot{\mu}}(\dot{\nu}t_{X_1}/g + \dot{m}t_E - \dot{\mu}(t_{X_1}/g - t_E))$. This presentation reveals that three parameters can be estimated from ${}_mY_{W_1}(\dot{\mu})$, namely $p_1 \equiv \dot{m}$, $p_2 \equiv \dot{\nu}t_{X_1}/g + \dot{m}t_E$, $p_3 \equiv t_{X_1}/g - t_E$. The expression for ${}_mY_{W_1}n_{*W_1}$ can be written out in the same way, where $*$ stands for H, O or N. This identifies $p_{4,5,6} \equiv n_{*X_1}\dot{\nu}t_{X_1}/g + n_{*E}\dot{m}t_E$ and $p_{7,8,9} \equiv n_{*X_1}t_{X_1}/g - n_{*E}t_E$ as estimatable parameters. These are rather complicated functions of the parameters of interest, but it is possible to change to functions of these identifiable parameters that relate in a simpler way to the parameters of interest. One possible choice is $p_1 = \dot{m}$, $p_2 + p_1p_3 = (\dot{m} + \dot{\nu})t_{X_1}/g$, $p_3 = t_{X_1}/g - t_E$, $\frac{p_{4,5,6}+p_1p_{7,8,9}}{p_2+p_1p_3} = n_{*X_1}$ and $p_{4,5,6} - p_3\frac{p_{4,5,6}+p_1p_{7,8,9}}{p_2+p_1p_3} = (n_{*X_1} - n_{*E})t_E$. This seems to be the set of estimatable parameters that relate to the original ones in the most simple way. If the maximum population growth rate $\dot{\mu}_m = \frac{\dot{\nu}-\dot{m}g}{1+g}$ and the scaled functional response at which no growth occurs, $f_0 = l_d = \dot{m}g/\dot{\nu}$, see (4.9), is known as well, all parameters can be estimated.

This discussion on estimation is given because if you try to estimate more parameters than is possible from data using, for instance, a non-linear regression technique, the computer program that you use will report trouble. The problems can have many sources and in the more complex situations it is not always easy to identify the troubles as an invariance problem. The standard way to obtain parameter values with this method is to start with 'dreaming up' values that are hopefully 'close' to the least squares estimates. Some values, however, are more accurately known in advance than others. The second step is to obtain least squares estimates of the lesser known parameters, keeping the values of the better known parameters fixed at their initially guessed values. Then one proceeds to involve more and more parameters in the estimation procedure. Troubles will start somewhere during this procedure, as soon as the set of parameters that is involved in the estimation contains an invariance problem. If the sum of squared deviations from model expectations is minimized with a Newton Raphson or related method, a domain error will occur because the iteration matrix has determinant zero, so that its inverse does not exist. Such an error will also occur if the initial guesses are totally out of range.

Examples

Examples of application are based on the work of Paul Hanegraaf [285], and are given in figure 5.18 for the prokaryote *Klebsiella* under aerobic conditions that does not make products and in figure 5.19 for the yeast *Saccharomyces* under anaerobic conditions that makes three types of product. Mortality is assumed to be unimportant in both examples. The DEB model describes the macro-chemical equation very well. Although the total number of parameters is large in both examples, the numbers of parameters per data set are only 1.1 and 1.5, respectively. Note that the abundance of nitrogen in the total biomass increases a bit for increasing population growth rates, which implies that, in this case, the energy reserves are richer in nitrogen than the structural biomass.

In the first data set, the values for $\dot{\nu}$ and g are poorly defined, while $\dot{m}$ is fixed accurately. This should not come as a surprise after the discussion on estimation. The relative abundance of the elements suggests that ribosomal RNA is an important component of the reserves in the prokaryote example, in view of the high population growth rate. This will be discussed further on {250}. Yeasts appear to be relatively rich in proteins when they grow fast, but their maximum growth rate is about half that of *Klebsiella.*

Application of (5.66) allows a partitioning of oxygen consumption and carbon dioxide production over the three energy fluxes:

	Klebsiella		*Saccharomyces*	
assimilation	$t_{OA} = -0.145$ h	$t_{CA} = 0.060$ h	$t_{OA} = 0$ h	$t_{CA} = 0.604$ h
maintenance	$t_{OM} = -0.287$ h	$t_{CM} = 0.310$ h	$t_{OM} = 0$ h	$t_{CM} = 0.065$ h
growth invest.	$t_{OG} = -0.028$ h	$t_{CG} = 0.068$ h	$t_{OG} = 0$ h	$t_{CG} = 0.005$ h

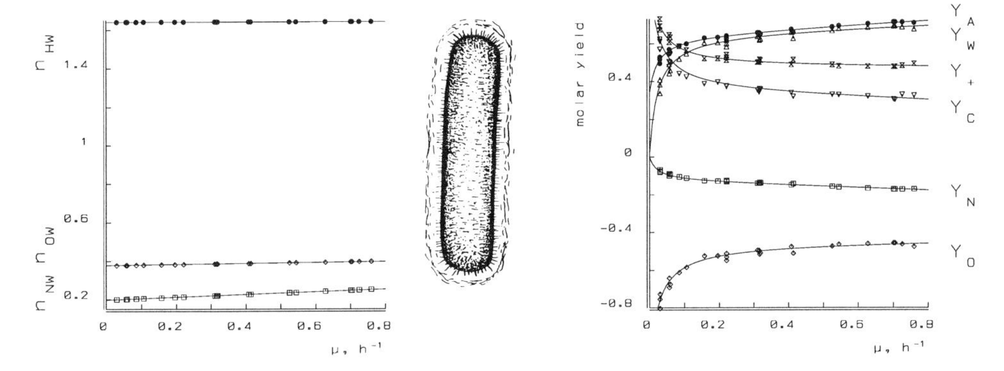

Figure 5.18: Relative abundances (left figure) of the elements H (∘), O (◇) and N (□) in the biomass and the molar yield (right figure) of biomass, △ ${}_mY_{W_1}$, carbon dioxide, ▽ ${}_mY_C$, water, ∘ ${}_mY_A$, oxygen, ◇ ${}_mY_O$, ammonia, □ ${}_mY_N$, and protons, ⋈ ${}_mY_+$, as functions of population growth rate for *Klebsiella aerogenes* growing on glycerol at 35 °C. Data from Esener *et al.* [207,208]. The curves are based on expectations of the DEB model for filaments. For the chosen values of $\dot{\mu}_m = 1.2$ h^{-1} and $f_0 = 0.01$, the estimated parameters and their standard deviations are

$\dot{\nu} = 2.7\ (3.7)$ h^{-1}	$g = 1.2\ (3.1)$	$\dot{m} = 0.022\ (0.00064)$ h^{-1}
$t_{X_1} = 0.29\ (0.34)$ h	$t_E = 0.31\ (0.34)$ h	
$n_{HX_1} = 1.641\ (0.0038)$	$n_{OX_1} = 0.38\ (0.0038)$	$n_{NX_1} = 0.195\ (0.0026)$
$n_{HE} = 1.646\ (0.008)$	$n_{OE} = 0.43\ (0.63)$	$n_{NE} = 0.36(0.17)$

Relative abundances of the elements H (∘), O (◇) and N (□) in the biomass.

Densities of substrate (glucose, ◇) and biomass (dry weight, □)

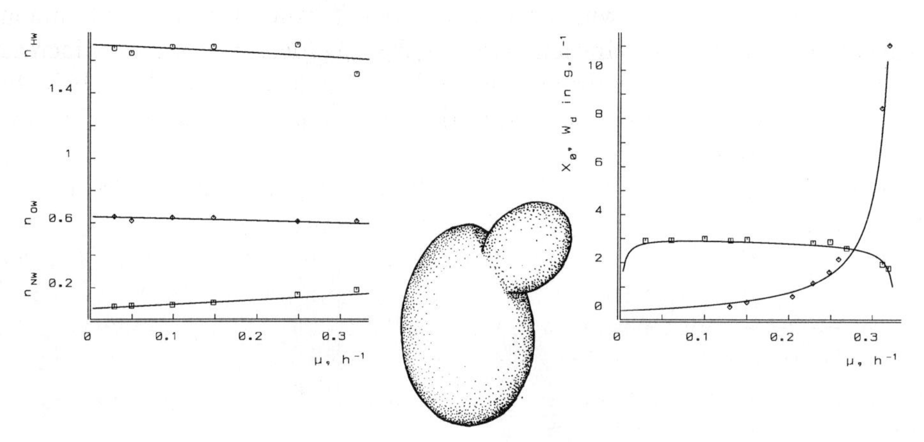

Densities of products ethanol, △, glycerol, ▽, pyruvate, ⋈

Weight-specific consumption/prod. rates of glucose, □, CO_2, ◇ and ethanol, △

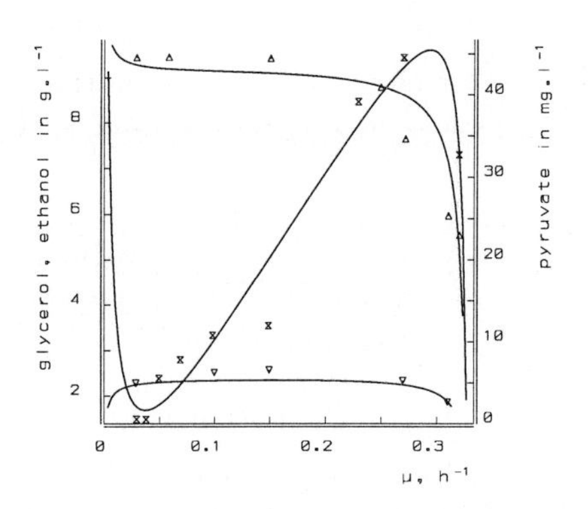

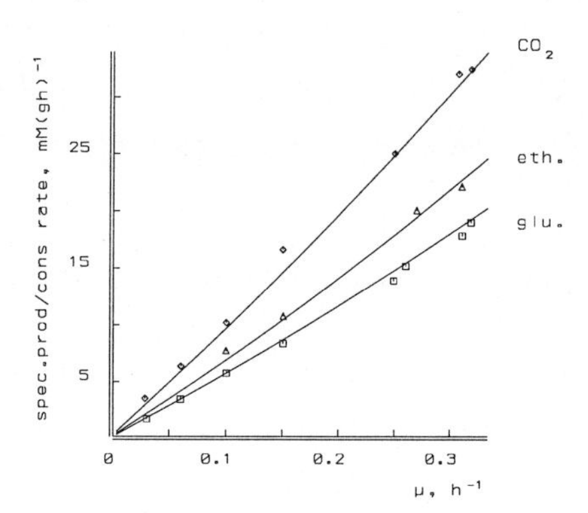

Figure 5.19: All these functions of population growth rate of *Saccharomyces cerevisiae* at 30 °C and a glucose concentration of 30 $\mathrm{g\,l^{-1}}$ in the feed have been fitted simultaneously. The observation that the maximum throughput rate is 0.34 $\mathrm{h^{-1}}$ has also been used. Data from Schatzmann [631]. The curves are based on expectations of the DEB model, with parameters (and standard deviations)

$\dot{\nu} = 0.461\ (0.008)\ \mathrm{h^{-1}}$	$g = 0.385\ (0.022)$	$\dot{m} = 0.0030\ (0.0007)\ \mathrm{h^{-1}}$
$t_{X_1} = 0.098\ (0.004)$ h	$t_E = 0.211\ (0.006)$ h	$K = 1.79\ \mathrm{g\,l^{-1}}$
$n_{HX_1} = 1.70\ (0.011)$	$n_{OX_1} = 0.637\ (0.011)$	$n_{NX_1} = 0.071\ (0.011)$
$n_{HE} = 1.55\ (0.022)$	$n_{OE} = 0.572\ (0.020)$	$n_{NE} = 0.205\ (0.021)$
ethanol	glycerol	pyruvate
$t_{P_1A} = 1.698\ (0.011)$ h	$t_{P_2A} = 1.561\ (0.022)$ h	$t_{P_3A} = 0.0066\ (0.00004)$ h
$t_{P_1M} = 0.637\ (0.011)$ h	$t_{P_2M} = 0.572\ (0.020)$ h	$t_{P_3M} = 0.0013\ (0.0018)$ h
$t_{P_1G} = 0.071\ (0.011)$ h	$t_{P_2G} = 0.205\ (0.021)$ h	$t_{P_3G} = -0.0077\ (0.00006)$ h

Comparisons in the vertical direction are relatively straightforward. Under aerobic conditions, maintenance contributes most in carbon dioxide production, while under anaerobic conditions assimilation is most important. Oxygen consumption associated with maintenance is 10 times that associated with growth investment, while the ratio only amounts to 4.5 for carbon dioxide. The respiration ratio, $-{}_mY_C/{}_mY_O$ decreases a bit for increasing population growth rate. Since both oxygen and carbon dioxide yields are hyperbolic functions of the population growth rate, the respiration ratio is also a hyperbolic function of the population growth rate.

In the second data set, three products are made by the yeast: glycerol ($n_{HP_1} = 8/3$, $n_{OP_1} = 1$), ethanol ($n_{HP_2} = 3$, $n_{OP_2} = 0.5$) and pyruvate ($n_{HP_3} = 4/3$, $n_{OP_3} = 1$). The description of product formation involves 3 time-parameters per product minus 3 constraints, making 2 parameters per product. A negative time parameter means that the product is consumed, rather than produced, in the corresponding energy flux. So it is possible that compounds are produced at a rate proportional to one energy flux and consumed at a rate proportional to another energy flux. No theoretical problems occur as long as there is an overall net production.

The curves in figure 5.19 are based on the following equations: for element composition n_{*W_1} see (5.68) with yields ${}_mY_{X_1}$ and ${}_mY_E$ in (5.57), for substrate density $X_0 = Kf(1-f)^{-1}$, with $f = g\frac{\dot{m}+\dot{\mu}}{\dot{\nu}-\dot{\mu}}$, for biomass W_1 see (5.58), for product densities $P_* = (X_r - X_0)\,{}_mY_{P_*}$ with ${}_mY_{P_*}$ in (5.57), for maximum throughput rate $\frac{\dot{\nu}-\dot{m}g(1+K/X_r)}{1+g(1+K/X_r)}$, for weight-specific uptake rate for glucose $\dot{\mu}(X_r - X_0)/W_1$ and for weight-specific production of carbon dioxide $\dot{\mu}(X_r - X_0)/W_1$ and of ethanol ${}_mY_{P_1}\dot{\mu}(X_r - X_0)/W_1$.

The experimental data do not obey the mass balance for carbon and oxygen in detail. Measurements of the volatile ethanol seem to be less reliable. The mass balance-based model fit of figure 5.19 suggests that the measured values represent 75% of the real ones, when the measurement error is considered as a free parameter. The saturation coefficient K was poorly fixed by the data, the chosen value should be considered as an educated guess.

Note that the maintenance rate coefficient $\dot{m}$ for *Klebsiella* at 35 °C is about ten times that for *Saccharomyces* at 30 °C. The maintenance rate coefficient for fungi is usually found to be much smaller in the literature [61], which Bulthuis [111] explained by the fact that fungi make a lot of protein at high population growth rates, which costs a lot of energy. As the maintenance rate coefficient is the ratio of maintenance and growth costs, this reduces the maintenance rate coefficient for fungi. Since protein density is coupled to the growth rate, however, the assumption of homeostasis dictates that most protein must be conceived as part of the reserves, so that the costs for synthesis of structural biomass is not higher for this reason.

5.4.3 Dissipating heat

In special situations, heat that dissipates from bioreactors can have a direct practical interest. The significance of dissipating heat for fundamental science is, however, in the access it gives to the free energy of structural biomass as well as of reserves. These free energies are required for an access to entropy of biomass. Although there are several ways to measure entropies of pure chemical compounds, the indirect route via dissipating heat is probably the only one available for living biomass. The setting within a thermodynamical framework is essential for various reasons, and a more rigorous definition of the concept 'energy' itself is just one of them. This subsection shows how the free energies of structural biomass and reserves can be obtained from measurements.

Many microbiological studies assume that the relative abundances of the elements are independent of feeding conditions. In the DEB theory this would mean that the relative abundances in the structural biomass and the energy reserves are the same. This allows the application of statistics such as 1 C-mole of biomass has a dry-weight of $W_d = 24.6$ g, the mean degree of reduction is 4.2, the mean Gibbs free energy is -67 kJ C-mol^{-1} (pH =7, 1 bar at 25 °C, thermodynamic reference) or $+474.6$ kJ C-mol^{-1} (pH=7, combustion reference) [305]. Since biomass composition is not constant, such crude statistics are of limited value and a more subtle approach is necessary to quantify dissipating heat.

The dissipating heat is usually related to oxygen consumption, by a fixed conversion of 519 ($\pm$13) kJ(mol O_2)$^{-1}$ [29]. This choice is not fully satisfactory for the lack of a mechanistic underpinning and because it is obviously not applicable to anaerobic conditions. The correlation between dissipating heat and carbon dioxide production has been found to be reduced by variations in type of substrate [136]. Heijnen [304] related dissipating heat to C-moles of formed biomass. This choice is problematic because of maintenance. If substrate density is low enough, no new biomass will be produced but heat will still dissipate. Within the context of the DEB model, the only satisfactory choice is to relate dissipating heat to the three energy fluxes assimilation, maintenance and growth investment, just as has been done for mass fluxes. The conservation law for energy tells exactly how heat dissipation is coupled to these energy fluxes, so no new model parameter shows up. The details are as follows.

The amount of dissipating heat depends on the chemical potentials of all compounds in the bioreactor. The first problem to consider is that the chemical potential of a compound itself depends on the concentration of that compound and on the concentration of all other compounds in the bioreactor. Although complex, this problem is well defined, except for living biomass. One of the reasons is that the concept 'concentration' does not extend to compounds tied into biomass, because these compounds are not homogeneously distributed in the environment, nor in the cell. This is the reason why I use the term 'density'. Figure 5.19 shows that biomass density hardly depends on the throughput rate. In practice, this also holds for most other compounds, except for the concentration of substrate. If changes in concentrations affect chemical potentials substantially, the chemical potential for substrate will be the first point to check. The extremes of the concentration of substrate are found for throughput rate $\dot{p} = 0$, where $X_0 = \frac{Kg\dot{m}}{\dot{\nu}-g\dot{m}}$ according to (5.51), and for throughput rate $\dot{p} = \dot{p}_m$, where $X_0 = X_r$ if death is negligible. The chemical potential of a compound depends on its concentration X as $\tilde{\mu} = \tilde{\mu}_{\text{ref}} + RT \ln X/X_{\text{ref}}$, where $R = 8.31441$ JK^{-1}mol^{-1} is the gas constant. It is just an unhappy coincidence that this standard notation for chemical potentials reminds us of population growth rates, with which it has nothing to do. The maximum relative effect of differences in concentrations of substrate on the chemical potential is

$$\frac{\tilde{\mu}_{X_0,\max} - \tilde{\mu}_{X_0,\min}}{\tilde{\mu}_{X_0,\mathrm{ref}}} = \frac{RT}{\tilde{\mu}_{X_0,\mathrm{ref}}} \ln\left\{\frac{\dot{\nu} - g\dot{m}}{g\dot{m}} \frac{X_r}{K}\right\}$$

In the example of figure 5.19, where the chemical potential of glucose is 2856 kJ mol^{-1} in the combustion frame of reference, the maximum relative effect amounts to 0.00777, which is negligibly small in view of many other uncertainties. Although the effect of changes in concentrations should be tested in each practical application, in this section I will not explicitly correct chemical potentials for differences in concentrations.

The second problem to consider is the free energy of structural biomass. I shall here treat it as some (unknown) constant, but the assumption that it is a constant which does not depend on the population growth rate is hard to substantiate. The mean age of the cells is inversely proportional to the population growth rate, cf. {177}, which means that the cumulated energy invested into development, cf. {97}, decreases with increasing population growth rate. This investment in development may reduce entropy during the cell cycle and so affects free energy.

Let ΔH denote the heat dissipation per C-mole of substrate that has been taken up, and collect the chemical potentials (free energies) of the various components in the vectors

$$\begin{aligned} \tilde{\boldsymbol{\mu}}_M^T &\equiv (\ \tilde{\mu}_C \ \ \tilde{\mu}_A \ \ \tilde{\mu}_O \ \ \tilde{\mu}_N \ \ \tilde{\mu}_+ \ ) && (5.69) \\ \tilde{\boldsymbol{\mu}}_D^T &\equiv (\ \tilde{\mu}_{X_0} \ \ \tilde{\mu}_{X_1} \ \ \tilde{\mu}_E \ \ \tilde{\mu}_{P_1} \ \ \tilde{\mu}_{P_2} \ \ \cdots \ ) && (5.70) \end{aligned}$$

just as for yields in (5.60) and (5.61). The chemical potentials of organic compounds are expressed in C-moles. The energy balance equation now reads

$$0 = \Delta H + \tilde{\boldsymbol{\mu}}_M^T \mathbf{Y}_M + \tilde{\boldsymbol{\mu}}_D^T \mathbf{Y}_D \tag{5.71}$$

Substitution of (5.63) gives for the heat yield

$$\Delta H = (\tilde{\boldsymbol{\mu}}_M^T \mathbf{un} - \tilde{\boldsymbol{\mu}}_D^T)\mathbf{Y}_D \tag{5.72}$$

which can again be decomposed into contributions from the three energy fluxes as has been done for the mass fluxes in (5.46). Let κ_{HA}, κ_{HM} and κ_{HG} denote the fraction of the energy fluxes of uptake, maintenance, and growth investment that is dissipated as heat; see figure 5.20. The heat yield can be written as the ratios of the energy fluxes and the substrate uptake rate in moles

$$\begin{aligned} \Delta H &= \kappa_{HA}\tilde{\mu}_{X_0} + \frac{\kappa_{HM}[\dot{M}]X_1}{d_{mx}[\dot{I}_m]fX_1} + \frac{\kappa_{HG}[G]\frac{d}{dt}X_1}{d_{mx}[\dot{I}_m]fX_1} && (5.73) \\ &= \kappa_{HA}\tilde{\mu}_{X_0} + \frac{\kappa_{HM}[\dot{M}]}{d_{mx}[\dot{I}_m]f} + \frac{\kappa_{HG}[G]}{d_{mx}[\dot{I}_m]f}\,\frac{\dot{\nu}e - \dot{m}g}{e+g} && (5.74) \\ \Delta H/\tilde{\mu}_E &= t_{HA}\dot{\nu} + t_{HM}\frac{\dot{m}g}{f} + t_{HG}\frac{g}{f}\,\frac{\dot{\nu}e - \dot{m}g}{e+g} && (5.75) \\ &= t_{HA}\dot{\nu} + (t_{HM}\dot{m} + t_{HG}\dot{\mu})\frac{\dot{\nu} - \dot{\mu}}{\dot{m} + \dot{\mu}} \quad \text{in equilibrium} && (5.76) \end{aligned}$$

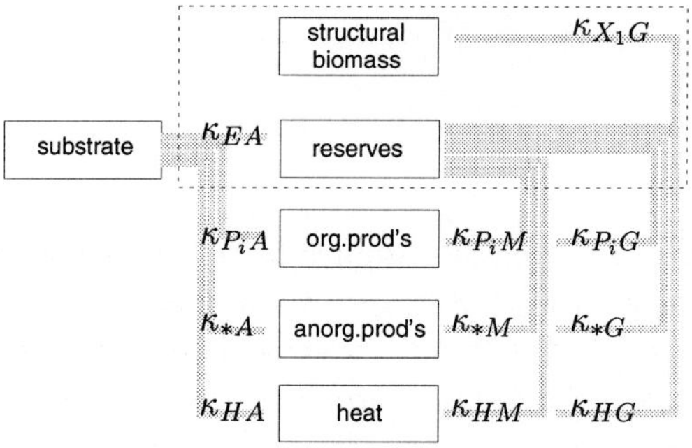

Figure 5.20: The partitionning of the energy fluxes of uptake, maintenance and growth investment in a filament. The size of the fluxes depends on substrate density, body size and amount of reserves, but the partitionning is fixed. The κ's in vertical direction add to 1. The stippled box indicates the organism.
For $*$ standing for C, A, O, N or $+$, we have

$$
\begin{array}{rclrclrcl}
\kappa_{EA} & = & -\frac{\tilde{\mu}_E t_{EA}}{\tilde{\mu}_{X_0} t_{X_0A}} & & & & \kappa_{X_1G} & = & -\frac{\tilde{\mu}_{X_1} t_{X_1G}}{\tilde{\mu}_E t_{EG}} \\
\kappa_{P_iA} & = & -\frac{\tilde{\mu}_{P_i} t_{P_iA}}{\tilde{\mu}_{X_0} t_{X_0A}} & \kappa_{P_iM} & = & -\frac{\tilde{\mu}_{P_i} t_{P_iM}}{\tilde{\mu}_E t_{EM}} & \kappa_{P_iG} & = & -\frac{\tilde{\mu}_{P_i} t_{P_iG}}{\tilde{\mu}_E t_{EG}} \\
\kappa_{*A} & = & -\frac{\tilde{\mu}_* t_{*A}}{\tilde{\mu}_{X_0} t_{X_0A}} & \kappa_{*M} & = & -\frac{\tilde{\mu}_* t_{*M}}{\tilde{\mu}_E t_{EM}} & \kappa_{*G} & = & -\frac{\tilde{\mu}_* t_{*G}}{\tilde{\mu}_E t_{EG}} \\
\kappa_{HA} & = & -\frac{\tilde{\mu}_E t_{HA}}{\tilde{\mu}_{X_0} t_{X_0A}} & \kappa_{HM} & = & -\frac{\tilde{\mu}_E t_{HM}}{\tilde{\mu}_E t_{EM}} & \kappa_{HG} & = & -\frac{\tilde{\mu}_E t_{HG}}{\tilde{\mu}_E t_{EG}}
\end{array}
$$

where $\tilde{\mu}_E = [E_m]/[d_{me}]$ is the chemical potential of the reserves and

$$(\; t_{HA} \;\; t_{HM} \;\; t_{HG} \;) \equiv t_E(\; \kappa_{HA}/\kappa_{EA} \;\; \kappa_{HM} \;\; \kappa_{HG} \;)$$

where $\kappa_{EA} \equiv [\dot{A}_m](d_{mx}[\dot{I}_m]\tilde{\mu}_{X_0})^{-1}$ denotes the fraction of energy that has been taken up as substrate that arrives in the reserves as assimilation energy.

The time parameters are given by

$$\tilde{\mu}_E(\; t_{HA} \;\; t_{HM} \;\; t_{HG} \;) = (\tilde{\boldsymbol{\mu}}_M^T \mathbf{un} - \tilde{\boldsymbol{\mu}}_D^T)\mathbf{t}_D \tag{5.77}$$

where the matrix $\mathbf{t}_D$ is defined in (5.66). The derivation is analogous to that of (5.66); substitute the yield coefficients (5.57) in (5.72) and equate (5.72) to (5.76), multiply by $\dot{m} + \dot{\mu}$ to convert the equation to second degree polynomials in $\dot{\mu}$ and equate the three polynomial coefficients to each other. So, given the chemical potentials of all compounds, (5.76) together with (5.77) gives the dissipating heat. Note that (5.77) represents three equations that are independent of $\dot{\mu}$, while (5.71) represents infinitely many equations, because it still depends on $\dot{\mu}$. The introduction of the time parameters, therefore, reduces complexity considerably.

Relationship (5.77) can be used to obtain the chemical potentials $\tilde{\mu}_{X_1}$ and $\tilde{\mu}_E$ from measurements of ΔH as function of $\dot{\mu}$. Non-linear regression will provide estimates for $\tilde{\mu}_E t_{HA}$, $\tilde{\mu}_E t_{HM}$ and $\tilde{\mu}_E t_{HG}$, since $\dot{\nu}$ and $\dot{m}$ in (5.76) can be obtained from mass fluxes, as has been demonstrated in the previous subsection. The practical implementation is to consider t_{HA}, t_{HM}, t_{HG}, $\tilde{\mu}_{X_1}$ and $\tilde{\mu}_E$ all as free parameters and the three equations (5.77) as constraints on the parameter values. Lagrange multipliers can be used to deal with these constraints. In the combustion frame of reference, we have $\boldsymbol{\mu}_M^T = \mathbf{0}$. If no products are formed, (5.77) reduces to $t_{HM} = t_E$ and $\tilde{\mu}_{X_1} = \tilde{\mu}_E(t_{HM} - t_{HG})g/t_{X_1}$ and $\tilde{\mu}_E = \tilde{\mu}_{X_0}\dot{\nu}^{-1}(t_{HA} + t_{HM})^{-1}$.

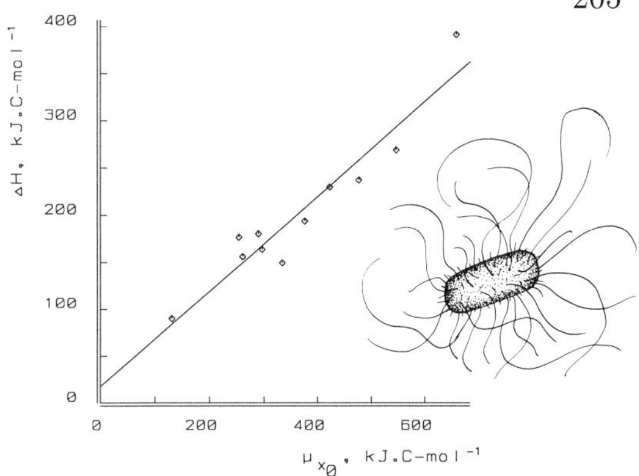

Figure 5.21: The amount of dissipating heat at maximum population growth rate is linear in the free energy per C-mole of substrate on the basis of combustion reference (pH = 7). Data from Rutgers [622] and Heijnen and van Dijken [305,304] for *Pseudomonas oxalaticus*, growing aerobically at 30 °C on a variety of substrates.

The chemical potential for the total biomass is $\tilde{\mu}_{W_1} = (\tilde{\mu}_{X_1\, m}Y_{X_1} + \tilde{\mu}_{E\, m}Y_E)\, {}_mY_{W_1}^{-1}$. This chemical potential depends on population growth rate via (5.55) and (5.65). Although the estimation of the chemical potential for biomass is laborious, application of these equations solves the problem in principle.

The estimation procedure can be simplified by making use of the empirical finding that heat dissipation is proportional to oxygen consumption for all $\dot{\mu}$. So $\Delta H\, {}_mY_O^{-1}$ is constant at value $\tilde{\mu}_h \simeq 519$ kJ(mol O_2)$^{-1}$. Substitution of the proposition that $\Delta H = \tilde{\mu}_{h\, m}Y_O$ into (5.72), which resulted from the energy balance equation, requires that

$$\tilde{\boldsymbol{\mu}}_D^T = (\tilde{\mu}_h \mathbf{u}_{O*} + \tilde{\boldsymbol{\mu}}_M^T \mathbf{u})\mathbf{n}$$

where $\mathbf{u}_{O*}$ denotes the third row of $\mathbf{u}$, i.e. the one that corresponds to oxygen. The chemical potentials $\tilde{\mu}_{X_1}$ and $\tilde{\mu}_E$ are elements 2 and 3 of $\tilde{\boldsymbol{\mu}}_D$. The first element, $\tilde{\mu}_{X_0}$, can be used to fix $\tilde{\mu}_h = (\tilde{\mu}_{X_0} - \tilde{\boldsymbol{\mu}}_M^T \mathbf{u}\mathbf{n}_{*X_0})(\mathbf{u}_{O*}\mathbf{n}_{*X_0})^{-1}$. Although these relationships are attractively simple, the basis is empirical only and should therefore be treated with some caution.

When different substrates are compared, the dissipating heat tends to increase with the free energy of substrate. The molar yield of biomass at a fixed population growth rate was found to be proportional to the free energy of substrate if the maximum volume-specific assimilation rate $[\dot{A}_m]$ and the maximum reserve capacity $[E_m]$ are proportional to the free energy per C-mole of substrate, see on {191}. The dissipating heat at maximum population growth rate is approximately linear in the free energy per C-mole of substrate if the combustion reference is used. This frame of reference is necessary because a high free energy of substrate corresponds with a high degree of reduction, which requires more oxygen to release the energy. In the combustion reference, this extra use of oxygen does not affect the relationship between free energy of substrate and heat dissipation. The approximate linearity of this relationship is easy to infer, in absence of product formation, when $\dot{\mu} = \dot{\mu}_m = \frac{\dot{\nu} - \dot{m}g}{1+g}$ is substituted in $\mathbf{Y}_D$ in (5.72); the result is

$$\mathbf{Y}_D^T = (\ -1 \quad t_E\dot{\mu}_m \quad t_{X_1}\dot{\mu}_m \quad t_{P_1A}\dot{\nu} + t_{P_1M}\dot{m}g + t_{P_1G}\dot{\mu}_m g \quad \cdots\ )$$

Since $\dot{\nu}$ is independent of the chemical potential, and g is inversely proportional to it, the maximum population growth rate will tend to $\dot{\nu}$ for increasing values for the chemical potential. If no products are formed, $\mathbf{Y}_D^T$ has only three elements; t_E is via $[\dot{d}_{me}]$ and $[E_m]$ proportional to the chemical potential of the substrate $\tilde{\mu}_{X_0}$, which itself is the first element of $\tilde{\boldsymbol{\mu}}_D$ in (5.72). This is why ΔH is linear in $\tilde{\mu}_{X_0}$. This is confirmed by the data of Rutgers [622]; see figure 5.21.

The idea that type of substrate and environmental conditions affect the substrate/energy conversion $[\dot{I}_m]/[\dot{A}_m]$ (and $[E_m]$) but nothing else, is consistent with analyses of data from Pirt [556], who plotted the inverse of the yield against the inverse of the population growth rate and obtained the linear relationship formulated by Marr [455]. According to the DEB theory for filaments with small reserve capacities $[E_m]$, this relationship is $\frac{1}{Y} = \frac{[\dot{I}_m]}{[\dot{A}_m]}[G] + \frac{[\dot{I}_m]}{[\dot{A}_m]}\frac{[\dot{M}]}{\dot{\mu}}$. As Pirt noted, this relationship is linear in $\dot{\mu}^{-1}$, but the slope depends on the substrate-energy conversion $[\dot{A}_m]/[\dot{I}_m]$. Pirt found a wide range of 0.083 to 0.55 h^{-1} on a weight basis for two species of bacteria growing on two substrates, aerobically and anaerobically. The ratio of the slope and the intercept equals the maintenance rate coefficient, $\dot{m} \equiv [\dot{M}]/[G]$, which does not depend on the substrate-energy conversion. Pirt's data fall in the narrow range of 0.0393 to 0.0418 h^{-1} [406]. These findings support the funnel concept, which states that a wide variety of substrates is decomposed to a limited variety of building blocks, which depends of course on the nature of the substrate and environmental conditions; these products are then built into biomass, which only depends on internal physiological conditions, subject to homeostasis.

Battley [41] argued that maintenance might follow from kinetic arguments, but that thermodynamics arguments lead to a rejection of the concept. This subsection and the previous ones show, however, that the coefficients of mass and energy balance equations can be written as functions of kinetic parameters and that such equations by no means exclude maintenance.

5.5 Computer simulations

As long as food density remains constant and stable age-distributions exist, it is possible to study most phenomena analytically, as illustrated in the preceding sections. For many purposes non-equilibrium situations should be considered, which requires computer simulation studies. Two strategies can be used to follow population dynamics: the family-tree method and the frequency method.

The family-tree method evaluates the changes of the state variables for each individual in the population at each time increment. For this purpose, the individuals are collected in a matrix, where each row represents an individual and each column the value of a state variable. At each time increment rows can be added and/or deleted and at regular time intervals population statistics such as the total volume of individuals are evaluated. The amount of required computer time is thus roughly proportional to the number of individuals in the population which must, therefore, be rather limited. This restricts the applicability of this method for analytical purposes, because at low numbers of individuals stochastic phenomena, such as those involved in survival, tend to dominate. The method is very flexible, however, which makes it easy to incorporate differences between individuals with respect to their parameter values. Such differences are realistic and appear to affect population dynamics substantially; see {210}. Kaiser [367,368] used the programming environment SIMULA successfully to study the population dynamics of individual dragon flies, mites and rotifers.

The frequency method is based on bookkeeping in terms of (hyperbolic) partial differential equations. Several strategies exist to integrate these equations numerically. The method of the escalator boxcar train, perfected by Andree de Roos [611], follows cohorts of individuals through the state space. The border of the state space where individuals appear at birth, is partitioned into cells, which are allowed to collect a cohort of neonates for a specified time increment. The reduction of the number of individuals in the cohorts is followed for each time increment, as the cohort moves through the state space. The amount of computer time required is proportional to the number of cohorts, which relates to the volume of the state space as measured by the size of the cells. The number of cells must be chosen by trial and error. The escalator boxcar train is just one method of integrating the partial differential equation, but it appears to be an efficient one compared with methods that use a fixed partitioning of the state space into cells that transfer numbers of individuals among them.

A nasty problem of the (partial) differential approach for the description of population dynamics is the continuity of the number of neonates if the reproduction rate is very small. This situation occurs in equilibrium situations, if the loss rate is small. The top predators especially are likely to experience very small loss rates. Details of the handling of energy reserves to produce or not produce a single young prove to have a substantial effect on population dynamics.

The formulation of population changes in terms of partial differential equations is as follows for dividing isomorphs in chemostats. Let $n(t,e,l)$ denote the density of individuals having scaled energy density e and scaled length l at time t, so $\int_{e_1}^{e_2}\int_{l_1}^{l_2} n(t,e,l)\,dl\,de$ is the number of individuals having a scaled energy density somewhere between e_1 and e_2 and a scaled length somewhere between l_1 and l_2. The total number of individuals equals $N(t) = \int_0^1\int_0^1 n(t,e,l)\,dl\,de$, the mean scaled surface area is $\mathcal{E}\underline{l}^2(t) = N(t)^{-1}\int_0^1\int_0^1 l^2 n(t,e,l)\,dl\,de$, and the total biovolume is $X_1(t) = N(t)V_m\mathcal{E}\underline{l}^3$, with $\mathcal{E}\underline{l}^3 = N(t)^{-1}\int_0^1\int_0^1 l^3 n(t,e,l)\,dl\,de$. The change of the density is given by the von Foerster equation [224,661] for two state variables:

$$\frac{\partial}{\partial t}n(t,e,l) = -\frac{\partial}{\partial l}\left(n(t,e,l)\frac{d}{dt}l\right) - \frac{\partial}{\partial e}\left(n(t,e,l)\frac{d}{dt}e\right) - (\dot{p} + \dot{h}(e,l))n(t,e,l) \qquad (5.78)$$

where $\frac{d}{dt}e$ and $\frac{d}{dt}l$ are given in (3.22) and (4.3) and the hazard rate $\dot{h}(e,l)$ in (3.61). This hyperbolic partial differential equation has the boundary condition

$$n(t,e,l_b)\left.\frac{d}{dt}l\right|_{l=l_b} = 2n(t,e,l_d)\left.\frac{d}{dt}l\right|_{l=l_d} \qquad (5.79)$$

For division into two equal parts, we take $l_b = l_d 2^{-1/3}$. Substitution of (4.2) and (4.3) gives

$$n(t,e,l_b) = 2n(t,e,l_d)\frac{\dot{v}e - g\dot{m}l_d}{\dot{v}e - g\dot{m}l_b} \qquad (5.80)$$

This formulation ties fission to the growth process. The formulations of Sinko and Streifer [661], and Metz and Diekmann [478], treat both processes as independent. Together with

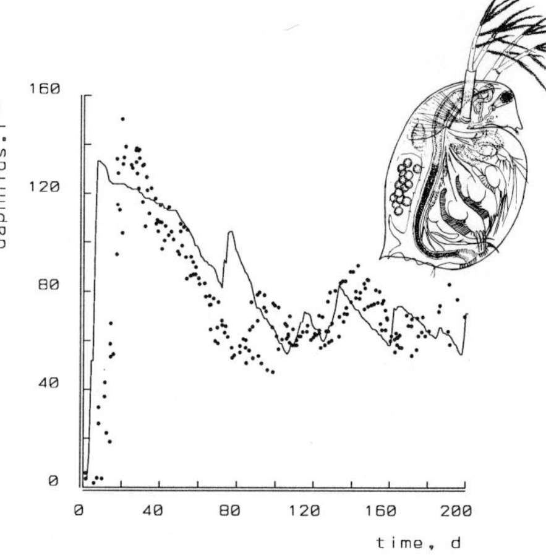

Figure 5.22: Computer simulation of a DEB-structured population of *Daphnia magna*, compared to a real laboratory population at 20 °C with a supply of 5×10^7 cells *Chlorella saccarophila* d^{-1}, starting from 5 individuals. Data from Fitsch [221]. The parameter values have been obtained independently from observations of individuals. Parameters:

$\{\dot{I}_m\}$	5×10^4 cells mm^{-2} h^{-1}	$\dot{m}g$	0.33 h^{-1}	g	0.033
K	3×10^5 cells ml^{-1}	l_b	0.133	l_p	0.417
$\ddot{p}_a$	1.1×10^{-6} h^{-2}	cv	0.5	q	0.9

the partial differential equation and its boundary condition, the differential equation

$$\frac{d}{dt}X_0 = \dot{p}(X_r - X_0) - \{\dot{I}_m\}V_m^{2/3} f N \mathcal{E} \underline{l}^2 \tag{5.81}$$

determines the fate of the food, which shows up in $\frac{d}{dt}e$ and $\frac{d}{dt}l$. Since these equations describe changes only, we also need the value of $X_0(0)$ and $n(0, e, l)$ for all e and l for some point $t = 0$ in time to evaluate the time path.

If the individuals propagate by reproduction rather than by division, (5.78) and (5.81) still apply, but the hazard rate $\dot{h}(e, l)$ is now given in (4.4) and the boundary condition reads

$$n(t, e, l_d) = \int_e \int_l \dot{R}(e, l) n(t - a_b, e, l)\, dl\, de \tag{5.82}$$

where the reproduction rate $\dot{R}$ is given in (4.5). A problem with this delay boundary condition is that we now need to specify $n(t, e, l)$ for all t between 0 and the incubation time a_b as the initial condition. A substantial simplification is obtained by just neglecting this time delay and hoping that this will not affect population dynamics too much.

Figure 5.22 demonstrates that computer simulations of DEB structured daphnids closely match the dynamics of laboratory populations. The strength of the argument is in the fact that the parameter values for individual performance have been obtained independently.

5.5.1 Synchronization

Computer simulations of fed-batch cultures of reproducing isomorphs reveal a rather unexpected property of the DEB model. In these simulations the food supply to the population is taken to be constant and the population is harvested by the process of aging and in a random way. To reduce complicating factors as much as possible, only parthenogenetically reproducing females are considered with parameter values that are realistic for *Daphnia magna* feeding on the green alga *Chlorella pyrenoidosa* at 20 °C. Reproduction in daphnids is coupled to moulting, which occurs every 2 to 3 days at 20 °C, irrespective of food availability. Just after moulting, the brood pouch is filled with eggs which hatch just before the next moult. So the intermoult period is beautifully adapted to the incubation time and

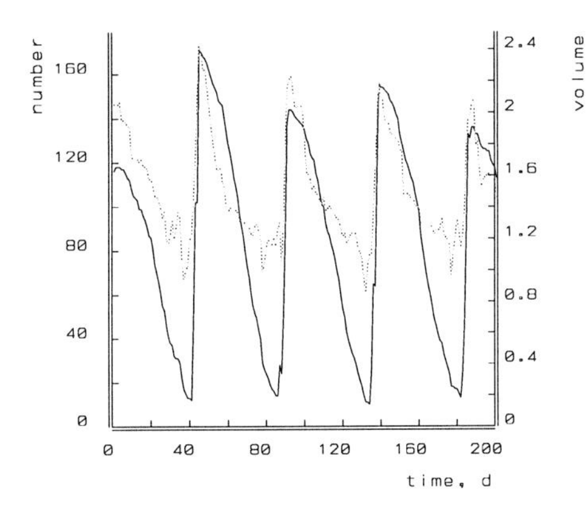

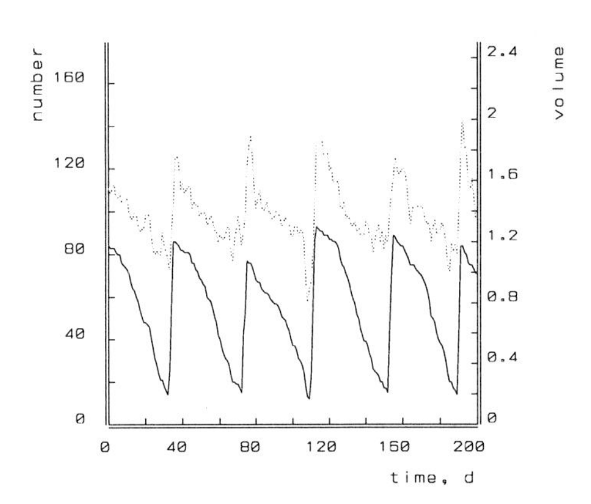

Figure 5.23: The number of individuals (—) and the total biovolume (···) in a simulated batch culture of daphnids subjected to aging as the only harvesting effort. The individuals accumulate reproductive effort during the incubation time in the left figure, while they reproduce egg by egg in the right one. The parameters are $\dot{p}X_r = 7$ units d^{-1}, $l_b = 0.133$, $l_p = 0.417$, $\dot{I}_m = 4.99$ units d^{-1}, $\kappa = 0.3$, $\dot{m} = 10\ \mathrm{d}^{-1}$, $g = 0.033$, $\ddot{p}_a = 2.5 \times 10^{-5}\ \mathrm{d}^{-2}$.

the buffer for energy allocated to reproduction stays open during the intermoult period. These details are followed in the simulation study because many species produce clutches rather than single eggs.

Figure 5.23 shows a typical result of the population trajectory: the numbers oscillate substantially at low random harvesting rates. Closer inspection reveals that the shape of the cycles of numbers closely follows the survival function of the aging process. The individuals appear to synchronize their life cycles, i.e. their ages, lengths and energy reserve densities, despite the fact that the founder population consisted of widely different individuals. This synchronization is reinforced by the accumulation of reproductive effort into clutches, but it also occurs with single-egg reproduction. The path individuals take in their state space closely follows the no-growth condition. Growth in these populations can only occur via thinning by aging and the resultant amelioration of food shortage. After reaching adult volumes, the individuals start to reproduce and mothers are soon outcompeted by their offspring, because they can survive at lower food densities. This has indeed been observed in experimental populations [268,709].

Having observed the synchronization of the individuals, it is not difficult to quantify population dynamics from an individual perspective when we now know that the scaled functional response cycles from $f = l_b$ to l_p. Starting from a maximum $N(0) = \frac{\dot{p}X_r g^2 \dot{m}^2}{\{\dot{I}_m\} l_b^3 \dot{v}^2}$ at time 0, the numbers drop according to $N(t) = N(0) \exp\{-\int_0^t \dot{h}(t_1)\, dt_1\}$, down to $N(t_n) = N(0)(l_b/l_p)^3$. The total biovolume is about constant at $X_1 = \frac{\dot{p}X_r V_m^{1/3}}{\kappa\{\dot{I}_m\}}$; see table 5.3. At the brief period of take-over by the next generation, the population deviates a little from this regime. It is interesting to note that growth and reproduction are fully determined by the aging process in this situation. Length-at-age curves do not resemble the satiation curve that is characteristic of the von Bertalanffy growth curve, they are more or less exponential.

Table 5.3: Oscillations can affect crude population statistics. This table compares statistics for computer simulations, on the assumption that either reproduction by clutches, or by eggs laid one at a time, with statistics that assume the stable age distribution.

statistic	clutch	single-egg	stable age
mean scaled functional response, f	0.355	0.340	0.452
mean scaled biomass density, x	1.095	0.99	0.943
mean number of individuals, N	87.0	55.3	18.3
scaled yield coefficient, y	0.214	0.186	0.120

Biovolume density and the yield are increased by the oscillatory dynamics, compared to expectation on the basis of the stable age distribution.

If the harvesting effort is increased, the population experiences higher food densities and the model details for growth and reproduction become important. The shape of the length-at-age curves switches from 'exponential' to von Bertalanffy, the cycle period shortens, the generations overlap for a longer period because competition between generations becomes less important, and the tendency to synchronize is reduced. All these changes result from the tendency of populations to grow to situations of food shortage if harvesting rates drop.

Similar synchronization phenomena are known for the bakers' yeast *Saccharomyces cerevisiae* [532,123]. It produces buds at soon the cell exceeds a certain size. This gives a synchronization mechanism that is closely related to the mechanism for *Daphnia*.

5.5.2 Variation between individuals

Although it is not unrealistic to have fluctuating populations at constant food input [665], the strong tendency of individuals to synchronize their life cycles seems to be unrealistic. Yet the model describes the input-output relationships of individuals rather accurately. A possible explanation is that at the population level some new phenomena play a role, such as slightly different parameter values for different individuals. This gives a stochasticity of a different type than that of the aging process, which is effectively smoothed out by the law of large numbers. This way of introducing stochasticity seems attractive because the replicatebility of physiological measurements within one individual generally tends to be better than between individuals. The exact source of variation in energy parameters, however, is far from obvious. This applies especially to parthenogenetically reproducing daphnids, where recombination is usually assumed not to occur. Hebert [298] however, has reported that (natural) populations of daphnids, which probably originate from a limited number of winter eggs, can have substantial genetic variation. Branta [86] was able to obtain a rapid response to selection in clones of daphnids, which could not be explained by the occurrence of spontaneous mutations. Cytoplasmic factors possibly play a much more important role in gene expression than is recognized at the moment. Koch [395] has

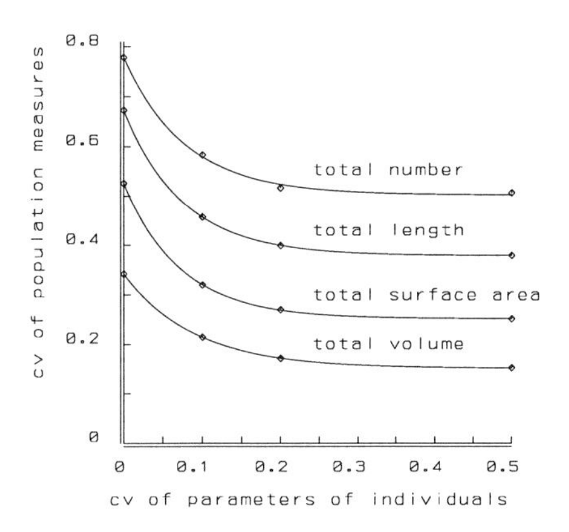

Figure 5.24: The coefficient of variation of the total number, length, surface area and volume of individuals in the population as functions of the coefficient of variation of the scatter parameter that operates on the parameters of individuals. The sharp initial reduction points to the limited realism of strictly deterministic models.

discussed individual variability among bacteria.

In principle, it is possible to allow all parameters to scatter independently, but this seems neither feasible nor realistic. High ingestion rates, for instance, usually go with high assimilation rates and storage capacities. The parameter values of the DEB model for different species appear to be linked in a simple way, as discussed in the section on parameter variation. We assume here that the parameters of the different individuals within a species are also linked in this way but vary within a narrow range. The parameters for a particular individual remain constant during its lifetime. In this way, we require only one simple individual-specific multiplier operating on (some of) the original parameters of the DEB model to produce the scatter. The way the scatter appears in the scaled parameters is even simpler [415].

Parameter variation between individuals has interesting effects on population dynamics: a log-normally distributed scatter with even a small coefficient of variation is enough to prevent death by starvation at the take-over of the new generation. Moreover, each generation becomes extinct only halfway through the period of the next generation and the amplitude of the population oscillations is significantly reduced; see figure 5.23. This may be quantified by its effect on the coefficient of variation for population measures, defined as

$$c_j = \left(\frac{1}{t_n} \int_0^{t_n} \sum_i l_i^j(t)\, dt \right)^{-1} \sqrt{\frac{1}{t_n} \int_0^{t_n} \left(\sum_i l_i^j(t) - \frac{1}{t_n} \int_0^{t_n} \sum_i l_i^j(t_1)\, dt_1 \right)^2 dt} \qquad (5.83)$$

for $j = 0, 1, 2, 3$. Integration is taken over one typical cycle of length t_n and the summation over all individuals in the population. For values larger than 0.2, the coefficient of variation of the scatter parameter barely depresses the variation coefficient of the population measures further; see figure 5.24.

The oscillations are also likely to be less if one accounts for spatial heterogeneities. This is realistic even for daphnids, because some of the algae adhere to the walls of the vessels and some (but not all) daphnids feed on them [238,336]. The general features of the dynamics of experimental populations are well captured by the DEB model. Emphasis is given to the competition for food, which Slobodkin considered to be the only type of interaction operative in his experimental food limited populations. He suggested that the competition between different age-size categories was responsible for the observed intrinsic oscillations, which is confirmed by this model analysis. Nelly van der Hoeven [327] has concluded, on the basis of a critical survey of the literature on experimental daphnid populations with constant food input, that some fluctuations are caused by external factors. Even populations that tend to stabilize do so, however, by way of a series of damped oscillations, while others seem to fluctuate permanently.

5.6 Chains

This section presents the first preliminary results that one has to obtain when aiming at community dynamics. The general strategy is to link populations of DEB-structured individuals into food chains, and in the future into food webs, where the parameters of the participating species are subjected to body size scaling relationships, as worked out in the next chapter. This reduces the problem of community dynamics in principle, to that of particle size distributions in taxon free communities, as reviewed by Damuth *et al.* [153].

In the previous section, the population was subjected to an experimentally controlled harvesting rate. This approach allows an extension to food chains, where the predator population, not the experimenter, sets the harvesting effort. We still have indirect access to the prey population via the harvesting effort exercised on the predator population. The total volumes of substrate, prey and predator will settle to equilibrium value (or a cyclic pattern therein), as a result of the (still constant) substrate flux to the three step food chain and the experimentally controlled harvesting rate. The coupling between the prey and predator populations is given by

$$(\dot{p}_1 - \dot{p})X_1 = \{\dot{I}_m\}_{1,2} f_{1,2} X_2 \frac{\mathcal{E}\underline{V}_2^{2/3}}{\mathcal{E}\underline{V}_2} \text{ with } f_{1,2} \equiv \frac{X_1}{K_2 + X_1} \tag{5.84}$$

where the indices refer to the species numbers and $\dot{p}_1$ denotes the individual-specific predation rate, which is defined by this equation. The coupling between the substrate and the prey population is given in exactly the same way. With only three trophic levels, we have $\dot{p}_2 = \dot{p}$.

Studying the dynamic behaviour of the food chain in the chemostat environment, Bob Kooi [401,402] constructed 'operating diagrams'. These diagrams give the dynamics of the system as a function of the operating parameters, the rate of dilution $\dot{p}$ and concentration of food X_r. The results presented in the operating diagrams are based upon local stability analysis based on standard techniques, which are not described here. The analyses assumed that the energy reserves are at their equilibrium value. Numerical studies produced the same results for filaments without constraints on energy reserves. The parameter values were chosen equal those given by Nisbet *et al.* [510], who based them on an actual substrate-bacterium-ciliate chain.

Nisbet *et al.* noted that the experimental system appears to be much more stable than is predicted by the 'Double Monod' model, i.e. a special case of the DEB model for filaments with $[\dot{M}]_i = 0$ and $[E_m] = 0$, for $i = 1, 2$. They concluded that the introduction of maintenance, as proposed by Pirt, increased the range of operation parameters that give stable chains; however, real world chains still appear to be more stable.

The results for filaments are summarized in figure 5.25, which compares the operating diagrams for a chain extended from length 2, 3 to 4, with and without maintenance. The results for dividing isomorphs are almost indistinguishable from these results for filaments. This does not come as a surprise because the yield coefficients are almost the same; see figure 5.14. The effect of an increase of energy reserves is that washout occurs at much lower dilution rates if the other parameters remain the same. The area of the control variables $\dot{p}$ and X_r where the chain is stable increases substantially. Note, however, that this way of comparing models is fit for theoretical analysis, but tells less about real systems. If models with and without energy reserves are fitted to real systems, they are likely to differ in parameters that do not relate to energy reserves. This is because they must produce about the same output as a result of the fitting procedure.

All these results are based upon local stability analysis. Global dynamics in unstable regions have been studied by numerical integration of the set of differential equations, by a Runge–Kutta method starting from perturbed equilibrium. Hastings and Powell [295] observed chaos in a three-species food chain that was closely related to the Double Monod model. The main differences were that the substrate was growing logistically and that the prey and predator died at widely different rates. The DEB model showed no chaos, only limit cycles for unstable regions, but we have no proof of its absence. We did, however, spot another source of problems: the prey density can become extremely low in the limit cycles, which severely complicates numerical analysis and hampers the biological interpretation.

The non-equilibrium dynamics of food chains can be rather complex and sensitively depends on the initial conditions. Figure 5.26 illustrates results for a substrate-bacteria-myxamoebae chain in a chemostat. Bob Kooi has been able to fit the experimental data to the DEB model for filaments with remarkable success. All parameters were estimated on the basis of a weighted least squares criterion. The main dynamic features are well described by the model. The myxamoebae decrease more rapidly in time than the throughput rate allows by shrinking during starvation. The type of equilibrium of this chain is known as a spiral sink, so that this chain ultimately stabilizes, and the period reduces with the amplitude. The numerical integration of the set of differential equations that describe the system has been a 4-th order Runge–Kutta method with a time step of 0.01 h. This particular data set has been used to illustrate the application of catastrophe theory by Saunders [630], who concluded that simple generalizations of the Lotka–Volterra model cannot fit this particular dataset and his analyses strongly suggest that the feeding rate for each individual myxamoeba is proportional to the product of the bacteria and the myxamoebae densities. This implies an interaction between the myxamoebae and Bazin and Saunders [47] suggested that the myxamoebae measure their own density via folic acid. Although interactions cannot be excluded, the goodness of fit of the DEB model makes clear that it is not necessary to include interactions. The significance of realistic descriptions without interaction is in the extrapolation to other systems; if species-specific interactions would dominate systems behaviour, there can be hardly any hope for the feasibility of community ecology.

no maintenance or reserves

maintenance and reserves

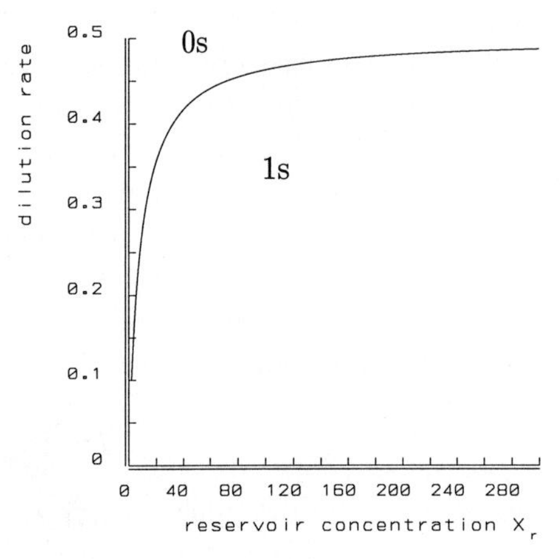

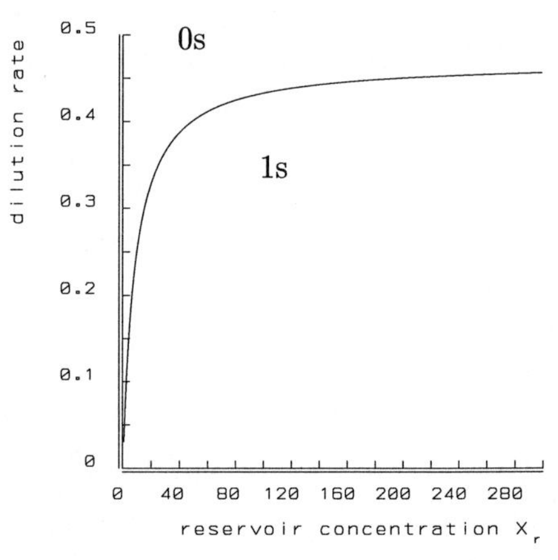

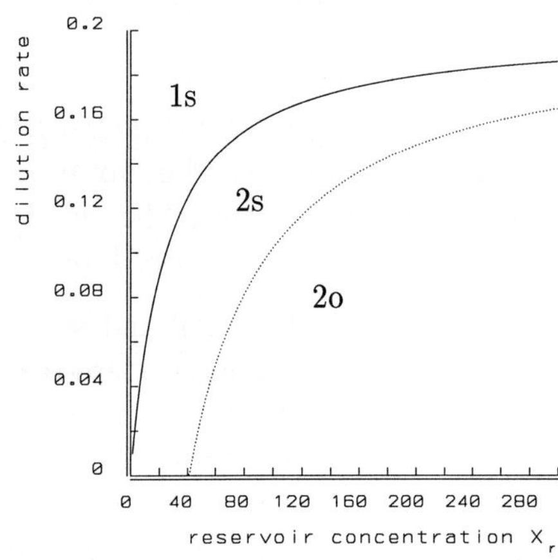

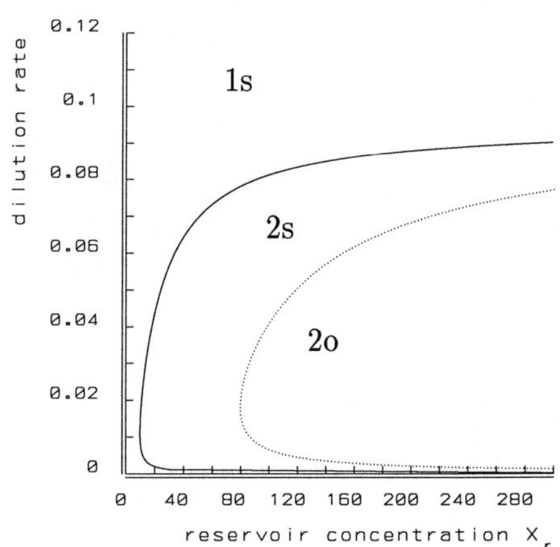

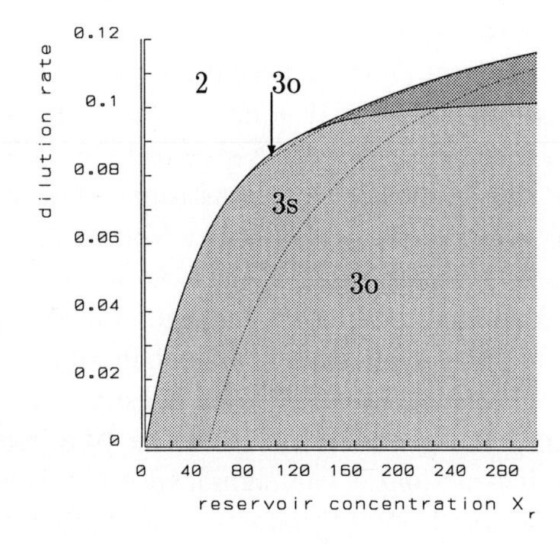

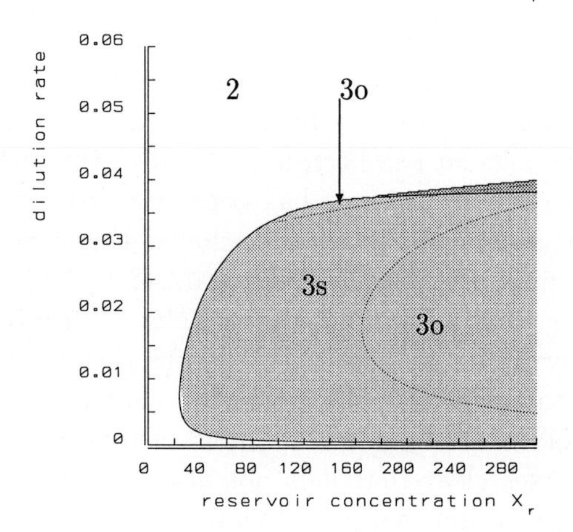

Figure 5.25: Operating diagrams for food chains of filaments of length 2, 3 and 4, in the Monod model (no maintenance, no reserves, left column) and the DEB model (right column). The numbers give the highest level that can exist given the operating conditions; 's' stands for stable coexistence and 'o' for unstable or oscillatory coexistence. The stippled curves represent bifurcation diagrams. The dark area indicates 2 solutions for the equilibrium, but the second solution is complex valued with a positive real component, which means that the chain of full length cannot exist in this solution. The parameter values are

$V_{d,1}^{1/3} = 0.63\,\mu\text{m}$	$y_{0,1} = 0.4$	$\dot{\nu}_1 = 40\,\text{h}^{-1}$	$K_{0,1} = 8\,\text{mg}\,\text{l}^{-1}$	$g_1 = 80$	$\dot{m}_1 = 0.025\,\text{h}^{-1}$
$V_{d,2}^{1/3} = 50.4\,\mu\text{m}$	$y_{1,2} = 0.6$	$\dot{\nu}_2 = 0.2\,\text{h}^{-1}$	$K_{1,2} = 9\,\text{mg}\,\text{l}^{-1}$	$g_2 = 1$	$\dot{m}_2 = 0.001\,\text{h}^{-1}$
$V_{d,3}^{1/3} = 100\,\mu\text{m}$	$y_{2,3} = 0.6$	$\dot{\nu}_3 = 0.0756\,\text{h}^{-1}$	$K_{2,3} = 10\,\text{mg}\,\text{l}^{-1}$	$g_2 = 0.504$	$\dot{m}_3 = 0.0075\,\text{h}^{-1}$

Figure 5.26: A chemostat with a three-step food chain of glucose X_0, the bacterium *Escherichia coli* X_1 and the cellular slime mold *Dictyostelium discoideum* X_2 at 25 °C, throughput rate $\dot{p} = 0.064\ \text{h}^{-1}$ and a glucose concentration of $X_r = 1\ \text{mg}\,\text{ml}^{-1}$ in the feed. Data from Dent *et al.* [164].

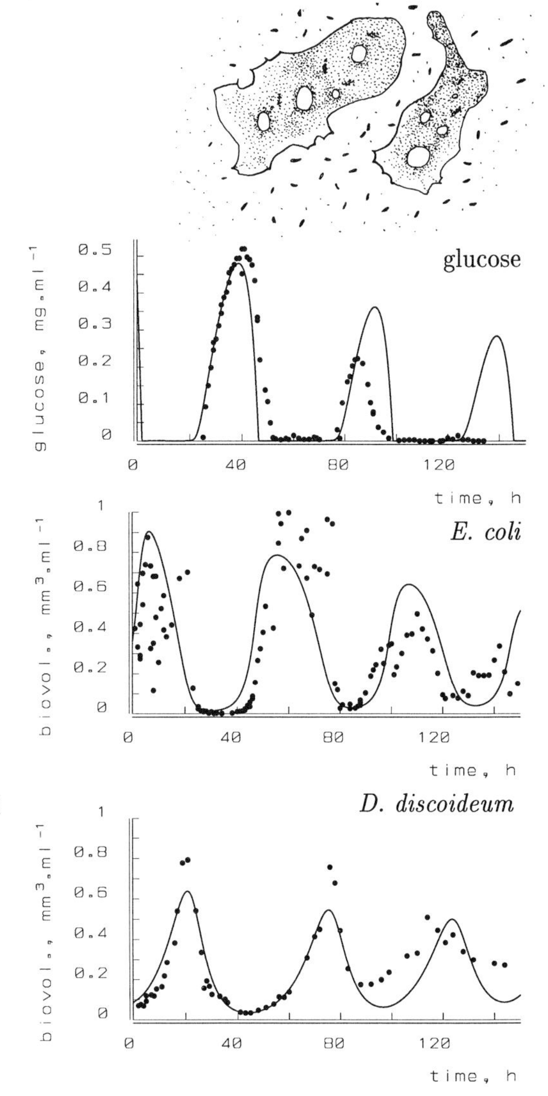

The parameter values and equations are

$X_0(0)$	0.433			$\text{mg}\,\text{ml}^{-1}$
$X_1(0)$	0.361	$X_2(0)$	0.084	$\text{mm}^3\,\text{ml}^{-1}$
$e_1(0)$	1	$e_2(0)$	1	-
K_1	1.312	K_2	0.173	$\frac{\mu\text{g}}{\text{ml}}, \frac{\text{mm}^3}{\text{ml}}$
g_1	0.84	g_2	4.07	-
$\dot{m}_1$	0.002	$\dot{m}_2$	0.159	h^{-1}
$\dot{\nu}_1$	0.652	$\dot{\nu}_2$	1.853	h^{-1}
$[\dot{I}_m]_1$	0.619	$[\dot{I}_m]_2$	0.261	$\frac{\text{mg}}{\text{mm}^3\,\text{h}}, \text{h}^{-1}$

$$\frac{d}{dt}X_0 = \dot{p}(X_r - X_0) - \frac{X_0 X_1 [\dot{I}_m]_1}{K_1 + X_0}$$

$$\frac{d}{dt}X_1 = \left(\frac{\dot{\nu}_1 e_1 - \dot{m}_1 g_1}{e_1 + g_1} - \dot{p}\right) X_1 - \frac{X_1 X_2 [\dot{I}_m]_2}{K_2 + X_1}$$

$$\frac{d}{dt}X_2 = \left(\frac{\dot{\nu}_2 e_2 - \dot{m}_2 g_2}{e_2 + g_2} - \dot{p}\right) X_2$$

$$\frac{d}{dt}e_1 = \dot{\nu}_1 \left(\frac{X_0}{K_1 + X_0} - e_1\right)$$

$$\frac{d}{dt}e_2 = \dot{\nu}_2 \left(\frac{X_1}{K_2 + X_1} - e_2\right)$$

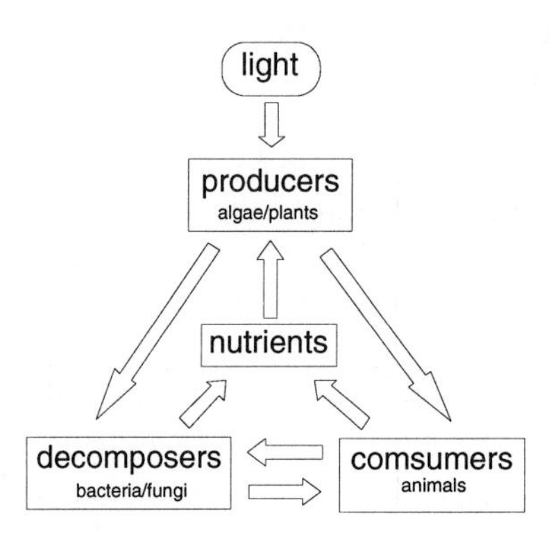

Figure 5.27: The most simple non-degenerated community consists of three components: primary producers that gain energy from light and take up nutrients independently to produce biomass, consumers that feed on producers and decomposers, to reconvert to consumers, and decomposers that recycle nutrients from producers and consumers. The community is rather closed for nutrients, but requires a constant supply of energy. Influx and efflux of nutrients largely determine the long term behaviour of the community.

5.7 Communities

Although the DEB theory has not yet been worked out to the extent that the dynamics of communities can be analysed, a brief discussion of the salient features of communities helps to clarify the aim and scope of such an exercise; as illustrated in figure 5.27 for the most simple non-degenerated community one can imagine. The coupling between energy and mass fluxes is most tight for the consumers, least tight for the producers, while the decomposers take an intermediate position. The coupling for the consumers is via the rather constant chemical composition of food. The consumers are usually organized in food webs, i.e. branched food chains, as discussed in the previous section. If it is necessary to distinguish groups within each component, competition must be considered, but the outcome is greatly affected by spatial heterogeneities. The flux of organic matter through food webs is mainly set by body size scaling relationships, to be discussed in the next chapter. The main nutrients that usually affect standing crops most drastically are nitrogen compounds and phosphate. Chemoautotrophy represents another energy source for some communities. In terrestrial environments, water plays an essential role, both directly as a 'nutrient' for producers and indirectly in the influx and efflux of nitrogen and phosporus.

The type of problems that can be analysed successfully by community modelling relate to a much larger scale in space and time than is relevant for individuals and populations. Processes on totally different space/time scales combine poorly into one model. Just as it has been a bad idea to model processes of growth on the basis of ATP fluxes, cf. {5}, it is a bad idea to model community processes on the basis of individuals. Generally, it even will be disappointing to try to trace particular species in community models. A useful field of applications of community models is in biogeochemistry on a regional or global scale. A promising approach for community modelling in relation to the DEB theory is via application of yield factors in the description of population behaviour, in combination with energy and mass conservation laws as on {192}. The gist of this approach is that the link between properties of individuals and communities is still preserved, because yield factors can be written as functions of energy parameters for individuals, as demonstrated in this chapter.

Chapter 6

Comparison of species

The range of body sizes is enormous. A bacterium with full physiological machinery has a volume of about 0.25×10^{-18} m^3. Some parasitic forms are much smaller. The blue whale has a volume of up to 135 m^3. A sequoia may even reach a volume of 2000 m^3, but one can argue that it is not all living material. Ironically enough, the organism with the largest linear dimensions is usually classified as a 'micro-organism': the fungus *Armillaria bulbosa* is reported to occupy at least 15 hectares and exceeds 10 Mg or 10 m^3 [669]. The factor between the volumes of bacterium and whale is 5.4×10^{20}, that between the volume a water molecule occupies in liquid water and that of a bacterium is 'only' 10^{10}.

These differences in size reflect differences in physiological processes, which the DEB theory tries to capture. The DEB model has structural body volume as a state variable. This implies that parameters which occur in the description of the process of energy uptake and use are independent of body volume of a particular individual. Ultimate body volume, and in particular the maximum body volume, can be written as a simple compound parameter. This is why (some) parameters of the DEB model must have a (simple) relationship with ultimate body volume. This powerful argument is so simple that it can easily be overlooked. A comparison of the energetics of different species, ranging from bacteria to whales reduces, in the DEB theory, to a comparison of sets of parameter values. This is different from comparison within a species, where we have only one set of parameter values, though different body volumes. This chapter deals with theory of parameter values, which includes body size scaling relationships, optimization problems and evolutionary aspects.

6.1 Body size scaling relationships

The standard way to study body size scaling relationships is allometric: apply linear regression to the logarithmically transformed quantity of interest as a function of the logarithm of total body weight [117,470,542,638]. I have already given my reservations with respect to the physical dimensions, {12}, but I also object to the application of regression methods. My objections have a deeper root than the presence of 'measurement error' in the independent variable, which is usually the whole body wet weight. The dependent variable, i.e. some quantity of interest, can be considered to be a compound parameter for a particular species, and this can hardly be conceived of as a random variable. Each species of the (limited) set living on earth happens to have a particular value for the quantity under consideration. This value is a result of evolutionary processes. Values of related species are thus likely to be dependent in a statistical sense. Moreover, evolutionary theory aims to explain a particular value while the application of regression methods implies that you leave the deviation in the black box. The random deviation from the (allometric) deterministic function, which regression analysis treats as 'measurement error', does not have a meaningful biological interpretation. A consequence of this point of view is that statistical tests on the 'exact' value of the scaling parameter must be considered to be misleading.

I prefer a different approach to the subject of body size scaling relationships which is implicit in the DEB theory. Although the relationships are mostly not of the allometric type if log-log plotted, straight lines approximate the result very well. To facilitate a comparison with the literature, I will refer frequently to the allometric (dimensionless) scaling parameter.

The tendencies discussed in the next few sections can be inferred on the basis of general principles of physical and chemical design. On top of these tendencies, species-specific adaptations occur that cause deviations from the expected tendencies. A general problem in body size scaling relationships is that large bodied species frequently differ from small bodied species in a variety of ways, such as behaviour, diet etc. These life styles require specific adaptations, which hamper simple inter-species comparison. MacMahon [469] applied elasticity arguments to deduce allometric scaling relationships for the shape of skeletal elements. Godfrey *et al.* [263] demonstrated, for mammals, that deviations from a simple geometrical upscaling of skeletal elements is due to size-related differences in life styles.

6.2 Primary scaling relationships

The parameter values of the DEB model tend to depend on maximum body volume in a predictable way that does not use any empirical argument. This makes it possible to predict how physiological quantities that can be written as functions of DEB parameters depend on maximum body volume. The core of the argument is that parameters that relate to the physical design of the organism are all proportional to volumetric maximum length, while the rest are size-independent. The latter parameters relate to molecular processes, which are thus essentially density-based. Reaction rates as described by the law of mass action depend on meeting frequencies between particles and do not depend on the (absolute) size of the organism. The difference between physical design and density-based parameters relates to the difference between intensive and extensive quantities.

The parameter values of a reference species number 1 with maximum body volume $V_{m,1}$ have an extra index to compare the parameter set with that of another species, number 2, with maximum body volume $V_{m,2}$. The primary scaling relationships are given in table 6.1 and are compared with the relationships from the invariance property of the DEB model. This property is that parameter sets that differ in a special way that involves an arbitrary factor z, result in identical energetics at strictly constant food densities, cf. {112}. A

Table 6.1: The relationship between the parameters of the DEB model for species 1 and 2 according to the invariance property (upper panel) and according to the primary scaling relationships (lower panel). The ratio of the ultimate volumetric body lengths of species 1 and 2 equals the zoom factor z

K_2	=	$K_1 z + X(z-1)$	$\{\dot{I}_m\}_2$	=	$\{\dot{I}_m\}_1 z$	$[\dot{M}]_2$	=	$[\dot{M}]_1$	$\{\dot{H}\}_2$	=	$\{\dot{H}\}_1$
$V_{b,2}^{1/3}$	=	$V_{b,1}^{1/3}$	$\{\dot{A}_m\}_2$	=	$\{\dot{A}_m\}_1 z$	$[G]_2$	=	$[G]_1$	$\ddot{p}_{a,2}$	=	$\ddot{p}_{a,1}$
$V_{p,2}^{1/3}$	=	$V_{p,1}^{1/3}$	$[E_m]_2$	=	$[E_m]_1 z$	κ_2	=	κ_1	q_2	=	q_1
K_2	=	$K_1 z$	$\{\dot{I}_m\}_2$	=	$\{\dot{I}_m\}_1 z$	$[\dot{M}]_2$	=	$[\dot{M}]_1$	$\{\dot{H}\}_2$	=	$\{\dot{H}\}_1$
$V_{b,2}^{1/3}$	=	$V_{b,1}^{1/3} z$	$\{\dot{A}_m\}_2$	=	$\{\dot{A}_m\}_1 z$	$[G]_2$	=	$[G]_1$	$\ddot{p}_{a,2}$	=	$\ddot{p}_{a,1}$
$V_{p,2}^{1/3}$	=	$V_{p,1}^{1/3} z$	$[E_m]_2$	=	$[E_m]_1 z$	κ_2	=	κ_1	q_2	=	q_1

striking resemblance exists between the relationships of parameter sets on the basis of the primary scaling relationships and the invariance property. The only deviations are in the parameters relating to physical design.

From the primary scaling relationships, other scaling relationships can be derived for all processes to which the DEB theory applies. Maximum volume itself is just one, though eye catching, compound parameter. Maximum volume is a result of energy uptake and use, not a factor determining these processes. Maximum volume can serve as a paradigm to compare species: $V_{m,2} = (\kappa_2\{\dot{A}_m\}_2/[\dot{M}]_2)^3 = (\kappa_1 z\{\dot{A}_m\}_1/[\dot{M}]_1)^3 = z^3 V_{m,1}$. One possible interpretation of the arbitrary zoom factor z is thus the ratio between the ultimate length measures of two species. Although it is usual and convenient to study how physiological quantities and life histories depend on body size, it is essential to realize that all, inclusive of body size itself, depend on the coupled processes of energy uptake and use. Before I discuss how many other compound parameters depend on body size, it is instructive to review the primary parameters first.

Molecular biology stresses again and again the similarities of cells, independent of the body size of the organism. It thus seems reasonable to assume that cells of equal size have about the same maintenance costs. Since the maintenance of cells is probably a major part of the maintenance of the whole individual, it seems natural that volume-specific maintenance is independent of body size. The same holds for the costs of growth. The values of $[\dot{M}]$, $\{\dot{H}\}$, $[G]$, κ and $\ddot{p}_a$ for a particular growing individual could in principle differ (a bit) for each time increment. The DEB model, however, assumes that these process parameters are constant. The assumption that they are independent of ultimate volume is the only one that is consistent with the structure of the DEB model.

Since κ and $[\dot{M}]$ are independent of maximum body size, $\{\dot{A}_m\}$ has to be proportional to the cubic root of the ultimate volume, because of the relationship $V_m^{1/3} = \kappa\{\dot{A}_m\}/[\dot{M}]$. This relationship makes $\{\dot{A}_m\}$ a physical design parameter. The parameters $\{\dot{I}_m\}$ and $[E_m]$ are also physical design parameters, because the ratio of them with $\{\dot{A}_m\}$ relates to density-based molecular processes. Since the DEB model in fact assumes that digestion is complete (else digestion efficiency would depend on feeding level, cf. {247}), and the ratio $\{\dot{A}_m\}/\{\dot{I}_m\}$ represents digestion efficiency, $\{\dot{I}_m\}$ has to be proportional to $\{\dot{A}_m\}$ and so to the cubic root of the ultimate volume as well. The same holds for the reserve capacity parameter $[E_m]$, because the ratio $\{\dot{A}_m\}/[E_m]$ stands for energy conductance. Like digestion efficiency, it could in principle change (a bit) for each time increment in a growing individual, but it is assumed to be constant in the DEB model. Both ratios could have been introduced as the primary parameters, which would turn maximum assimilation rate and storage capacity into compound parameters. This is mathematically totally equivalent. Such a construction would leave V_b, V_p, $\{\dot{I}_m\}$ and K as the only parameters relating to the physical design of the organism.

The body size dependence of the saturation coefficient is less easy to see, because species differ so much in their feeding behaviour. At low food densities this constant can be interpreted as the ratio of the maximum ingestion and filtering rates in a filter feeder such as *Daphnia*. If maximum beating rate is size independent, as has been observed, the filtering rate is proportional to surface area. Since the maximum specific ingestion rate $\{\dot{I}_m\}$ scales with a length measure, the saturation coefficient K should scale with a length measure as well. More detailed modelling of the beating rate would involve 'molecular' density-based formulations for the filtering process, which turns the saturation coefficient into a derived compound parameter. This is not attempted here, because the formulations would only apply to filtering, while many species do not filter.

The life stage parameters, V_b and V_p, show an extremely wide range of variation among different taxa. Huge fishes can lay very small eggs and thus have small values for V_b. For example, the ocean sunfish *Mola mola* can reach a length of 4 m and can weigh more than 1500 kg, it can produce clutches of 3×10^{10} tiny eggs. The other extreme within the bony fishes is the oviviparous coelacanth *Latimeria chalumnae*, which can reach a length of 2 m and a weight of 100 kg. It produces eggs with a diameter of 9 cm in clutches of some 26. (If we include the cartilaginous fish, the whale shark *Rhincodon typus* wins with a 12–18 m length, more than 8165 kg weight and eggs of some 30 cm.) The tendency of egg size to be proportional to ultimate size only holds for related species at best, as within the squamate reptiles [655]. This matter will be discussed further under r and K strategies, {239}.

The gist of the argument for primary scaling relationships is that they can be derived from the structure of the DEB model and they do not involve empirical arguments.

6.3 Secondary scaling relationships

This section gives examples of the derivation of body size scaling relationships of a variety of eco-physiological phenomena that can be written as a compound parameter of the DEB model. The derivation follows the same path time and again and is worked out in detail for respiration. Food density will be assumed to be constant or high.

Since the independent variable in body size scaling relationships is standard wet weight, we should first consider how wet weight relates to the primary parameter values. From (2.9) it follows for $[E] = [E_m]$ that $W_w = ([d_{wv}] + [d_{we}](1 + E_{\dot{R}}/[E_m]))V$. The ultimate volumetric length is $(\kappa\{\dot{A}_m\} - \{\dot{H}\})/[\dot{M}]$, see (3.17) for $f = 1$, so that the ultimate wet weight equals

$$W_{w,\infty} \simeq ([d_{wv}] + [d_{we}])(\kappa\{\dot{A}_m\} - \{\dot{H}\})^3[\dot{M}]^{-3} \tag{6.1}$$

at abundant food for isomorphs that have a negligibly small amount of reserves allocated to reproduction. Wet weights are sensitive to body chemistry. The structural body mass and in particular water content and type of reserve materials are different in unrelated species. This hampers comparisons that include species as different as jelly fish and elephants. If comparisons are restricted to related species, for example among mammals, the structural volume-weight conversion $[d_{wv}]$ will be independent of body volume, while $[d_{we}]$ increases with volumetric length, because it includes the maximum reserve capacity $[E_m]$ in its definition; see below (2.9). Since $\{\dot{A}_m\}$ increases with a volumetric length, while κ, $\{\dot{H}\}$ and $[\dot{M}]$ are independent of body volume, this means that the volume-specific wet weight $[W_w] \equiv W_w/V$ increases with body volume for two reasons. The first is the increasing contribution of energy reserves, the second is the decreasing effect of volume reduction due to heating. The last reason only applies to endotherms, of course.

To simplify the argument, the ultimate weight of ectotherms at high food densities is in the rest of this section taken to approximate the maximum body volume after division by $[d_w] = 1\,\mathrm{g\,cm^{-3}}$. This has also been done for endotherms, and although this is even less correct, as explained, the tendencies are crude enough to legitimise this approximation.

6.3.1 Respiration

Respiration rate should be discussed first for historical reasons; see the chapter 1. The scaling parameter has been found to be 0.66 for unicellulars, 0.88 for ectotherms and 0.69 for endotherms [548]. The exact value differs among authors taking their data from the literature. The variations are due, in part, to differences in the species included and in the experimental conditions under which respiration rates were measured. For crustaceans Vidal and Whitledge [739] present values of 0.72 and 0.85, and Conover [134] gives 0.74. If species ranging from bacteria to elephants are included, the value 0.75 emerges. It has become an almost magic number in body size scaling relationships. Few authors make the distinction between *intra*species, cf. {103}, and *inter*species relationships. Many explanations have been proposed; some are based on muscle tension [469] or running speed [570], for instance. As for intraspecies relationships, I find these explanations not completely satisfactory, because mechanics plays only a minor role in energy budgets and the argument is too specific, since it applies to a very much restricted group of species. As mentioned, respiration rates are usually thought to reflect routine metabolic costs. It is no wonder that the explanation for the scaling parameter being less than one for ectotherms is still a hot issue. I hope to have made it clear by now that costs other than routine metabolic costs contribute to respiration as well, which makes it possible that respiration scales with a parameter less than one, while routine metabolic costs scale with a parameter one.

If we compare individuals with the same parameter set, (3.42) shows that the respiration rate is proportional to a weighted sum of surface area and volume: $(\dot{v} + \dot{m}V_h^{1/3})V^{2/3} + \dot{m}V$ at constant food density. This is nothing new. If we compare different parameter sets the construction is as follows. Step one is to set the energy density and the body volume equal to their maximum values $[E_m]$ and V_m, and write the quantity of interest out using the primary parameters. Step two is to multiply each primary parameter value that relates to physical design by the zoom factor; see table 6.1. Step three is to study how the respiration rate depends on maximum body volume $\mathcal{V} \equiv z^3 V_{m,1}$ by variation of the zoom factor z. For this purpose, the substitution $z = (\mathcal{V}/V_{m,1})^{1/3}$ is made. Starting from (3.42), the three steps result in

$$\begin{aligned}
\dot{C}_m &= \frac{[G]/\kappa}{[G]/\kappa[E_m]+1}\left(\left(\frac{\{\dot{A}_m\}}{[E_m]} + \frac{[\dot{M}]}{[G]}V_h^{1/3}\right)\left(\frac{\{\dot{A}_m\}}{[\dot{M}]/\kappa}\right)^2 + \frac{[\dot{M}]}{[G]}\left(\frac{\{\dot{A}_m\}}{[\dot{M}]/\kappa}\right)^3\right) \\
\dot{C}_{m,2} &= \frac{[G]_1/\kappa_1}{[G]_1/\kappa_1[E_m]_1 z+1}\left(\left(\frac{\{\dot{A}_m\}_1 z}{[E_m]_1 z} + \frac{[\dot{M}]_1}{[G]_1}V_{h,1}^{1/3}\right)\left(\frac{\{\dot{A}_m\}_1 z}{[\dot{M}]_1/\kappa_1}\right)^2 + \frac{[\dot{M}]_1}{[G]_1}\left(\frac{\{\dot{A}_m\}_1 z}{[\dot{M}]_1/\kappa_1}\right)^3\right) \\
&= \frac{[\dot{M}]_1}{\kappa_1}\frac{(\dot{v}_1/\dot{m}_1 + V_{h,1}^{1/3})z^2 V_{m,1}^{2/3} + z^3 V_{m,1}}{g_1/z+1} \\
\dot{C}_{m,\mathcal{V}} &= \frac{[\dot{M}]_1}{\kappa_1}\frac{(\dot{v}_1/\dot{m}_1 + V_{h,1}^{1/3})\mathcal{V}^{2/3} + \mathcal{V}}{\mathcal{V}^{-1/3}\dot{v}_1/\dot{m}_1+1}
\end{aligned}$$

The interpretation of $\mathcal{V}$ is the maximum body volume of a species and is a compound parameter, not a state variable. The respiration rate is thus approximately proportional to a weighted sum of volume$^{2/3}$ and volume^1 if $g_1 \ll z$. So it relates to body volume in almost, but not exactly, the same way as it does within a species.

The respiration rate will appear almost as a straight line in a double-log plot against body volume, the slope being somewhere between 0.66 and 1, depending on the species that have been included. Surface-bound heating costs dominate in endotherms, so a plot that includes them will be close to a line with slope 0.66. The slope for the *Bathyergidae*, a family of rodents that are practically ectothermic, is found to be close to 1 [442], as expected.

6.3.2 Feeding

Maximum ingestion rate

The maximum ingestion rate for an individual of volume V is $\dot{I}_m = \{\dot{I}_m\}V^{2/3}$, so $\dot{I}_{m,\mathcal{V}} = \{\dot{I}_m\}_1 V_{m,1}^{-1/3}\mathcal{V}$. The maximum ingestion thus scales allometrically with body volume, but with a scaling parameter of .66 for intra-species comparisons and 1 for inter-species comparisons. Farlow [211] gives an empirical scaling parameter of 0.88, but value 1 also fits the data well. For endotherms especially, a scaling parameter of somewhat less than 1 is expected for weight as the independent variable, because of the increase in volume-specific weight, as explained. In a thorough study of scaling relationships, Calder [117] coupled the inter-specific ingestion rate directly to the respiration rate, without using an explicit model for energy uptake and use. The present DEB-based considerations force one to deviate from intuition.

Gut capacity

Within a species, isomorphy implies a gut capacity that is a constant fraction of body volume. This must also hold for inter-species comparisons, as long as body design and diet are comparable and this has been found for birds and mammals [117]. The mean gut residence time of food particles is thus independent of body size as a consequence, because ingestion rate is proportional to body size, while it was found to be proportional to a length measure for intra-species comparisons.

Maximum filtering rate

The filtering rate is maximal at low food densities. If particle retention is complete, it is given by $\dot{F}_m = \dot{I}_m/K = V^{2/3}\{\dot{I}_m\}/K$. So, $\dot{F}_{m,\mathcal{V}} = \mathcal{V}^{2/3}\{\dot{I}_{m,1}\}/K_1$. This was found by Brendelberger and Geller [90].

Speed

Since biomechanics is not part of the DEB theory, this is not the right place for a detailed discussion on Reynolds and Froude numbers, although interesting links are possible. Speed of movement has only a rather indirect relationship with feeding or other aspects that bear on energy budgets. A few remarks are, therefore, made here.

McMahon and Bonner [470] found that the speed of sustained swimming for species ranging from larval anchovy, via salmon, to blue whales scales with the square root of volumetric length; they underpinned this finding with mechanical arguments. Since the energy costs for swimming are proportional to squared speed and to surface area, cf. {63}, the total costs for movement would scale with cubed length, or $\mathcal{V}$, for a common travelling time. This is consistent with the DEB theory, where the costs for travelling are taken to be a fixed fraction of the maintenance costs.

A similar result appears to hold for the speed of flying, but by a somewhat different argument. The cruising speed, where the power to fly is minimal, is proportional to the square root of the wing loading [708]. If a rough type of isomorphy applies, comparing insects, bats and birds, wing loading, i.e. the ratio of body mass and wing area, scales with length, so that cruising speed scales with the square root of length [470].

Arguments for why standard cruising rate for walking tends to be proportional to length, have been given on {59,63}. If energy invested in movement is proportional to volume and taken to be part of the maintenance costs, the intra- and inter-species scaling relationships work out in the same way.

Maximum diving depth

Birkhead [93] found that the maximum diving depth for auks and penguins tend to be proportional to volumetric length. This can be understood if diving depth is proportional to the duration of the dive; the latter is proportional to length, cf. {59} by the argument that respiration rate of these endotherms is about proportional to surface area and oxygen reserves to volume.

Minimum food density

The minimum food density at which an isomorph of body volume V can live for a long time is found from the condition that energy derived from ingested food just equals the maintenance costs, so $\dot{I}\{\dot{A}_m\}/\{\dot{I}_m\} = \dot{M}$, or $f_\dagger = \frac{X_\dagger}{K+X_\dagger} = \frac{[\dot{M}]}{\{\dot{A}_m\}}V^{1/3}$. The solution is $X_\dagger = \frac{V^{1/3}K[\dot{M}]/\{\dot{A}_m\}}{1-V^{1/3}[\dot{M}]/\{\dot{A}_m\}}$. At this food density, the individual can only survive, not reproduce. For different species, we obtain the condition $X_{\dagger,\mathcal{V}} = \frac{\mathcal{V}^{1/3}K_1[\dot{M}]_1/\{\dot{A}_m\}_1}{1-V_{m,1}^{1/3}[\dot{M}]_1/\{\dot{A}_m\}_1}$. Minimum food density, also called the threshold food density, is thus proportional to volumetric length. An important ecological consequence is that at a given low food density, small individuals can survive, while the large ones can not. This explains, for instance, why bacteria in oligotrophic seas are so small.

This result only applies to situations of constant food density. If it is fluctuating, storage capacity becomes important, which tends to increase with body size; see {128}. The possibility of surviving in dynamic environments then works out to be rather complex. Stemberger and Gilbert [681] found that threshold food density tends to increase with body size for rotifers, as expected, but Gliwicz [261] found the opposite for cladocerans. This result can be explained, however, by details of the experimental protocol. The threshold food density was obtained by plotting the growth rate against food density and selecting the value where growth is nil. Growth at the different food densities was measured from two-day old individuals exposed to a constant food density for 4 days. The reserves at the start of the growth experiment, which depend on culture conditions, will contribute substantially to the result.

6.3.3 Growth

Maximum growth

As follows from (4.8), the maximum growth rate for different species equals

$$\frac{4}{27}\frac{\dot{m}_1 g_1}{(\mathcal{V}/V_{m,1})^{1/3} + g_1}(\mathcal{V}^{1/3} - V_{h,1}^{1/3})^3$$

and is thus about proportional to volume$^{2/3}$. This fits Calow and Townsend's data very well [119].

Von Bertalanffy growth rate

The von Bertalanffy growth rate at high food density is $\dot{\gamma}_\mathcal{V} = (3/\dot{m}_1 + 3\mathcal{V}^{1/3}/\dot{v}_1)^{-1}$ for different species. It decreases almost linearly with volumetric length. This is consistent with empirical findings; see figure 6.1. The parameters and data sources are listed in table 6.2. This table is so extensive because the fit with the von Bertalanffy growth curve is used to support the argument that it is possible to formulate a theory that is not species-specific, {1}. If one collects growth data from the literature, an amazingly large fraction fits the von Bertalanffy curve despite the fact that most data sets are from specimens collected in the field. Since it is hard to believe that food density has been constant during the growth period, this suggests that food has been abundant; this is relevant for population dynamics.

If the von Bertalanffy growth rate is plotted against maximum volumetric length, the scatter is so large that it obscures their relationship. This is largely due to differences in body temperature. A fish in the North Sea with a yearly temperature cycle between 3 and 14 °C grows much slower than a passerine bird with a body temperature of 41 °C. This is not due to fundamental energy differences in their physiology. If corrected to a common body temperature according to the Arrhenius relationship with an Arrhenius temperature of 12500 K, the expected relationship is revealed and the differences between fishes and birds disappear. Since temperature had not been measured in most cases, I had to estimate them in a rather crude way. For most molluscs and fish data I used general information on local climate and guessed water temperatures (which depend on the, frequently unknown, depth). The body temperatures of birds and mammals have also been guessed. Uncertainties about temperature doubtlessly contributed the most to the remaining scatter. The corrected rates are not meant as predictions for actual growth rates at this body temperature because most North Sea fish and birds will die almost instantaneously if the temperature was realized. The average energy conductance, $\dot{v}$, of 261 species at 37 °C appears to be 5.49 mm d^{-1}, 0.885 mm d^{-1} at 25 °C, or 0.433 mm d^{-1} at 20 °C. This is the best evidence that the maximum storage capacity increases with volumetric length, just as the maximum surface-specific assimilation rate does.

The contribution of maintenance in the von Bertalanffy growth rate is small for large bodies, which explains that the von Bertalanffy growth rate is about proportional to $\mathcal{V}^{-1/3}$, as Ricklefs [598] found for birds for instance.

Table 6.2: The von Bertalanffy parameters and their standard deviations as calculated by non-linear regression. The shape coefficient converts the size measure used to volumetric length. For shape coefficient 1, the data refer to wet weight, except for *Saccharomyces*, *Actinophrys* and *Asplanchna*, where volumes were measured directly. The data for *Mnemiopsis* and *Calanus* refer to dry weight. The other data are length measures, mostly total body length. Where the standard deviation is not given, the parameters of the authors are given. Temperatures between brackets were inferred from the location on earth. Where two temperatures are given, a sinusoidal fluctuation between these extremes is assumed. In the column 'sex': f=female, m=male.

species	sex	length mm L_∞	s.d. mm	shape coeff d_m	rate a^{-1} $\dot{\gamma}$	s.d. a^{-1}	location NS	location EW	temp °C	source
Ascomyceta										
Saccharomyces carlsbergensis		4.59e-3	2.16e-5	0.806	11830	318	lab	lab	30	[56]
Heliozoa										
Actinophrys spec.		0.0043	2.2e- 5	1	2891	368	lab	lab		[701]
Rhizopoda										
Amoeba proteus		2.79	0.016	0.01	832.2	56.9	lab	lab	23	[568]
Ciliata										
Paramecium caudatum		2.969	0.062		1638	210	lab	lab	17	[636]
Ctenophora										
Pleurobrachia pileus	fm	15.04	0.436	0.702	33.27	2.49	lab	lab	20	[274]
Mnemiopsis mccradyi	fm	8.851	0.927	3.90	11.61	1.88	lab	lab	26	[589]
Rotifera										
Asplanchna girodi	f	0.2400	7.32e-4	1	193.7	4.92	lab	lab	20	[606]
Annelida										
Dendrobeana veneta	fm	14.5	0.24	1	12.04	0.73	lab	lab	20	Bos, pc
Mollusca										
Aplysia californica	fm	112.2	6.05	1	4.840	0.871	lab	lab	18-20	[541]
Urosalpinx cinerea	fm	30.94	1.31	0.397	0.8116	0.11	31S	152E	-1-25	[229]
Achatina achatina	fm	106.5	2.45	0.543	1.121	0.0770	5N	0E	(25)	[326]
Helix aspera	fm	25.06	0.498	0.68	1.098	0.0960	lab	lab	(18-20)	[154]
Patella vulgata	fm	46.93	0.306	0.310	0.4296	7.91e-3	54N	4.40W	(4-17)	[783]
Monodonta lineata	fm	21.92	0.130	0.716	0.6213	0.0171	52.25N	4.05W	(4-17)	[772]
Biomphalaria pfeifferi	fm	7.538	0.0497	1	4.879	0.201	lab	lab	25	[480]
Lymnaea stagnalis	fm	15.37	0.0584	1	10.81	0.204	lab	lab	20	[666]
Helicella virgata	fm	9.888	0.215	1	3.316	0.163	35S	139E	11-16	[561]
Macoma baltica		21.57	0.154	0.423	3.00	0.0869	41.31N	70.39W	10.56	[259]
Cerastoderma glaucum		29.24	1.86	0.558	2.221	0.380	40.50N	14.10E	13-30	[351]
Venus striatula		37.76	25.1	0.471	0.1961	0.210	55.50N	4.40W	6-13	[17]
Ensis directus		142.2		0.187	0.5830		54.35N	8.45E	4-17	[700]
Mytilus edulis		95.92	2.02	0.394	0.1045	5.109e-3	53.36N	9.50W	7-17	[607]
Placopecten magellanicus		162.3	1.01	0.388	0.1671	2.842e-3	47.10N	53.36W	0-18	[447]
Perna canaliculus		191.2	10.6	0.394	0.3555	0.0342	36.55S	174.47E	17	[317]
Hyridella menziesi		74.62	2.05	0.400	0.1331	8.38e-3	(36.55S)	(147.47E)		[354]
Mya arenaria		91.31		0.407	0.1866		41.39N	70.42W	(4-17)	[101]
Loligo pealei	f	455.3	39.5	0.398	0.4201	0.0551	41.31N	70.39W	(4-17)	[697]
Loligo pealei	m	918.2	111	0.398	0.2122	0.0315	41.31N	70.39W	(4-17)	[697]
Brachiopoda										
Terebratalia transversa		48.39	1.09	0.640	0.3140	0.0163	47.30N	122.5W	(4-17)	[526]
Crustacea										
Daphnia pulex	f	2.366	0.0192	0.526	44.25	2.10	lab	lab	20	[595]
Daphnia longispina	f	2.951	0.0260	0.520	61.32	2.92	lab	lab	25	[349]
Daphnia magna	f	5.136	0.0970	0.526	35.04	1.83	lab	lab	20	[407]
Daphnia magna	m	2.813	0.0440	0.526	66.80	5.11	lab	lab	20	[407]
Daphnia cucullata	f	1.049	0.0214	0.480	58.25	9.71	lab	lab	20	[740]
Daphnia hyalina	f	1.717	0.0399	0.520	47.52	5.93	lab	lab	20	[740]
Ceriodaphnia pulchella	f	0.7503	0.0122	0.520	39.89	5.04	lab	lab	20	[740]
Ceriodaphnia reticulata	f	1.038	0.0210	0.520	49.28	3.30	lab	lab	20	[407]
Chydorus sphaericus	f	0.4115	1.10e-3	0.560	52.63	0.969	lab	lab	20	[740]

Diaphanosoma brachyurum	f	1.380	0.0198	0.520	46.50	3.72	lab	lab	20	[740]
Leptodora kindtii	f	8.632	0.204	0.300	26.96	2.64	lab	lab	20	[740]
Bosmina longirostris	f	0.5289	0.0215	0.520	38.73	6.50	lab	lab	20	[740]
Bosmina coregoni	f	0.4938	0.0104	0.520	66.90	9.59	lab	lab	20	[740]
Calanus pacificus		6.295	1.02	0.215	8.863	1.89	lab	lab	12	[522]
Dissodactylus primitivus	f	11.02	0.410	0.635	1.025	0.0732	lab	lab	(18)	[559]
Dissodactylus primitivus	m	9.013	0.212	0.635	1.362	0.0742	lab	lab	(18)	[559]
Euphasia pacifica		12.91	2.35	0.197	1.008	0.369	lab	lab	10	[459]
Homarus vulgaris		186.6	6.99	0.939	0.05543	3.36e-3	lab	lab	10	[315]
Cancer pagurus	f	9.707	0.385	1	0.2711	0.0122	50.30N	2.45W	(5-18)	[54]
Cancer pagurus	m	115.6	0.513	1	0.3513	0.0174	50.30N	2.45W	(5-18)	[54]
Dichelopandalus bonnieri		25.73	1.97	0.882	0.4795	0.0824	54N	4.40W	(4-17)	[6]
Gammarus pulex	m	4.355	0.0570	1	3.300	0.177	lab	lab	15	[698]
Gammarus pulex	f	4.089	0.0554	1	2.218	0.123	lab	lab	15	[698]
Calliopius laeviusculus		15.27	0.699	0.262	13.52	1.96	lab	lab	15	[151]
Uniramia										
Tomocerus minor		3.903	0.0848	0.351	6.600	0.379	lab	lab	20	[364]
Orchesella cincta		3.652	0.0858	0.351	4.948	0.351	lab	lab	20	[364]
Isotomata viridis		3.034	0.0751	0.351	6.52	0.469	lab	lab	20	[364]
Entomobrya nivalis		1.981	0.0830	0.351	3.416	0.418	lab	lab	20	[364]
Lepidocyrtus cyaneus		1.181	0.0666	0.351	9.840	2.17	lab	lab	20	[364]
Orchesella cincta		1.281	0.0151	1	6.817	0.354	lab	lab	20	[356]
Phaenopsectra coracina		1.745	0.147	1	2.388	0.779	63.14N	10.24E	4	[1]
Diura nanseni		2.782	0.0460		6.328	0.536	60.15N	6.15E	0-20	[26]
Capnia pygmaea		1.024	0.0967		2.493	0.663	60.15N	6.15E	1-20	[26]
Locusta migratoria		10.82	0.237	1	44.82	7.36	lab	lab	23-36	[440]
Chironomus plumosus	f	4.053	0.272	1	21.88	5.50	lab	lab	15	[348]
Chironomus plumosus	m	3.211	0.0415	1	52.74	4.77	lab	lab	15	[348]
Cheatognata										
Sagitta hispida	fm	9.431	0.150	0.15	44.80	5.25	lab	lab	21	[588]
Echinodermata										
Lytechenus variegatus		46.10	0.147	0.70	3.913	0.199	18.26N	77.12W	26-29	[376]
Echinocardium cordatum		34.50	0.425	0.696	0.4590	0.0232	53.10N	4.15E	5-12	[189]
Echinocardium cordatum		36.70	0.375	0.696	0.5320	0.0259	53.40N	4.30E	5-14	[189]
Echinocardium cordatum		44.90	0.405	0.696	0.4960	0.0212	54.15N	4.30E	5-16	[189]
Tunicata										
Oikopleura longicauda	fm	0.829	0.049	0.520	56.56	6.62	lab	lab	20	[215]
Oikopleura dioica		0.952	0.327	0.560	63.97	37.3	lab	lab	20	[215]
Chondrichthyes										
Raja montaqui	fm	695.9	11.0	0.184	0.1874	0.0140	52-54N	3-7E	(4-17)	[331]
Raja brachyura		1589	213	0.184	0.1018	0.0261	52-54N	3-7E	(4-17)	[331]
Raja clavata	f	1303	107	0.184	0.09297	0.0163	52-54N	3-7E	(4-17)	[331]
Raja clavata	m	952.7	29.8	0.184	0.1557	0.0145	52-54N	3-7E	(4-17)	[331]
Raja erincea		542.9	32.6	0.184	0.2787	0.0542	41.05N	73.10W	1-19.1	[593]
Prionace glauca		4230		0.165	0.1100		48N	7W	(5-18)	[683]
Osteichthyes										
Accipenser stellatus		2120	30.5	0.198	0.05396	1.46e-3	(45.10N)	(28.30E)	(4-23)	[63]
Clupea sprattus		157.0	0.557	0.200	0.5847	4.60e-3	52.30N	2E	(4-17)	[347]
Coregonus lavaretus		397.3	8.39	0.203	0.3295	0.0221	54.35N	2.50W	(5-15)	[28]
Salvelinus willughbii	f	385.4	72.9	0.225	0.2495	0.0973	54.20N	2.57W	(5-15)	[236]
Salvelinus willughbii	m	328.9	12.7	0.224	0.3545	0.0366	54.20N	2.57W	(5-15)	[236]
Salmo trutta		585.8	18.0	0.216	0.4769	0.0411	53.15N	4.30W	(4-17)	[343]
Salmo trutta		576.2	20.6	0.240	0.2921	0.0253	57.40N	5.10W	5-12.8	[121]

Salmo trutta		420.2	3.13	0.240	0.4157	0.0107	54.20N	2.57W	(5-15)	[143]
Oncorhynchus tschawytscha		155.2	11.9	1	0.9546	0.217	36S	147E	(11-16)	[115]
Thymallus thymallus		459.6	8.44	0.240	0.4656	0.0224	52.09N	2.41W	(5-15)	[307]
Esox lusius	f	948.7	88.3	0.209	0.2101	0.0718	50.17N	3.39W	(5-15)	[87]
Esox lusius	m	703.6	13.0	0.209	0.4016	0.0455	50.17N	3.39W	(5-15)	[87]
Esox masquinongy		2091	848	0.199	0.04503	0.0263	44N	79W	(5-15)	[495]
Rutilus rutilus		441.6	15.8	0.258	0.1661	0.0116	52.30N	0.30E	(5-15)	[142]
Leuciscus leuciscus		252.6	2.32	0.258	0.3329	0.0131	52.30N	0.30E	(5-15)	[142]
Barbus grypus		1036	25.2	0.206	0.1265	6.59e-3	35.75N	44.7E	(17-30)	[7]
Abramis brama		546.0		0.225	0.1142		53.15N	2.30W	(5-15)	[265]
Gambusia holbrookii	f	61.72	2.34	0.250	0.9366	0.216	38.40N	9.40W	(5-25)	[228]
Poecilia reticulata	f	50.58	1.14	0.252	1.667	0.0690	lab	lab	21	[730]
Merluccius merluccius		1265	78.4	0.222	0.2075	0.0184	55.45N	5W	(8-12)	[27]
Lota lota		1009	60.3	0.193	0.09768	0.0103	53N	98W	(5-15)	[316]
Gadus merlangus	f	898.6	12.2	0.222	0.08626	2.07e-4	54N	4.40W	(8-12)	[84]
Gadus merlangus	m	772.8	9.03	0.222	0.08626	2.07e-4	54N	4.40W	(8-12)	[84]
Gadus morhua		1089	43.2	0.222	0.1308	9.26e-3	40N	60W	10	[400]
Gadus aeglefinus		106.5		1	0.2000		53-57N	0-7E	(4-17)	[69]
Atherina presbyter		124.0	3.20	0.238	1.091	0.109	51.55N	1.20W	(5-18)	[727]
Gasterosteus aculeatus		52.41	2.62	0.250	1.019	0.249	52.20N	3W	(4-17)	[363]
Pygosteus pungitius		41.28	1.03	0.200	1.777	0.468	52.20N	3W	(4-17)	[363]
Nemipterus marginatus		232.8	35.8	0.243	0.5047	0.227	6N	116E	(26-30)	[535]
Labrus bergylta		509.2	8.64	0.258	0.07170	3.30e-3	54N	4.40W	(4-17)	[172]
Ellerkeldia huntii		152.1	10.8	0.319	0.3350	0.0791	35.30S	174.40E	(12-22)	[361]
Lepomis gibbosus		61.86	9.04	1	0.1415	0.0342	45.40N	89.30W	(5-15)	[490]
Lepomis macrochirus		71.62	16.8	1	0.1292	0.0467	45.40N	89.30W	(5-15)	[490]
Perca fluviatilis		317.9	22.5	0.25	0.1615	0.0242	56.10N	4.45W	8-14	[649]
Tilapia species		129.6	20.7	1	3.542	1.10	31.30N	35.30E	(37)	[443]
Liza vaigiensis		746.3	31.8	0.258	0.1758	0.0147	17S	145E	(18-27)	[271]
Mugil cephalus		595.0	27.2	0.258	0.3350	0.0370	17S	145E	(18-27)	[272]
Valamugil seheli		635.3	35.0	0.258	0.2725	0.0291	17S	145E	(18-27)	[272]
Seriola dorsalis		1373	30.7	0.231	0.1155	5.72e-3	33N	118W	(15-20)	[43]
Ammodytes tobianus		140.9	1.98	0.147	0.7305	0.0595	50.47N	1.02W	5-18	[586]
Thunnus albacares		2745	636	0.266	0.1481	0.0509	0-10N	165E	(26-30)	[514]
Thunnus thynnus		3689	448	0.266	0.06623	0.0144	53-57N	0-7E	(4-17)	[721]
Coryphoblennius galerita		69.55	2.72	0.250	0.4011	0.0598	50.20N	4.10W	(5-18)	[488]
Pomatoschistus norvegicus		48.80	0.770	0.252	2.466	0.305	56.20N	5.45W	(8-14)	[257]
Gobio gobio		154.9	15.9	0.250	0.7519	0.495	51N	2.15W		[451]
Gobio gobio		174.8	3.84	0.250	0.4165	0.0321	51.50N	8.30W	(4-17)	[379]
Gobius cobitis		213.9	14.9	0.295	0.2082	0.0385	48.45N	4W	(5-18)	[254]
Gobius paganellus		79.89	1.94	0.200	0.4790	0.0463	54N	4.40W	(4-17)	[487]
Lesueurigobius friesii		65.82	0.623	0.252	0.5628	0.0349	55.45N	5W	8-12	[501]
Lesueurigobius friesii		63.72	0.409	0.252	0.6826	0.0322	56.20N	5.45W	(8-14)	[255]
Blennius pholis		150.5	3.36	0.250	0.2464	0.0176	50.20N	4.10W	(5-18)	[488]
Arnoglossus laterna		93.55	3.06	0.200	0.4544	0.0895	56.15N	5.40W	(8-14)	[256]
Hypoglossus hypoglossus		632.7	54.7	1	0.04797	6.04e-3	59N	152W	(3-14)	[674]
Scophthalmus maximus	f	669.4	14.2	0.266	0.2165	0.0298	53-57N	0-7E	(3-14)	[360]
Scophthalmus maximus	m	495.3	6.93	0.272	0.3247	0.0222	53-57N	0-7E	(3-14)	[360]
Pleuronectes platessa		142.1		1	0.09500		53-57N	0-7E	(4-17)	[69]
Solea vulgaris		78.41		1	0.4200		53-57N	0-7E	(4-17)	[69]
Amphibia										
Rana tigrina	l	12.79	0.670	1	15.75	1.88	lab	lab	30-33	[155]
Rana sylvatica		8.201	0.154	1	30.97	6.64	36.05N	81.50W	21-26	[770]

Triturus vulgaris	l	26.40		0.353	3.960		59.30N	10.30E	-5-14	[174]
Triturus cristatus	l	40.40		0.353	4.080		59.30N	10.30W	-5-14	[174]
Reptilia										
Emys orbicularis	f	182.1	1.98	0.500	0.2707	0.0124			(22)	[132]
Emys orbicularis	m	161.8	1.56	0.500	0.3453	0.0172			(22)	[132]
Vipera berus		539.0	33.0	0.075	0.3734	0.0657			(20)	[234]
Eunectes notaeus	f	3283	50.9	0.075	0.2552	0.0165	lab	lab	(20)	[547]
Eunectes notaeus	m	2946	94.5	0.075	0.2030	0.0251	lab	lab	(20)	[547]
Aves										
Eudyptula minor nov.		114.7	5.67	1	15.60	2.69			39.5	[386]
Pygoscelis papua		191.8	3.35	1	15.31	0.965			39.5	[744]
Pygoscelis antarctica		163.6	5.29	1	16.88	2.12			39.5	[744]
Pygoscelis adeliae		159.9	7.77	1	15.47	2.81			39.5	[744]
Pygoscelis adeliae		188.7	3.47	1	14.32	0.698			39.5	[705]
Aptenodytes patagonicus		250.0		1	8.508	0.164			39.5	[687]
Pterodroma cahow		63.16	0.465	1	62.96	1.55			39.5	[773]
Pterodroma phaeopygia		79.2	0.93	1	20.08	3.43			39.5	[290]
Puffinus puffinus		83.90	0.069	1	41.55	2.87			39	[99]
Diomedea exulans		229.1	1.02	1	5.541	0.176			39.5	[720]
Oceanodroma leucorhoa		41.53	0.282	1	26.37	1.58			39.5	[599]
Oceanodroma furcata		44.73	0.339	1	23.28	1.16			39.5	[75]
Phalacrocorax auritus		149.5	6.31	1	18.18	1.81			39.5	[190]
Phaethon rubricaudata		101.1	1.45	1	13.03	0.923			39.5	[166]
Phaethon lepturus		72.79	1.12	1	18.77	2.03			39.5	[166]
Sula sula		80.01	1.18	1	11.82	1.53			39.5	[166]
Sula bassana		172.7	2.50	1	12.41	0.639			39.5	[505]
Cionia cionia		158.0	6.10	1	18.36	2.35			39.5	[144]
Phoeniconaias minor		116.8	3.01	1	11.31	1.30			39.5	[62]
Florida caerulea		68.19	1.16	1	42.63	3.61			39.5	[762]
Anas platyrhynchos		117.3	0.330	1	17.75	0.410			39.5	[313]
Anas platyrhynchos		151.3	0.353	1	17.04	0.307			39.5	[313]
Anas platyrhynchos		145.5	1.94	1	10.26	0.680			39.5	[485]
Anas platyrhynchos		154.8	1.65	1	13.14	4.56			39.5	[615]
Anser anser		181.5	2.99	1	7.895	0.626			39.5	[485]
Buteo buteo	f	103.7	1.17	1	27.57	1.34			39.5	[564]
Buteo buteo	m	95.99	1.11	1	27.90	1.45			39.5	[564]
Falco subbuteo		66.16	0.689	1	46.77	3.57			39.5	[70]
Meleagris gallopavo		256.1	9.89	1	4.340	0.782			39.5	[129]
Meleagris gallopavo		296.2	26.2	1	3.657	1.18			39.5	[129]
Phasianus colchicus	f	100.3	1.86	1	6.610	0.738			39.5	[485]
Phasianus colchicus	m	118.8	4.25	1	5.004	0.746			39.5	[485]
Gallus domesticus	f	136.5	1.24	1	4.625	0.209			39.5	[530]
Gallus domesticus	m	153.5	2.22	1	4.522	0.305			39.5	[530]
Bonasia bonasia		85.17	2.68	1	7.807	0.740			39.5	[59]
Colinus virginianus		56.90	0.328	1	10.81	0.427			39.5	[613]
Coturnix coturnix		55.41	0.761	1	14.94	0.784			39.5	[95]
Rallus aquaticus		51.66	0.730	1	14.45	0.0882			39.5	[657]
Gallinula chloropus		67.05	1.20	1	20.00	1.72			39.5	[204]
Philomachus pugnax	f	47.41	1.04	1	39.46	2.75			39.5	[633]
Philomachus pugnax	m	59.94	2.18	1	29.09	2.97			39.5	[633]
Haematopus moquini		103.4	5.69	1	10.63	1.40			39.5	[325]
Chlidonias leucopterus		42.76	0.502	1	66.39	4.08			39.5	[370]
Sterna fuscata		57.94	0.364	1	22.21	1.07			39.5	[103]

Sterna dougalli		50.15	1.12	1	33.97	3.77	39.5	[430]
Sterna hirundo		46.74	1.10	1	35.29	4.76	39.5	[430]
Rissa tridactyla		76.07	0.715	1	32.98	1.79	39.5	[460]
Larus argentatus		115.1	1.70	1	16.53	0.791	39.5	[675]
Catharacta skua		131.3	4.64	1	17.42	2.37	39.5	[686]
Catharacta skua		100.5	0.610	1	40.69	3.12	39	[241]
Catharacta maccormicki		104.8	0.310	1	60.29	3.18	39	[241]
Stercorarius longicaudus		83.90	0.069	1	41.55	2.87	30	[241]
Ptychoramphus aleuticus		59.66	0.373	1	23.73	0.913	39.5	[734]
Cuculus canoris		45.49	0.884	1	49.29	4.00	39.5	[784]
Cuculus canoris		50.26	1.45	1	38.56	3.68	39.5	[784]
Cuculus canoris		52.02	7.20	1	42.11	2.12	39.5	[784]
Cuculus canoris		52.44	1.40	1	39.91	3.60	39.5	[784]
Glaucidium passerinum	f	42.36	0.309	1	46.98	2.02	39.5	[643]
Glaucidium passerinum	m	41.86	0.484	1	41.57	2.51	39.5	[643]
Asio otus		64.94	0.596	1	36.54	1.77	39.5	[768]
Tyto alba		68.25	1.18	1	21.68	2.70	39.5	[276]
Strix nebulosa		98.26	0.960	1	16.43	0.730	39.5	[484]
Steatornis capensis		94.59	5.24	1	12.96	2.39	39.5	[673]
Apus apus		37.44	0.274	1	45.55	2.88	39.5	[760]
Selasphorus rufus		16.33	0.475	1	58.44	9.88	$\leq$ 41	[135]
Amazilia fimbriata		16.12	0.110	1	69.86	3.54	$\leq$ 41	[296]
Ramphastos dicolorus		70.11	1.89	1	28.52	4.01	39.5	[88]
Sturnus vulgaris		40.83	0.332	1	82.71	5.04	41	[764]
Bombycilla cedrorum		34.16	0.392	1	73.37	4.31	41	[596]
Petrochelidon pyrrhonota		31.19	0.520	1	69.64	6.40	41	[596]
Toxostoma curvirostre		36.62	0.695	1	49.82	3.62	41	[596]
Tyrannus tyrannus		35.53	0.673	1	59.43	4.60	41	[497]
Sylvia atricapilla		25.59	0.142	1	108.2	11.7	41	[67]
Garrulus glandarius		52.34	2.85	1	39.82	8.52	41	[382]
Campylorhynchus brunneicap.		32.79	0.200	1	65.85	6.70	41	[597]
Emberiza schoeniclus		25.88	0.238	1	138.7	12.1	41	[74]
Troglodytes aedon		22.29		1	105.9		41	[20]
Phylloscopus trochilus		22.41	0.576	1	76.78	8.86	41	[642]
Parus major		27.47	0.207	1	59.90	2.33	41	[31]
Parus ater		23.40	0.232	1	75.74	3.88	41	[438]
Montacilla flava		9.910	0.298	2.913	55.19	4.42	41	[173]
Agelaius phoeniceus	f	35.94	0.951	1	75.16	7.58	41	[146]
Agelaius phoeniceus	m	40.66	0.529	1	65.28	2.74	41	[146]
Gymnorhinus cyanocephalus		44.84	0.596	1	49.68	2.97	41	[40]
Eremophila alpestris		30.81	1.24	1	75.98	10.9	41	[48]
Mammalia								
Macropus parma		148.6	0.615	1	2.736	0.0942	35.5	[462]
Macropus fuliginosus		261.6	34.8	1	2.397	0.910	35.5	[563]
Trichosurus caninus		137.8	1.06	1	1.754	0.561	35.5	[340]
Trichosurus vulpecula		139.3	1.34	1	3.715	0.184	35.5	[444]
Perameles nasuta		100.5	0.967	0.961	4.743	0.175	35.5	[444]
Setonix brachyurus		116.6		1	1.728	0.117	35.5	[728]
Suncus murinus	f	26.58	0.160	1	30.92	1.37	37	[185]
Suncus murinus	m	29.88	0.267	1	20.64	1.27	37	[185]
Sorex minutus		65.00		0.294	32.97	0.674	36	[344]
Desmodus rotundus		30.68	0.175	1	8.775	0.277	(35.5)	[637]
Homo sapiens	m	1648	58.5	0.244	0.1490	0.0158	37	[120]

Lepus europaeus		148.3	1.60	1	5.034	0.530	37	[98]
Oryctolagus cuniculus		116.6	1.11	1	6.507	0.272	37	[728]
Notomys mitchellii		27.09	0.412	1	21.54	1.64	38	[145]
Notomys cervinus		23.85	0.456	1	23.94	3.00	38	[145]
Notomys alexis		27.43	0.382	1	20.03	1.24	38	[145]
Pseudomys novaehollandiae		24.88	0.101	1	13.00	0.386	38	[377]
Castor canadensis		234.4	1.64	1	5.117	0.365	38	[8]
Mus musculus		34.24	0.474	1	15.09	0.924	38	[530]
Mus musculus	f	31.87	0.129	1	22.33	1.31	38	[530]
Mus musculus	m	33.98	0.118	1	26.66	1.28	38	[530]
Rattus fuscipes		171.5	4.08	0.280	9.333	0.843	38	[704]
Rattus norvegicus		75.23	0.301	1	9.286	0.279	38	[530]
Tachyoryctes splendens		64.87	0.992	1	8.231	0.680	38	[576]
Balaenoptera musculus		37810	5420	0.188	0.05884	0.0208	37	[667]
Balaenoptera musculus	f	26200		0.188	0.2240		37	[435]
Balaenoptera musculus	m	25000		0.188	0.2160		37	[435]
Balaenoptera physalus	f	22250		0.180	0.2220		37	[435]
Balaenoptera physalus	m	2.1000		0.180	0.2221		37	[435]
Balaenoptera borealis	f	15300		0.197	0.1337		37	[435]
Balaenoptera borealis	m	14800		0.197	0.1454		37	[435]
Delphinapterus leucas	f	3056	54.4	0.254	0.2700	0.0399	37	[253]
Delphinapterus leucas	m	3589	86.5	0.254	0.1876	0.0227	37	[253]
Canus domesticus		387.2	1.46	1	4.168	0.120	37	[530]
Lutra lutra	f	178.1	1.32	1	2.870	0.156	37	[682]
Lutra lutra	m	197.7	1.38	1	2.692	0.143	37	[682]
Pagaphilus groenlandicus		486.4	7.44	1	0.4787	0.0673	37	[427]
Mirounga leonina	m	5580	356	0.254	0.1492	0.0265	37	[428]
Mirounga leonina	f	2933	42.7	0.254	0.3094	0.0480	37	[428]
Mirounga leonina	m	1799	149	1	0.1185	0.0278	37	[106]
Mirounga leonina	f	704.0	20.4	1	0.3661	0.0982	37	[106]
Leptonychotes weddelli		685.4		1	0.3001	0.0184	37	[106]
Loxodonta a.africana	f	1392	14.5	1	0.1016	8.16e-3	37	[429]
Loxodonta a.africana	m	1723	45.4	1	0.07173	7.81e-3	37	[429]
Rangifer tarandus	f	470.2	1.84	1	1.263	0.0589	37	[465]
Rangifer tarandus	m	534.4	4.39	1	1.000	0.0617	37	[465]
Bos domesticus	f	815.4	3.66	1	0.9957	1.73e-3	38.5	[530]
Alces alces		712.6	12.7	1	0.5930	0.159	37	[309]

Minimum embryonic period

Because the DEB model is volume-structured rather than age-structured, the length of the various life stages is closely tied to growth. The gestation time is proportional to volume$^{1/3}$, excluding any delay in implantation. Weasels and probably armadillos are examples of species that usually observe long delays, possibly to synchronize the juvenile period with favourable environmental conditions. Figure 6.2 illustrates that the expected scaling relationship is appropriate for 250 species of eutherian mammals. The mean energy conductance was found to be 2 mm d^{-1} at some 37 °C. This is less than half the mean temperature corrected value found from the von Bertalanffy growth rates of juveniles and adults, a difference that must be left unexplained at this moment.

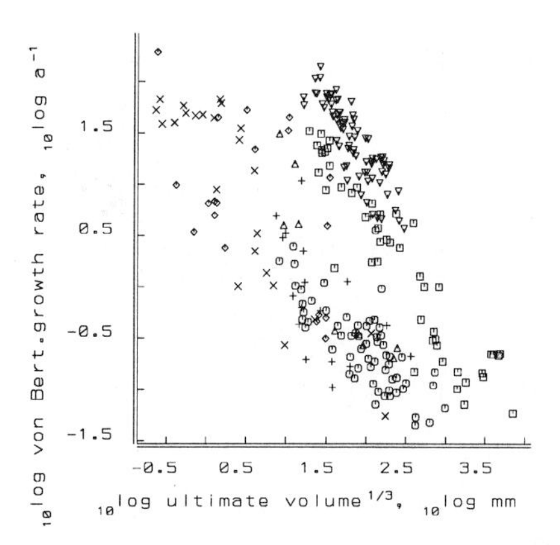

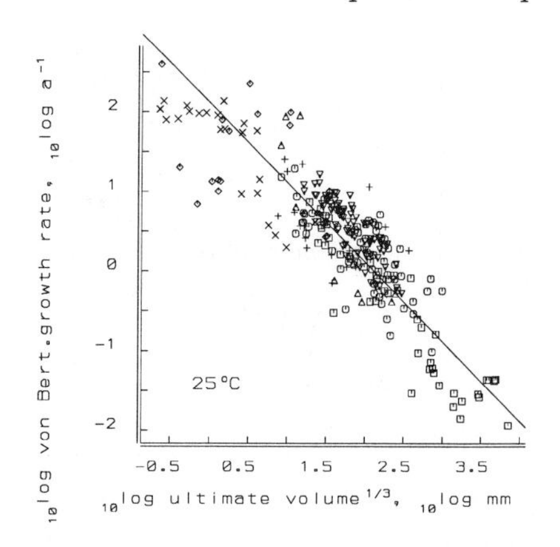

Figure 6.1: The von Bertalanffy growth rate as a function of maximum volumetric length. The left figure shows the rate as estimated from the original data, while the right figure gives the rates corrected to a common body temperature of 25 °C. The markers refer to ▽ birds, □ mammals, △ reptiles and amphibians, ∘ fishes, × crustaceans, + molluscs, ◇ others. The line has slope −1, which is expected on the basis of the DEB theory.

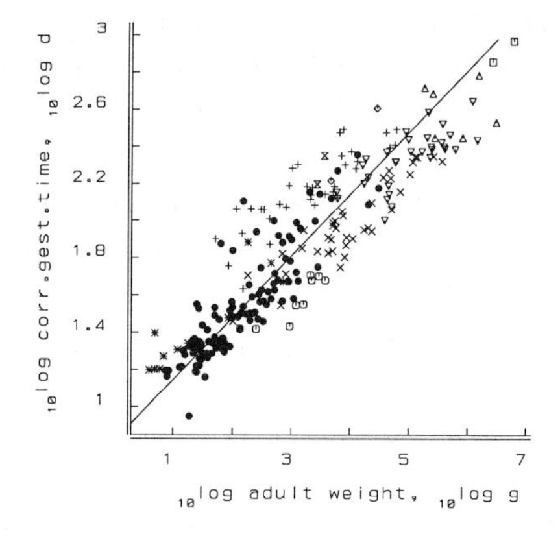

Figure 6.2: The gestation time of eutherian mammals tends to be proportional to volumetric length (line). Data from Millar [486]. The times have been corrected for differences in relative birth weight, i.e. birth weight as a fraction of adult weight, by multiplication of the ratio of the mean relative birth weight$^{1/3}$, 0.396, and the actual relative birth weight$^{1/3}$. The symbols refer to ∗ *Insectivora*, + *primates*, ◇ *Edentata*, ∘ *Lagomorpha*, • *Rodentia*, × *Carnivora*, □ *Proboscidea*, ⋈ *Hyracoidea*, △ *Perissodactyla*, ▽ *Artiodactyla*.

Incubation time (3.29) depends on volume in a more complex way, but it is also approximately proportional to body volume$^{1/3}$, or alternatively to egg volume$^{1/4}$; the scaled egg costs e_0 do not depend on body size, so that egg costs themselves $E_0 = e_0[E_m]V_m$ scale with $\mathcal{V}^{4/3}$ or $\mathcal{V} \propto E_0^{3/4}$ so that $a_b \propto \mathcal{V}^{1/3} \propto E_0^{1/4}$. Figure 6.3 gives the log-log plot for the species that breed in Europe. These data are very similar to those of Rahn and Ar [577], who included species from all over the world. Although the scatter is considerable, the data are consistent with the expectation. Note that within a species, large eggs hatch earlier than small ones, though one needs to look for species with egg dimorphism to find a large enough difference between egg weights.

The tube noses *Procellariiformes*, incubate longer, while they also have relatively heavy eggs, and so relatively large chicks. If corrected for this large volume at birth, their incubation time falls within the range of other species. This correction has been done

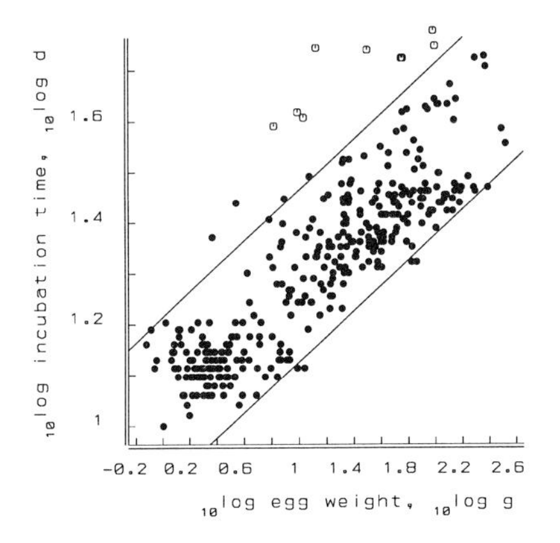

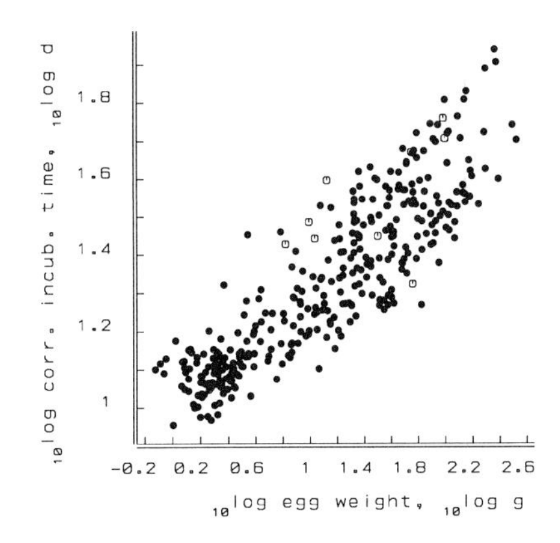

Figure 6.3: The incubation time for european breeding birds as a function of egg weight (left figure). Data from Harrison [291]. The lines have a slope of 0.25. The tube noses (∘) sport long incubation times. If corrected for a common relative volume at birth (right figure), this difference largely disappears.

by calculating the egg weight first, from $([d_w]\pi/6)$(egg length)(egg breadth)^2. (Data from Harrison [291].) The weight at birth is about 0.57 times the initial egg weight [742]. The scaled length at birth is about $(W_b/W_\infty)^{1/3}$. (This is not 'exact' because of the weight-volume conversion and the volume reduction due to heating.) Bergmann and Helb [58] give adults weights. The incubation time is then corrected for differences in scaled length at birth on the basis of (3.29) for small values of the investment ratio g and a common value for the scaled length at birth, 0.38.

The application of the DEB model has been useful in identifying the proper question, which is not why the incubation time of tube noses is that long, but why they lay so large an egg. The bird champion in this respect is the kiwi *Apteryx*, which produces eggs of 350 g to 400 g , while the adult weight is only 2200 g. It has an incubation period of a respectable 78 days. The relatively low incubation temperature of 35.4 °C extends incubation in comparison to other birds, which usually incubate at 37.7 °C [116,117]. This accounts for some 17-20 days extension with an Arrhenius temperature of 10000–12500 K, however, most of this long incubation relates to the very large relative size of the egg. The relative size of the egg itself is a result of the energy uptake and use pattern. This matter will be taken up again in the discussion on strategies, {239}.

If one or more primary parameters are known, the value of a certain compound parameter such as the (minimum) incubation time can be predicted with much more accuracy. On the basis of growth data for the cassin's auklet during the juvenile phase, I predicted an incubation period of 40 days [408], not knowing that it has been measured and actually found to be 37–42 days [453]. It is more difficult to verify my prediction of an incubation period for the hadrosaur of 145 days if it was ectothermic. This calculation accounted for a birth length of 35 cm with an adult length of 700 cm [334], while the Nile crocodile has a birth length of 20 cm, an adult length of 700 cm and an incubation period of 80–90 days [279]. If true, the hadrosaur must have been a very patient animal! This consideration supports the hypothesis of e.g. Bakker [30] and Desmond [165], that dinosaurs

Figure 6.4: The striped tenrec *Hemicentetes semispinosus* is a curious 'insectivore' of 110 grams from the rainforests of Madagascar that feeds on arthropods and earthworms and finds its way about using sonar. Walking in the forest, you can spot it easily by its head shaking, not unlike that of an angry lizard. Its juvenile period of 35 days is the shortest among mammals. The gestation period is 58 days [197].

were endothermic. The high body temperature doubtlessly reduces the incubation time considerably.

Reptilian champion in incubation time is the tuatara *Sphenodon punctatus* where the 4 g hatchling leaves the egg after 15 months. The low temperature, 20–25 °C, contributes to this record.

The European cuckoo is a breeding parasite which parks each of its many eggs into the nest of a 'host', which has an adult body weight of only 10% of that of the cuckoo. The eggs of the host are one half to three quarters the size of that of the cuckoo. On the basis of egg size alone, therefore, the cuckoo egg should hatch later than the eggs of the host, while in fact it usually hatches earlier despite the later date of laying. If the relative size of the egg with respect to the adult is taken into account, the DEB theory correctly predicts the observed order of hatching. The essence of the reasoning is that, since the cuckoo is much larger than the host, the cuckoo uses the reserves at a higher rate (i.e. $\{\dot{A}_m\}$ is larger), and, therefore, it grows faster in the absolute sense. Growth is so much faster that the difference in birth weight with the chicks of the host is more than compensated. In non-parasitic species of the cuckoo family, the eggs are much larger [784], which indicates that the small egg size is an adaptation to the parasitic way of life. The extra bonus for the European cuckoo is that it can produce many small eggs (about 20-25), which helps to overcome the high failure rate of this breeding strategy.

Minimum juvenile period

The juvenile period at high food density for different species is

$$a_{p,\mathcal{V}} = \frac{1}{\dot{\gamma}_{\mathcal{V}}} \ln \frac{V_{m,1}^{1/3} - (V_{h,1}V_{m,1}/\mathcal{V})^{1/3} - V_{b,1}^{1/3}}{V_{m,1}^{1/3} - (V_{h,1}V_{m,1}/\mathcal{V})^{1/3} - V_{p,1}^{1/3}}$$

It increases almost linearly with length. This relationship fits Bonner's data, as given in Pianka [78,550] very well, however, this data set uses actual lengths, rather than the more appropriate volumetric ones.

The Guinness book of world records mentions the striped tenrec *Hemicentetes semispinosus*, see figure 6.4, as the mammal with the shortest juvenile period [425]. The cuis *Galea musteloides*, a 300–600 g South American hystricomorph rodent, usually ovulates at some 50 days, but sometimes does so within 11 days of birth [656,747]. Many smaller mammals have a longer juvenile period, which points to the fact that body scaling relationships only give tendencies and not reliable predictions.

6.3.4 Reproduction

Energy investment into an egg

For small values of the energy investment ratio g, the scaled energy investment e_0 into a single egg, as given in (3.27), is independent of maximum body volume, so that for the unscaled energy investment E_0 holds: $E_{0,\mathcal{V}} = E_{0,1}(\mathcal{V}/V_{m,1})^{4/3}$. This does not necessarily translate into the egg weight being proportional to body weight$^{4/3}$, because the energy content, i.e. the chemical composition, may also show scaling relationships. The larger species also have to observe mechanical constraints, and small species can have problems with heating themselves during development. This may cause deviations from expected tendencies. The volume of the hatching young is proportional to the maximum volume of the adult (if corrected for the volume reduction due to heating in endotherms), according to primary scaling relationships. The European birds have egg weights approximately proportional to adult weights. Calder [117] and Rahn *et al.* [580] obtained egg weights proportional to adult weights$^{0.77}$; Birkhead [93] found that egg weight of auks is proportional to adult weight$^{0.72}$.

Water loss in eggs

The use of energy (stored in lipids etc.) relates to water that will evaporate from bird eggs. Part of this water is formed by the oxidation of energy-rich compounds, and part of it consists of the watery matrix in which the compounds are embedded for the purpose of giving enzymes a correct environment and for transport of the products. The total loss of water during the incubation period, therefore, reflects the total use of energy $E_0 - E_b$. Since, like the energy investment into a single egg E_0, the amount of energy at birth $E_b = [E_m]V_b$ is also proportional to $\mathcal{V}^{4/3}$, the loss of water must be a fixed proportion of egg volume. Rahn, Ar and Paganelli [578,19] found that it is some 15% of the initial egg weight. If the use of energy relates to water loss directly, one would expect that the initial loss rate is small and builds up gradually. The egg usually decreases linearly in weight, as Gaston [243] found for the ancient murrelet *Synthliboramphus antiqua*. This is to be expected on physical grounds, of course. The specific density of an egg can be used to determine the length of time it has been incubated. This process of water loss implies that the water content of the reserves changes during incubation, but its range is rather restricted. The functional and physical aspects of water loss in eggs thus coincide beautifully.

Maximum reproductive rate

The maximum reproductive rate, as given in (3.51) is $\dot{R}_{m,\mathcal{V}} = \dot{R}_{m,1}(V_{m,1}/\mathcal{V})^{1/3}$ for the different species. This is a beautiful example showing that the size relationships within a species work out differently from those between species. Within species comparisons show that large individuals reproduce at a higher rate than small ones, while the reverse holds for between species comparisons. Like most of the other scaling relationships mentioned in this chapter, this only reflects tendencies allowing substantial deviations. The trade-off between a small number of large young and a large number of small young is obvious.

The partition coefficient κ does not depend on body size, thus a small species spends the same fraction of energy that it utilizes from its reserves on reproduction as a large species. (That is, if the energy required for the maintenance of maturity is negligibly small.) Most studies do not deal with dynamic models for energy allocation, however, but with static ones. Such studies aim to describe the (instantaneous) allocation of resources to the various end points, given an individual of a certain size. If we express the energy spent on reproduction as a fraction of the energy taken up from the environment (at constant food density), this fraction decreases with increasing body volume. This is because ingestion rate increases with volume, see {222}, and utilized energy (respiration rate) with a weighted sum of surface area and volume. This illustrates once again the importance of explicit theories for the interpretation of data.

6.3.5 Survival

Starvation

In the section on prolonged starvation {128}, the time till death by starvation for an individual with an initial scaled energy density of $e(0) = l$ was found to be $t_\dagger = \dot{v}^{-1}V^{1/3}\ln\kappa^{-1}$ or $t_\dagger = \dot{v}^{-1}V^{1/3}\kappa^{-1}$ depending on its storage dynamics during starvation. In the first expression the individual does not change its storage dynamics, and in the second one it spends energy on maintenance only. The corresponding survival times for different species are thus $t_{\dagger,\mathcal{V}} = \dot{v}_1^{-1}\mathcal{V}^{1/3}\ln\kappa_1^{-1}$ or $t_{\dagger,\mathcal{V}} = \dot{v}_1^{-1}\mathcal{V}^{1/3}\kappa_1^{-1}$. They are thus proportional to volume$^{1/3}$. Threlkeld [717] found a scaling parameter of 1/4, but 1/3 also fits the data well.

Constant food densities thus select for small body volume, because small volume aids sur… densities select for large body volume, be… l over prolonged starvation. Brook and Do… , the larger species of zooplankton domina… tion does not lie in the size dependence of … erate the other way round), but in the len… sufficient food. This has been confirmed ex…

Li…

Gr… ation of the DEB model, but it is practical to … eeds $(1-\epsilon)^3 V_\infty$, as the end of the growth pe… say. The length of the growth period at co… nts to $\dot{\gamma}^{-1}\ln\epsilon(1-l_b/f)$. It thus increases wi… st as the juvenile period. The mean life sp… period that die from aging is found from

(4.48) to be $\frac{1}{3}\Gamma(\frac{1}{3})(\frac{1}{6}\ddot{p}_a\dot{m})^{-1/3}$ for $l = f$. The mean life span is thus indepent of maximum body volume of a species. Finch [219] concluded that the scanty data on life spans of ectotherms do not reveal clear-cut relationships with body volume. Large variations in life spans exist, both within and between taxa. The ratio of the growth period and the mean life span is $5.55\ddot{p}_a^{1/3}\dot{m}^{-2/3}(1+f/g)\ln\epsilon(1-l_b/f)$ and increases with volumetric length. If this ratio approaches 1, life span tends to increase with maximum body volume in a sigmoid manner.

In the section on aging, {105}, I discussed the coupling between effectiveness of antioxidants, life spans and genetical flexibility. If aging allows long life spans, individuals are likely to possess effective means of dealing with a threatening environment such as avoidance behaviour for dangerous situations (learning), physiological regulation to accommodate changes in diet, temperature and so on. This is likely to involve large brain size and thus an indirect coupling between brain size and life span. The brain may also be involved in the production of antioxidants or the regulation thereof, which makes the link between brain size and life span more direct. Birds have larger brain to body weight ratios than mammals and live twice as long. The life spans of both mammals and birds tend to scale empirically with weight$^{0.2}$ [117,219], which is close to volume$^{1/3}$. Although I have not worked out aging for endotherms quantitatively, this is consistent with the DEB-based expectation, because surface bound heating costs dominate respiration, and thus aging. Brain size is found, empirically, to be approximately proportional to surface area in birds and mammals [117]. Mammals tend to have higher volume-specific respiration rates than birds [759], which contributes to the difference in mean life span and jeopardizes easy explanations.

It must be stressed that these life span considerations relate to aging, though it is doubtful that aging is a major cause of death under field conditions. Suppose that size and age independent of death dominate under those conditions and that food web interactions work out such that the population remains at the same level while food is abundantly available. To simplify the argument, let us focus on species that have a size at first maturation close to the ultimate size. The death rate can then be found from the characteristic equation (5.22) for $\dot{\mu} = 0$ and $\text{Prob}\{\underline{a}_\dagger > a\} \simeq \exp\{-\dot{p}a\}$ and $\dot{R}(a) \simeq (a > a_p)\dot{R}_{m,\mathcal{V}}$. Substitution gives $\exp\{-\dot{p}_\mathcal{V} a_{p,\mathcal{V}}\} = \dot{p}_\mathcal{V}/\dot{R}_{m,\mathcal{V}}$. I have shown already that the age at first maturation $a_{p,\mathcal{V}}$ increases almost linearly with length, {234}, and the maximum reproduction rate $\dot{R}_{m,\mathcal{V}}$ decreases with length, {235}. The death rate $\dot{p}_\mathcal{V}$ must, therefore, decrease with length, so that the life span $\dot{p}_\mathcal{V}^{-1}$ increases with length.

These considerations help to explain the results of Shine and Charnov [655] that the product of the von Bertalanffy growth rate and the life span, $\dot{\gamma}_\mathcal{V}/\dot{p}_\mathcal{V}$, is independent of body size for snakes and lizards. Charnov and Berrigan [126] argued that the ratio of the juvenile period and the life span is also independent of body size. They tried to understand this empirical result from evolutionary arguments. Since the juvenile period is appoximately proportional to length as well, {234}, the ratio with the life span is roughly independent of body size. The present derivation also specifies the conditions under which the result is likely to be found, without using evolutionary arguments.

6.4 Tertiary scaling relationships

Primary and secondary scaling relationships follow directly from the invariance property of the DEB model. The class of tertiary scaling relationships invokes indirect effects via the population level. The assumptions that lead to the DEB model, table 4.1, must for tertiary scaling relationships be supplemented with assumptions on individual interactions. The chapter on 'living together' considers the most simple one: interaction is via the resource only, {159}. This makes tertiary scaling relationships a weaker type. Body size scaling relations are usually much less obvious at the community level [149], due to a multitude of complicating factors. Nonetheless, they can be of interest for certain applications.

6.4.1 Abundance

Geographical distribution areas are frequently determined by temperature tolerance limits; see {44}. Temperature and food abundance also determine species abundance in more subtle ways.

Since both the maximum ingestion rate and maintenance costs are proportional to body volume, abundance is likely to be inversely proportional to body volume, so $N \propto \mathcal{V}^{-1}$. This has been found by Peters [542], but Damuth [152] gives a scaling of -0.76. This relationship can only be an extremely crude one. Abundances depend on primary production levels, positions in the food web, etc. Nee *et al.* [502] point to the relationships between phylogenetic position, position in food webs and abundances in birds.

6.4.2 Distribution

High food densities go with large ultimate body sizes within a species. If different geographical regions which differ systematically in food availability are compared, geographical races can develop in which these size differences are genetically fixed. Since high food densities occur more frequently towards the poles and low food densities in the tropics, body sizes between these races follow a geographical pattern known as the Bergmann rule; see {132}.

It is tempting to extend this argument to different species feeding on comparable resources. This is possible to some extent, but another phenomenon complicates the result. Because of the yearly cycle of seasons, which are more pronounced towards the poles, food tends to be more abundant towards to poles, but at the same time the length of the good season tends to shorten. The time required to reach a certain size (for instance the one at which migration is possible) is proportional to a volumetric length. This implies that maximum size should be expected at the polar side of the temperate regions, depending on parameter values, migratory behaviour, endothermism, etc. This probably holds for species such as geese, that migrate to avoid bad seasons. Geist [246] reported a maximum body weight at some 60° latitude and smaller weights both at higher and lower latitudes for New World deer and races of wolves. He found a maximum body size for sheep at some 50° latitude. Ectotherms that stay in the region can 'choose' the lower boundary of the temperature tolerance range such that they switch to the torpor state as soon as the temperature drops to a level at which food becomes sparse. This reduces the growth rate, of course, but not the ultimate body size. It then depends on harvesting mechanisms whether or not the mean body size in a population is affected.

6.4.3 Population growth rate

Since the (maximum) reproduction rate decreases with a length measure and the juvenile period increases with a length measure, the maximum population growth rate decreases somewhat faster than a length measure, especially for the small species. A crude approximation is the implicit equation obtained from (5.23):

$$\exp\{-\dot{\mu}_{\mathcal{V}} a_{p,1} (\mathcal{V}/V_{m,1})^{1/3}\} = (\mathcal{V}/V_m)^{1/3} \dot{\mu}_{\mathcal{V}} / \dot{R}_{m,1}$$

For dividing isomorphs, the population growth rate is inversely proportional to the division interval, which corresponds with a juvenile period from an energetics point of view. This gives $\dot{\mu}_{\mathcal{V}} = \dot{\mu}_1 (\mathcal{V}/V_{m,1})^{-1/3}$. Fenchel [216] obtained an empirical scaling of weight$^{-1/4}$ for protozoa.

6.5 Evaluation of strategies

Several comparisons of strategies have already been made to support statements during model development and analysis. See for instance on {173,181}. This section presents some additional strategies, that would have disrupted the flow of arguments if discussed in other chapters.

6.5.1 r *vs* K strategy

The ecological literature is full of references to what is known as r and K strategies, as introduced by MacArthur and Wilson [446]. The symbol r refers to the population growth rate and K to the carrying capacity; these two parameters occur in the logistic growth equation, which plays a central role in ecology. Under the influence of Pianka [549], organisms are classified relative to each other with respect to a number of coupled traits, the extremes being an 'r-strategist' and a 'K-strategist'. Many of the coupled traits mentioned by Pianka can now be recognized as direct results of body size scaling relationships for eco-physiological characteristics. The search for factors in the environment selecting for r or K strategies can as a first approximation be translated into that for factors selecting for a small or large body size.

6.5.2 Small *vs* large eggs

Most optimization arguments lead to the uninspiring result that reproduction rate or population growth rate is maximized by producing an infinitely large number of infinitesimally small young. No energy argument seems to forbid this possibility. It is hard to understand

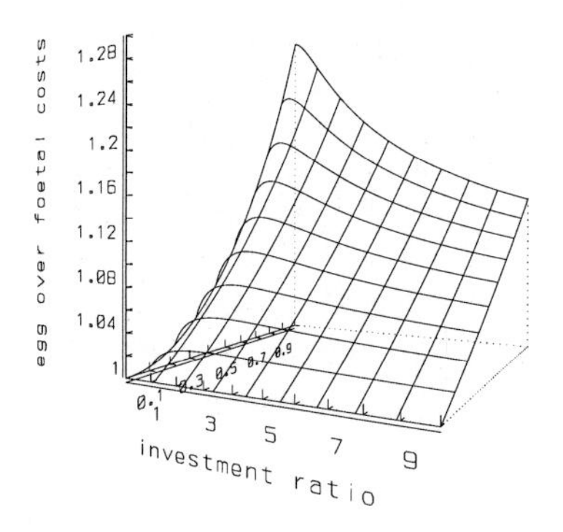

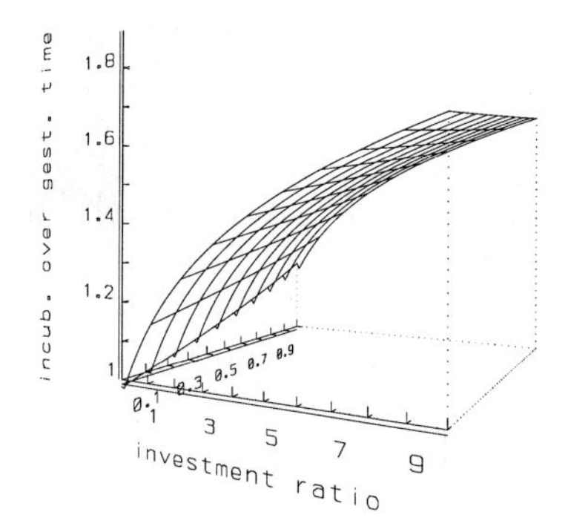

Figure 6.5: The energy costs of the production of an egg relative to that of a foetus (left) and incubation time relative to gestation time (right), as a function of the investment ratio g and scaled length at birth l_b (plotted on the y-axis) at high energy density at birth, $e_b = 1$.

why it pays to produce (few) large eggs. One possibility is in accounting for a changing spatially heterogeneous environment. Reproduction is usually synchronized with a favourable season, which is usually short. The reason why the crossbill breeds in midwinter in Scotland, for instance, is that it feeds its young with spruce seeds, which are mature early in spring. This habitat is not always favourable for them; if the seeds are finished, they have to move out. The same holds for ducks breeding in Iceland, where the adult starts to incubate while there is still snow. When the chicks hatch, food is available, but not for long; soon after they are able to fly, the conditions grow worse and they are forced to migrate to the sea. These examples are obvious, but the principle is probably quite common. The selection constraint is, therefore, a maximum period to complete development up to a stage allowing for migration.

It is consistent with the structure of the DEB model, that such a stage can be tied to a certain body volume. That the time needed to reach such a volume is strongly reduced by laying large eggs is obvious from the expression for the juvenile period. The fact that birds with large eggs, such as shearwaters and the kiwi, also have long incubation times does not devalue the argument. The DEB model shows that the time till the chick reaches a certain size would be even longer if the eggs were smaller. This insight is one of the gains of formalized reasoning, where all relevant variables can be considered at the same time. Another aspect to consider for endotherms is that small young have a hard time maintaining a high body temperature.

6.5.3 Egg *vs* foetus

The ratio of the energy costs for egg and foetus production is shown in figure 6.5 in case of high reserve density at birth, $e_b = 1$. This figure also shows the ratio of the incubation and gestation time. For very small investment ratios, g, the latter ratio becomes

$$\frac{\sqrt{2}e_b u^3}{l_b}\left(\frac{1}{2}\ln\frac{u^2+u\sqrt{2}+1}{u^2-u\sqrt{2}+1}+\arctan\frac{u\sqrt{2}}{1-u^2}\right)$$

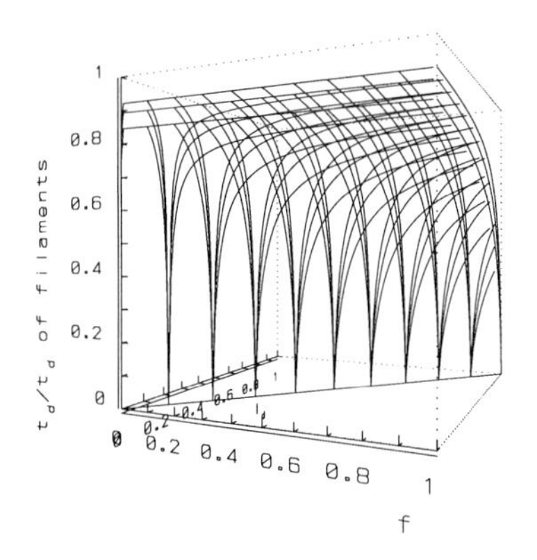

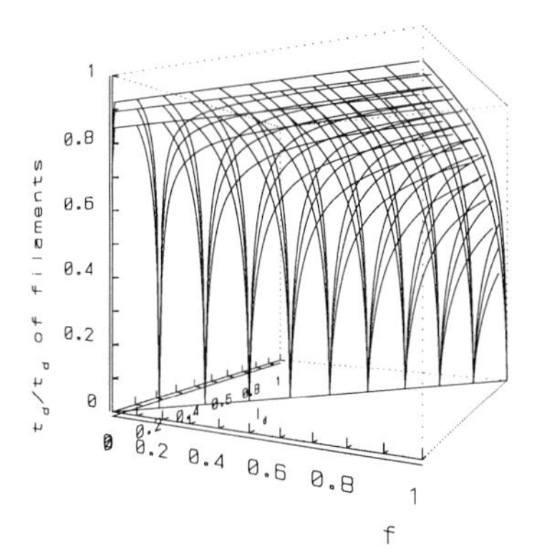

Figure 6.6: Stereo view of the division interval (z-axis) of cocci (upper surface), 2D-isomorphs (middle surface) and 3D-isomorphs (lower surface) as a fraction of that of filaments and as a function of the scaled functional response (f, x-axis) and scaled length at division (l_d, y-axis).

with $u \equiv (4e_b/l_b - 1)^{-1/4}$. For very small scaled lengths at birth, this ratio becomes $B_{x_b}(\frac{1}{3}, 0)x_b^{-1/3}/3$, with $x_b \equiv \frac{g}{e_b+g}$. The development of the embryo in an egg is somewhat retarded at the end of incubation, due to the diminishing reserves. This means that the incubation period is somewhat longer than the corresponding gestation period and that the cumulative costs at birth of an egg are somewhat higher than that of a foetus. This comparison assumes that all parameters are equal. Another difference is that, when breeding, the incubating individual is more restricted in its freedom than the pregnant mother.

6.5.4 Changing shape and growth

The change in shape during growth has been taken into account by multiplying parameters that involve surface areas in the growth equation by the appropriate shape correction function, defined as the ratio of the real surface area and that of an isomorph, such that the function has value 1 for the volume at division. The comparison of the effects of a change in shape is on the basis of equal energetics at the moment of division, i.e. equal uptake rate, equal maintenance costs, equal energy costs per unit of structural body volume, etc. This basis allows a comparison of division intervals and so of population growth rates. If $\dot{\gamma} \equiv \frac{\dot{m}g}{3(f+g)}$ denotes the von Bertalanffy growth rate, substitution of $V = V_d$ in (3.15), (3.35), (3.39), (4.37) and (4.41) results for $l_b = 2^{-1/3}l_d$ in

$$\dot{\gamma}t_d = \ln\frac{f-l_d/2}{f-l_d} \quad \text{for 0D-isomorphs} \qquad \dot{\gamma}t_d = \frac{2}{3}\ln\frac{f-l_d2^{-1/2}}{f-l_d} \quad \text{for 2D-isomorphs}$$

$$\dot{\gamma}t_d = \frac{l_d\ln 2}{3(f-l_d)} \quad \text{for 1D-isomorphs} \qquad \dot{\gamma}t_d = \ln\frac{f-l_d2^{-1/3}}{f-l_d} \quad \text{for 3D-isomorphs}$$

$$\dot{\gamma}t_d = \frac{l_d}{3(l_d-0.8f)}\ln\frac{0.6f-0.5l_d}{f-l_d} \quad \text{for cocci}$$

Rods are between cocci and filaments (i.e. 1D-isomorphs), depending on the aspect ratio δ. These scaled division intervals only depend on the scaled functional response f and the scaled (volumetric) length at division l_d. Figure 6.6 shows that the division intervals can be ordered into the sequence 3D-isomorphs, 2D-isomorphs, cocci, 1D-isomorphs, where 3D-isomorphs have the shortest and 1D-isomorphs the longest division interval. This does not come as a surprise because a baby 1D-isomorph has the same surface area/volume ratio as the mother, while a baby 3D-isomorph is better off. The reason to choose the mother as a reference rather than the baby is because it links up more closely with the trigger of DNA replication. The volume at the start of replication is the only one that is fixed, although the growth rate dependence of the volume at division is not taken into account here. The differences in division intervals hardly depend on the values of f and l_d, as long as the scaled length at division is small in comparision with the scaled functional response. For small l_d and $f = 1$, the division intervals relate as

$$\text{0D-isom.} : \text{1D-isom.} : \text{cocci} : \text{2D-isom.} : \text{3D-isom.} = 2.16 : 1 : 0.92 : 0.89 : 0.85$$

6.5.5 Deletion of disused DNA

Bacteria as a group are much more diverse in their metabolism than eukaryotes. Within the α-subgroup of the purple non-sulphur bacteria, a wide variety of complex metabolic pathways occurs, each involving a considerable number of genes [689]. This can only be understood by assuming that the ancestor of this group possessed all the pathways for, e.g. denitrification, aerobic and anaerobic photosynthesis, methylotrophy etc. During evolution, most species lost one or more of these traits; This brings us to the problem of to understand why it can be beneficial for species to cut out DNA that is not used in a particular environment rather than leaving it unused.

As shown in figure 6.7, the DEB model offers an explanation; the population growth rate decreases for increasing DNA duplication time t_D, particularly at high substrate levels. As the growth process continues during DNA duplication, the cell becomes larger the longer the DNA duplication period, if DNA duplication is triggered once the cell reaches a certain specific size. Since the uptake of substrate relates to surface area, and the surface area-volume ratio grows worse the larger the cell, the cell is better off reducing the time required to duplicate DNA. The effect of the DNA duplication time on the population growth rate is less at low substrate levels, because the division intervals are extended under these circumstances.

Cutting out disused DNA is just one way to reduce the DNA duplication time [691]. Another possibility is to maintain two chromosomes that are duplicated simultaneously, as in *Rhodobacter sphaeroides* [699], or more frequently, to maintain megaplasmids [233, 359,689].

The evolutionary significance of a high population growth rate should probably be found in the spatial and temporal heterogeneity of the environment. Useful substrates for heterotrophs are usually rare. If a plant or animal dies, the locally present microbes will grow at a high rate over a short period. If the subsequent selection processes thin randomly, the most abundant species has the best opportunity of surviving till the next event at which substrate becomes available. Since the ratio of the numbers grows exponentially at a rate equal to the difference in the population growth rates, small differences can be significant for long growth periods.

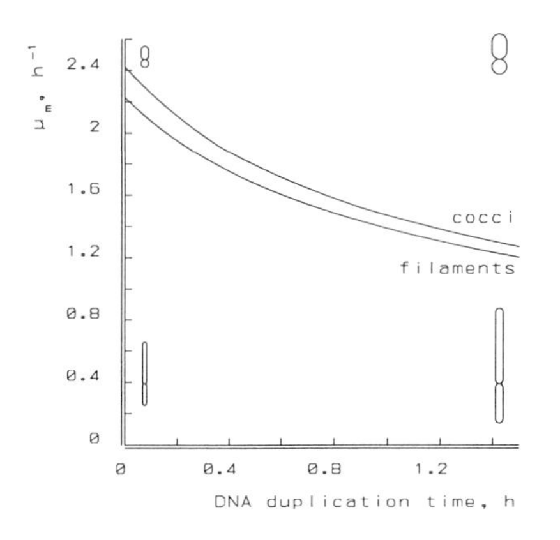

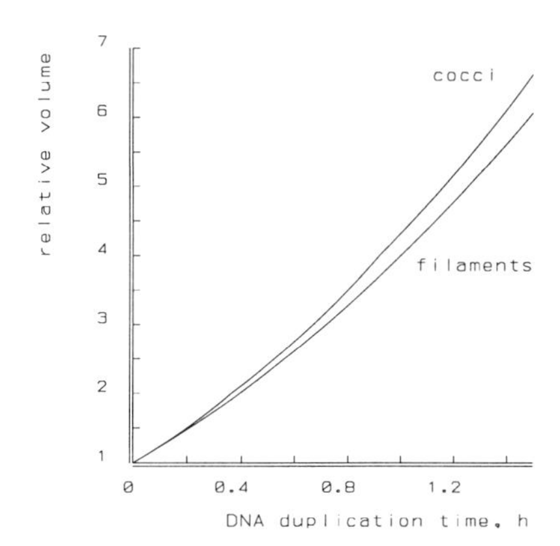

Figure 6.7: Maximum population growth rate decreases for increasing DNA duplication times. The curves are for aspect ratio $\delta = 0$, and 0.6. The aspect ratio is specified just prior to division and is fixed. Cell shape and relative size are indicated just before and after division for $\delta = 0.1$ and 0.6, at a doubling time of 0 and 1.5 h. Cell volume at division relative to the volume that triggers DNA duplication, V_d/V_p, is given in the right figure. Numerical studies show that the figure is independent of parameter values for l_p, g and $\dot{m}$, given maximum population growth rate.

6.5.6 Fitness

The relative amount of effort spent on reproduction differs from one species to another. Even within a species, it can depend on environmental conditions, as discussed for the light cycle in the pond snail. A change in the partition coefficient κ not only affects the energy allocated to reproduction, but also growth rate and, therefore, indirectly, ingestion rate, the length of the embryo and juvenile stage, aging acceleration and, as a result, survival probability. The intimate coupling between all these traits makes it essential to have rather comprehensive models for energy uptake and use when in the evaluation the evolutionary aspects of reproduction allocation.

This is one of the reasons why I am rather sceptical of evolutionary discussions that neglect coupling of traits. Another reason is that it is not at all obvious to me what traits precisely are favoured by evolution. Is it (mean) reproduction rate, the total number of offspring an individual produces over a life time, the population growth rate, or efficiency of conversion of food into biomass? The last criterion is seldom mentioned because this efficiency is usually taken to be constant. Given a limited amount of available food, this efficiency determines population numbers and is, therefore, most relevant in an evolutionary perspective. I doubt that one criterion exists that applies in every case.

A necessary, but not sufficient, condition for a genome to propagate is that the species survives. Thus, it must not be out-competed by other species in a dramatic way. Population growth rate may be relevant here, if environmental conditions are sufficiently homogeneous in space and time and the competitors more or less mixed. If, however, potential competitors live separated with relatively short periods of local mixing followed by random and drastic thinning, conversion efficiency from food to biomass is more relevant. Within

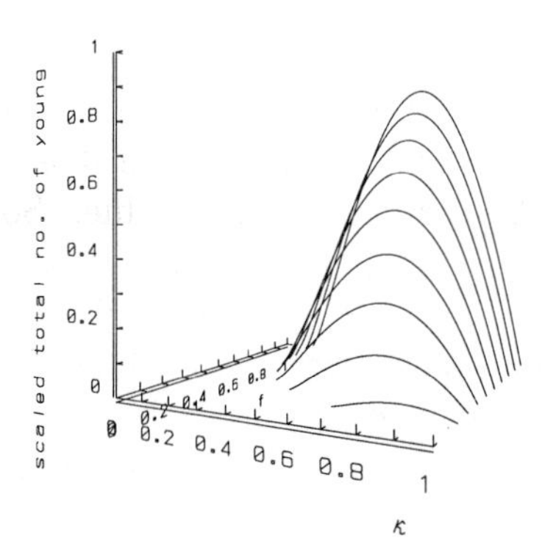

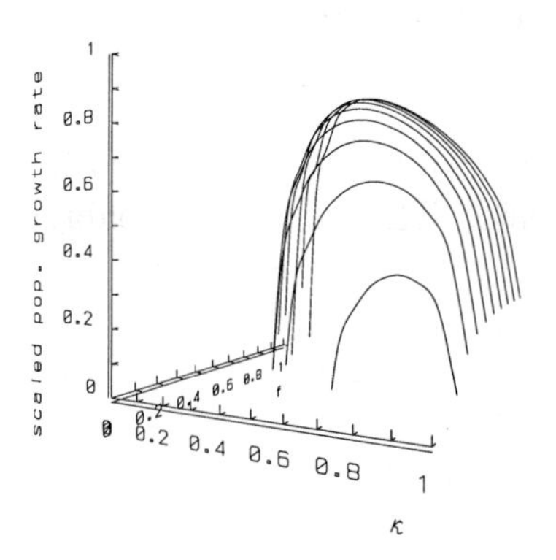

Figure 6.8: The effect of different values for the fraction κ of catabolic energy spent on maintenance plus growth in 3D-isomorphs at different but constant food densities on the total number of young (left) and the population growth rate (right). The scaled functional response f is plotted on the y-axis.

a population, the total amount of offspring from one individual seems most important as long as everything else can be considered to be constant. This, however, cannot be true, because offspring will affect resources. It might very well be, of course, that a large number of offspring spoils the environment and its resources to such an extent that extinction becomes probable. How do we measure evolutionary success? Is the evolution of the always rare and 200 Ma old tuatara less successful than the over-abundant human species which will probably reproduce itself into extinction well before 3 Ma of age as a species? Must the potential to generate new species be included into a measure for evolutionary success?

The aspects of propagation that make evolutionary sense depend on the scales of time and space. A proper setting of the problem involves the ecosystem and global levels of organization and are thus well beyond reach at the moment. Although I mistrust the evolutionary relevance of most optimization considerations, the results are frequently intriguing and inspiring. I will, therefore, evaluate how κ relates to various fitness measures for ectothermic isomorphs.

The general idea is to take the set of parameters, appropriate for a certain species, and study what mean number of total offspring and what population growth rate it would have at constant food density, if the species had another value for κ. Not all parameters are relevant. The minimum set is $\{\kappa_1, l_{b1}, l_{p1}, g_1, \dot{v}, \dot{m}, \ddot{p}_{a1}\}$, where the index 1 indicates that the (compound) parameter depends on the value κ_1. Figure 6.8 illustrates that at high food levels, the optimal κ is about 0.6 for the total number of young and 0.2 for the population growth rate. The difference is in the appearance of the first few young, which dominates the population growth rate, but not the total number of young. For lower food densities, the optimal value for κ is larger.

6.6 Origin of life

Comparing species ultimately leads to speculations about the origin of life. Such speculations are relatively straightforward in the context of the DEB theory, because it is not species-specific. Therefore, it probably also applies to the very first forms of life. So it does not suffer from the problem that collections of species-specific models for energetics have: if model 1 applies to species A and model 2 to species B, what model would apply to the common ancestor of species A and B if changes during evolution are gradual? This problem only has a solution if models 1 and 2 can be converted to each other in a continuous way. This poses severe constraints on the structure of models that make sense in an evolutionary context.

The very first cells probably did not have an advanced structure, so that they are likely to have been isomorphically growing spheres. The cells probably also did not have an advanced system for dividing into two equal parts. The surface tension of the (outer) membrane prohibits the separation of very small daughter cells. In turbulent environments protocells cannot grow to a large size before being torn apart into daughter cells that are not very different in size. In less turbulent environments cells can grow to larger sizes, while the daughter cells are able to differ more in size.

Homeostasis is a rather advanced trait which makes the job of regulation a much easier one. It is likely that it took a long time to achieve this trait, so that the composition of the protocells followed chemical changes in the environment more closely. Maintenance costs consist primarily of maintaining concentration gradients across the membrane and of replacing proteins and were probably low or absent in protocells.

The formation of bilayered membranes can occur and still does occur abiotically, especially on agitated surfaces of water, such as in coastal areas. Modern cells have phospholipid membranes that are impermeable to most compounds and exchange material with the environment via ion channels, which are complex proteins (in organisms alive today). Such exchange, therefore, requires a rather advanced machinery for protein synthesis which was probably RNA based. (Ling [434] argues, however, that uptake is largely determined by properties of the cytoplasm and that membranes are not that impermeable.) Since the discovery that RNA can catalyze its own splicing in the absence of proteins [124], most authors now agree that RNA appeared earlier than proteins [513], even though the abiotic synthesis of such complex RNA has not been demonstrated. It is hard to see how RNA in the environment could have been of much significance. Its concentration as well as that of its substrates (amino acids) were doubtlessly extremely low. The accumulation of the products inside membranes of protocells could hardly have been of significance. The situation is obviously much better for RNA molecules captured in protocells with a membrane, provided that the membrane is permeable for amino acids and other substrates. The quantitative aspects of RNA activity will be discussed in the next chapter on {250}. DNA appeared later to fulfill the function of an archive for RNA.

Growth can occur abiotically via the accumulation of compounds in the membranes from the environment. Originally these compounds were probably rarely subjected to chemical transformation. If RNA was present in the protocells that catalyzed transformations such that accumulation was enhanced, positive selection of such protocells would be a fact. The catalytic role of proteins then comes as a next step. The energy required for transformations could obviously not have been derived from the respiratory chain because of its complexity. According to de Duve [191], endogenic reactions were originally fuelled by thioesters. Most authors assume an early occurrence of photochemical processes and think that modern heterotrophs evolved from photo-autotrophs. The accumulation rate of substrate at low concentrations was probably proportional to the surface area of the membrane just as it still is. Protocells will have grown in the 'cube root' phase, see {147}, because maintenance processes were relatively unimportant.

I realize of course that this account is not very specific. A biochemical discussion is outside the scope of this book. If this account is realistic, however, it gives weight to the concept of the 'individual' being basic to eco-energetics.

Chapter 7

Suborganismal organization

The DEB model has its roots in suborganismal organization, thus this level of organization has been discussed already at several places in this book. The topics discussed related directly to the model assumptions. It is tempting to explore the consequences of the model for suborganismal processes that relate only indirectly to the model assumptions. To what extent is it possible to build models for elements of suborganismal organization that satisfy three requirements simultaneously: they must have a mechanistic basis, they must be realistic and they must be consistent with the DEB model. The general idea is that, although the DEB model appears to survive tests for the performance of individuals and populations, the primary assumptions of the DEB model may turn out to be inconsistent with molecular phenomena.

A lot of work still has to be done to penetrate the lower levels of organization within the context of the DEB model, and in this chapter I report results that have been obtained so far.

7.1 Digestion

Many studies of energy transformations assume that the energy gain from a food item does not depend on the size of the individual or on the ingestion rate. The usefulness of this assumption in ecological studies is obvious, and the DEB model uses it as well. In view of the relationship of gut residence time to both size and ingestion rate, this assumption needs further study. The nutritional gain from a food particle may depend on gut residence time, as has been observed by Richman and Schindler [595,635]. These findings are suspect for two reasons. The first reason is that assimilation efficiencies are usually calculated per unit of dry weight of consumer, while the metabolically inert energy reserves, which contribute substantially to dry weight, tend to increase with food density. The second reason is that, while the nutritional value of faecal pellets may decrease with increasing gut residence time, it is not obvious whether the animal or the gut microflora gained from the difference. I will discuss here to what extent a model for digestion based on described feeding behaviour can be made consistent with the 'constant gain' assumption (see assumption 4 of table 4.1).

When animals such as daphnids are fed with artificial resin particles mixed through their algal food, the appearance of these particles in the faeces supports the plug flow type of model for the digestion process, as proposed by Penry and Jumars [209,538,539]. The shape of the digestive system also suggests plug flow. The basic idea is that materials enter and leave the system in the same sequence and that they are perfectly mixed radially. Mixing or diffusion along the flow path is assumed to be negligible. (This is at best a first approximation, because direct observation shows that particles sometimes flow in the opposite direction.)

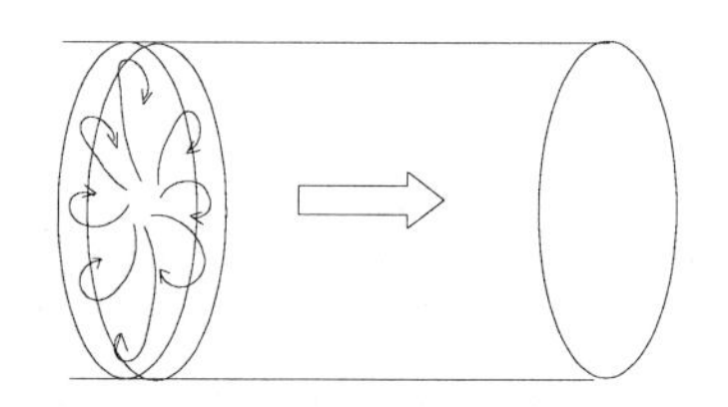

Suppose that a thin slice of gut contents can be followed during its travels along the cylinder-like digestive tract under conditions of constant ingestion rate. The small changes in the size of the slice during the digestion process are ignored. The gut content of a 4 mm *D. magna* is about 0.1 mm^3, while the capacity is about 6.3×10^5 cells of *Scenedesmus*, see figure 3.10, of some 58 μm^3 per cell, which gives a total cell volume of 0.0367 mm^3. The cells occupy some 37% of the gut volume, which justifies the neglect of volume changes for the slice. The volume of the slice of thickness L_λ is $V_s = \pi L_\lambda L_\phi^2/4$, where L_ϕ is the diameter of the gut, and $\pi L_\lambda L_\phi$ is the surface area of the gut.

Suppose that the metabolizable energy in the food must be freed before it can be absorbed through the gut wall, which generally involves some chemical transformations. The rate of freeing energy from the gut contents is taken to be proportional to the concentration of enzymes which have been secreted by or via the gut wall. The activity period of the different types of enzyme will probably vary. If the whole gut wall is involved in the secretion of enzymes and deactivation follows a simple first order process, the amount S_g of enzyme in the slice will follow $\frac{d}{dt}S_g = \{\dot{S}_g\}\pi L_\lambda L_\phi - \dot{k}_e S_g$, where $\{\dot{S}_g\}$ is the (constant) secretion rate of enzyme per unit of surface area of gut wall and $\dot{k}_e$ is the decay rate of enzyme activity. The equilibrium concentration of enzyme is thus $[S_g] \equiv S_g/V_s = 4\{\dot{S}_g\}(L_\phi \dot{k}_e)^{-1}$. So the enzyme concentration is larger in smaller individuals because of the more favourable surface area/volume ratio of the slice.

A simple Michaelis–Menten kinetics for the change in metabolizable energy density locked in the food, E_f, gives $\frac{d}{dt}E_f = -\dot{k}_f f_f 4\{\dot{S}_g\}(L_\phi \dot{k}_e)^{-1}$, where $f_f = E_f/(K_f + E_f)$. The parameter $\dot{k}_f$ is a rate constant for digestion. If the absorption of absorbable energy through the gut wall again follows Michaelis–Menten kinetics, the change of absorbable energy density in the slice, E_g, is given by $\frac{d}{dt}E_g = \dot{k}_f f_f 4\{\dot{S}_g\}(\dot{k}_e L_\phi)^{-1} - \dot{k}_g f_g 4\{C\}/L_\phi$ where $f_g \equiv E_g/(K_g + E_g)$ and $\{C\}$ is the number of carriers in a unit surface area of gut wall. The parameter $\dot{k}_g$ is a rate constant for absorption. This two-step double Michaelis–Menten kinetics for digestion with plug flow has been proposed independently by Dade *et al.* [150].

The conservation law for energy can be used to deduce that the total amount of energy taken up from the slice equals the slice volume times $(E_f(0) - E_f(t_g) - E_g(t_g))$, where $E_f(0)$ denotes the metabolizable energy density at ingestion and t_g the gut residence time.

To evaluate to what extent food density and the size of the organism affect digestion, it is helpful to define a digestion and uptake efficiency of metabolizable energy, $U = E_f(0) - E_f(t_g) - E_g(t_g)$ and to let t_g be the unit of time. So for $t^* = t/t_g$, $E_f^*(t^*) = E_f(t^*)/E_f(0)$ and $E_g^*(t^*) = E_g(t^*)/E_g(0)$, the equations for the change in energy locked into food and in absorbable energy become

$$\frac{d}{dt^*}E_f^* = -\frac{\dot{k}_f 4\{\dot{S}_g\}t_g}{\dot{k}_e E_f(0) L_\phi}\frac{E_f^*}{K_f/E_f(0)+E_f^*} \tag{7.1}$$

$$\frac{d}{dt^*}E_g^* = \frac{\dot{k}_f 4\{\dot{S}_g\}t_g}{\dot{k}_e E_g(0) L_\phi}\frac{E_f^*}{K_f/E_f(0)+E_f^*} - \frac{\dot{k}_g 4\{C\}t_g}{E_g(0) L_\phi}\frac{E_g^*}{K_g/E_g(0)+E_g^*} \tag{7.2}$$

The efficiency now becomes $U = 1 - E_f^*(1) - E_g^*(1)$. For isomorphs, where gut diameter L_ϕ is proportional to whole body length L, these equations imply that energy uptake from food is independent of body size. Shorter gut residence time in small individuals is exactly compensated by higher enzyme concentration. This is because the production of short living enzymes is taken to be proportional to the surface area of the gut. An obvious alternative would be a long living enzyme that is secreted in the anterior part of the digestive system. If this part is a fixed proportion of the whole gut length the result of size independence is still valid.

Efficiency depends on food density as long as digestion is not complete, i.e. if $E_f^*(1)$ and $E_g^*(1)$ are not negligibly small. If absorption is a rapid process, so that $E_g^*(1) \simeq 0$, $E_f^*(1)$ can be solved implicitly from (7.1) via separation of variables. Substitution of $U = 1 - E_f^*(1)$ into this solution gives

$$U\frac{E_f(0)}{K_f} - \ln\{1-U\} = \frac{\dot{k}_f 4\{\dot{S}_g\}t_g}{\dot{k}_e K_f L_\phi} \tag{7.3}$$

For a chosen value for efficiency U close to 1, (7.3) provides a constraint on parameter values in order to achieve almost complete digestion. More specifically, it relates the rate of enzyme secretion to the ingestion rate of food items with metabolizable energy density $E_f(0)$. So it relates ingestion rate to food quality.

If the saturation coefficient of the freeing process, K_f, is negligibly small, (7.1) reduces to the zero-th order process $\frac{d}{dt^*}E_f^* = -\frac{\dot{k}_f 4\{\dot{S}_g\}t_g}{\dot{k}_e E_f(0) L_\phi}$, giving

$$E_f^*(t^*) = \left(1 - \frac{\dot{k}_f 4\{\dot{S}_g\}t_g}{\dot{k}_e E_f(0) L_\phi}t^*\right)_+ \tag{7.4}$$

The energy density of gut contents thus decreases linearly with time (and distance). This has been proposed by Hungate [342], who modelled the 42 hour digestion of alfalfa in ruminants. Digestion is complete if $\dot{k}_f 4\{\dot{S}_g\}t^* t_g > \dot{k}_e E_f(0) L_\phi$.

The above model can be extended easily to cover a lot of different enzymes in different sections of the gut, without becoming much more complicated, as long as the additivity assumptions of their mode of action and their products hold. Food usually consists of many components that differ in digestibility. Digestion can only be complete for the animal in question if the most resistent component is digested. If the digestion of each component follows zero-order kinetics, the flux across the anterior gut wall is larger than across the posterior gut wall. This situation is worked out for the freshwater oligochaete *Nais elinguis*, which propagates through division into two parts, the anterior part growing faster than the posterior one. This is to be expected if longitudinal mixing of compounds in the worm is limited; see the section on segmented individuals, {155}.

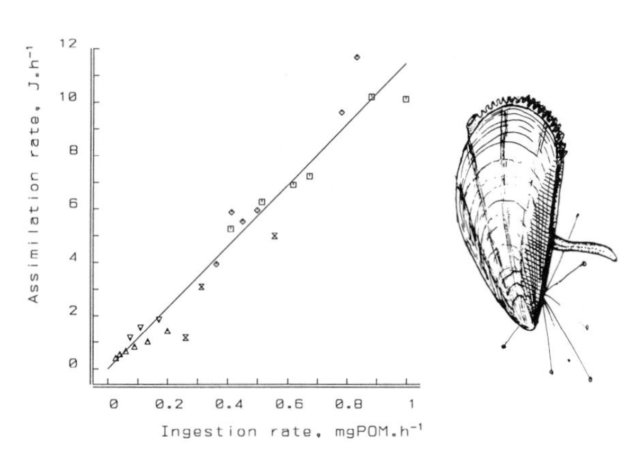

Figure 7.1: The assimilation rate as a function of ingestion rate for mussels (*Mytilus edulis*) ranging from 1.75 to 5.7 cm. Data from [44,45,81,387,297]. All rates are corrected to 15 °C. The fitted line is $\dot{A} = \dot{I}\{\dot{A}_m\}/\{\dot{I}_m\}$ with $\{\dot{A}_m\}/\{\dot{I}_m\}$ = 11.5 (s.d. 0.34) J mg POM^{-1}.

Microflora is likely to play an important role in the digestion process of all herbivores, including daphnids. It can provide additional nutrients by fermenting carbohydrates and by synthesizing amino acids and essential vitamins. Daphnids are able to derive structural body components and lipids from the cellulose of algal cell walls [640], though it is widely accepted that daphnids, like almost all other animals, are unable to produce cellulase. Endogenous cellulase production is only known from some snails, wood-boring beetles, shipworms and thysanurans [441]. The leaf cutting ant *Atta* specifically cultures fungi, probably to obtain cellulase [457]. Bacteria have been found in the guts of an increasing number of crustaceans [500], but not yet in daphnids [640]. In view of the short gut residence times for daphnids, it is improbable that the growth of the daphnid's gut flora plays an important role. Digestion of cellulose is a slow process and the digestive caecum is situated in the anterior part of gut, it is thus probable that daphnids produce enzymes that can pass through cell walls, because they do not have the mechanics to rupture them.

The existence of a maximum ingestion rate implies a minimum gut residence time. With a simple model for digestion, it is possible to relate the digestive characteristics of food to the feeding process, on the assumption that the organism aims at complete digestion. The energy gain from ingested food is then directly proportional to the ingestion rate, if prolonged feeding at constant, different, food densities is considered. See figure 7.1 from [286]. Should temperature affect feeding in a different way than digestion, the close harmony between both processes would be disturbed, which would lead to incomplete digestion under some conditions.

7.2 Protein synthesis

RNA, mainly consisting of rRNA, is an example of a compound known to be more abundant in cells growing at a high rate [393]. This property is used to measure the growth rate of fish, for example [108,339]. In prokaryotes, which can grow much faster, the increase in rRNA is much stronger. This section will, therefore, focus on prokaryotes. Within the DEB model, we can only account for this relationship when (part of the) RNA is included in the energy reserves. This does not seem unrealistic, because when the cell experiences a decline in substrate density and thus a decline in energy reserves, it is likely to gain energy

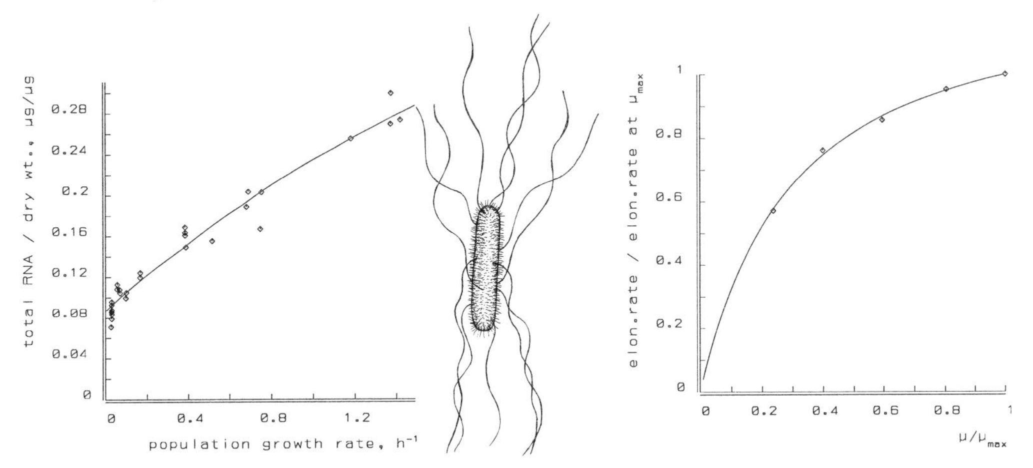

Figure 7.2: The concentration of RNA as a function of the population growth rate in *E. coli.* Data from Koch [393]. The least squares estimates of the parameters (with s.d.) are $\alpha_e = 0.44(0.05)$, $\alpha_v = 0.087(0.005)$ and $[d_{de}]/[d_{dv}] = 20.7(5.4)$.

Figure 7.3: Elongation rate in *E. coli* for $\delta = 0.3$, $l_d = 0.24(0.019)$, $g = 32.4(91.9)$. Data from Bremer and Dennis [89]. Both elongation rate and population growth rate are expressed as fractions of their maximum value of $\dot{\mu} = 1.73$ h^{-1} with an elongation rate of 21 aa s^{-1}rib^{-1}.

through the degradation of ribosomes [158]. It also makes sense, because the kinetics of reserve energy density is assumed to be a first order one, which implies that the use of reserves increases with their density. The connection between the abundance of RNA, i.e. the apparatus for protein synthesis, and energy density is, therefore, a logical one. No assumption of the DEB model implies that the energy reserves should be inert materials with no other function than being reserves. The analysis of the data from Esener in the section on mass-energy coupling, {192}, also points to the conclusion that rRNA can be a significant part of the reserves in bacteria.

The dynamics of RNA is most easy to describe when RNA constitutes a fixed fraction of the energy reserves. This is also the simplest condition under which homeostasis for energy reserves holds in sufficient detail to apply to RNA. The rate of RNA turnover is completely determined by this assumption. It also has strong implications for the translation rate and the total number of translations made from a particular RNA molecule.

RNA as a fraction of dry weight is given in figure 7.2. If the weight of RNA is a fraction α_v of the dry weight of structural biomass and a fraction α_e of the dry weight of the energy reserves, the fraction of dry weight that is RNA equals

$$W_{RNA}/W_d = \frac{\alpha_v[d_{dv}]V + \alpha_e[d_{de}]fV}{[d_{dv}]V + [d_{de}]fV} = \frac{\alpha_v + \alpha_e f[d_{de}]/[d_{dv}]}{1 + f[d_{de}]/[d_{dv}]}$$

The parameters of figure 5.11 were used to relate $\dot{\mu}$ to f. This indicates that at least in prokaryotes almost all RNA is part of the reserves and about half the energy reserves consist of RNA.

The mean translation rate of a ribosome, known as the elongation rate, is proportional to the ratio of the rate of protein synthesis and the energy reserves, E. The rate of protein synthesis is proportional to the growth rate plus part of the maintenance rate which is higher, the lower the growth rate in bacteria [690]. The peptide elongation rate is plotted in figure 7.3 for *E. coli* at 37 °C. If we neglect the contribution of maintenance to protein synthesis, the elongation rate at constant substrate density is proportional to the ratio of the growth rate $\frac{d}{dt}V$ and the stored energy $[E_m]fV$. For a typical rod, i.e. a rod of size $\mathcal{E}\underline{V}$ given in table 5.1, substitution of (3.37) shows that the elongation rate should be proportional to $\dot{\mu}/f$ at population growth rate $\dot{\mu}$. It allows the estimation of the parameter l_b, which is hard to obtain in another way.

The life time of a compound in the reserves is exponentially distributed with a mean residence time of $(\dot{\nu}(\frac{\delta}{3}\frac{V_d}{V} + 1 - \frac{\delta}{3}))^{-1}$; see (3.36). The mean residence time thus increases during the cell cycle. At division it is $\dot{\nu}^{-1}$, independent of the (population) growth rate. The total number of transcriptions of a ribosome, in consequence, increases with the population growth rate. Outside the cell, RNA is rather stable. The fact that the RNA fraction of dry weight depends on feeding conditions indicates that a RNA molecule has a restricted life span inside the cell.

7.3 Allometric growth and regulation

The bill of the guillemot *Uria aalge* is just one example of non-isomorphic growth. Although of little energetical significance, the κ-rule provides the structure to describe such deviations.

In the development of the DEB theory, only somatic and reproductive tissue have been distinguished for the sake of simplicity. The assumption of isomorphy covers other tissue as fixed fractions of the somatic tissue, conceived as a lumped sum. The elaboration below makes explicit that the mechanism behind the κ-rule implies a particular type of growth regulation. It also reveals the intimate connection between the κ-rule and allometric growth.

In a bit more detail, the κ-rule (3.11) can be rephrased as

$$\kappa_i \frac{V_i}{V_+}\dot{C} = [G_i]\frac{d}{dt}V_i + [M_i]V_i \tag{7.5}$$

where V_i denotes tissue (or organ or part of body) i, and $V_+ \equiv \sum_{i=1}^{k} V_i$ is the total body volume. Since blood flow is space-filling, the fraction V_i/V_+ stands for the relative length

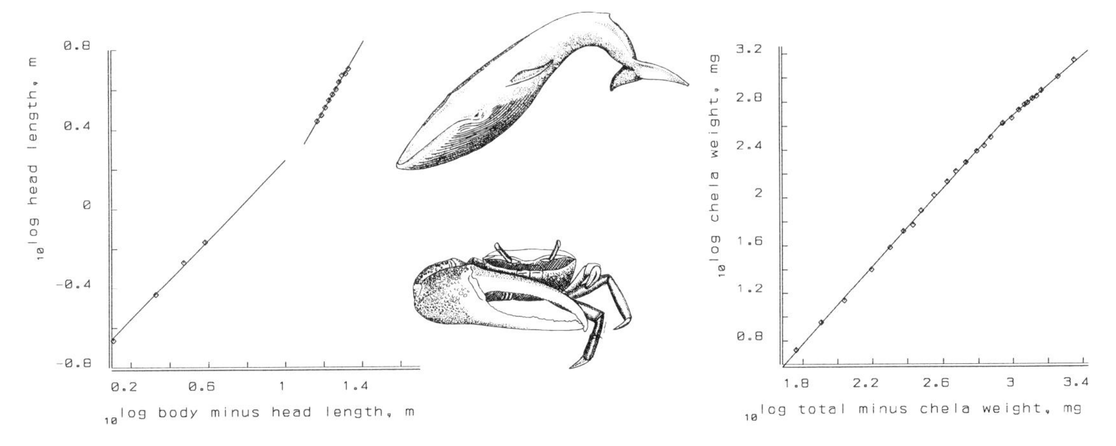

Figure 7.4: Examples of allometric growth: $\log y = a + b \log x$. Left: The head length (from the tip of the nose to the blow hole), with respect to total body length minus the head length in the male blue whale, *Balaenoptera musculus*. The first 4 data points are from foetuses, where growth is isomorphic ($b = 1$). Thereafter the head extends more rapidly ($b = 1.65$). Right: The weight of the large chela with respect to that of the rest of the body in the male fiddler crab *Uca pugnax*. Initially the chela grows rapidly ($b = 1.63$) until a rest of body weight of 850 mg, thereafter it slows down a little ($b = 1.23$). Data from Huxley [345].

of the track followed by blood as it flows through tissue i. In the above, only somatic and reproductive tissue were distinguished, so $k = 2$. Isomorphism implies that V_i/V_+ remains fixed, so $\kappa = \kappa_1 V_1/V_+$ has been taken, while $\frac{d}{dt}V_1 = \frac{d}{dt}V_2$. The extra uptake by reproductive tissue did not result in enhanced growth of the reproductive tissue, but in production that becomes lost for the body. If isomorphism is dropped as a condition and if more types of tissue are to be distinguished, (7.5) can be written as $[G_i]\frac{d}{dt}V_i = \frac{\kappa_i}{V_+}\dot{C}(1 - \frac{[\dot{M}_i]V_+}{\kappa_i \dot{C}})V_i$. Allometrical growth of tissue i with respect to tissue j results, that is $\frac{dV_i}{dV_j} = \frac{\kappa_i [G_j]}{\kappa_j [G_i]} \frac{V_i}{V_j}$, if $\frac{[\dot{M}_i]V_+}{\kappa_i \dot{C}}$ is small.

Allometric growth with respect to total body volume occurs if the contribution of tissue i to total body volume is insignificant, because $V \neq \sum_i \alpha_i V^{\beta_i}$ if $\beta_i \neq 1$ for some i, whatever the values of positive α_i's. Absolute growth requires specification of how feeding and digestion (and heating for endotherms) depend on the volume and shape of the different tissues. It is likely to become complex. Allometric growth of extremities and skeletal elements frequently occurs, as illustrated in figure 7.4. Houck *et al.* [337] used this growth as an argument to delineate taxa in *Archaeopteryx*. It is improbable, however, that whole-organism energetics is seriously affected by these relative changes. This paragraph only serves to illustrate that the mechanism behind allometric growth (of appendages) is intimately connected to the κ-rule. Note that the dimensional problems that are usually connected to allometric relationships, {12}, do not apply in this implementation of allometric growth.

Isomorphs thus require growth regulation over the different body parts. Without control, allometric growth results. For isomorphs $[V_i] \equiv V_i/V_+$ must remain fixed, so that

$\frac{d}{dt}V_i = [V_i]\frac{d}{dt}V_+$ must hold. This implies for the DEB model that the organism must accelerate or retard the growth of organ/tissue/part i by a factor $[V_i]\frac{dV_+}{dV_i} \simeq g_i \sum_j [V_j]/g_j$, with $g_i \equiv \frac{[G_i]}{\kappa_i[E_m]}$. (The approximation holds for $\dot{M}_i << \kappa_i \dot{C}$.) The mechanism of control may be via the density of carriers that transfer resources from the blood to the tissue. The carrier density in membranes of large tissues/parts, should be less than that in small tissues/parts for a particular value.

The acceleration/retardation factor demonstrates that the carrier density does not have to change during growth. Other types of growth regulation are also possible. This discussion is only about the effects of regulation, rather than its mechanism.

Chapter 8

Ecotoxicity

The first purpose of this chapter is to show how the DEB model help us solve ecotoxicological problems; in particular, I will show how it can be used to build physiologically underpinned models for the uptake of (xenobiotic) compounds, for the characterization of their effects on individual performance and for the translation of these effects to the population and ecosystem level.

The need for physiological underpinning of the kinetics of compounds is felt in the ever increasing size of the data base that is being built up at the moment for experimental data on the kinetics of a wide variety of compounds in a wide variety of organisms. The search for structures in this data set is essential for the interpretation of experimental results, for predictions concerning unknown combinations of compounds and species, for the design of informative experiments and for the filtering of less reliable data. The domain of application also extends to the field of drugs: pharmaco-kinetics.

The need for a characterization of effects of compounds and their extrapolation to the population and ecosystem level originates in the huge problems man is creating in his environment at the moment. It is relatively easy to determine effects on individuals in the laboratory, whereas the real problems are at the ecosystem level. This calls for models to link them. The purpose is to analyse how bad a particular pollution event or situation is, to set priorities for cleaning up the mess and to set norm values for maximum emission of compounds into the environment for industry and transport. We can only be extremely modest in our claims to understand long-term effects at the ecosystem level. This book attempts to contribute to one particular aspect of what is needed. A lot of work still has to be done.

The position I take with respect to the application of the DEB theory to ecotoxicological problems is that you can agree or disagree with the details of the DEB theory, but you will need something like it to enable the translation of effects on individuals to the population level. A lot of attempts have been made and doubtlessly will be made to circumvent the translation problem and to experiment directly with populations to measure effects at this level. However, if you want to understand what is going on, you still need translation. For a purely descriptive approach, you can do without translation to some extent, but a lot of problems lie in wait for you. This is not the right place to discuss them at length; see e.g. [409,412], but an important aspect is the substantial probability of what is known as an error of the second kind in statistics: there are effects, but you fail to spot them. In this case, it is due to the erratic behaviour of populations and our limited biological knowledge of how to experiment with them in a meaningful way.

The DEB model has just the proper level of resolution for ecotoxicological applications. This is no coincidence, I designed it specifically for this purpose. The treatment of ecotoxicological application in this book can be justified for two other reasons. The first reason is that the kinetics of a xenobiotic compound do not differ from that of any other compound that is not regulated physiologically. The kinetics of physiologically regulated compounds can only be studied satisfactorily in the light of the non-regulated ones. It therefore has a bearing on mass exchange between organism and environment in general. Moreover, a lot of plant-animal relationships closely resemble chemical warfare. The second reason is that toxic compounds can affect particular energy allocations. This provides the opportunity to use toxicological experiments as a tool to study the mechanisms behind energy allocation patterns.

This chapter first discusses simple one-compartment uptake/elimination behaviour in a variable environment as relevant, for example, to monitoring programs. It restates and modestly extends the work of Thomann and Mueller [712] on this point. The physiological change of the organism is at first neglected. This serves as an introduction to the next section, which enlarges on physiological change with respect to size, energy reserves and reproductive output and invokes the DEB model. Then follows a link to lethal and sublethal effects for individuals and effects at the population level.

8.1 One-compartment kinetics

To monitor the concentration of a pollutant in waterways, it sometimes makes sense to determine the concentration of pollutant in mussels, $[Q](t)$, which have been exposed in such waterways, rather than to determine the concentration in the water, $c(t)$. The first argument is that of bio-availability. Not all of the chemically determined pollutant in the water is actually available to organisms, due to the variety of chemical forms in which the pollutant is present (e.g. ligands). Concentrations in tissues provide information which is, therefore, more relevant to the problem of pollution. Another argument is that the concentration of pollutant in the water may have some sharp peaks, which are easy to miss with infrequent determinations. Mussels, in some way, integrate the concentration in the environment in time. It should, therefore, still be possible to observe a trace of such peaks in the concentrations found in the tissue by infrequent sampling, provided the mussels do not close their valves during such peaks. Let us study this argument in more detail, assuming that uptake is from the dissolved compound in the water and elimination is from the aqueous fraction of the body.

Suppose that the concentration in the tissue follows a simple one-compartment process, i.e.

$$\frac{d}{dt}[Q] = [\dot{k}_{da}]c(t) - \dot{k}_a[Q] \tag{8.1}$$

where $\dot{k}_a$ is the transfer rate from the aqueous fraction of the body to the dissolved pool in the environment, called the elimination rate, and $[\dot{k}_{da}]$ the reverse rate, the uptake rate. The ratio $[\dot{k}_{da}]/\dot{k}_a$ is known as the bioconcentration coefficient and can be less than 1. Although many texts treat the bioconcentration coefficient as a dimensionless one, it actually has dimension environmental volume$\times$(body volume)$^{-1}$ because the sum of both types of volume does not have a useful role to play. Most texts in fact use environmental volume$\times$(body dry weight)$^{-1}$, or for soils environmental dry weight$\times$(body dry weight)$^{-1}$. The uptake kinetics is the same as in the Lotka–Volterra model for the uptake of food, and can thus be considered as a linear approximation of the hyperbolic functional response for low concentrations.

If we know $[Q](t)$ in sufficient detail, and if it is sufficiently smooth so that it can be described by a cubic spline function for instance, we can reconstruct $c(t)$ through

$$c(t) = (\frac{d}{dt}[Q](t) + \dot{k}_a[Q])/[\dot{k}_{da}] \tag{8.2}$$

We need an explicit expression of $[Q](t)$ in terms of $c(t)$, however, because we want to study the effects of rapidly changing concentrations in the water. The solution is found from (8.1) to be

$$[Q](t) = [Q](0)\exp\{-t\dot{k}_a\} + [\dot{k}_{da}]\int_0^t \exp\{-(t-t_1)\dot{k}_a\}c(t_1)\,dt_1 \tag{8.3}$$

If $c(t)$ is actually constant, (8.3) reduces to

$$[Q](t) = [Q](0)\exp\{-t\dot{k}_a\} + \left(1 - \exp\{-t\dot{k}_a\}\right)\frac{[\dot{k}_{da}]}{\dot{k}_a}c \tag{8.4}$$

which is known as the accumulation curve.

8.1.1 Random increment input

The concentration in the environment is usually not constant. In experimental setups, it can be evaluated using mass balances. In the environment, it follows some stochastic process, which involves many factors. Some simple choices for such processes are discussed in this and the next few subsections. These subsections can be skipped without loss of continuity.

The purpose of the following subsections is to derive statistical properties of one-compartment models with a stochastic input, as is realistic for an individual exposed in the field to a xenobiotic. Different stochastic inputs will result in very similar statistical properties. Knowledge of these properties can be helpful for the interpretation of data and for the setup of monitoring programs.

The mathematically most simple type of stochastic input is when the continuous function $c(t)$ is approximated by a step function which changes only at discrete time points t_i, $i = 1, 2, \cdots$, that are a fixed time interval t_d apart. That is, $c(t)$ is constant over a time interval $(t_i, t_{i+1}]$, at value c_i. The concentration in the tissue at the end of the interval is given by

$$[Q]_{i+1} = [Q](t_i + t_d) = r[Q]_i + (1-r)c_i[\dot{k}_{da}]/\dot{k}_a \tag{8.5}$$

with $[Q]_i \equiv [Q](t_i)$ and $r \equiv \exp\{-t_d \dot{k}_a\}$. Now we assume that the values c_i represent trials taken independently from some probability density function. This is reasonable for the situation in a river, where the well-mixed water surrounding a mussel is completely replaced in an interval of length t_d. This input process is called a random increment process. The schedule (8.5) is known as a (first-order) stochastic difference equation or an auto-regression process, because the new value for $[Q]$ is a weighted sum of the old values and an independent random variable. As these processes play such a central role in the theory of stochastic processes, the description is first given in terms of auto-regression processes, and afterwards in less conventional continuous time formulations. This reveals differences in discrete time and continuous time formulations.

If we let the interval t_d shrink down to an infinitesimally short one, the stochastic difference equation transforms into a stochastic differential equation. This can be seen by subtracting $[Q]_i$ on both sides from (8.5), letting $t_d \to 0$ and noting that $\lim_{t_d \to 0}(1-r)/t_d = \dot{k}_a$. It then reduces to

$$\frac{d}{dt}[\underline{Q}] \equiv \lim_{t_d \to 0} \frac{[\underline{Q}]_{i+1} - [\underline{Q}]_i}{t_d} = [\dot{k}_{da}]\underline{c}(t) - \dot{k}_a[\underline{Q}] \tag{8.6}$$

where $\underline{c}(t)$ follows what is known as a random increment process. When $t_d \to 0$, we have to reduce var $\underline{c}_i$ at the proper rate to arrive at the variance of the increments, otherwise the process does not have interesting properties. This remark only serves to point to the relationship between auto-regressive processes and stochastic differential equations. The discussion will be restricted to auto-regressive schemes to avoid measure theoretical problems associated with that limiting process. These problems originate from details in the definition of integration of functions, such as the realization of the process $\underline{c}(t)$ that is discontinuous at infinitely many points.

The auto-regressive scheme can be expressed alternatively as a moving average scheme:

$$[\underline{Q}]_{i+1} = r^{i+1}[\underline{Q}]_0 + (1-r)\frac{[\dot{k}_{da}]}{\dot{k}_a}\sum_{j=0}^{i} r^j \underline{c}_{i-j} \tag{8.7}$$

The expected value for $[\underline{Q}]_{i+1}$ will be

$$\mathcal{E}[\underline{Q}]_{i+1} = r^{i+1}\mathcal{E}[\underline{Q}]_0 + (1-r)\frac{[\dot{k}_{da}]}{\dot{k}_a}\mathcal{E}\underline{c}_i \sum_{j=0}^{i} r^j \tag{8.8}$$

Ultimately, i.e. for large i, this reduces to $\mathcal{E}[\underline{Q}]_i = \mathcal{E}\underline{c}_i[\dot{k}_{da}]/\dot{k}_a$. Thus, the ultimate mean concentration in the tissue is just the mean concentration in the environment times the bioconcentration coefficient $[\dot{k}_{da}]/\dot{k}_a$. This result corresponds with the deterministic situation if $\underline{c}$ is constant. Then $[Q](\infty) = c[\dot{k}_{da}]/\dot{k}_a$, as can be inferred directly from (8.4).

The variance of the concentrations in the tissue is obtained by taking variances at both sides of (8.7), which leads to

$$\text{var}\,[\underline{Q}]_i = \text{var}\,\underline{c}_i \left(\frac{[\dot{k}_{da}]}{\dot{k}_a}\right)^2 \frac{1-r}{1+r} \tag{8.9}$$

Similarly, the covariance equals $\text{cov}\,([\underline{Q}]_i, \underline{c}_i) = 0$, so that both concentrations are uncorrelated.

The subsequent concentrations in the tissue are correlated, as opposed to the concentrations in the water. This is expressed by the auto-covariance function $\text{cov}\,([\underline{Q}]_{i+h}, [\underline{Q}]_i)$, or the auto-correlation function $\text{cor}\,([\underline{Q}]_{i+h}, [\underline{Q}]_i)$ (both considered as functions in h), given from (8.7) by

$$\text{cov}\,([\underline{Q}]_{i+h}, [\underline{Q}]_i) = r^{|h|}\text{var}\,[\underline{Q}]_i \tag{8.10}$$

and

$$\text{cor}\,([\underline{Q}]_{i+h}, [\underline{Q}]_i) = \frac{\text{cov}\,([\underline{Q}]_{i+h}, [\underline{Q}]_i)}{\text{var}\,[\underline{Q}]_i} = r^{|h|} \tag{8.11}$$

We can also study how $[\underline{Q}]_{i+h}$ depends on $\underline{c}_i$ by looking at the (cross-)correlation function $\text{cor}\,([\underline{Q}]_{i+h}, \underline{c}_i)$ which in this case can easily be derived to be

$$\text{cor}\,([\underline{Q}]_{i+h}, \underline{c}_i) \equiv \frac{\text{cov}\,([\underline{Q}]_{i+h}, \underline{c}_i)}{\sqrt{\text{var}\,[\underline{Q}]_i\,\text{var}\,\underline{c}_i}} = \sqrt{1-r^2}\,r^{h-1} \quad \text{for}\, h > 0 \tag{8.12}$$

The cross-correlation function in figure 8.1 shows how the concentration in the tissue lags behind concentration fluctuations in the water. The value $r = 0.8$ has been chosen in (8.11) and (8.12). This smoothing also results in a gradually decreasing auto-correlation function of $[\underline{Q}]$.

So far, we have studied the concentrations at time points ht_d, $h = 0, 1, \ldots$. Since the uptake/elimination process is in continuous time, the mean and variance of $[\underline{Q}](t)$ are of more practical interest than those of $[\underline{Q}]_i$. These statistics are defined as $\mathcal{E}[\underline{Q}](t) \equiv \lim_{t\to\infty} t^{-1}\int_0^t [\underline{Q}](t_1)\,dt_1$ and $\text{var}\,[\underline{Q}](t) \equiv \lim_{t\to\infty} t^{-1}\int_0^t ([\underline{Q}](t_1) - \mathcal{E}[\underline{Q}](t_1))^2\,dt_1$. To evaluate them, it is helpful to note that $[\underline{Q}](t) = [\underline{Q}]_i \frac{\exp\{-(t-t_i)\dot{k}_a\}-r}{1-r} + [\underline{Q}]_{i+1}\frac{1-\exp\{-(t-t_i)\dot{k}_a\}}{1-r}$ for $t \in (t_i, t_{i+1}]$. From this expression, it is easy to see that $t_d^{-1}\int_{t_i}^{t_{i+1}}[\underline{Q}](t)\,dt = \frac{[\underline{Q}]_{i+1}-r[\underline{Q}]_i}{1-r} + \frac{[\underline{Q}]_i - [\underline{Q}]_{i+1}}{t_d\dot{k}_a}$. The second term disappears in the mean, so that the mean of $[\underline{Q}](t)$ equals that of $[\underline{Q}]_i$. A similar evaluation of the variance results in

$$\text{var}\,[\underline{Q}](t) = \text{var}\,\underline{c}_i\left(\frac{[\dot{k}_{da}]}{\dot{k}_a}\right)^2\left(1 - \frac{1-r}{t_d\dot{k}_a}\right) \tag{8.13}$$

Comparison with (8.9) reveals the effect of the inclusion of the time points between $0, t_d, 2t_d, \ldots$ in the calculation of the variance. So the means of the discrete and the continuous time process are the same, but the variances differ. Since the concentrations in the water are constant within each time interval, the mean and variance of $\underline{c}(t)$ equal that of $\underline{c}_i$. The auto-correlation function of $\underline{c}(t)$ equals

$$\text{cor}\,(\underline{c}(t+t_h), \underline{c}(t)) = (|t_h| < t_d)(1 - |t_h|/t_d) \tag{8.14}$$

It thus decreases linearly to zero at $|t_h| = t_d$.

The correlation coefficient between $\underline{c}(t)$ and $[\underline{Q}](t)$ proves to be

$$\text{cor}\,([\underline{Q}](t), \underline{c}(t)) = \sqrt{1 - \frac{1-r}{t_d\dot{k}_a}} \equiv \rho \tag{8.15}$$

while $\underline{c}_i$ and $[\underline{Q}]_i$ are uncorrelated.

The cross-correlation function turns out to be

$$\begin{aligned}
\text{cor}\,([\underline{Q}](t+t_h), \underline{c}(t)) &= 0 \quad \text{for } t_h < -t_d \\
&= \frac{1}{\rho} + \frac{t_h}{\rho t_d} + \frac{r\exp\{-t_h\dot{k}_a\} - 1}{\rho t_d \dot{k}_a} \quad \text{for } -t_d \leq t_h < 0 \\
&= \frac{1}{\rho} - \frac{t_h}{\rho t_d} + \frac{1 - (2-r)\exp\{-t_h\dot{k}_a\}}{\rho t_d \dot{k}_a} \quad \text{for } 0 \leq t_h < t_d \\
&= \frac{(1-r)^2}{\rho r t_d \dot{k}_a}\exp\{-t_h\dot{k}_a\} \quad \text{for } t_h \geq t_d
\end{aligned} \tag{8.16}$$

It has a maximum at $t_h = \dot{k}_a^{-1}\ln(2-r)$.

The auto-correlation function of $[\underline{Q}](t)$ can best be obtained via the relationship

$$\frac{2\dot{k}_a}{[\dot{k}_{da}]}\text{cov}\,([\underline{Q}](t), [\underline{Q}](t+t_h)) = \text{cov}\,([\underline{Q}](t+t_h), \underline{c}(t)) + \text{cov}\,([\underline{Q}](t), \underline{c}(t+t_h)) \tag{8.17}$$

which can be obtained directly by applying (8.1) and using the property that, ultimately, we must have $\frac{d}{dt}\text{cov}\,([\underline{Q}](t), [\underline{Q}](t+t_h)) = 0$. The result is

$$\begin{aligned}
\text{cor}\,([\underline{Q}](t), [\underline{Q}](t+t_h)) &= \frac{1}{\rho^2} - \frac{|t_h|}{\rho^2 t_d} + \frac{r\exp\{|t_h|\dot{k}_a\} - (2-r)\exp\{-|t_h|\dot{k}_a\}}{2\rho^2 t_d \dot{k}_a} \quad \text{for } |t_h| \leq t_d \\
&= \frac{(1-r)^2}{2r\rho^2 t_d \dot{k}_a}\exp\{-|t_h|\dot{k}_a\} \quad \text{for } |t_h| > t_d
\end{aligned} \tag{8.18}$$

An interesting consequence of (8.17) is that for the linear system (8.6), with any stochastic input, holds:

$$\text{var}\,[\underline{Q}](t) = \text{cov}\,([\underline{Q}](t), \underline{c}(t))[\dot{k}_{da}]/\dot{k}_a \tag{8.19}$$

This implies that the correlation coefficient between the concentrations in tissue and water equals the ratio of their variation coefficients, thus

$$\text{cor}\,([\underline{Q}](t), \underline{c}(t)) = \frac{\text{cv}\,[\underline{Q}](t)}{\text{cv}\,\underline{c}(t)} \tag{8.20}$$

If the correlation coefficient between both concentrations is low, tissue smoothes out fluctuating concentrations in the water effectively, so that a few samples from the tissue tell as much as a large amount of samples from the water, as far as statistics is concerned. If the scatter in the concentration in the water is high, it is easy to miss peaks, while traces of such peaks are still measurable in the tissue. In practice however, behaviour of individuals, such as closing of valves by mussels, may complicate the analysis.

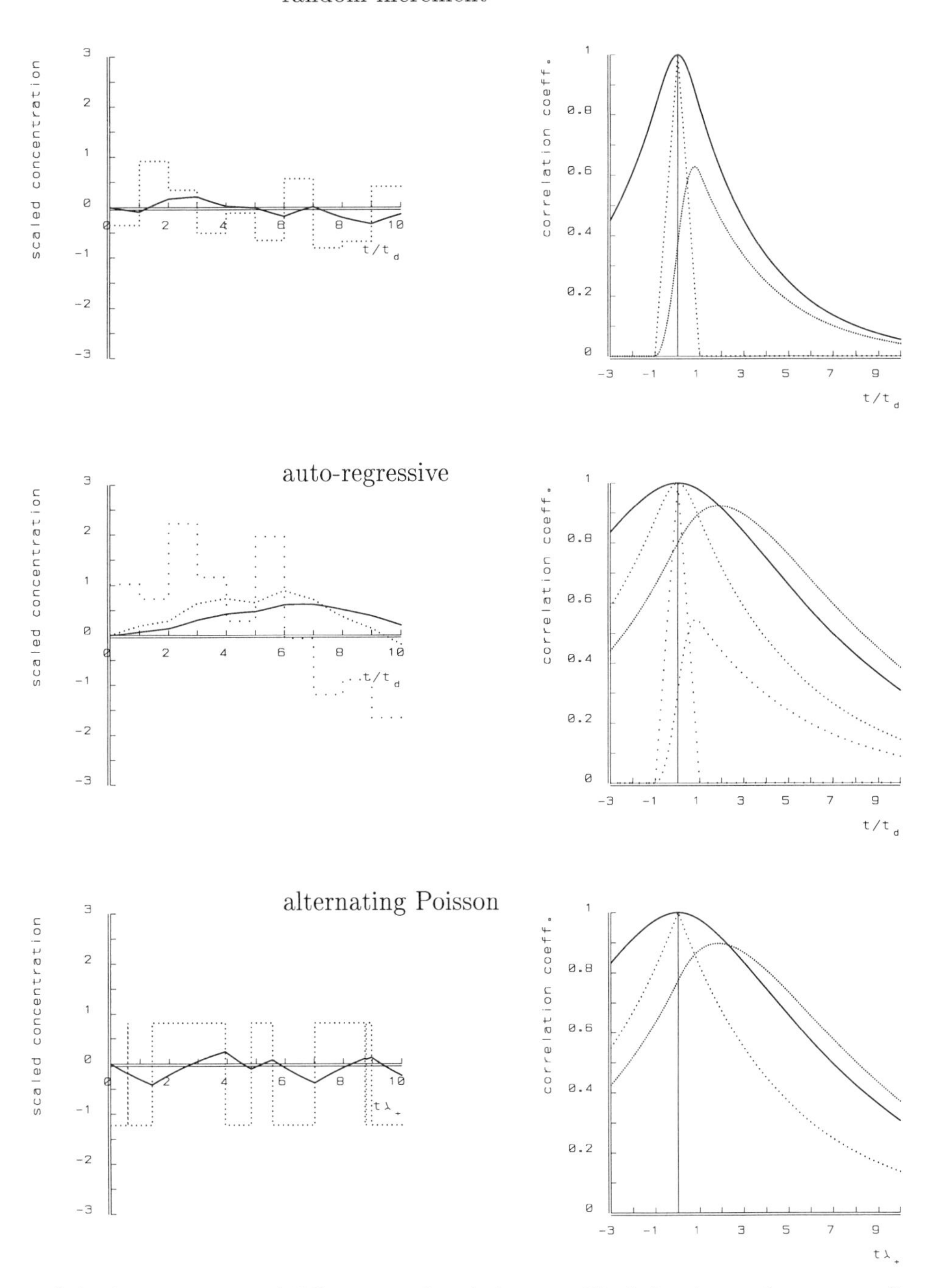

Figure 8.1: A comparison of different stochastic inputs. The left column shows a small part of a sample path of the concentrations in water and tissue, and the right shows the corresponding correlation functions. The three input processes are described in the text. The concentrations are normalized for zero mean and unit variance in the water (river). The line types of the figures in both columns correspond. Drawn: tissue, dotted: water, sparsely dotted: inflowing water. The dot-density of the cross-correlation curves is the mean of both variables.

8.1.2 Auto-regressive input

The behaviour of concentrations in the water usually shows more 'memory'. Suppose, we have a lake of constant volume V_c, with an inflow and an outflow of water at rate $\dot{k}_c$ per unit of time. The quotient $t_c \equiv V_c/\dot{k}_c$ is known as the residence time of water in the lake. If mixing is perfect, the concentration c in the lake depends on b in the inflowing water according to a one-compartment process:

$$\frac{d}{dt}c = (b(t) - c)/t_c \tag{8.21}$$

The solution, in analogy with (8.3), is:

$$c(t) = c(0)\exp\{-t/t_c\} + \frac{1}{t_c}\int_0^t \exp\{-(t-t_1)/t_c\}b(t_1)\,dt_1 \tag{8.22}$$

We will approximate $b(t)$ the same way we did $c(t)$. That is, b is constant over an interval $(t_i, t_i + t_d]$ at value b_i. Then we have

$$\begin{aligned} \underline{c}_{i+1} &= s\underline{c}_i + (1-s)\underline{b}_i \quad \text{with } s \equiv \exp\{-t_d/t_c\} \\ &= s^{i+1}\underline{c}_0 + (1-s)\sum_{j=0}^{i} s^j \underline{b}_{i-j} \end{aligned} \tag{8.23}$$

This is again an auto-regression scheme if the $\underline{b}_i$'s are independent and identically distributed.

In steady state we have, analogously with (8.7) and (8.9), $\mathcal{E}\underline{c}_i = \mathcal{E}\underline{b}_i$ and $\text{var}\,\underline{c}_i = \text{var}\,\underline{b}_i(1-s)/(1+s)$. The main difference with the river situation is that subsequent values for $\underline{c}_i$ are now correlated. The auto-covariance function (see (8.10)) is given by

$$\text{cov}\,(\underline{c}_{i+h}, \underline{c}_i) = s^{|h|}\text{var}\,\underline{c}_i \tag{8.24}$$

or, in analogy with (8.18),

$$\begin{aligned} \frac{\text{cov}\,(\underline{c}(t), \underline{c}(t+t_h))}{\text{var}\,\underline{b}_i} &= 1 - \frac{|t_h|}{t_d} + \frac{s\exp\{|t_h|/t_c\} - (2-s)\exp\{-|t_h|/t_c\}}{2t_d/t_c} \quad \text{for } |t_h| \le t_d \\ &= \frac{(1-s)^2}{2st_d/t_c}\exp\{-|t_h|/t_c\} \quad \text{for } |t_h| > t_d \end{aligned} \tag{8.25}$$

so that $\text{var}\,\underline{c}(t) = \rho_{AR}^2 \text{var}\,\underline{b}_i$ with $\rho_{AR} \equiv \sqrt{1 - \frac{1-s}{t_d/t_c}}$ representing the correlation coefficient $\text{cor}([\underline{Q}](t), \underline{c}(t))$.

The concentrations in the tissue can again be expressed as an auto-regressive or a moving average scheme, via substitution of

$$c(t - t_i) = c_i \exp\{-(t-t_i)/t_c\} + b_i(1 - \exp\{-(t-t_i)/t_c\})$$

in (8.3). The resulting scheme is:

$$\begin{aligned}
\underline{[Q]}_{i+1} &= \underline{[Q]}_i r + \frac{[\dot{k}_{da}]}{\dot{k}_a - 1/t_c}(\underline{c}_i - \underline{b}_i)(s-r) + \frac{[\dot{k}_{da}]}{\dot{k}_a}\underline{b}_i(1-r) \\
&= \underline{[Q]}_0 r^{i+1} + \frac{[\dot{k}_{da}]}{\dot{k}_a - 1/t_c}(s^{i+1} - r^{i+1})\underline{c}_0 + \\
&\quad + (1-r)\frac{[\dot{k}_{da}]}{\dot{k}_a}\sum_{j=0}^{i}\underline{b}_{i-j}r^j\left(1 - \frac{\dot{k}_a}{\dot{k}_a - 1/t_c}\left(1 - \frac{1-s}{1-r}\left(\frac{s}{r}\right)^j\right)\right)
\end{aligned} \tag{8.26}$$

The first two terms only refer to the initial concentration in the tissue and lake. Their contribution becomes negligible in the long run.

The mean concentration in the tissue again equals $\mathcal{E}\underline{[Q]}_i = \mathcal{E}\underline{[Q]}(t) = [\dot{k}_{da}]\dot{k}_a^{-1}\mathcal{E}\underline{c}_i$. The correlation function with the concentrations in the water can, for linear systems in general, be obtained directly from the auto-correlation function of the input (the concentration in the water in this case). From the definition of the correlation function and the general solution (8.3), it follows for large t that

$$\text{cor}\,(\underline{[Q]}(t+t_h), \underline{c}(t)) = [\dot{k}_{da}]\int_0^\infty \exp\{-t_1\dot{k}_a\}\text{cor}\,(\underline{c}(t), \underline{c}(t+t_h-t_1))\,dt_1 \tag{8.27}$$

The cross-covariance function between output and input is thus completely determined by the auto-covariance function of the input for linear systems. The evaluation of (8.27) is straightforward and results in:

$$\begin{aligned}
&\text{cov}\,(\underline{[Q]}(t+t_h), \underline{c}(t))\dot{k}_a/[\dot{k}_{da}] \\
&= \frac{(1-s)^2\exp\{t_h/t_c\}}{2s(1+1/t_c\dot{k}_a)t_d/t_c} \quad \text{for } t_h < -t_d \\
&= 1 + \frac{t_h}{t_d} - \frac{1}{t_d\dot{k}_a} - \frac{(2-s)\exp\{t_h/t_c\}}{2(1+1/t_c\dot{k}_a)t_d/t_c} + \frac{s\exp\{-t_h/t_c\}}{2(1-1/t_c\dot{k}_a)t_d/t_c} + \\
&\quad + \frac{\exp\{-t_h\dot{k}_a\}}{1-t_c^2\dot{k}_a^2}\frac{r}{t_d\dot{k}_a} \quad \text{for } -t_d \le t_h < 0 \\
&= 1 - \frac{t_h}{t_d} + \frac{1}{t_d\dot{k}_a} + \frac{s\exp\{t_h/t_c\}}{2(1+1/t_c\dot{k}_a)t_d/t_c} - \frac{(2-s)\exp\{-t_h/t_c\}}{2(1-1/t_c\dot{k}_a)t_d/t_c} + \\
&\quad - \frac{\exp\{-t_h\dot{k}_a\}}{1-t_c^2\dot{k}_a^2}\frac{2-r}{t_d\dot{k}_a} \quad \text{for } 0 \le t_h < t_d \\
&= \frac{(1-s)^2\exp\{-t_h/t_c\}}{2s(1-1/t_c\dot{k}_a)t_d/t_c} + \exp\{-t_h\dot{k}_a\}\frac{(1-r)^2}{rt_d\dot{k}_a}\alpha \quad \text{for } t_h \ge t_d \\
&\text{cov}\,(\underline{[Q]}(t+t_h), \underline{[Q]}(t))\dot{k}_a^2/[\dot{k}_{da}]^2 \\
&= 1 - \frac{|t_h|}{t_d} - \frac{(2-s)\exp\{-|t_h|/t_c\}}{2(1-t_c^{-2}\dot{k}_a^{-2})t_d/t_c} + \frac{s\exp\{|t_h|/t_c\}}{2(1-t_c^{-2}\dot{k}_a^{-2})t_d/t_c} + \\
&\quad - \frac{(2-r)\exp\{-|t_h|\dot{k}_a\}}{2(1-t_c^2\dot{k}_a^2)t_d\dot{k}_a} + \frac{r\exp\{|t_h|\dot{k}_a\}}{2(1-t_c^2\dot{k}_a^2)t_d\dot{k}_a} \quad \text{for } |t_h| \le t_d
\end{aligned}$$

$$= \frac{(1-s)^2 \exp\{-|t_h|/t_c\}}{2s(1-t_c^{-2}\dot{k}_a^{-2})t_d/t_c} + \frac{(1-r)^2 \exp\{-|t_h|\dot{k}_a\}}{2r(1-t_c^2\dot{k}_a^2)t_d\dot{k}_a} \quad \text{for } |t_h| > t_d$$

These expressions illustrate that even the most simple stochastic process results in considerable complexity. The cross-covariance function is not symmetrical in the delay t_h, as illustrated in figure 8.1. It has a peak at $t_h = (\dot{k}_a - 1/t_c)^{-1} \ln\left(\frac{1-r}{1-s}\right)^2 \frac{s}{r} \frac{2}{t_c\dot{k}_a(t_c\dot{k}_a+1)}$ if it happens to be larger than t_d.

8.1.3 Alternating Poisson input

The auto-regressive process has, because of its linearity, beautiful mathematical properties. It is possible to evaluate important properties of the concentration in the tissue, without specification of the distribution of the concentration in the inflowing water. The time interval over which the concentration in the inflowing water is taken to be constant is, however, a somewhat arbitrary element. It is, therefore, a good idea to compare the results with those of a different stochastic process for concentration in the environment: the alternating Poisson process.

Suppose that the environment can be partitioned into two types of patches. In one type, the concentration of the compound is c_0, in the other it is c_1. Suppose that the animal stays a period $\underline{t}_0$ in a patch with concentration c_0, then a period $\underline{t}_1$ in a patch with concentration c_1, then again a time $\underline{t}_0$ in a patch with concentration c_0, etc. Suppose next that the residence times follow an exponential distribution, with parameters $\dot{\lambda}_0$ and $\dot{\lambda}_1$ respectively. So Prob$\{\underline{t}_i > t\} = \exp\{-t\dot{\lambda}_i\}$, $i = 0, 1$. Think of it as an animal walking randomly over a patchy soil, or an animal only exposed to the compound when feeding, or a mussel exposed to a flow of badly mixed batches of water. The rates $\dot{\lambda}$ relate to the size of the patches, relative to the travelling rate of the animal. In particular, $\dot{\lambda}_i^{-1}$ is the mean residence time in a patch with concentration c_i. Such a process is called an alternating Poisson process or a random telegraph process (if $c_0 = 0$).

The mean of the alternating Poisson process is $\mathcal{E}\underline{c}(t) = c_0\dot{\lambda}_1/\dot{\lambda}_+ + c_1\dot{\lambda}_0/\dot{\lambda}_+$, where $\dot{\lambda}_+ \equiv \dot{\lambda}_0 + \dot{\lambda}_1$. The variance is var $\underline{c}(t) = (c_1 - c_0)^2\dot{\lambda}_0\dot{\lambda}_1\dot{\lambda}_+^{-2}$ and the auto-correlation function

$$\text{cor}\ (\underline{c}(t), \underline{c}(t+t_h)) = \exp\{-|t_h|\dot{\lambda}_+\} \tag{8.28}$$

The mean concentration in the tissue is again $\mathcal{E}[\underline{Q}](t) = [\dot{k}_{da}]\dot{k}_a^{-1}\mathcal{E}\underline{c}(t)$. This can be seen directly from (8.1) by taking the expectation at both sides and using the property that $\frac{d}{dt}\mathcal{E}Q(t) = 0$ at steady state. The probability density function of the concentration in the tissue at steady state turns out to be the beta distribution

$$\phi_{[\underline{Q}]}([Q]) = \frac{\Gamma(\dot{\lambda}_+/\dot{k}_a)}{\Gamma(\dot{\lambda}_0/\dot{k}_a)\Gamma(\dot{\lambda}_1/\dot{k}_a)}([Q] - c_0[\dot{k}_{da}]/\dot{k}_a)^{\dot{\lambda}_0/\dot{k}_a - 1}(c_1[\dot{k}_{da}]/\dot{k}_a - [Q])^{\dot{\lambda}_1/\dot{k}_a - 1} \tag{8.29}$$

Jaques Bedaux derived this result in [50]. The variance equals var $[\underline{Q}](t) = \left(\frac{[\dot{k}_{da}]}{\dot{k}_a}\right)^2 \frac{\text{var}\ \underline{c}(t)}{1+\dot{\lambda}_+/\dot{k}_a}$.

The correlation coefficient between the concentrations in the environment and in the tissue turns out to be

$$\text{cor}\ ([\underline{Q}](t), \underline{c}(t)) = \sqrt{1 + \dot{\lambda}_+/\dot{k}_a} \equiv \rho_{AP} \tag{8.30}$$

The auto-variance function of the concentrations in the tissue and the cross-covariance function with those in the environment are

$$\frac{\text{cov}\ ([\underline{Q}](t+t_h), \underline{c}(t))}{[\dot{k}_{da}]\dot{k}_a^{-1}\text{var}\ \underline{c}(t)} = \frac{\exp\{t_h\dot{\lambda}_+\}}{1+\dot{\lambda}_+/\dot{k}_a} \quad \text{for } t_h < 0 \tag{8.31}$$

$$= \frac{\exp\{-t_h\dot{\lambda}_+\} - \exp\{-t_h\dot{k}_a\}}{1-\dot{\lambda}_+/\dot{k}_a} + \frac{\exp\{-t_h\dot{k}_a\}}{1+\dot{\lambda}_+/\dot{k}_a} \quad \text{for } t_h \leq 0$$

$$\frac{\text{cov}\ ([\underline{Q}](t), [\underline{Q}](t+t_h))}{[\dot{k}_{da}]^2\dot{k}_a^{-2}\text{var}\ \underline{c}(t)} = \frac{\exp\{-|t_h|\dot{\lambda}_+\} - \exp\{-|t_h|\dot{k}_a\}\dot{\lambda}_+/\dot{k}_a}{(1+\dot{\lambda}_+/\dot{k}_a)(1-\dot{\lambda}_+/\dot{k}_a)} \tag{8.32}$$

The cross-covariance function has a peak at $t_h = (\dot{k}_a - \dot{\lambda})^{-1}\ln 2\rho_{AP}^{-2}$.

For special choices for $\mathcal{E}\underline{b}_i$ and var $\underline{b}_i$ of the auto-regressive process of the last subsection, the mean and variance of the concentration in the environment can be made identical. The auto-covariance functions of both processes are close to each other for $t_c = \dot{\lambda}_+^{-1}$. Figure 8.1 illustrates the remarkable similarity of the auto- and cross-correlation functions in case of auto-regressive and alternating Poisson inputs, dispite the substantial differences between these stochastic processes.

8.2 DEB-based kinetics

One-compartment models do not always give a satisfactory fit with experimental data. For this reason more-compartment models have been proposed [148,262,353,624]; because of their larger number of parameters, the fit is better, but an acceptable physical identification of the compartments is usually not possible. These models, therefore, contribute little to our understanding of kinetics as a process. A more direct link with the physiological properties of the organism and with the lipophilicity of the compound seems an attractive alternative, which does not, however, exclude more-compartmental models. As usual, the problem is not so much in the formulation of those complex models but in the useful application. Too many parameters can easily become a nuisance if few, scattered, data are available. Changes in physiological conditions can easily lead to introduction of many parameters. One reason to account for these changes is that they actually occur in many uptake experiments. It is practically impossible to feed a cohort of blue mussels in a two-month uptake/elimination experiment adequately in the laboratory; at the end of the experiment, the lipid content is reduced substantially.

I will focus in this section on non-metabolized compounds. For terrestrial animals, the usual uptake of xenobiotics from the environment is via food. Sometimes, uptake is via the lung or directly via the surface. In the aquatic environment uptake directly from water is especially important for hydrophilic organic compounds [104], and metals [80,81,602]. In aquatic animals that are chemically isolated from their environment, such as aquatic insects, birds and mammals, the common uptake route is via food. Walker [746] gives a discussion of uptake routes. Excretion can be through the surface directly, via excretion products and via gametes.

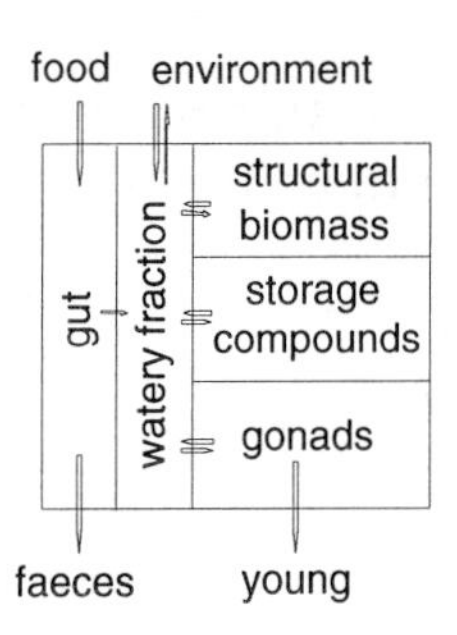

Figure 8.2: Uptake of a xenobiotic is via food and/or directly from the environment. Elimination is directly into the environment, possibly via faeces, and via reproductive output. Internal partitioning into four compartments is taken to be instantaneous. The exchange rate with the environment is taken to be proportional to surface area, apart from reproductive elimination.

Accumulation of lipophilic compounds and partitioning between different organs can be explained by the occurrence of stored lipids. Schneider [639] found large differences of PCB concentrations in different organs of the cod, but the concentrations did not differ when based on the phospholipid-free fraction of extractable lipids. Models for feeding condition dependent kinetics have been proposed [283,284,426], but they have a large number of parameters. The application of the DEB model involves relatively few parameters, due to one-compartment kinetics and instantaneous partitioning of the compound in the organism, as proposed by Barber *et al.* [33] and Hallam *et al.* [283]. The assumption of instantaneous partitioning of compounds is supported by the study of the elimination rate of 4,4′-dichlorobiphenyl (PCB15) in the pond snail *Lymnaea stagnalis* by Wilbrink *et al.* [769], who found that elimination rates are equal for different organs. The fact that structural biomass consists of organs differing in partition coefficients for the xenobiotic, is covered by the assumptions of isomorphism, homeostasis and instantaneous partitioning. The combination of these three assumptions implies that the concentration-time curve in one organ can be obtained from that in another organ by application of a fixed multiplication factor.

The basic idea is that changes in feeding conditions lead to changes in lipid content and thus in the uptake rate of the compound from the environment. The physics of the uptake process strongly suggests that it is proportional to the surface area of the organism; it thus links up beautifully with the structure of the DEB model. Indeed, the uptake and elimination processes can still be described by a one-compartment model, but with varying coefficients, rather than fixed ones as discussed in the last section. The changes in physiological conditions thus result in changes in the uptake and elimination rates. One of the side results is the relationship between the kinetics and size of the organism, which is useful for the interpretation of data and the design of experiments. Figure 8.2 gives the diagram of the flows of the compounds that are thought to exist.

8.2.1 Uptake and elimination

The presentation assumes that the exchange of the compound with the environment is via the aqueous fraction of the organism. In combination with an instantaneous internal partitioning of the compound, any other compartment as well as a combination of compartments can be selected for the exchange without any need to change the model. For simplicity's sake, suppose that the water content of the energy reserves contributes little to the wet weight. This means that $\frac{[d_{we}]-[d_{de}]}{[d_{wv}]-[d_{dv}]}(e+e_{\dot{R}})$ in (2.11) is small. The wet weight per unit of volume can now be partitioned as

$$[W_w] = [d_{dv}] + [d_{wa}] + [d_{we}](e + e_{\dot{R}}) \tag{8.33}$$

The contribution of reproduction-energy to wet weight obviously depends on the species. Mussels, for instance, spawn once in a year, and have such a pronounced cycle in volume-specific dry weight, that the reproductive energy cannot be neglected. If the water content of the energy reserves is not negligibly small, an alternative assumption that preserves simplicity is that reserves replace water; this has been worked out in [414], but I expect this difference to be of secondary interest only.

The amount of compound can likewise be partitioned as $Q = Q_v + Q_a + Q_e + Q_{\dot{R}}$ over the dry weight fraction of body volume, aqueous fraction of body volume, energy reserves and reproductive energy, respectively. The instantaneous partitioning of the compound over the four compartments can be symbolized as $\frac{[d_{wa}]}{[d_{dv}]}\frac{Q_v}{Q_a} = P_{va}$ and $\frac{[d_{wa}]}{[d_{we}]e}\frac{Q_e}{Q_a} = \frac{[d_{wa}]}{[d_{we}]e_{\dot{R}}}\frac{Q_{\dot{R}}}{Q_a} = P_{ea}$, where P_{va} and P_{ea} are taken to be fixed partition coefficients, which only depend on the nature of the compound and not on its concentration. Note that the concentrations in these partition coefficients are based on compartment weight, not volume. Since the energy reserves will have a relatively high lipid content in most species of animal, $P_{ea} > P_{va}$ will hold for lipophilic compounds and P_{ea} will be close to the octanol-water partition coefficient, which is known for many compounds.

We can now relate the total number of moles of the compound in the organism, Q, to that in the aqueous fraction, Q_a:

$$Q = Q_a\left(\frac{[d_{dv}]}{[d_{wa}]}P_{va} + 1 + \frac{[d_{we}]}{[d_{wa}]}P_{ea}(e + e_{\dot{R}})\right) \quad \text{or} \tag{8.34}$$

$$\frac{[Q]}{[Q_a]} = \frac{[d_{dv}]}{[d_{wa}]}P_{va} + 1 + \frac{[d_{we}]}{[d_{wa}]}P_{ea}(e + e_{\dot{R}}) \equiv P_{wa} \tag{8.35}$$

Uptake is now taken to follow two routes, directly from the environment, where the compound is present at concentration c_d, and via food, where it is at concentration c_x. Uptake is taken to be proportional to surface area. The argumentation for food has been given on {55}, and for direct uptake, the argumentation is extremely similar. The nature of the uptake can be passive or active, but the rate is taken to be proportional to the concentration in the environment and/or to food uptake. The reasoning is similar to the Lotka–Volterra model for feeding; see {161}. If the elimination rate is proportional to the concentration in the aqueous fraction, uptake kinetics follow one-compartment dynamics which reads

$$\frac{d}{dt}Q = V^{2/3}(\{\dot{k}_{da}\}c_d + \{\dot{k}_{xa}\}fc_x - \dot{k}_{ad}[Q_a]) \tag{8.36}$$

The application of the chain rule for differentiation gives for the concentration $[Q] = Q/V$:

$$\frac{d}{dt}[Q] = \frac{\{\dot{k}_{da}\}}{V^{1/3}}c_d + \frac{\{\dot{k}_{xa}\}}{V^{1/3}}fc_x - [Q]\left(\frac{\dot{k}_{ad}}{V^{1/3}P_{wa}} + \frac{d}{dt}\ln V\right) \tag{8.37}$$

where the partition coefficient P_{wa} is given in (8.35). It is instructive to compare this formulation with the standard one, given by (8.1). The role of $\{\dot{k}_{da}\}V^{-1/3}$ in (8.37) is most similar to that of $[\dot{k}_{da}]$, while $\dot{k}_{ad}V^{-1/3}$ now plays the role of $\dot{k}_a$ in (8.1). The interpretation of this elimination rate $\dot{k}_{ad}$ is thus different from $\dot{k}_a$, and has dimension length time^{-1}, rather than time^{-1}; so it is in fact a conductance. This is the reason for giving them a different index. The term $\frac{d}{dt}\ln V$ in (8.37) represents the dilution rate due to growth, which was omitted in the most simple formulation (8.1).

Since most measurements are done on the basis of weights, the kinetics of the variable $\langle Q\rangle_w = [Q]/[W_w]$ is of practical interest; it represents the number of moles per unit of wet weight. Since $\frac{d}{dt}\langle Q\rangle_w = (\frac{d}{dt}[Q])/[W_w] - \langle Q\rangle_w[d_{we}](\frac{d}{dt}e + \frac{d}{dt}e_{\dot{R}})/[W_w]$, its dynamics is

$$\frac{d}{dt}\langle Q\rangle_w = \frac{\{\dot{k}_{da}\}c_d + \{\dot{k}_{xa}\}fc_x}{[W_w]V^{1/3}} - \langle Q\rangle_w\left(\frac{\dot{k}_{ad}}{V^{1/3}P_{wa}} + \frac{d}{dt}\ln V + \frac{[d_{we}]}{[W_w]}(\frac{d}{dt}e + \frac{d}{dt}e_{\dot{R}})\right) \quad (8.38)$$

Apart from the initial conditions, this specifies the dynamics in the period between the moments of spawning or reproduction. At such moments, (wet) weight as well as the contents of xenobiotic compounds have a discontinuity, because the buffer of energy allocated to reproduction is emptied, possibly together with its load of xenobiotic compound. The most simple assumption is to let the compound in that buffer transfer to the egg. If reproduction occurs at time $t_{\dot{R}}$, and if $t_{\dot{R}}^-$ denotes a moment just before $t_{\dot{R}}$, and $t_{\dot{R}}^+$ just after, the ratio of the concentrations of compound equals

$$\frac{\langle Q\rangle_w(t_{\dot{R}}^+)}{\langle Q\rangle_w(t_{\dot{R}}^-)} = \frac{[d_{dv}]P_{va} + [d_{wa}] + [d_{we}]P_{ea}(e + e_{\dot{R}})}{[d_{dv}]P_{va} + [d_{wa}] + [d_{we}]P_{ea}e} \frac{[d_{dv}] + [d_{wa}] + [d_{we}]e}{[d_{dv}] + [d_{wa}] + [d_{we}](e + e_{\dot{R}})} \quad (8.39)$$

The first factor corresponds with the ratio of xenobiotic weights, the second factor with the ratio of body weights. This result can be larger as well as smaller than 1, depending primarily on the partition coefficient P_{ea}. If the moments of reproduction are frequent enough to neglect the contribution of $e_{\dot{R}}$ to wet weight and compound load, $\frac{d}{dt}e_{\dot{R}}$ can be replaced by $e_0\dot{R}$, which can be left out if the reproductive output is negligibly small. The elimination route via reproduction can be very important for rapidly reproducing species.

It is also possible that no compound is transduced through the reproduction process, as has been found for 4,4′-DCB in *Lymnaea* [769]. This implies a (sudden) increase of the concentration at reproduction, which can induce toxic effects.

The change of concentration at reproduction has of course an intimate relationship with the initial conditions for the offspring, which depend on the feeding conditions and the loading of the mother. Experience with chronic toxicity tests shows that most effects occur at hatching, which means that an egg must be considered to be rather isolated, chemically, from its environment apart of course, from gas exchange. An extreme consequence is that the amount of compound at egg formation is the same as that at hatching. This means that the concentration at hatching relates to that of the mother just after reproduction as

$$\langle Q\rangle_w(a_b) = \langle Q\rangle_w(t_{\dot{R}}^+)\frac{P_{ea}V_m}{P_{wa}V_b}e_0 \quad (8.40)$$

where the ratio P_{wa} is given in (8.35) and should now be evaluated at $e_{\dot{R}} = 0$.

The number of parameters that relate to the kinetics of the compound amounts to three exchange rates $\dot{k}_{ad}$, $\{\dot{k}_{da}\}$ and $\{\dot{k}_{xa}\}$, which are present in even the most simple one-compartment kinetics with two inputs, and two additional partition coefficients P_{va} and P_{ea}. On top of that, a number of parameters shows up that relate volumes to weights. The third class of parameters is from the DEB model via the expressions for $\frac{d}{dt}V$, $\frac{d}{dt}e$ and $\frac{d}{dt}e_{\dot{R}}$. Not all parameters are required to fit the model to experimental data, but it is obvious that additional physiological knowledge will help us interpret experimental results, especially if the physiological condition changes during the experiment. Although some of the physiological parameters can be estimated from uptake/elimination curves in principle, an independent and more direct estimation is to be preferred. The least restrictive assumption is that the change in size is small, so $\frac{d}{dt}V \simeq 0$ and V is constant, which depends on the length of the experiment relative to the growth rate. Then usually the assumption $\frac{d}{dt}e_{\dot{R}} \simeq 0$ follows for restrictiveness. The most restrictive variant will usually be that $\frac{d}{dt}e = 0$, in which case (8.38) reduces to a simple one-compartment kinetics with constant coefficients. It shows that the exchange rates are inversely proportional to the volumetric length of the organism.

Figures 8.3 and 8.4 illustrate the performance of the model to describe the uptake/elimination behaviour of the compounds hexachlorobenzene (octanol/water partition coefficient $\log K_{ow} = 5.45$ [620]) and 2-monochloronaphthalene ($\log K_{ow} = 3.90$ [521]). The mussels and fish were not fed during the experiment, which implies that their energy reserves decreased during the experiment. The fish depleted its energy reserves faster, because it was smaller than the mussel and its temperature was higher. As a result of the decrease in reserves, the fish started to eliminate during the accumulation phase of the experiment. The model successfully describes this phenomenon. The experiments were short enough to assume that the size of the test animals did not change and that the energy allocation to reproduction was negligibly small during the experiment. The concentration of xenobiotic compounds in the water changed during accumulation. A cubic spline was, therefore, fitted to these concentrations and used to obtain the concentrations in the wet weight.

8.2.2 Bioconcentration coefficient

The bioconcentration coefficient, BC, is an important concept in the kinetics of xenobiotics. It is used among other things as a crude measure to compare xenobiotic compounds and species and to predict effects. For aquatic species and hydrophilic compounds, it is usually defined as the ratio of the concentration in the organism and in the water, which are both taken to be constant. For terrestrial species and/or lipophilic compounds, it is usually defined as the ratio of the concentration in the organism and in the food. The application of the concept BC is a bit complicated in the present context, because the concentration in the organism does not become stationary, due to growth and reproduction, even if the concentration in the environment is constant, i.e. in water, food and at constant food density. If the growth rate is low in comparison to the exchange rates, the compound can be in pseudo-equilibrium, but its concentration still depends, generally, on the size of the organism. On top of this, reproduction causes a cyclic change in concentration. The oscillations become larger if the organism accumulates its reproductive output over a

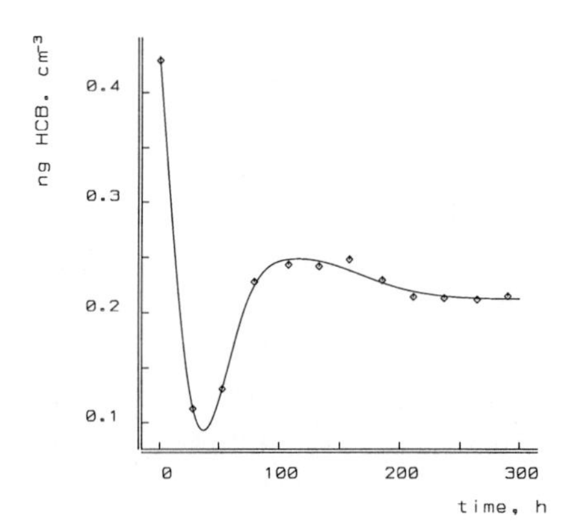

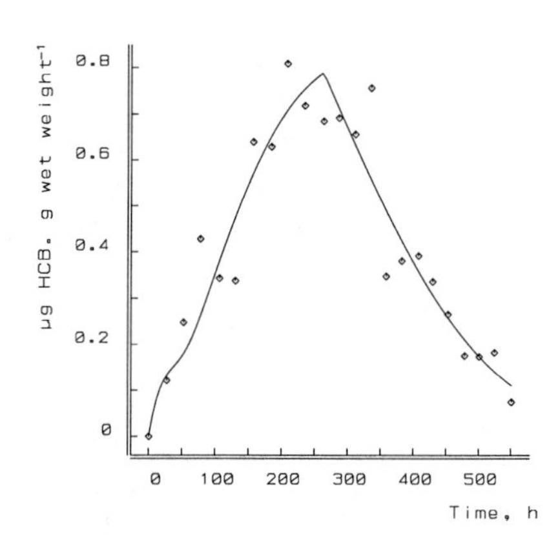

Figure 8.3: Measured concentration of hexachlorobenzene in water and in a starving 6.03 cm^3 freshwater mussel *Elliptio complanata* at 20 °C during a 264 h uptake/elimination experiment. Data from Russel and Gobas [620]. The least squares fitted curves are the cubic spline function for concentrations in the water and the model-based expectation for the concentration in the wet weight. From [414].

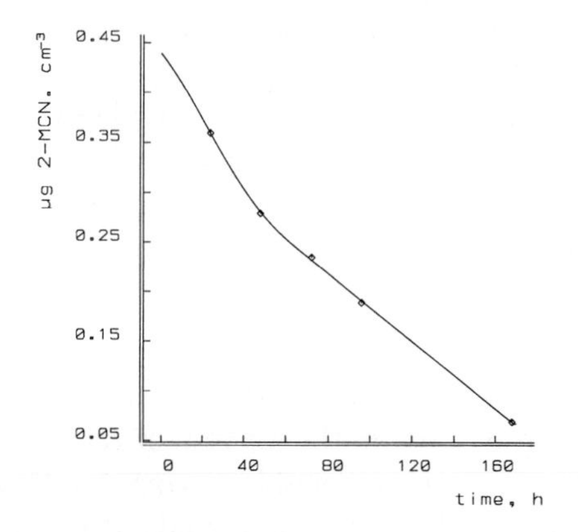

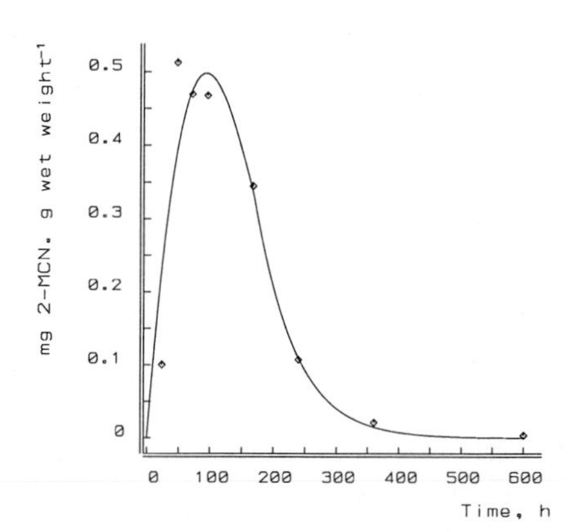

Figure 8.4: Measured concentration of 2-monochloronaphthalene in water and in a starving 0.22 cm^3 female guppy *Poecilia reticulata* at 22 °C during a 168 h uptake/elimination experiment. Data from Opperhuizen [521]. The least squares fitted curves are the cubic spline function for the concentrations in the water and the model-based expectation for the concentration in the wet weight. From [414].

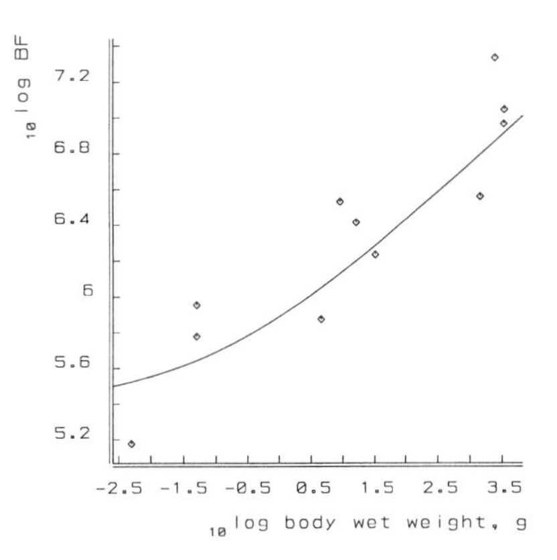

Figure 8.5: Bioconcentration coefficients for PCB153 in aquatic organisms in the field, as given in [414]. Data from Oliver and Niimi [508,519] and from the Dutch Ministry of Public Works and Transport. The curve represents the least squares fit of the linear relationship between the BC and the volumetric length. $\{\dot{k}_{xa}\}P_{xd}\{\dot{k}_{da}\}^{-1}\mathcal{V}^{1/3}$= 46 mm.

longer time period. If food density is constant for a long enough period, we have $e = f$ and $\frac{d}{dt}e = 0$. The ultimate concentration on the basis of wet weight then reduces for low growth and reproduction rates to

$$\langle Q\rangle_w \rightarrow \frac{\{\dot{k}_{da}\}c_d + \{\dot{k}_{xa}\}fc_x}{[W_w]\dot{k}_{ad}} \left(\frac{[d_{dv}]}{[d_{wa}]}P_{va} + 1 + \frac{[d_{we}]}{[d_{wa}]}P_{ea}f\right) \tag{8.41}$$

This expression can be used to predict how the bioconcentration coefficient depends on body size if species are compared. Since $\{\dot{k}_{xa}\}$ is proportional to $\{\dot{I}_m\}$, BC is expected to be linear in the volumetric length. The trend in $[E_m]$ almost cancels out. If the ratio of concentration in the food and that in the water equals some fixed partition coefficient, P_{xd} say, then we should expect that BC $\propto \{\dot{k}_{da}\} + \{\dot{k}_{xa}\}P_{xd}$ is almost linear in the volumetric length. Figure 8.5 illustrates that the BC for the highly lipophilic compound 2,4,5,2′,4′,5′ hexachlorobiphenyl (PCB153) for aquatic animals indeed depends on body size, and that this can be explained on the basis of the present reasoning.

This expectation is thus based solely on differences in the uptake of the amount of food. Especially in terrestrial habitats and more debatably in aquatic ones, accumulation in the food chain occurs. Since top predators tend to have the largest body size, it can be difficult to distinguish food chain effects from body size effects. Food chain effects operate via the partition coefficient for food/water, P_{xd}, body size effects via the uptake rate via food, $\{\dot{k}_{xa}\}$.

8.2.3 Metabolic transformations

If compounds are metabolized, the usual effect is that the products are less lipophilic than the original compound, so P_{ea} is reduced. In this way, the product will be eliminated at a higher rate. Since the toxic effects of compounds can be tied to the concentration in the organism, such a reduction obviously makes sense. It is tempting to couple the rate of metabolic transformation to the overall catabolic rate of the organism, $\dot{C}$. Since the catabolic rate is proportional to a weighted sum of a surface area and a volume, the bioconcentration coefficient is likely to increase with body volume more steeply for such compounds.

I do not know of data that can be used to test this idea, so there seems no point in working it out quantitatively. If the kinetics of the products and the original compound all require the full set of parameters, one will need a lot of measurements. Data will hopefully indicate which simplifying assumptions can be made. I mention this to point to the general lack of data on kinetics in relation to physiological information and nutritional status.

8.3 Biological effects

Only two types of effects are of primary, ecotoxicological, interest: those that affect survival and those that affect reproduction. These effects determine population dynamics, and thus production and existence. Due to the coupling between the various processes of energy uptake and use, many other effects of compounds have an indirect effect on reproduction. For instance, the conservation law for energy implies that a reduction of food uptake has indirect effects on reproduction. The DEB model describes the routes that translate these effects into an effect on reproduction. One is rather direct, via reduced reserves and the κ-rule. Another is more indirect via reduced growth. Small individuals eat less than large ones, so less energy is available for reproduction. Effects on growth and maintenance can also be translated into an effect on reproduction on the basis of the DEB model. These types of effects relate directly to energetics. In steady state, their consequences can be evaluated by changing one or more parameter values of the DEB model. Such a study is not very different from a more general one on the evolutionary implications of parameter settings.

The environmental relevance of mutagenic effects is still in debate. A frequently heard opinion from some industrialists is that mutagenic effects have no environmental impact at all, stating that the direct effect on survival is negligibly small and the loss of gametes does not count from an ecological point of view. The way aging is treated within the DEB model closely links up with mutagenic effects, particularly if the free radical mechanism is correct. Mutagenic compounds have about the same effect on organisms as free radicals. As a consequence, mutagenic effects can be studied by changing aging acceleration (in the case of metazoans). The DEB model offers the possibility of evaluating the consequences of mutagenic effects along the same lines as the effects on energy fluxes. I have already mentioned the setting of aging acceleration as a compromise between the life span of individuals and the evolutionary flexibility of the genome. The effects of changes in aging acceleration must then be found over a time scale of many generations and involve interspecies relationships. This makes such effects extremely hard to study, both experimentally and theoretically. The lack of reliable models for this time scale makes it difficult to draw firm conclusions. The fact that mutagenic compounds tend to be rather reactive and, therefore, generally have a short life in the environment is part of the problem, perhaps makes them less relevant to the problem of environmental pollution if emissions are incidentically only.

The significance of mutagenic effect on human health is widely recognized, particularly in relation to the occurrence of tumours and cancer. The Ames test is frequently applied to test compounds for mutagenic effects. The DEB model offers a framework for interpreting the sometimes unexpected results from these tests. The Ames test is discussed, for this reason, in a subsection of the section on effects on populations, as it is basically aimed at this level of organization.

The environmental significance of teratogenic effects, i.e. effects on the development of organisms, is even less recognized than the significance of mutagenic effects. Fortunately, only a few compounds seem to have a teratogenic effect as their primary effect and these fall outside the scope of this book.

8.3.1 Steady state effects on individuals

For many practical purposes of evaluating expected effects in natural or semi-natural systems where the concentration of a compound is at a certain (mean) level, it suffices to model the effect by a change in one or more of the primary DEB parameters. This is because the concentration in the organism will not change too much. The mode of action of the compound determines which parameters are affected. The mode of action itself, unfortunately, depends on the concentration; for example most compounds affect reproduction directly or indirectly at low concentrations, but survival is affected at high concentrations. This complicates the matter considerably. In the section on effects on populations, this approach will be continued.

The analysis of effects as they show up in toxicity tests, and natural systems where the concentration of compound changes considerably, should be carried out on a dynamic basis. Most American work deals with river systems with discharges. Traveling down the river, individuals experience a rapid change from an unpolluted environment to a polluted one. The occurrence of effects relates directly to the uptake kinetics of the compound. This process will be studied in the next subsection.

8.3.2 Dynamic effects on individuals

This subsection discusses the standard view on the characterization of lethal effects of xenobiotics, and then proposes an alternative that aims to solve some of the problems that are connected with this standard view.

An approach that has proven to be rather successful is to tie the occurrence of effects to the concentration in the tissue. In combination with the idea of an instantaneous partitioning of the compound over the different body fractions, as has been discussed in the previous section, it no longer matters if the effects originate from the disfunctioning of one or more particular organs, or of the whole body. If the concentration in one particular organ exceeds some threshold, it will at the same time exceed another threshold in another organ. This is of course no longer true if partitioning is a slow process with respect to the uptake and elimination rate of the compound.

For some reason, not all individuals show effects of the same intensity at the same time, if exposed to the compound in a certain concentration or at a certain dose. Part of the differences can be explained by differences in physiological condition, lipid content and size. It is possible to remove most but not all differences by strict standardization of the test organisms. The standard view is that the individuals differ in threshold values. So

Figure 8.6: A description of the occurrence of effects (here survival) in terms of the uptake kinetics of the compound and a (fixed) distribution of threshold values for the individuals. At a certain exposure period, the concentration in the tissue reaches a certain level, which translates into an effect in a certain fraction of the individuals

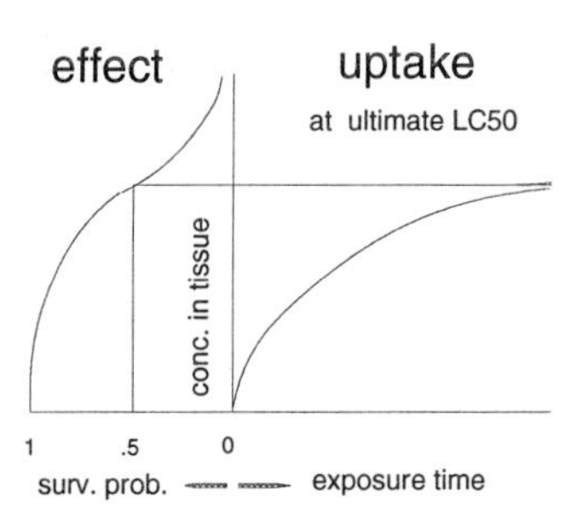

the process is a fully deterministic one for a single individual, but a stochastic process for a cohort. Figure 8.6 illustrates the concept.

Usual choices for the description of the distribution of the threshold values with respect to effects on survival are the log-normal or the log-logistic distribution. The motivation is empirical only. Since the threshold value remains unknown in practice, the survival probability is related directly to the dose or the concentration in the environment. For the log-logistic distribution of the concentration $\underline{c}_\dagger$ that causes death, this is

$$\text{Prob}\,\{\underline{c}_\dagger > c\} = q_0 \left(1 + \left(\frac{c}{c_{L50}}\right)^{1/\beta}\right)^{-1} \tag{8.42}$$

where c_{L50} is known as the LC50, the lethal concentration for 50% of the individuals, or more precisely, the concentration at which the survival probability is half the survival probability in the blank for a given exposure period. (This notation will appear unusual in the eyes of toxicologists, but I want to observe the mathematical standard of using only one character for one symbol.) The parameter β is a measure of the variance. The log-normal distribution is most similar to the log-logistic one. If the variance for the normally distributed variable $\ln\{\underline{c}_\dagger/c_{L50}\}$ is σ^2, the mapping $\sigma = \beta\pi/\sqrt{3} \simeq 1.8\beta$ results in equal variances for both distributions and the mapping $\sigma = \beta 4/\sqrt{2\pi} \simeq 1.6\beta$ in equal slopes of the plot of the survival probability against the concentration.

The parameter q_0 stands for the survival probability in the blank, i.e. if $c = 0$. It is assumed to be close to 1 and accounts for 'accidental losses'. Usually, too little attention is given to the actual nature of these losses. In many cases it reflects the condition of the test organisms, in which case it is misleading to treat such a cause of death as independent from the effects of the compound. The survival probability cannot be factorized into components relating to either the blank mortality or the compound induced mortality. It is frequently possible to obtain extremely low LC50 values by imposing some form of 'stress', such as oxygen depletion.

Variations in the conditions of the test organisms are substantial, even in the best standardized cultures. The chronic (21 days) toxicity test with daphnids (*D. magna*) for dichromate is considered to be the best standardized test used on a routine basis to check the condition of cultures in many toxicological laboratories. The standardization includes almost all aspects of culture and test conditions. The ratio of the upper and the lower boundary of the 95% confidence interval for the LC50.21d estimate is usually less than 1.1. If data from the best skilled technicians are compared throughout the year in a single laboratory, the ratio between extreme values is about 2. If the results of round robin tests are compared, the ratio is 5 to 10 for a compound as stable as dichromate. I mention these figures to point out the practical limitations of modelling in this field. A discussion about the statistical analysis of toxicity tests is outside the scope of this book; see [49,403,404,413].

The standard description of the concentration-response relation can be extended to the time domain by subjecting the concentration of toxicant in the tissue to a simple uptake/elimination kinetics such as in (8.4) [403]. If a no-effect level c_0 is introduced as well, the survival probability at exposure time t and concentration c becomes in absence of blank mortality

$$q(c,t) = \left(1 + \left(\frac{((1 - \exp\{-t\dot{k}_a\})c - c_0)_+}{c_{L50.\infty} - c_0}\right)^{1/\beta}\right)^{-1} \tag{8.43}$$

where $c_{L50.\infty}$ stands for the ultimate LC50. The LC50-time behaviour of this model is

$$c_{L50}(t) = c_{L50.\infty}(1 - \exp\{-t\dot{k}_a\})^{-1}$$

so the LC50 decreases exponentially in time at a rate that equals the elimination rate of the compound.

Figure 8.7 illustrates that the coupling of uptake kinetics to effects really works. The example is for cadmium chloride solutions in fresh water with algae as the food for daphnids. On alternate days, the solutions were refreshed and new algal food was supplied. The cadmium had the tendency to adsorb to the algal cells, which settled slowly to the bottom of the vessels and became no longer available to the daphnids. The concentration of available cadmium thus follows a saw-like pattern as illustrated, and the concentration in the tissue follows this pattern in a smooth way. If the settling kinetics is a simple first order one, the ratio of the concentrations before and after refreshment, after a fixed period, should be independent of the concentration after refreshment. The experimental confirmation is illustrated in figure 8.7. The crucial point is that the disappearance rate of the available cadmium could be estimated from the survival pattern of the daphnids, which is a strong argument in favour of the coupling between uptake and effects.

A weak point in this approach is the empirical nature of the threshold distribution and its existence in well standardized cultures. One would expect that full standardization leads to fully deterministic effects. What are the causes of the variation between individuals? It is unlikely that they differ that much in uptake/elimination kinetics. They are also too similar in biochemistry to explain how the threshold concentrations can differ so much in the rare cases of extreme standardization. Another weak point is that if survival probability is plotted against the concentration, the slope of the sigmoid graph usually tends to increase with exposure time. This is not possible with a fixed threshold distribution, which implies that the slope does not change.

A solution to these problems can be found in a stochastic approach to the occurrence of effects on a single individual, much along the same lines as is done for modelling of aging. The survival of a single individual is then described in terms of a hazard rate that depends

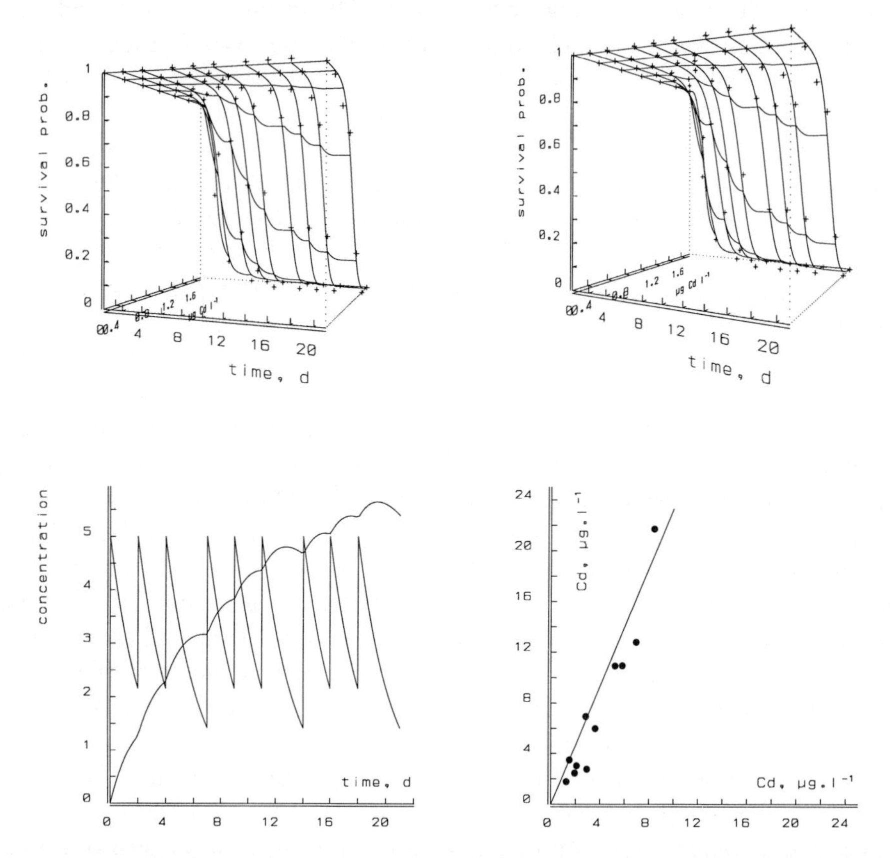

Figure 8.7: Stereo view of the survival probability of daphnids (*D. magna*, z-axis) as a function of exposure time (x-axis) and the concentration of cadmium chloride, plotted $_{10}\log$ transformed in μg l^{-1} on the y-axis. The concentrations of available cadmium followed a saw-like pattern, due to adsorption to settling algae. The lower left figure illustrates a first order settling kinetics and a first order uptake/elimination behaviour for the concentrations in the tissue. The lower right figure gives the analyses of the concentration in the water just before (x-axis) and 48 h after (y-axis) refreshment. The line is not based on these points, but on an analysis of the mortality pattern of the daphnids [403].

on the concentration of the compound in the organism, i.e. the hazard rate is

$$\dot{h}_c \propto ([Q]_1 - [Q])_+ \quad \text{and/or} \quad \dot{h}_c \propto ([Q] - [Q]_0)_+ \tag{8.44}$$

where $[Q]_1$ and $[Q]_0$ stand for the lower and the upper boundary of the concentrations of compound that do not affect survival. The background of such boundaries is that evolution takes place in a chemically varying environment. Organisms somehow managed to cope with a concentration range for each component. If the concentration leaves this tolerance range, effects will show up, just as with temperature. For xenobiotics, such as cadmium, the lower boundary of the range is simply $[Q]_1 = 0$. For copper, the lower boundary is some value larger than zero, because organisms need a small amount of copper. Effects are likely to occur at both low and high concentrations. As early as in 1969, Sprague [678] pointed to the similarity of effects of oxygen shortage and the presence of toxic compounds in fish. The consequences of this similarity did not receive proper attention, unfortunately.

The proportionality constant to describe the effect on the hazard rate probably differs for shortages and excesses. This relates to differences in mechanisms. If the concentration exceeds the tolerance range substantially, it is likely that death will strike via other mechanisms than for small excesses. This would restrict the applicability of the model to relatively small ranges of concentration. In practice, however, very wide concentration ranges are frequently used, as in range-finding tests on a routine basis. In the rest of this subsection, I will assume that $[Q]_1 = 0$ for simplicity's sake.

For comparison, the standard approach in fact assumes that the hazard rate is proportional to $([Q] > [Q]_\dagger)$, where $[Q]_\dagger$ is the threshold and the proportionality constant is extremely large. In the present proposal, the individuals do not have to differ any longer to understand why they do not all die exactly at the same moment, because stochasticity is in each of them. It remains possible that the individuals actually differ in parameter values, which become visible as soon as the cultures are less standardized. In this respect the problem is most similar to that of energetics.

The idea for hazard modelling can be worked out quantitatively as follows for a constant concentration in the environment.

Because of the general lack of knowledge about relevant concentrations in tissue, those in the environment will be used to specify the hazard rate. If the initial concentration in the tissue is negligibly small and if the concentration of compound in the tissue follows a simple first order (i.e. one-compartment) kinetics, the hazard rate at constant concentration c in the environment is

$$\dot{h}_c = \dot{k}_\dagger((1 - \exp\{-t\dot{k}_a\})c - c_0)_+$$

The proportionality constant $\dot{k}_\dagger$ has the interpretation of a killing rate with dimension (concentration time)$^{-1}$. It is a measure for the toxicity of the compound with respect to survival. The no-effect concentration $c_0 \equiv [Q]_0\dot{k}_a/[\dot{k}_{da}]$ in the environment can be interpreted as the highest concentration that will never result in an effect, if the concentration is constant. If $c > c_0$, but constant, and if the initial concentration in the tissue is 0, effects start to show up at $t_0 = -\dot{k}_a^{-1}\ln\{1 - c_0/c\}$, the moment at which the concentration in the tissue exceeds the no-effect concentration. In the absence of 'natural' mortality, the survival probability q for $c > c_0$ and $t > t_0$ is

$$q(c,t) = \exp\left\{\dot{k}_\dagger \dot{k}_a^{-1} c(\exp\{-t_0 \dot{k}_a\} - \exp\{-t\dot{k}_a\}) - \dot{k}_\dagger (c - c_0)(t - t_0)\right\} \tag{8.45}$$

Although this equation looks rather massive, it only has three parameters which are of all of practical interest: the no effect level, the killing rate and the elimination rate. The standard approach (8.43) has four parameters. The slope parameter of the standard approach for the analysis of dose-response relationships is missing, and it is generally ignored in practical applications. This parameter only occurs in the hazard model after we conceive one or more parameters as a random trial from some distribution (for instance the log-logistic one) to allow for variation between individuals. Of course, this would complicate the access to the parameters of interest. The models (8.43) and (8.45) are compared in figure 8.8. The more elaborate description of the DEB-based kinetics could be used to describe survival patterns in more detail. Practical limitations are likely to ruin such an attempt if no measurements for the concentration in the tissue are available. An appropriate experimental design can usually avoid such complications. Figure 8.9 illustrates the application of (8.45) to the results of some standard toxicity tests. Note that this formulation implies that the concentration-response relationships become steeper for longer exposure periods.

An interesting special case concerns extremely small elimination rates, so $\dot{k}_a \to 0$. The accumulation process reduces to $\frac{d}{dt}[Q] = [\dot{k}_{da}]c$, so that $[Q](t) = [\dot{k}_{da}]ct$ if the initial concentration in the tissue is negligibly small. The no-effect level (in the environment) is now 0, because a very small concentration in the environment will result ultimately in a very high concentration in the tissue. A no-effect level in the tissue, i.e. the upper boundary of the tolerance range, still exists, of course, and is exceeded at $t_0 = [Q]_0([\dot{k}_{da}]c)^{-1}$. The hazard rate amounts to $\dot{h}_c = \ddot{k}_\dagger c(t - t_0)_+$. The relationship between the killing acceleration $\ddot{k}_\dagger$ and the killing rate $\dot{k}_\dagger$, in the case that $\dot{k}_a \neq 0$, is $\ddot{k}_\dagger = \dot{k}_\dagger \dot{k}_a$. The survival probability is

$$q(c,t) = \exp\{-\ddot{k}_\dagger c(t^2/2 - t_0^2/2 - t_0 t)\} \tag{8.46}$$

For small no-effect levels in the tissue, so $t_0 \to 0$, this represents a Weibull distribution with shape parameter 2. The only difference with the survival probability related to aging, cf. {152}, is the extra accumulation step of products made by affected DNA, which resulted in a Weibull distribution with shape parameter 3.

In this special case, the full response surface in the concentration-exposure time-plane is described by just one parameter, the killing acceleration $\ddot{k}_\dagger$. One step towards more complex models is the introduction of the upper boundary of the tolerance range, via $[Q]_0/[\dot{k}_{da}]$ in t_0. Next comes the introduction of the elimination rate $\dot{k}_a$, which allows a new parameter basis: $\dot{k}_\dagger$, $\dot{k}_a$ and c_0. Then follow changes in the chemical composition (and size) of the animal by introduction of the partition coefficients P_{va} and P_{ea}, and/or a separation of uptake routes via the dissolved fraction $\{\dot{k}_{da}\}$ or via food $\{\dot{k}_{xa}\}$. Finally, we should allow for metabolic transformations, which have not been worked out in this book. So the level of complexity of the model can be fully trimmed to the need and/or practical limitations. The more complex the model is, the more one needs to know (and measure) about the behaviour of the compound in the environment, changes in the nutritional status of the animals, growth, reproduction, etc. If experimental research and model-based analysis

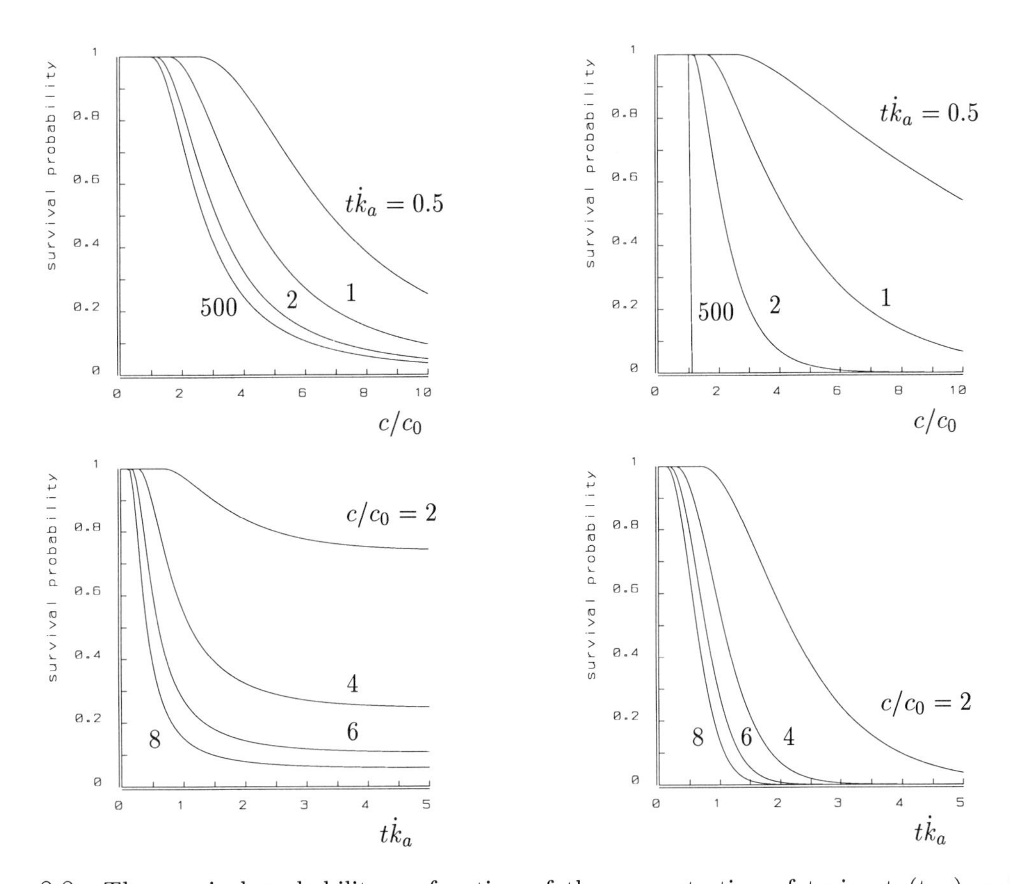

Figure 8.8: The survival probability as function of the concentration of toxicant (top) and exposure time (bottum) for the standard log-logistic model (left) and the hazard model (right). The log-logistic model is extended to include a no-effect level c_0, while uptake/elimination follows a simple one-compartment model with elimination rate $\dot{k}_a$ in both models. The parameters $c_{L50.\infty}$ and β of the log-logistic model have been chosen such that both models have the same c_{L50} and c_{L25} values for exposure time $\dot{k}_a^{-1}$ for the choice $\dot{k}_\dagger = \dot{k}_a/c_0$. This is the case when $c_{L50.\infty} = 2.69c_0$ and $\beta = 0.51$. In the log-logistic model, some individuals will survive forever, even if the concentration exceeds the no-effect level. In the hazard model, all individuals will eventually die if the concentration exceeds the no-effect level.

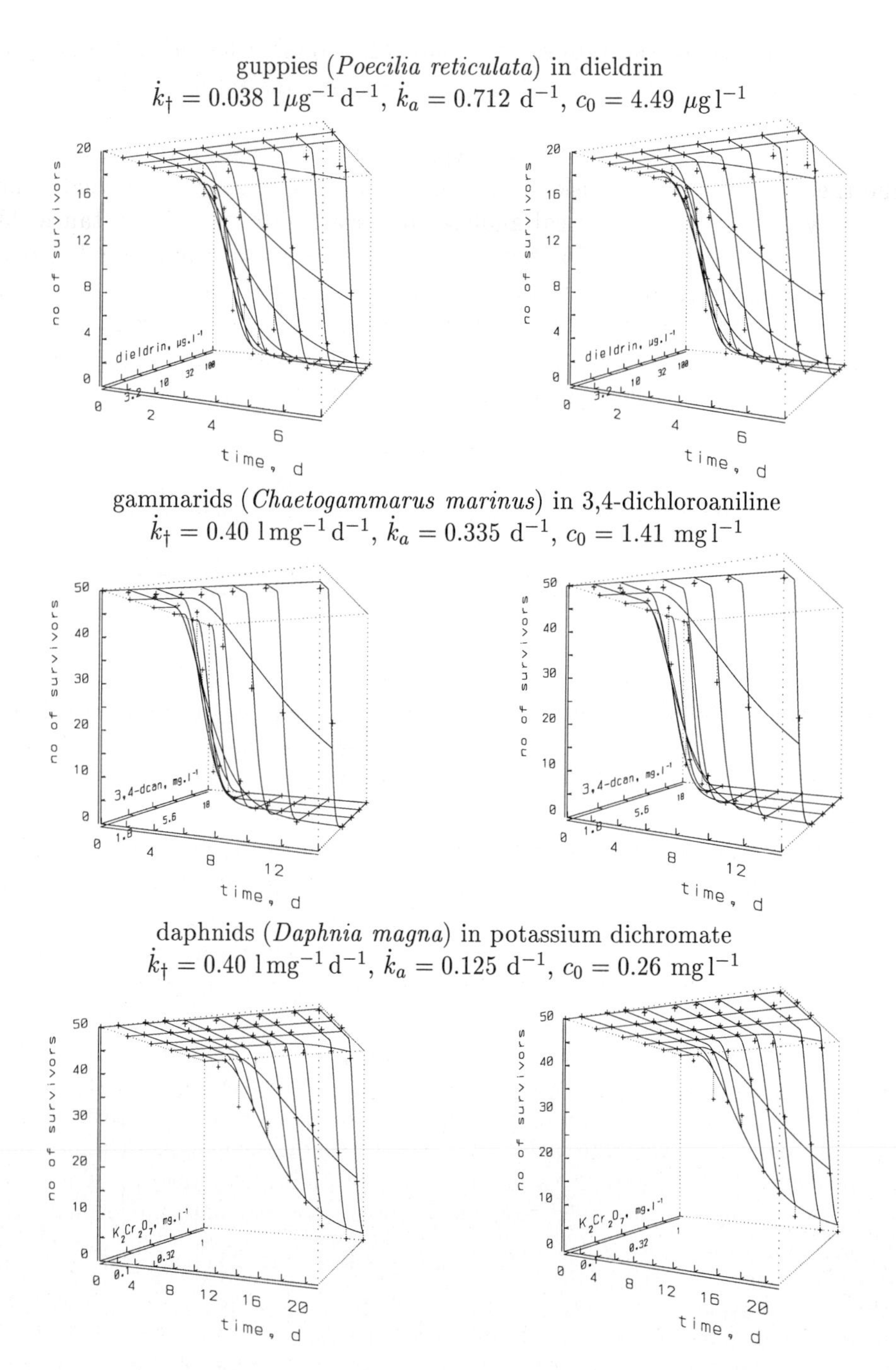

Figure 8.9: Stereo view of the number of surviving individuals, z-axis, as a function of exposure time, x-axis, to toxic compounds, y-axis. The expected number of surviving individuals is based on the idea that the hazard rate is proportional to the concentration in the tissue that exceeds the no-effect level under first order kinetics. Unpublished data, kindly provided by Ms Adema (IMW–TNO laboratories)

of results are combined in the proper way, one will probably feel an increasing need to define precise experimental conditions and avoid complicating factors, such as uncontrolled changes in exposure.

Besides a higher degree of theoretical elegance, this description of effects on survival makes the theory on competing risks available for direct application to toxicity and links up smoothly with standard statistical analyses of hazard rates; see for instance [39,141, 156,369,464]. The significance of a toxic stress for a particular individual depends on other risks, such as aging and starvation. Similar to the treatment of yields, {123}, an instantaneous and a non-instantaneous measure for this significance can be of practical value. If $\dot{h}$ stands for the hazard tied to aging as before, and $\dot{p}$ for other risks, such as predation, an obvious instantaneous measure for the significance of the toxic stress is

$$\dot{h}_c(\dot{h}_c + \dot{h} + \dot{p})^{-1}$$

A useful non-instantaneous measure is the effect on the expected life span

$$1 - \frac{\mathcal{E}\underline{a}_\dagger}{\mathcal{E}\underline{a}_\dagger|\dot{h}_c = 0} = 1 - \frac{\int_0^\infty \exp\left\{-\int_0^t(\dot{h}_c(t_1) + \dot{h}(t_1) + \dot{p}(t_1))\,dt_1\right\}\,dt}{\int_0^\infty \exp\left\{-\int_0^t(\dot{h}(t_1) + \dot{p}(t_1))\,dt_1\right\}\,dt} \tag{8.47}$$

In this way, both dimensionless measures are scaled between 0 and 1.

A consequence of the hazard-based model is that the hazard rate is higher than in the blank as soon as the concentration in the tissue exceeds the no-effect level. This implies that under those conditions the compound will kill the individual for sure if the exposure is long enough and if the individual lives for a long enough period. This means that the ultimate LC50 equals the no-effect level, which degrades the value of the LC50 as a characteristic for the toxicity of a compound. The LC50-time curve can be obtained by equating (8.45) to 0.5 and solving c numerically for choices of exposure times t. Figure 8.10 shows how the LC50 depends on the no-effect level and the exposure time. For increasing no-effect levels, it takes longer for the LC50 to approach the no-effect level.

If a no-effect level is incorporated into the standard sigmoid concentration-response relationship with a free slope parameter, estimation problems usually show up. The sigmoid responses with and without a no-effect level are generally too similar to be told apart on the basis of experimental results. The no-effect level is much better fixed by the hazard-based model than by the standard model. The hazard-based model also has sigmoid concentration-response relationships, but the slope is fully determined by the elimination and killing rates. Contrary to the standard sigmoid model, the slope therefore contains information about the no-effect level. Since the no-effect level has a much higher environmental significance than the LC50, the overall conclusion can only be that (8.45) should be preferred above the standard model. It should be realized however that it is an extremely simple model. Compounds with more complex kinetics or metabolic effects will deviate from this expectation. Application on a routine basis can still be valuable to trace this more complex behaviour. In view of the objectives of routine testing of chemicals, minor deviations do not matter.

The comparison of LC50 values for different species after a fixed exposure time to some compound suffers from differences in the kinetics that are unrelated to the actual

Figure 8.10: Stereo view of the LC50, c_{L50} on the z-axis, as a function the exposure time t on the x-axis, and the no-effect level c_0 on the y-axis. The hazard rate here equals $\dot{k}_\dagger$ times the difference between the concentration in the tissue and the no-effect level, while the compound follows a first-order kinetics with elimination rate $\dot{k}_a$. The three variables are made dimensionless in the way indicated in the figure. Note that the exposure time does not start from 0 since the LC50 is theoretically infinitely large there, apart from boundary conditions set by the chemical properties of the compound.

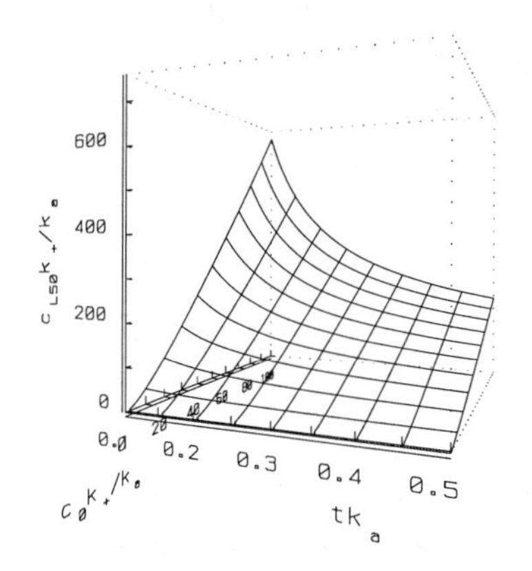

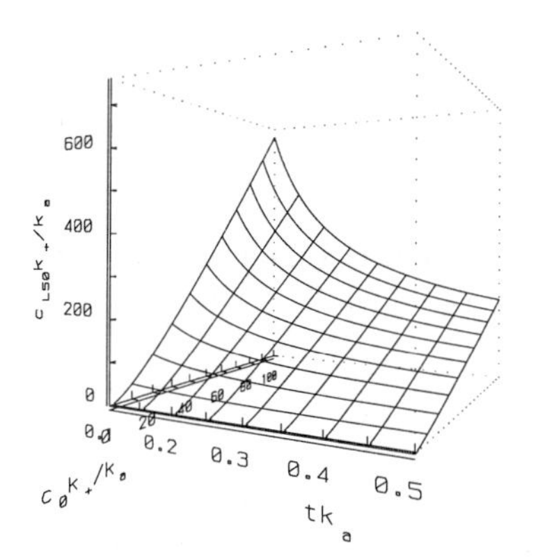

toxicity. Large bodied species generally have higher LC50 values, not because they are less sensitive, but because uptake is less rapid. This complicates the search for patterns in toxicity. Killing rates suffer much less from these problems. The model in which the LC50 is introduced is primarily inspired by experimental design and not by biological mechanisms. The two models beautifully illustrate the importance of preconceptions in research.

8.3.3 Effects on populations

In the chapter on population dynamics, it has been argued that many phenomena can be understood on the assumption that individuals only interact via resources. This extremely simple assumption on interactions allows, as a consequence, that population dynamics can be deduced directly from the energetics of the individuals as a first approximation. Consistent with this idea, we can try to understand effects at the population level in terms of consequences of effects on individuals. This points to the relationship between individual performance and population dynamics and it will lead to some unexpected conclusions about the way effects on individuals show up at population level.

An extremely useful approach for understanding population effects is to focus on steady states, where food density is constant. It implies that populations will (eventually) grow exponentially at rate $\dot{\mu}$, say. As explained, this follows from the property that the offspring repeat the survival and reproduction pattern of the parents. The relationship between food density and $\dot{\mu}$ is given by the characteristic equation (5.22). The survival probability and reproduction rate is given for DEB isomorphs in (3.57) with (3.56) and (3.47) with (3.26), respectively. Given the set of energy parameters, $\dot{\mu}$ can vary from 0 to $\dot{\mu}_m$, depending on the food density. It cannot become negative in steady state, because the population will then become extinct due to the discrete nature of numbers of individuals.

Suppose that the xenobiotic compound in a certain concentration affects just one energy parameter. The compound causes a rise in the maintenance costs, for instance, or in the costs for growth, etc. Suppose next, that the concentration of the compound is such that the population growth rate at high food densities is reduced to some fraction of that in the blank situation, say a fraction of 0.9. This thus corresponds with a particular change in one parameter. Now we can study how $\dot{\mu}$ depends on food density, given the new set of parameters, which differ from the original ones by a change in that single parameter, and compare it with the corresponding $\dot{\mu}$ in the blank situation. This approach makes it possible to compare effects of compounds with different modes of action. Figure 8.11 illustrates the result for ectothermic DEB isomorphs. For effects on growth or reproduction, the population growth rate is just a fixed fraction of that in the blank. For effects on maintenance and survival, the effect becomes much more severe at low population growth rates, thus at low food densities. Figure 8.12 illustrates that these differences can be observed in experimental populations. The compound 3,4-dichloroaniline primarily affects reproduction and the effect on $\dot{\mu}$ is practically independent of food density, while vanadium primarily affects maintenance and $\dot{\mu}$ is much more affected at low food levels. For other examples; see [405].

The difference in population effects can be understood intuitively as follows. If the population is at its carrying capacity, $\dot{\mu} = 0$, and reproduction and loss rates are both very low, food availability completely governs the reproduction rate. All resources are used for maintenance. Effects on maintenance, therefore, show up directly in this situation, but effects on growth and reproduction remain hidden, unless the effect is so strong that replacement is impossible. If the population is growing at a high rate, energy allocation to maintenance is just a small fraction of available energy. Even considerable changes in this small fraction will, therefore, remain hidden, but effects on production rates are now revealed. This implies that at a constant concentration of compound in the environment, the effect at the population level depends on food availability and thus is of a dynamic nature. This reasoning does not yet use the more subtle effects of uptake via food as opposed to those via the environment directly.

The effects at low population growth rates can be studied if the population is at its carrying capacity. If food supply to a fed batch culture is constant, the number of individuals at carrying capacity is proportional to the food supply rate, cf. figure 5.12. If the loss rate, and so the reproduction rate, is small, the ratio of the food supply rate and the number of individuals is a good measure of the maintenance costs. Figure 8.13 illustrates that some compounds, such as vanadium and bromide, affect these maintenance costs while others don't and 'only' cause death in this situation. It also shows that the effect is almost linear in the concentration, as are the effects on survival, aging and mutagenicity, as we will see in the next section.

Figure 8.11: Population growth rate in a stressed situation is plotted against that in a blank situation, when only one energy parameter is affected at the same time for reproducing isomorphs (left) and dividing filaments (right). The effect of compounds with different mode of action is standardized such that the maximum population growth rate is 0.9 times that in the blank. Food density is assumed to be constant. Relative effects in isomorphs on growth $[G]$, reserve capacity $[E_m]$ and reproduction q are almost independent of the feeding conditions, while those on assimilation $\{\dot{A}_m\}$, maintenance $[\dot{M}]$ and survival $\ddot{p}_a$, are much stronger under poor feeding conditions. The effect on the partitioning fraction κ is different from the rest and probably does not correspond with an effect of a toxic compound. The relative effects in filaments are largely comparable to those in isomorphs for growth and maintenance. Effects on assimilation $[\dot{A}_m]$, coincide with effects on survival $\dot{p}_a$.

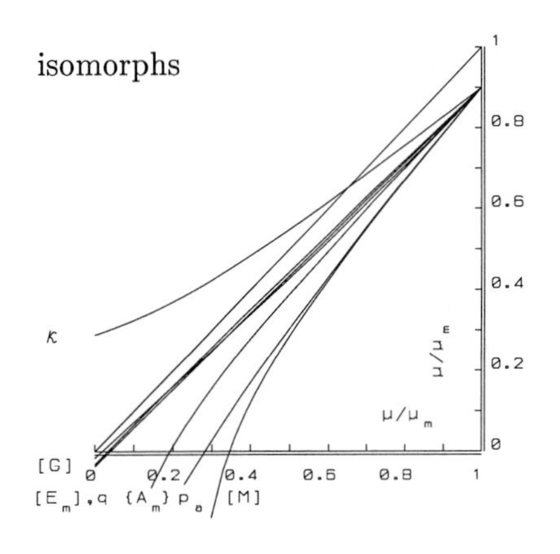

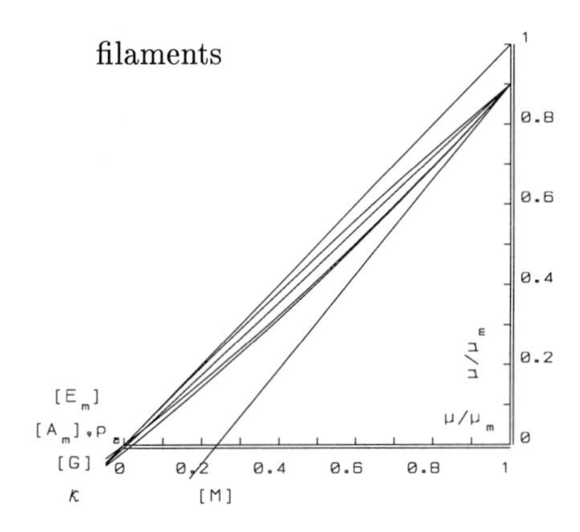

8.3.4 Mutagenicity

The *Salmonella* test, also known as the Ames test, is a popular test for the mutagenic properties of a compound [15,454]. It is discussed here because the results of the test can sometimes only be understood if energy side effects are taken into account, for which the DEB model gives a useful framework. This section relies heavily on [328,418,419].

The test is carried out as follows. Bacteria (mutants of *Salmonella typhimurium*) that cannot produce the amino acid histidine are grown on an agar plate with a small amount of histidine but otherwise large amounts of all sorts of nutrients. When the histidine becomes depleted, these histidine auxotrophs stop growing at a colony size of typically 8 to 32 cells. Histidine auxothrophic bacteria can undergo a mutation enabling them to synthesize the necessary histidine themselves, as can the wild strain. They become histidine-prototrophic and continue to grow, even if the histidine on the plate is depleted. (They only synthesize histidine if it is not available in the environment.) Colonies that contain histidine-prototrophs are called revertant colonies and can eventually be observed with the naked eye when the colony size has thousands of cells. The number of revertant colonies relates to the concentration of the compound that has been added to the agar plate and its mutagenic capacity.

Liver homogenate of metabolically stimulated rats is sometimes added to simulate mu-

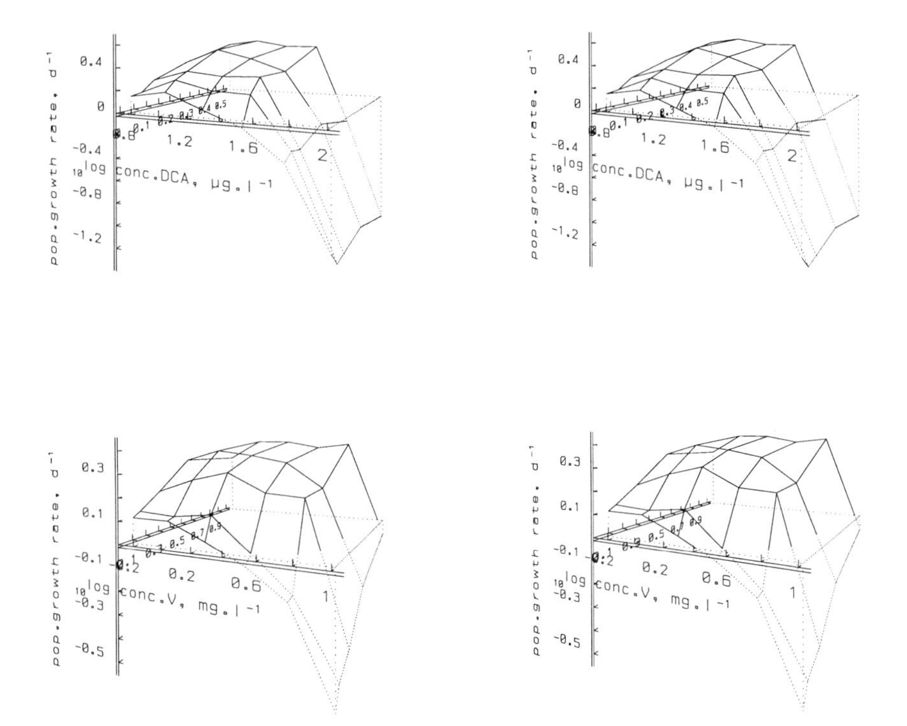

Figure 8.12: Stereo view of the population growth rate of the rotifer *Brachionus rubens* (z-axis) as a function of food density (y-axis) and concentration of toxic compound (x-axis): 3,4 dichloroaniline (above) and potassium metavanadate (below). Food density is in 1.36×10^9 cells *Chlorella pyrenoidosa* per litre, temperature is 20 °C. The difference in shape of the response surfaces is due to differences in mode of action of the compounds, as predicted by the DEB theory.

Figure 8.13: The ratio of the food supply rate to a population of daphnids and the number of individuals at carrying capacity in fed-batch cultures as a function of the concentration of compound at 20 °C. The crosses + refer to the occurence of mortality. Only compounds that affect maintenance give a positive response.

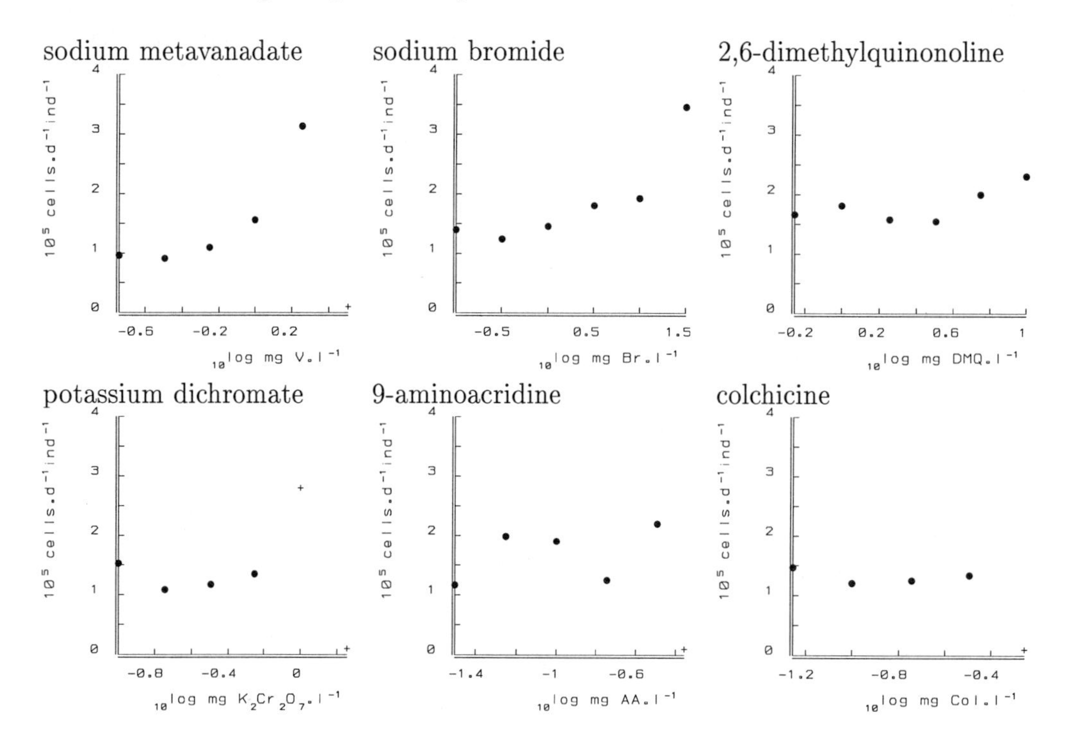

tagenicity for vertebrates. The primary interest for mutagenicity is in human health problems, as explained. Vertebrates have many metabolic pathways that prokaryotes do not have. Enzymes in this homogenate sometimes transform non-mutagenic compounds into mutagenic ones, sometimes they do the opposite or have no effects at all.

Some initial histidine is necessary, because bacteria that do not grow and divide do not seem to mutate, or, at least, the mutation is not expressed. This ties mutation frequency to energetics. It is a most remarkable observation, with many consequences. Since maintenance processes also involve some protein synthesis, one would think that mutations should also be expressed if growth ceases, but observation teaches otherwise. If a compound is both mutagenic and reduces growth, the moment of histidine depletion is postponed, so that effective exposure time to the mutagenic compound is increased. Some brands of agar contain small amounts of compounds that become (slightly) mutagenic after autoclavating. This gives a small mutagenic response in the blank. If a test compound only affects growth and is not mutagenic at all, the number of revertant colonies will increase with the concentration of test compound. Such responses make it necessary to model the combined mutation/growth process for the interpretation of the test results.

The rest of this section gives a simple account, appropriate for DEB filaments from a culture that resembles the (initial) growth conditions on the agar plate. For a more detailed account; see [328]. A description for DEB rods would be more accurate but also more complex and would hide the message.

Suppose that the initial amount of histidine on a plate is just enough for the synthesis of N_h cells. Figure 5.5 shows that the histidine reserves are small enough to be neglected. If the inoculum size on the plate is N_0, the number of cells develops initially as $N(t) = N_0 \exp\{\dot{\mu}t\}$. Histidine thus becomes depleted at time $t_h = \dot{\mu}^{-1} \ln\{1 + N_h/N_0\}$. If the mutation rate per unit of DNA is constant, say at value $\dot{p}_m$, the probability of at least one mutation occurring in the descendants of one auxotrophic cell becomes for low mutation rates

$$1 - \exp\left\{-\dot{p}_m \int_0^{t_h} (N(t)/N_0)\, dt\right\} = 1 - \exp\left\{-\frac{\dot{p}_m N_h}{\dot{\mu} N_0}\right\} \simeq \frac{\dot{p}_m N_h}{\dot{\mu} N_0} \tag{8.48}$$

The probability of back mutation is small enough to be neglected. The expected number of revertant colonies is N_0 times (8.48), so that the number of revertant colonies is hardly affected by the inoculum size. The effect of an increase in the number of micro-colonies on the plate is cancelled by the resulting reduction of exposure time.

A consequence of the assumption that the mutation frequency per unit of DNA is constant is that the mutations are independent of each other. This means that the number of revertant colonies on a plate follows a binomial distribution, which is well approximated by the Poisson distribution for low mutation rates. (There are typically less than 100 revertant colonies with a typical inoculum size of 10^8 per plate.)

The significance of this expression is that the effect of inoculum size and the amount of histidine become explicit. Variations in these variables, which are under experimental control, translate directly into extra variations in the response. If a compound affects the population growth rate, it also affects the expected number of revertant colonies. I refer to the subsection on population growth rates, {282}, for a discussion of how individual performance (substrate uptake, maintenance, growth) relates to population growth rates. This defines how effects on individual performance translate into effects on population growth rates. This remark not only applies to effects of the test compound, but also to the nutritional quality of the agar.

The mutation rate is usually found to be proportional to the concentration of test compound. This means that each molecule has a certain probability to cause a mutation. Deviations from this relationship can usually be related to changes in the stability of the compound on the plate. Many mutagenic compounds are rather reactive, so that the concentration usually decreases substantially before t_h. Others, such as nitrite, diffuse to the deeper layers of the agar plate and become less available for the bacteria in the upper layer. It is easy to circumvent this problem by adding the compound to the (thick) nutritive bottom layer when it is still liquid, rather than to the (thin) top layer. However, this would increase the financial costs of the test. If metabolic activation is applied, the concentrations of the original compound and the products are likely to become complex compound-specific functions of time. One strategy for interpreting the test results is to analyse and model time stability of compounds in the Ames test. A better strategy would be to change the experimental procedure in such a way that these complexities do not occur.

Bibliography

[1] K. Aagaard. Profundal chironimid populations during a fertilization experiment in Langvatn, Norway. *Holarct. Ecol.*, pages 325–331, 1982.

[2] R.A. Ackerman. Growth and gas exchange of embryonic sea turtles (*Chelonia, Caretta*). *Copeia*, 1981:757–765, 1981.

[3] R.A. Ackerman. Oxygen consumption by sea turtle (*Chelonia, Caretta*) eggs during development. *Physiol. Zool.*, 54:316–324, 1981.

[4] R.A. Ackerman, G.C. Whittow, C.V. Paganelli, and T.N. Pettit. Oxygen consumption, gas exchange, and growth of embryonic wedge-tailed shearwaters (*Puffinus pacificus chlororhynchus*). *Physiol. Zool.*, 53:201–221, 1980.

[5] D.L. Aksnes and J.K. Egge. A theoretical model for nutrient uptake in phytoplankton. *Mar. Ecol. Prog. Ser.*, 70:65–72, 1991.

[6] A.H.Y. Al-Adhub and A.B. Bowers. Growth and breeding of *Dichelopandalus bonnieri* in Isle of Man Waters. *J. Mar. Biol. Ass. U. K.*, 57:229–238, 1977.

[7] A.W.H. Al-Hakim, M.I.A. Al-Mehdi, and A.H.J. Al-Salman. Determination of age, growth and sexual maturity of *Barbus grypus* in the Dukan reservoir of Iraq. *J. Fish Biol.*, 18:299–308, 1981.

[8] M. Aleksiuk and I.M. Cowan. The winter metabolic depression in Arctic beavers (*Castor canadensis* Kuhl) with comparisons to California beavers. *Can. J. Zool.*, 47:965–979, 1969.

[9] R.M. Alexander. *Exploring biomechanics; Animals in motion.* Scientific American Library, New York, 1992.

[10] V.Y. Alexandrov. *Cells, molecules and temperature.*, volume 21 of *Ecol. Stud.* Springer-Verlag, 1977.

[11] A.L. Alldredge. Abandoned larvacean houses. A unique food source in the pelagic environment. *Science (New York)*, 177:885–887, 1972.

[12] A.L. Alldredge. Discarded appendicularian houses as sources of food, surface habitats and particulate organic matter in planktonic environments. *Limnol. Oceanogr.*, 21:14–23, 1976.

[13] A.L. Alldredge. House morphology and mechanisms of feeding in *Oikopleuridae* (*Tunicata, Appendicularia*). *J. Zool. Lond.*, 181:175–188, 1977.

[14] T.F.H. Allen and T.B. Starr. *Hierarchy. Perspectives for ecological complexity.* University Press of Chicago, 1982.

[15] B.N. Ames, J. McCann, and E. Yamasaki. Methods for detecting carcinogens and mutagens with *Salmonella*/mammalian microsome mutagenicity test. *Mutat. Res.*, 31:347–364, 1975.

[16] A.W. Angulo y González. The prenatal growth of the albino rat. *Anat. Rec.*, 52:117–138, 1932.

[17] A.D. Ansell. Reproduction, growth and mortality of *Venus striatula* (Da Costa) in Kames Bay, Millport. *J. Mar. Biol. Ass. U. K.*, 41:191–215, 1961.

[18] R.S. Appeldoorn. Variation in the growth rate of *Mya arenaria* and its relationship to the environment as analyzed through principal components analysis and the ω parameter of the von Bertalanffy equation. *J. Fish Biol.*, 81:75–84, 1983.

[19] A. Ar and H. Rahn. Interdependence of gas conductance, incubation length, and weight of the avian egg. In J. Piiper, editor, *Respiration function in birds.* Springer-Verlag, 1978.

[20] E.A. Armstrong. *The wren.* The New Naturalist. Collins, 1955.

[21] N. Armstrong, R.E. Chaplin, D.I. Chapman, and B. Smith. Observations on the reproduction of female wild and park Fallow deer (*Dama dama*) in southern England. *J. Zool. Lond.*, 158:27–37, 1969.

[22] D.K. Arrowsmith and C.M. Place. *An introduction to dynamical systems.* Cambridge University Press, 1990.

[23] D.K. Arrowsmith and C.M. Place. *Dynamical systems. Differential equations, maps and chaotic behaviour.* Chapman & Hall, 1992.

[24] P.W. Atkins. *Physical chemistry.* Oxford University Press, 1989.

[25] M. Atlas. The rate of oxygen consumption of frogs during embryonic development and growth. *Physiol. Zool.*, 11:278–291, 1938.

[26] T. Baekken. Growth patterns and food habits of *Baetis rhodani*, *Capnia pygmaea* and *Diura nanseni* in a west Norwegian river. *Holarct. Ecol.*, 4:139–144, 1981.

[27] T.B. Bagenal. The growth rate of the hake, *Merluccius merluccius* (L.), in the Clyde and other Scottish sea areas. *J. Mar. Biol. Ass. U. K.*, 33:69–95, 1954.

[28] T.B. Bagenal. Notes on the biology of the schelly *Coregonus lavaretus* (L.) in Haweswater and Ullswater. *J. Fish Biol.*, 2:137–154, 1970.

[29] J.E. Bailey and D.F. Ollis. *Biochemical engineering fundamentals.* McGraw-Hill, 1986.

[30] R. Bakker. *The dinosaur heresies.* Longman, 1986.

[31] J.H.van Balen. A comparative study of the breeding ecology of the great tit *Parus major* in different habitats. *Ardea*, 61:1–93, 1973.

[32] C.A.M. Baltus. De effekten van voedselregime en lood op de eerste reproduktie van *Daphnia magna.* Technical report, Dienst DBW/RIZA, 1990.

[33] M.C. Barber, L.A. Suárez, and R.R. Lassiter. Modeling bioconcentration of nonpolar organic pollutants by fish. *Environ. Toxicol. Chem.*, 7:545–558, 1988.

[34] J. Barcroft. *Researches on pre-natal life.*, volume 1. Blackwell Sci. Publ., 1946.

[35] J. Barcroft, R.H.E. Elliott, L.B. Flexner, F.G. Hall, W. Herkel, E.F. McCarthy, T. McClurkin, and M. Talaat. Conditions of foetal respiration in the goat. *J. Physiol. (Lond.)*, 83:192–214, 1935.

[36] J. Barcroft, L.B. Flexner, W. Herkel, E.F. McCarthy, and T. McClurkin. The utilization of oxygen by the uterus in the rabbit. *J. Physiol. (Lond.)*, 83:215–221, 1935.

[37] R.D.J. Barker, T.G. Cartledge, F. Dewhurst, and R.O. Jenkins. *Principles of cell energetics.* Butterworth-Heinemann Ltd., Oxford, 1992.

[38] G.A. Bartholomew and D.L. Goldstein. The energetics of development in a very large altricial bird, the brown pelican. In R.S. Seymour, editor, *Respiration and metabolism of embryonic vertebrates.*, pages 347–357. Dr. W. Junk Publishers, Dordrecht, 1984.

[39] I.V. Basawa and B.L.S.P. Rao. *Statistical inference for stochastic processes.* Academic Press, 1980.

[40] G.C. Bateman and R.P. Balda. Growth, development, and food habits of young Pinon jays. *Auk*, 90:39–61, 1973.

[41] E.H. Battley. *Energetics of microbial growth.* J. Wiley & Sons, Inc., 1987.

[42] E.H. Battley. Calculation of entropy change accompanying growth of *Escherichia coli* K-12 on succinic acid. *Biotechnol. Bioeng.*, 41:422–428, 1993.

[43] J.L. Baxter. A study of the yellowtail *Seriola dorsalis* (Gill). *Fish. Bull. (Dublin)*, 110:1–96, 1960. State Calif. Dept. Nat. Res., Div. Fish. Game.

[44] B.L. Bayne, A.J.S. Hawkins, and E. Navarro. Feeding and digestion by the mussel *Mytilus edulis* L. (*Bivalvia, Mollusca*) in mixtures of silt and algal cells at low concentrations. *J. Exp. Mar. Biol. Ecol.*, 111:1–22, 1987.

[45] B.L. Bayne, A.J.S. Hawkins, E. Navarro, and I.P. Iglesias. Effects of seston concentration on feeding, digestion and growth in the mussel *Mytilus edulis. Mar. Ecol. Prog. Ser.*, 55:47–54, 1989.

[46] B.L. Bayne and R.C. Newell. Physiological energetics of marine molluscs. In A.S.M.

Saleuddin and K.M. Wilbur, editors, *Physiology*, volume 4 of *The Mollusca*, chapter 1. Academic Press, 1983.

[47] M.J. Bazin and P.T. Saunders. Determination of critical variables in a microbial predator-prey system by catastrophe theory. *Nature (Lond.)*, 275:52–54, 1978.

[48] R.C. Beason and E.C. Franks. Development of young horned larks. *Auk*, 90:359–363, 1983.

[49] J.J.M. Bedaux and S.A.L.M. Kooijman. Hazard-based analysis of bioassays. *J. Environ. Stat.*, 1993. to appear.

[50] J.J.M. Bedaux and S.A.L.M. Kooijman. Stochasticity in deterministic models. In C.R. Rao, G.P. Patil, and N.P. Ross, editors, *Handbook of Statistics 12: Environmental Statistics.*, volume 12. North Holland, 1993.

[51] J.R. Beddington and D.B. Taylor. Optimum age specific harvesting of a population. *Biometrics*, 29:801–809, 1973.

[52] D.J. Bell. *Mathematics of linear and nonlinear systems.* Clarendon Press, Oxford, 1990.

[53] G.I. Bell and E.C. Anderson. Cell growth and division. *Biophys. J.*, 7:329–351, 1967.

[54] D.B. Bennett. Growth of the edible crab (*Cancer pagurus* L.) off south-west England. *J. Mar. Biol. Ass. U. K.*, 54:803–823, 1974.

[55] A.A. Benson and R.F. Lee. The role of wax in oceanic food chains. *Sci. Am.*, (3):77–86, 1975.

[56] T.G.O. Berg and B. Ljunggren. The rate of growth of a single yeast cell. *Biotechnol. Bioeng.*, 24:2739–2741, 1982.

[57] C. Bergmann. Über die Verhältnisse der Wärmeökonomie der Tiere zu ihrer Grösse. *Goett. Stud.*, 1:595–708, 1847.

[58] H. Bergmann and H. Helb. *Stimmen der Vögel Europas.* BLV Verlagsgesellschaft, München, 1982.

[59] H.H. Bergmann, S. Klaus, F. Muller, and J. Wiesner. *Das Haselhuhn.* Die Neue Brehm-Bücherei. A. Ziemsen-Verlag, Wittenberg Lutherstadt, 1978.

[60] J. Bernardo. Determinants of maturation in animals. *Trends Ecol. & Evol.*, 8:166–173, 1993.

[61] D.R. Berry, editor. *Physiology of industrial fungi.* Blackwell Sci. Publ., 1988.

[62] H.H. Berry. Hand-rearing lesser flamingos. In J. Kear and N. Duplaix-Hall, editors, *Flamingos.* T. & A.D.Poyser, Berkhamsted, 1975.

[63] L.von Bertalanffy. A quantitative theory of organic growth. *Hum. Biol.*, 10:181–213, 1938.

[64] L.von Bertalanffy. Quantitative laws in metabolism and growth. *Q. Rev. Biol.*, 32:217–231, 1957.

[65] L.von Bertalanffy. *General systems theory.* George Braziller, New York, 1968.

[66] L.von Bertalanffy. Basic concepts in quantitative biology of metabolism. *Helgol. Wiss. Meeresunters.*, 9:5–34, 1969.

[67] P. Berthold. Über den Einfluss der Nestlingsnahrung auf die Jugendentwickling, insbesondere auf das Flugelwachstum, bei der Mönchsgrasmücke (*Sylvia atricapilla*). *Vogelwarte*, 28:257–263, 1976.

[68] J. Best. The influence on intracellular enzymatic properties for kinetics data obtained from living cells. *J. Cell. Comp. Physiol.*, 46:1–27, 1955.

[69] R.J.H. Beverton and S.J. Holt. On the dynamics of exploited fish populations. *Fish. Invest. Ser. I*, 2, 1957.

[70] R. Bijlsma. *De boomvalk.* Kosmos vogelmonografieën. Kosmos, Amsterdam, 1980.

[71] D.G. Blackburn. Convergent evolution of viviparity, matrotrophy and specializations for fetal nutrition in reptiles and other vertebrates. *Am. Zool.*, 32:313–321, 1992.

[72] F.F. Blackman. Optima and limiting factors. *Ann. of Bot.*, 19:281–295, 1905.

[73] K. Blaxter. *Energy metabolism in animals and man.* Cambridge University Press, 1989.

[74] H. Blumel. *Die Rohrammer.* Die Neue Brehm-Bücherei. A. Ziemsen-Verlag, Wittenberg Lutherstadt, 1982.

[75] P.D. Boersma, N.T. Wheelwright, M.K. Nerini, and E.S. Wheelwright. The breeding biology of the fork-tailed storm-petrel (*Oceanodroma furcata*). *Auk*, 97:268–282, 1980.

[76] S. Bohlken and J. Joosse. The effect of photoperiod on female reproductive activity and growth of the fresh water pulmonate snail *Lymnaea stagnalis* kept under laboratory conditions. *Int. J. Invertebr. Reprod.*, 4:213–222, 1982.

[77] C. Bohr and K.A. Hasselbalch. Über die Kohlensäureproduction des Hühnerembryos. *Skand. Arch. Physiol.*, 10:149–173, 1900.

[78] J.T. Bonner. *Size and cycle: An essay on the structure of biology.* Princeton University Press, 1965.

[79] T.A. Bookhout. Prenatal development of snowshoe hares. *J. Wildl. Manage.*, 28:338–345, 1964.

[80] T. Borchardt. Influence of food quantity on the kinetics of cadmium uptake and loss via food and seawater in *Mytilus edulis*. *Mar. Biol.*, 76:67–76, 1983.

[81] T. Borchardt. Relationships between carbon and cadmium uptake in *Mytilus edulis*. *Mar. Biol.*, 85:233–244, 1985.

[82] F.van den Bosch. *The velocity of spatial population expansion.* PhD thesis, Rijksuniversiteit Leiden, 1990.

[83] F.van den Bosch, J.A.J. Metz, and O. Diekmann. The velocity of spatial expansion. *J. Math. Biol.*, 28:529–565, 1990.

[84] A.B. Bowers. Breeding and growth of whiting (*Gadus merlangus* L.) in Isle of Man Waters. *J. Mar. Biol. Ass. U. K.*, 33:97–122, 1954.

[85] A.E. Brafield and M.J. Llewellyn. *Animal energetics.* Blackie, Glasgow, 1982.

[86] A.M. Branta. Studies on the physiology, genetics and evolution of some *Cladocera*. Technical report, Carnegie Institute, Washington, 1939.

[87] P.R. Bregazzi and C.R. Kennedy. The biology of pike, *Esox lucius* L., in southern eutrophic lake. *J. Fish Biol.*, 17:91–112, 1980.

[88] W.W. Brehm. Breeding the green-billed toucan at the Walsrode Bird park. *Int. Zoo Yearb.*, 9:134–135, 1969.

[89] H. Bremer and P.P. Dennis. Modulation of chemical composition and other parameters of the cell by growth rate. In F.C. Neidhardt, editor, *Escherichia coli and Salmonella typhimurium.*, pages 1527–1542. Am. Soc. Microbiol., Washington, 1987.

[90] H. Brendelberger and W. Geller. Variability of filter structures in eight *Daphnia* species: Mesh sizes and filtering areas. *J. Plankton Res.*, 7:473–486, 1985.

[91] J.R. Brett. The relation of size to rate of oxygen consumption and sustained swimming speed of sockeye salmon (*Oncorhynchus nerka*). *J. Fish. Res. Board Can.*, 22:1491–1497, 1965.

[92] G.E. Briggs and J.B.S. Haldane. A note on the kinetics of enzyme action. *Biochem. J.*, 19:338–339, 1925.

[93] T. Brikhead. *Great auk islands; A field biologist in the Arctic.* T. & A.D.Poyser, London, 1993.

[94] F.H.van den Brink. *Zoogdierengids.* Petersons veldgidsenserie. Elsevier, 1955.

[95] I.L. Brisbin Jr and L.J. Tally. Age-specific changes in the major body components and caloric value of growing Japanese quail. *Auk*, 90:624–635, 1973.

[96] M.S. Brody and L.R. Lawlor. Adaptive variation in offspring size in the terrestrial isopod *Armadillium vulgare*. *Oecologia (Berlin)*, 61:55–59, 1984.

[97] S. Brody. *Bioenergetics and growth.* Rheinhold, New York, 1945.

[98] S. Broekhuizen. *Hazen in Nederland.* PhD thesis, Landbouw Hogeschool, Wageningen, 1982.

[99] M. Brooke. *The manx shearwater.* T. & A.D.Poyser, London, 1990.

[100] J.L. Brooks and S.I. Dodson. Predation, body size, and composition of plankton. *Science (New York)*, 150:28–35, 1965.

[101] D.J. Brousseau. Analysis of growth rate in *Mya arenaria* using the von Bertalanffy equation. *Mar. Biol.*, 51:221–227, 1979.

[102] G.H. Brown and J.J. Wolken. *Liquid crystals and biological structures.* Academic Press, 1979.

[103] W.Y. Brown. Growth and fledging age of sooty tern chicks. *Auk*, 93:179–183, 1976.

[104] W.A. Bruggeman, L.B.J.M. Martron, D. Kooiman, and O. Hutzinger. Accumulation and elimination kinetics of di-, tri-and tetra chlorobiphenyls by goldfish after dietary and aqueous exposure. *Chemosphere*, 10:811–832, 1981.

[105] B. Bruun. *Gids voor de vogels van Europa.* Elsevier, 1970.

[106] M.M. Bryden. Growth of the south elephant seal *Mirounga leonina* (Linn.). *Growth*, 33:69–82, 1969.

[107] T.L. Bucher. Parrot eggs, embryos, and nestlings: Patterns and energetics of growth and development. *Physiol. Zool.*, 56:465–483, 1983.

[108] L.J. Buckley. RNA–DNA ratio: an index of larval fish growth in the sea. *Mar. Biol.*, 80:291–298, 1984.

[109] J.J. Bull. Sex determination mechanisms: An evolutionary perspective. In S.C. Stearns, editor, *The evolution of sex and its consequences.*, pages 93–114. Birkhäuser, Basel, 1987.

[110] P.C. Bull and A.H. Whitaker. The amphibians, reptiles, birds and mammals. In G. Kuschel, editor, *Biogeography and ecology in New Zealand.*, pages 231–276. W. Junk b.v., Den Haag, 1975.

[111] B.A. Bulthuis. *Stoichiometry of growth and product-formation by Bacillus licheniformis.* PhD thesis, Vrije Universiteit, Amsterdam, 1990.

[112] C.W. Burns. The relationship between body size of filterfeeding cladocera and the maximum size of particles ingested. *Limnol. Oceanogr.*, 13:675–678, 1968.

[113] R. Burton. *Egg: Nature's miracle of packaging.* W. Collins, 1987.

[114] D.K. Button. Biochemical basis for whole-cell uptake kinetics: Specific affinity, oligotrophic capacity, and the meaning of Michaelis constant. *Appl. Environ. Microbiol.*, 57:2033–2038, 1991.

[115] P.L. Cadwallader and A.K. Eden. Food and growth of hatchery-produced chinook-salmon, *Oncorhynchus tschawytscha* (Walbaum), in landlocked lake Purrumbete, Victoria, Australia. *J. Fish Biol.*, 18:321–330, 1981.

[116] W.A. Calder III. The kiwi and egg design: Evolution as a package deal. *Bioscience*, 29:461–467, 1979.

[117] W.A. Calder III. *Size, function and life history.* Harvard University Press, 1984.

[118] H.B. Callen. *Thermodynamics.* J. Wiley & Sons, Inc., 1960.

[119] P. Calow and C.R. Townsend. Resource utilization in growth. In C.R. Townsend and P. Calow, editors, *Physiological ecology.*, pages 220–244. Blackwell Sci. Publ., 1981.

[120] N. Cameron. *The measurement of human growth.* Croom Helm, London, 1984.

[121] R.N. Campbell. The growth of brown trout, *Salmo trutta* L. in northern Scottish lochs with special reference to the improvement of fisheries. *J. Fish Biol.*, 3:1–28, 1971.

[122] M. Cashel and K.E. Rudd. The stringent response. In F.C. Neidhardt, editor, *Escherichia coli and Salmonella typhimurium.*, volume 2, pages 1410–1438. Am. Soc. Microbiol., 1987.

[123] L. Cazzador. Analysis of oscillations in yeast continuous cultures by a new simplified model. *Bull. Math. Biol.*, 53:685–700, 1991.

[124] T.R. Cech, A.J. Zaug, and P.J. Grabowski. In vitro splicing of the ribosomal RNA precursor of *Tetrahymena*: involvement of a guanosine nucleotide in the excision of the intervening sequence. *Cell*, 27:487–496, 1981.

[125] B. Charlesworth. *Evolution in age-structured populations.* Cambridge University Press, 1980.

[126] E.L. Charnov and D. Berrigan. Dimensionless numbers and the assembly rules for life histories. *Philos. Trans. R. Soc. Lond. B Biol. Sci.*, 332:41–48, 1991.

[127] K. Christoffersen and A. Jespersen. Gut evacuation rates and ingestion rates of *Eudiaptomus graciloides* measured by means of gut fluorescence method. *J. Plankton Res.*, 8:973–983, 1986.

[128] A. Clarke. Temperature and embryonic development in polar marine invertebrates. *Int. J. Invertebr. Reprod.*, 5:71–82, 1982.

[129] G.A. Clayton, C. Nixey, and G. Monaghan. Meat yield in turkeys. *Br. Poult. Sci.*, 19:755–763, 1978.

[130] J.H.L. Cloethe. Prenatal growth in the merino sheep. *Onderstepoort J. Vet. Sci. Anim. Ind.*, 13:417–558, 1939.

[131] J.F. Collins and M.H. Richmond. Rate of growth of *Bacillus cereus* between divisions. *J. Gen. Microbiol.*, 28:15–33, 1962.

[132] A. Comfort. *The biology of senescence.* Churchill Livingstone, Edinburgh, 1979.

[133] J.D. Congdon, D.W. Tinkle, and P.C. Rosen. Egg components and utilization during development in aquatic turtles. *Copeia*, 1983:264–268, 1983.

[134] R.J. Conover. Transformation of organic matter. In O. Kinne, editor, *Marine Ecology.* J. Wiley & Sons, Inc., 1978.

[135] G.D. Constanz. Growth of nestling rufous hummingbirds. *Auk*, 97:622–624, 1980.

[136] C.L. Cooney, D.I.C. Wang, and R.I. Mateles. Measurement of heat evolution and correlation with oxygen consumption during microbial growth. *Biotechnol. Bioeng.*, 11:269–281, 1968.

[137] S. Cooper. The constrained hoop: An explanation of the overschoot in cell length during a shift-up of *Escherichia coli.* *J. Bacteriol.*, 171:5239–5243, 1989.

[138] S. Cooper. *Bacterial growth and division.* Academic Press, 1991.

[139] J.D. Coronios. Development of behaviour in fetal cat. *Genet. Psychol. Monogr.*, 14:283–330, 1933.

[140] R.R. Coutts and I.W. Rowlands. The reproductive cycle of the skomer vole (*Clethrionomys glareolus skomerensis*). *J. Zool. Lond.*, 158:1–25, 1969.

[141] D.R. Cox and D. Oakes. *Analysis of survival data.* Monographs on statistics and applied probability. Chapman & Hall, 1984.

[142] D. Cragg-Hine and J.W. Jones. The growth of dace *Leuciscus leuciscus* (L.), roach *Rutilus rutilus* (L.) and chub *Sqalius cephalus* (L.) in Willow Brook, Northhamptonshire. *J. Fish Biol.*, 1:59–82, 1969.

[143] J.F. Craig. A note on growth and mortality of trout, *Salmo trutta* L., in afferent streams of Windermere. *J. Fish Biol.*, 20:423–429, 1982.

[144] G. Creutz. *Der Weiss-storch.* Die Neue Brehm-Bücherei. A. Ziemsen-Verlag, Wittenberg Lutherstadt, 1985.

[145] E.G. Crichton. Aspects of reproduction in the genus *Notomys* (*Muridae*). *Aust. J. Zool.*, 22:439–447, 1974.

[146] J.R. Cronmiller and C.F. Thompson. Experimental manipulation of brood size in red-winged blackbirds. *Auk*, 97:559–565, 1980.

[147] K.W. Cummins and J.C. Wuycheck. Caloric equivalents for investigations in ecological energetics. *Mitt. Int. Ver. Theor. Angew. Limnol.*, 18:1–95, 1971.

[148] E.H. Curtis, J.J. Beauchamp, and B.G. Blaylock. Application of various mathematical models to data from the uptake of methyl mercury in bluegill sunfish (*Lepomis macrochirus*). *Ecol. Modell.*, 3:273–284, 1977.

[149] H. Cyr and M.L. Pace. Allometric theory: Extrapolations from individuals to communities. *Ecology*, 74:1234–1245, 1993.

[150] W.B. Dade, P.A. Jumars, and D.L. Penry. Supply-side optimization: maximizing absorptive rates. In R.N. Hughes, editor, *Behavioural mechanisms of food selection.*, volume 20 of *NATO ASI*, pages 531–555. Springer-Verlag, 1990.

[151] M.J. Dagg. Complete carbon and nitrogen budgets for the carnivorous amphipod, *Calliopius laeviusculus* (Kroyer). *Int. Rev. Gesamten Hydrobiol.*, 61:297–357, 1976.

[152] J.D. Damuth. Of size and abundance. *Nature (Lond.)*, 351:268–269, 1991.

[153] J.D. Damuth. Taxon-free characterization of animal comminuties. In A.K. Behrensmeyer, J.D. Damuth, W.A. DiMichele, R. Potts, H.-D. Sues, and S.L. Wing, editors, *Terrestrial ecosystems through time; Evolutionary paleoecology of terrestrial plants amd animals.*, pages 183–203. University of Chicago Press, 1992.

[154] N. Dan and S.E.R. Bailey. Growth, mortality, and feeding rates of the snail *Helix aspersa* at different population densities in the laboratory, and the depression of activity of helicid snails by other individuals, or their mucus. *J. Molluscan Stud.*, 48:257–265, 1982.

[155] M.C. Dash and A.K. Hota. Density effects on the survival, growth rate, and metamorphosis of *Rana tigrina* tadpoles. *Ecology*, 61:1025–1028, 1980.

[156] H.A. David and M.L. Moeschberger. *The theory of competing risks.*, volume 39 of *Griffin's Statistical Monographs & Courses.* C. Griffin & Co Ltd, 1978.

[157] P.J. Davoll and M.W. Silver. Marine snow aggregates: life history sequence and microbial community of abandoned larvacean houses from Monterey Bay, California. *Mar. Ecol. Prog. Ser.*, 33:111–120, 1986.

[158] E.A. Dawes. The constrained hoop: An explanation of the overschoot in cell length during a shift-up of *Escherichia coli.* In T.R. Gray and J.R. Postgate, editors, *The survival of vegetative microbes.*, pages 19–53. Cambridge University Press, 1976.

[159] D.L. DeAngelis and L.J. Gross, editors. *Individual-based models and approaches in ecology.* Chapman & Hall, 1992.

[160] D.C. Deeming and M.W.J. Ferguson. Effects of incubation temperature on growth and development of embryos of *Alligator mississippiensis. J. Comp. Physiol.*, B159:183–193, 1989.

[161] T.E. DeLaca, D.M. Karl, and J.H. Lipps. Direct use of dissolved organic carbon by agglutinated benthic foraminifera. *Nature (Lond.)*, 289:287–289, 1981.

[162] M.W. Demment and P.J.van Soest. A nutritional explanation for body size patterns of ruminant and nonruminant herbivores. *Am. Nat.*, 125:641–672, 1985.

[163] W.R. Demott. Feeding selectivities and relative ingestion rates of *Daphnia* and *Bosmina. Limnol. Oceanogr.*, 27:518–527, 1982.

[164] V.E. Dent, M.J. Bazin, and P.T. Saunders. Behaviour of *Dictyostelium discoideum* amoebae and *Escherichia coli* grown together in chemostat culture. *Arch. Microbiol.*, 109:187–194, 1976.

[165] A.J. Desmond. *The hot-blooded dinosaurs.* Futura Publications Ltd., London, 1975.

[166] A.W. Diamond. The red-footed booby on Aldabra atoll, Indian Ocean. *Ardea*, 62:196–218, 1974.

[167] J.M. Diamond and R.K. Buddington. Intestinal nutrient absorption in herbivores and carnivores. In P. Dejours, L. Bolis, C.R. Taylor, and E.R. Weibel, editors, *Comparative physiology: Life in water and on land.* IX-Liviana Press, 1987.

[168] O. Diekmann, H.J.A.M. Heijmans, and H.R. Thieme. On the stability of the cell size distribution. *J. Math. Biol.*, 19:227–248, 1984.

[169] O. Diekmann, H.J.A.M. Heijmans, and H.R. Thieme. On the stability of the cell size distribution II: Time periodic developmental rates. *Comp. & Math. with Appls.*, 1985.

[170] O. Diekmann, H.A. Lauwerier, T. Aldenberg, and J.A.J. Metz. Growth, fission and the stable size distribution. *J. Math. Biol.*, 18:135–148, 1983.

[171] C. Dijkstra. *Reproductive tactics in the kestrel Falco tinnunculus; a study in evolutionary biology.* PhD thesis, University of Groningen, The Netherlands, 1988.

[172] F.A. Dipper, C.R. Bridges, and A. Menz. Age, growth and feeding in the ballan wrasse *Labrus bergylta* Ascanius 1767. *J. Fish Biol.*, 11:105–120, 1977.

[173] H. Dittberner and W. Dittberner. *Die Schafstelze.* Die Neue Brehm-Bücherei. A. Ziemsen-Verlag, Wittenberg Lutherstadt, 1984.

[174] D. Dolmen. Growth and size of *Triturus vulgaris* and *T. cristatus* (*Amphibia*) in different parts of Norway. *Holarct. Ecol.*, 6:356–371, 1983.

[175] W.D. Donachie. Relationship between cell size and time of initiation of DNA replication. *Nature (Lond.)*, 219:1077–1079, 1968.

[176] W.D. Donachie, K.G. Begg, and M. Vicente. Cell length, cell growth and cell division. *Nature (Lond.)*, 264:328–333, 1976.

[177] F. Dörr. Energetic and statistical relations. In W. Hoppe, W. Lohmann, H. Markl, and H. Ziegler, editors, *Biophysics.*, chapter 8, pages 316–371. Spinger-Verlag, 1983.

[178] P.G. Doucet and N.M.van Straalen. Analysis of hunger from feeding rate observations. *Anim. Behav.*, 28:913–921, 1980.

[179] W.L. Downing and J.A. Downing. Molluscan shell growth and loss. *Nature (Lond.)*, 362:506, 1993.

[180] R.L. Draper. The prenatal growth of the guinea-pig. *Anat. Rec.*, 18:369–392, 1920.

[181] P.G. Drazin. *Nonlinear systems.* Cambridge University Press, 1992.

[182] R.H. Drenth. Functional aspects of incubation in the herring gull. *Behaviour, Suppl.*, 17:1–132, 1970.

[183] M.R. Droop. Some thoughts on nutrient limitation in algae. *JPLAJ*, 9:264–272, 1973.

[184] M.R. Droop. 25 years of algal growth kinetics. *Bot. Mar.*, 26:99–112, 1983.

[185] G.L. Dryden. Growth and development of *Suncus murinus* in captivity on Guam. *J. Mammal.*, 49:51–62, 1968.

[186] E. Duclaux. Vie aérobie et anaérobie. In *Traité de microbiologie.*, volume 1, pages 208–212. Masson et cie., Paris, 1898.

[187] W.E. Duellman. Reproductive strategies of frogs. *Sci. Am.*, 7:58–65, 1992.

[188] W.E. Duellman and L. Trueb. *Biology of amphibians.* McGraw-Hill, 1986.

[189] G.C.A. Duineveld and M.I. Jenness. Difference in growth rates of the sea urchin *Echiocardium cordatum* as estimated by the parameter ω of the von Bertalanffy equation to skeletal rings. *Mar. Ecol.*, 19:65–72, 1984.

[190] E.H. Dunn. Growth, body components and energy content of nestling double-crested cormorants. *Condor*, 77:431–438, 1975.

[191] C. de Duve. *Blueprint for a cell: The nature and origin of life.* Neil Patterson Publishers, Burlington, North Carolina, 1991.

[192] M. Dworkin. *Developmental biology of the bacteria.* Benjamin-Cummings Publ. Co., California, 1985.

[193] O.N. Eaton. Weight and length measurements of fetuses of karakul sheep and of goats. *Growth*, 16:175–187, 1952.

[194] B. Ebenman and L. Persson. *Size-structured populations. Ecology and evolution.* Springer-Verlag, 1988.

[195] L. Edelstein-Keshet. *Mathematical models in biology.* Random House, New York, 1988.

[196] D.A. Egloff and D.S. Palmer. Size relations of the filtering area of two *Daphnia* species. *Limnol. Oceanogr.*, 16:900–905, 1971.

[197] J.F. Eisenberg. *The mammalian radiations. An analysis of trends in evolutions, adaptation, and behaviour.* University of Chicago Press, 1981.

[198] R.C. Elandt-Johnson and N.L. Johnson. *Survival models and data analysis.* J. Wiley & Sons, Inc., 1980.

[199] R.H.E. Elliott, F.G. Hall, and A.St.G. Huggett. The blood volume and oxygen capacity of the foetal blood in the goat. *J. Physiol. (Lond.)*, 82:160–171, 1934.

[200] G.W. Elmes, J.A. Thomas, and J.C. Wardlaw. Larvae of *Maculinea rebeli*, a large-blue butterfly, and their *Myrmica* host ants: wild adoption and hehaviour in ant-nests. *J. Zool. Lond.*, 223:447–460, 1991.

[201] S. Emerson. The growth phase in *Neurospora* corresponding to the logarithmic phase in unicellular organisms. *J. Bacteriol.*, 60:221–223, 1950.

[202] J.M. Emlen. *Population biology: The coevolution of population dynamics and behaviour.* MacMillan Publ. Co., N.Y., 1984.

[203] L.H. Emmons and F. Feer. *Neotropical rainforest mammals.* The University of Chicago Press, 1990.

[204] H. Engler. *Die Teichralle.* Die Neue Brehm-Bücherei. A. Ziemsen-Verlag, Wittenberg Lutherstadt, 1983.

[205] G. Ernsting and J.A. Isaaks. Accelerated ageing: a cost of reproduction in the carabid beetle *Notiophilus biguttatus* F. *Funct. Ecol.*, 5:229–303, 1991.

[206] G. Ernsting and J.A. Isaaks. On egg size variation in *Notiophilus biguttatus* F. (*Coleoptera*; *Carabidae*). In *Carabid beetles: Ecology and evolution.* Proc. 8th European Carabidologists; Meeting 1992., Kluwer, 1993. in press.

[207] A.A. Esener, J.A. Roels, and N.W.F. Kossen. Theory and applications of unstructured growth models: Kinetic and energetic aspects. *Biotechnol. Bioeng.*, 25:2803–2841, 1983.

[208] A.A. Esener, T. Veerman, J.A. Roels, and N.W.F. Kossen. Modeling of bacterial growth: Formulation and evaluation of a structured model. *Biotechnol. Bioeng.*, 24:1749–1764, 1982.

[209] E.G. Evers and S.A.L.M. Kooijman. Feeding and oxygen consumption in *Daphnia magna*; A study in energy budgets. *Neth. J. Zool.*, 39:56–78, 1989.

[210] N. Fairall. Prenatal development of the impala *Aepyceros melampus*. *Koedoe*, 12:97–103, 1969.

[211] J.O. Farlow. A consideration of the trophic dynamics of a late Cretaceous large-dinosaur community (Oldman Formation). *Ecology*, 57:841–857, 1976.

[212] H.S. Fawcett. The temperature relations of growth in certain parasitic fungi. *Univ. Calif. Berkeley Publ. Agr. Sci.*, 4:183–232, 1921.

[213] M.A. Fedak and H.J. Seeherman. Reappraisal of energetics of locomotion shows identical cost in bipeds and quadrupeds including ostrich and horse. *Nature (Lond.)*, 282:713–716, 1979.

[214] R. Fenaux. Rhythm of secretion of oikopleurid's houses. *Bull. Mar. Sci.*, 37:498–503, 1985.

[215] R. Fenaux and G. Gorsky. Cycle vital et croissance de l'appendiculaire *Oikopleura longicauda*. *Ann. Inst. Oceanogr.*, 59:107–116, 1983.

[216] T. Fenchel. Intrinsic rate of natural increase: The relationship with body size. *Oecologia (Berlin)*, 14:317–326, 1974.

[217] M.W.J. Ferguson. Palatal shelf elevation in the Wistar rat fetus. *J. Anat.*, 125:555–577, 1978.

[218] S. Fernandez-Baca, W. Hansel, and C. Novoa. Embryonic mortality in the alpaca. *Biol. Reprod.*, 3:243–251, 1970.

[219] C.E. Finch. *Longevity, senescence, and the genome.* University of Chicago Press, 1990.

[220] R.A. Fisher and L.H.C. Tippitt. Limiting forms of the frequency distribution of the largest or the smallest memeber of a sample. *Proc. Cambridge Phil. Soc.*, 24:180–190, 1928.

[221] V. Fitsch. *Laborvcrsuchc und Simula tionen zur kausalen Analyse der Populationsdynamik von Daphnia magna.* PhD thesis, Rheinisch-Westfälischen Technischen Hochschule Aachen, 1990.

[222] L.B. Flexner and H.A. Pohl. The transfer of radioactive sodium across the placenta of the white rat. *J. Cell. Comp. Physiol.*, 18:49–59, 1941.

[223] P.R. Flood. Architecture of, and water circulation and flow rate in, the house of the planktonic tunicate *Oikopleura labradoriensis*. *Mar. Biol.*, 111:95–111, 1991.

[224] H.von Foerster. Some remarks on changing populations. In F. Stohlman, editor, *The kinetics of cellular proliferation.* Grune and Stratton, 1959.

[225] J. Forsberg and C. Wiklund. Protandry in the green-veined white butterfly, *Pieris napi* L. (*Lepidoptera*; *Pieridae*). *Funct. Ecol.*, 2:81–88, 1988.

[226] B.L. Foster Smith. The effect of concentration of suspension and inert material on the assimilation of algae by three bivalves. *J. Mar. Biol. Ass. U. K.*, 55:411–418, 1975.

[227] B.L. Foster Smith. The effect of concentration of suspension on the filtration rates and pseudofaecal production for *Mytilus edulis* L., *Cerastoderma edule* L. and *Venerupsis pullastra* (Montagu). *J. Exp. Mar. Biol. Ecol.*, 17:1–22, 1975.

[228] P. da Franca. Determinacao da idade em *Gambusia holbrookii* (Girard). *Arq. Mus. Bocage*, 24:87–93, 1953.

[229] D.R. Franz. Population age structure, growth and longevity of the marine gastropod *Uropalpinx cinerea* Say. *Biol. Bull. (Woods Hole)*, 140:63–72, 1971.

[230] J.C. Frauenthal. Analysis of age-structured models. In T.G. Hallam and S.A. Levin, editors, *Mathematical ecology*, pages 117–147. Springer-Verlag, 1986.

[231] J.J.R. Fraústo da Silva and R.J.P. Williams. *The biological chemistry of the elements. The inorganic chemistry of life.* Clarendon Press, Oxford, 1993.

[232] A.G. Fredrickson, R.D. Megee III, and H.M. Tsuchiy. Mathematical models for fermentation processes. *Adv. Appl. Microbiol.*, 13:419–465, 1970.

[233] B. Friedrich. Genetics of energy converting systems in aerobic chemolithotrophs. In

H.G. Schlegel and E. Bowien, editors, *Autotrophic bacteria.*, pages 415–436. Springer-Verlag, 1989.

[234] E. Frommhold. *Die Kreuzotter.* Die Neue Brehm-Bücherei. A. Ziemsen-Verlag, Wittenberg Lutherstadt, 1969.

[235] B.W. Frost. Effects of size and concentration of food particles on the feeding behaviour of the marine copepod *Calanus pacificus. Limnol. Oceanogr.*, 17:805–815, 1972.

[236] W.E. Frost and C. Kipling. The growth of charr, *Salvinus willughbii* Gunther, in Windermere. *J. Fish Biol.*, 16:279–289, 1980.

[237] F.E.J. Fry. The effect of environmental factors on the physiology of fish. In W.S. Hoar and D.J. Randall, editors, *Fish physiology.*, volume 6, pages 1–87. Academic Press, 1971.

[238] G. Fryer. Evolution and adaptive radiation in *Macrothricidae* (*Crustacea*: *Cladocera*): A study in comparative functional morphology and ecology. *Proc. R. Soc. Lond. B Biol. Sci.*, 269:142–385, 1974.

[239] E. Fuller, editor. *Kiwis.* Swan-Hill Press, Shrewsbury, England, 1991.

[240] T.W. Fulton. *The Sovereignty of the seas.* Edinburgh & London., 1911.

[241] R.W. Furness. *The skuas.* T. & A.D.Poyser, Calton, 1987.

[242] V.F. Galliucci and T.J. Quinn. Reparameterizing, fitting, and testing a simple growth model. *Trans. Am. Fish. Soc.*, 108:14–25, 1979.

[243] A.J. Gaston. *The ancient murrelet.* T. & A.D. Poyser, London, 1992.

[244] R.E. Gatten, K. Miller, and R.J. Full. Energetics at rest and during locomotion. In M.E. Feder and W.W. Burggren, editors, *Environmental physiology of the amphibians.*, chapter 12, pages 314–377. The University of Chicago Press, 1992.

[245] M. Gatto, C Ricci, and M. Loga. Assessing the response of demographic parameters to density in a rotifer population. *Ecol. Modell.*, 62:209–232, 1992.

[246] V. Geist. Bergmann's rule is invalid. *Can. J. Zool.*, 65:1035–1038, 1987.

[247] W. Geller. Die Nahrungsaufname von *Daphnia pulex* in Aghängigkeit von der Futterkonzentration, der Temperatur, der Köpergrösse und dem Hungerzustand der Tiere. *Arch. Hydrobiol./ Suppl.*, 48:47–107, 1975.

[248] W. Geller and H. Müller. Seasonal variability in the relationship between body length and individual dry weight as related to food abundance and clutch size in two coexisting *Daphnia. J. Plankton Res.*, 7:1–18, 1985.

[249] R.B. Gennis. *Biomembranes. Molecular structure and function.* Springer-Verlag, 1989.

[250] M. Genoud. Energetic strategies of shrews: ecological constraints and evolutionary implications. *Mammal Review*, 4:173–193, 1988.

[251] R.D. Gettinger, G.L. Paukstis, and W.H.N. Gutzke. Influence of hydric environment on oxygen consumption by embryonic turtles *Chelydra serpentina* and *Trionyx spiniferus. Physiol. Zool.*, 57:468–473, 1984.

[252] W.M. Getz and R.G. Haight. *Population Harvesting.*, volume 27 of *Monographs in population biology.* Princeton University Press, 1989.

[253] W. Gewalt. *Der Weisswal (Delphinapterus leucas).* Die Neue Brehm-Bücherei. A. Ziemsen-Verlag, Wittenberg Lutherstadt, 1976.

[254] R.N. Gibson. Observations on the biology of the giant goby *Gobius cobitis* Pallas. *J. Fish Biol.*, 2:281–288, 1970.

[255] R.N. Gibson and I.A. Ezzi. The biology of a Scottish population of Fries' goby, *Lesueurigobius friesii. J. Fish Biol.*, 12:371–389, 1978.

[256] R.N. Gibson and I.A. Ezzi. The biology of the scaldfish, *Arnoglossus laterna* (Walbaum) on the west coast of Scotland. *J. Fish Biol.*, 17:565–575, 1980.

[257] R.N. Gibson and I.A. Ezzi. The biology of the Norway goby, *Pomatoschistus norvegicus* (Collett), on the west coast of Scotland. *J. Fish Biol.*, 19:697–714, 1981.

[258] A. Gierer. *Hydra* as a model for the development of biological form. *Sci. Am.*, (12):44–54, 1974.

[259] M.A. Gilbert. Growth rate, longevity and maximum size of *Macoma baltica* (L.). *Biol. Bull. (Woods Hole)*, 145:119–126, 1973.

[260] S. Glasstone, K.J. Laidler, and H. Eyring. *The theory of rate processes.* McGraw-Hill, 1941.

[261] Z.M. Gliwicz. Food thresholds and body size in cladocerans. *Nature (Lond.)*, 343:638–640, 1990.

[262] K. Godfrey. *Compartment models and their application.* Academic Press, 1983.

[263] L. Godfrey, M. Sutherland, D. Boy, and N. Gomberg. Scaling of limb joint surface areas in anthropoid primates and other mammals. *J. Zool. Lond.*, 223:603–625, 1991.

[264] J.C. Goldman and E.J. Cerpenter. A kinetic approach of temperature on algal growth. *Limnol. Oceanogr.*, 19:756–766, 1974.

[265] C.R. Goldspink. A note on the growth-rate and year-class strength of bream, *Abramus brama* (L.), in three eutrophic lakes, England. *J. Fish Biol.*, 19:665–673, 1981.

[266] M. Gophen and W. Geller. Filter mesh size and food particle uptake by *Daphnia*. *Oecologia (Berlin)*, 64:408–412, 1984.

[267] S.J. Gould. *Ontogeny and phylogeny.* Belknap Press, Cambridge, 1977.

[268] C.E. Goulden, L.L. Henry, and A.J. Tessier. Body size, energy reserves, and competitive ability in three species of cladocera. *Ecology*, 63:1780–1789, 1982.

[269] C.E. Goulden and L.L. Hornig. Population oscillations and energy reserves in planktonic cladocera and their consequences to competition. *Proc. Nat. Acad. Sci. U. S. A.*, 77:1716–1720, 1980.

[270] B.W. Grant and W.P. Porter. Modelling global macroclimatic constraints on ectotherm energy budgets. *Am. Zool.*, 32:154–178, 1992.

[271] C.J. Grant and A.V. Spain. Reproduction, growth and size allometry of *Liza vaigiensis* (Quoy & Gaimard) (*Pisces*: *Mugilidae*) from North Queensland inshore waters. *Aust. J. Zool.*, 23:475–485, 1975.

[272] C.J. Grant and A.V. Spain. Reproduction, growth and size allometry of *Valamugil seheli* (Forskal) (*Pisces*: *Mugilidae*) from North Queensland inshore waters. *Aust. J. Zool.*, 23:463–474, 1975.

[273] J. Gray. The growth of fish I. the relationship between embryo and yolk in *Salmo fario*. *J. Exp. Biol.*, 4:215–225, 1926.

[274] W. Greve. Ökologische Untersuchungen an *Pleurobrachia pileus* 1. Freilanduntersuchungen. *Helgol. Wiss. Meeresunters.*, 22:303–325, 1971.

[275] W. Greve. Ökologische Untersuchungen an *Pleurobrachia pileus* 2. Laboratoriumuntersuchungen. *Helgol. Wiss. Meeresunters.*, 23:141–164, 1972.

[276] R.S.de Groot. Origin, status and ecology of the owls in Galapagos. *Ardea*, 71:167–182, 1983.

[277] R. Grundel. Determinants of nestling feeding rates and parental investment in the mountain chickadee. *Condor*, 89:319–328, 1987.

[278] J. Guckenheimer and P. Holmes. *Nonlinear oscillations, dynamical systems, and bifurcations of vector fields.*, volume 42 of *Applied Mathematical Sciences.* Springer-Verlag, 1983.

[279] C.A.W. Guggisberg. *The crocodiles.* David and Charles, Newton Abbot, 1980.

[280] R. Günther. *Die Wasserfrösche Europas.* Die Neue Brehm-Bücherei. A. Ziemsen-Verlag, Wittenberg Lutherstadt, 1990.

[281] R. Haase. *Thermodynamics of irreversible processes.* Dover Publications, 1990.

[282] U. Halbach. Einfluss der Temperatur auf die Populationsdynamik der planktischer Radertieres *Brachionus calyciflorus* Pallas. *Oecologia (Berlin)*, 4:176–207, 1970.

[283] T.G. Hallam, R.R. Lassiter, and S.A.L.M. Kooijman. Effects of toxicants on aquatic populations. In S.A. Levin, T.G. Hallam, and L.F. Gross, editors, *Mathematical Ecology*, pages 352–382. Springer-Verlag, 1989.

[284] T.G. Hallam and J.L.de Luna. Effects of toxicants on population: A qualitative approach III. environmental and food chain pathways. *J. Theor. Biol.*, 109:411–429, 1984.

[285] P.P.F. Hanegraaf. *Coupling of mass and energy yields in micro-organisms. Application of the Dynamic Energy Budget model and an experimental approach.* PhD thesis, Vrije Universiteit, Amsterdam, 1993. in prep.

[286] R.J.F.van Haren and S.A.L.M. Kooijman. Application of a dynamic energy budget model to *Mytilus edulis*. *Neth. J. Sea Res.*, 32, 1993. to appear.

[287] D. Harman. Role of free radicals in mutation, cancer, aging and maintenance of life. *Radiat. Res.*, 16:752–763, 1962.

[288] D. Harman. The aging process. *Proc. Nat. Acad. Sci. U. S. A.*, 78:7124–7128, 1981.

[289] F.M. Harold. *The vital force. A study of bioenergetics.* Freeman, 1986.

[290] M.P. Harris. The biology of an endangered species, the dark-rumped petrel (*Pterodroma phaeopygia*), in the Galapagos Islands. *Condor*, 72:76–84, 1970.

[291] C. Harrison. *A field guide to the nests, eggs and nestlings of European birds.* Collins, London, 1975.

[292] R. Harrison and G.G. Lunt. *Biological membranes. Their structure and function.* Blackie, Glasgow, 1980.

[293] R.J. Harvey, A.G. Marr, and P.R. Painter. Kinetics of growth of individual cells of *Escherichia coli* and *Azotobacter agilis*. *J. Bacteriol.*, 93:605–617, 1967.

[294] K.A. Hasselbalch. Über den respiratorischen Stoffechsel des Hühnerembryos. *Skand. Arch. Physiol.*, 10:353–402, 1900.

[295] A. Hastings and T. Powell. Chaos in a three-species food chain. *Ecology*, 72:896–903, 1991.

[296] F. Haverschmidt. Notes on the life history of *Amazilia fimbriata* in Surinam. *Wilson Bull.*, 61:69–79, 1952.

[297] A.J.S. Hawkins and B.L. Bayne. Seasonal variations in the balance between physiological mechanisms of feeding and digestion in *Mytilus edulis* (*Bivalvia, Mollusca*). *Mar. Biol.*, 82:233–240, 1984.

[298] P.D.N. Hebert. The genetics of *Cladocera*. In W.C. Kerfoot, editor, *Evolution and ecology of zooplankton communities.*, pages 329–336. University Press of New England, 1980.

[299] P.D.N. Hebert. Genotypic characteristic of cyclic parthenogens and their obligately asexual derivates. In S.C. Stearns, editor, *The evolution of sex and its consequences.*, pages 175–195. Birkhäuser, Basel, 1987.

[300] H. Heesterbeek. R_0. PhD thesis, Rijksuniversiteit Leiden, 1992.

[301] H.J.A.M. Heijmans. Holling's 'hungry mantid' model for the invertebrate functional response considered as a Markov process. Part III: Stable satiation distribution. *J. Math. Biol.*, 21:115–1431, 1984.

[302] H.J.A.M. Heijmans. On the stable size distribution of populations reproducing by fission into two unequal parts. *Math. Biosci.*, 72:19–50, 1984.

[303] H.J.A.M. Heijmans. *Dynamics of structured populations.* PhD thesis, University of Amsterdam, 1985.

[304] J.J. Heijnen. A new thermodynamically based correlation of chemotrophic biomass yields. *Antonie van Leeuwenhoek*, 60:235–256, 1991.

[305] J.J. Heijnen and J.P.van Dijken. In search of a thermodynamic description of biomass yields for the chemotrophic growth of micro organisms. *Biotechnol. Bioeng.*, 39:833–858, 1992.

[306] J.J. Heijnen and J.A. Roels. A macroscopic model describing yield and maintenance relationships in aerobic fermentation. *Biotechnol. Bioeng.*, 23:739–761, 1981.

[307] J.M. Hellawell. Age determination and growth of the grayling *Thymallus thymallus* (L.) of the River Lugg, Herefordshire. *J. Fish Biol.*, 1:373–382, 1969.

[308] M.A. Hemminga. *Regulation of glycogen metabolism in the freshwater snail Lymnaea stagnalis.* PhD thesis, Vrije Universiteit, Amsterdam, 1984.

[309] W.G. Heptner and A.A. Nasimowitsch. *Der Elch (Alces alces).* Die Neue Brehm-Bücherei. A. Ziemsen-Verlag, Wittenberg Lutherstadt, 1974.

[310] D. Herbert. Some principles of continuous culture. In G. Tunevall, editor, *Recent progress in microbiology, Symposium of the 7th International Congress for Microbiology.*, volume 7, pages 381–396, Stockholm, 1958. Almqvist and Wiksell.

[311] C.F. Herreid II and B. Kessel. Thermal conductance in birds and mammals. *Comp. Biochem. Physiol.*, 21:405–414, 1967.

[312] H.C. Hess. The evolution of parental care in brooding spirorbid polychaetes: the effect of scaling constraints. *Am. Nat.*, 141:577–596, 1993.

[313] D.J.S. Hetzel. The growth and carcass characteristics of crosses between alabio and tegal ducks and moscovy and pekin drakes. *Br. Poult. Sci.*, 24:555–563, 1983.

[314] H.R. Hewer and K.M. Backhouse. Embryology and foetal growth of the grey seal, *Halichoerus grypus*. *J. Zool. Lond.*, 155:507–533, 1968.

[315] C.J. Hewett. Growth and moulting in the common lobster (*Homarus vulgaris* Milne-Edwards). *J. Mar. Biol. Ass. U. K.*, 54:379–391, 1974.

[316] L.C. Hewson. Age, maturity, spawning and food of burbot, *Lota lota*, in Lake Winnipeg. *J. Fish. Res. Board Can.*, 12:930–940, 1955.

[317] R.W. Hickman. Allometry and growth of the green-lipped mussel *Perna canaliculus* in New Zealand. *Mar. Biol.*, 51:311–327, 1979.

[318] R. Hile. Age and growth of the cisco *Leuchichthys artedi* (Le Sueur) in lake of the north-eastern High Land Eisconsian. *Bull. Bur. Fish., Wash.*, 48:211–317, 1936.

[319] R. Hill and C.P. Whittingham. *Photosynthesis*. Methuen & Co., London, 1955.

[320] T.L. Hill. *An introduction to statistical thermodynamics*. Dover, New York, 1986.

[321] M.W. Hirsch and S. Smale. *Differential equations, dynamical systems, and linear algebra*. Academic Press, 1974.

[322] P.W. Hochachka. *Living without oxygen*. Harvard University Press, 1980.

[323] P.W. Hochachka and M. Guppy. *Metabolic arrest and the control of biological time*. Harvard University Press, 1987.

[324] P.W. Hochachka and G.N. Somero. *Biochemical adaptation*. Princeton University Press, 1984.

[325] P.A.R. Hockey. Growth and energetics of the African black oystercatcher *Haematopus moquini*. *Ardea*, 72:111–117, 1984.

[326] J.K.M. Hodasi. The effects of different light regimes on the behaviour biology of *Achatina (Achatina) achatina* (Linne). *J. Molluscan Stud.*, 48:283–293, 1982.

[327] N.van der Hoeven. The population dynamics of daphnia at constant food supply: A review, re-evaluation and analysis of experimental series from the literature. *Neth. J. Zool.*, 39:126–155, 1989.

[328] N.van der Hoeven, S.A.L.M. Kooijman, and W.K.de Raat. *Salmonella*-test: relation between mutagenicity and number of revertant colonies. *Mutat. Res.*, 234:289–302, 1990.

[329] M. Hoffman. Yeast biology enters a surprising new phase. *Science (New York)*, 255:1510–1511, 1992.

[330] M.A. Hofman. Energy metabolism, brain size and longevity in mammals. *Q. Rev. Biol.*, 58:495–512, 1983.

[331] M.J. Holden. The growth rates of *Raja brachyura*, *R. clavata* and *R. montaqui* as determined from tagging data. *J. Physiol. (Lond.)*, 34:161–168, 1972.

[332] C.S. Holling. Some characteristics of simple types of predation and parasitism. *Can. Entomol.*, 91:385–398, 1959.

[333] S.C. Holt. Anatomy and chemistry of spirochetes. *Microbiol. Rev.*, 42:114–160, 1978.

[334] J.R. Horner. The nesting behaviour of dinosaurs. *Sci. Am.*, (4):92–99, 1984.

[335] H.J. Horstmann. Sauerstoffverbrauch und Trockengewicht der Embryonen von *Lymnaea stagnalis* L. *Z. Vgl Physiol.*, 41:390–404, 1958.

[336] R.A. Horton, M. Rowan, K.E. Webster, and R.H. Peters. Browsing and grazing by cladoceran filter feeders. *Can. J. Zool.*, 57:206–212, 1979.

[337] M.A. Houck, J.A. Gauthier, and R.E. Strauss. Allometric scaling in the earliest fossil bird, *Archaeopteryx lithographica*. *Science (New York)*, 247:195–198, 1990.

[338] S.E.L. Houde and M.R. Roman. Effects of food quality on the functional ingestion response of the copepod *Acartia tonsa*. *Mar. Ecol. Prog. Ser.*, 48:69–77, 1987.

[339] J. Hovekamp. *On the growth of larval plaice in the North Sea*. PhD thesis, University of Groningen, 1991.

[340] R.A. How. Reproduction, growth and survival of young in the mountain possum, *Trichosurus caninus* (*Marsupialia*). *Aust. J. Zool.*, 24:189–199, 1976.

[341] A.St.G. Huggett and W.F. Widdas. The relationship between mammalian foetal weight and conception age. *J. Physiol. (Lond.)*, 114:306–317, 1951.

[342] R.E. Hungate. The rumen microbial ecosystem. *Ann. Rev. Ecol. Syst.*, 6:39–66, 1975.

[343] P.C. Hunt and J.W. Jones. Trout in Llyn Alaw, Anglesey, North Wales II. Growth. *J. Fish Biol.*, 4:409–424, 1972.

[344] R. von Hutterer. Beobachtungen zur Geburt und Jugendentwicklung der Zwergspitzmaus, *Sorex minutus* L. (*Soricidae- Insectivora*). *Z. Saeugetierkd*, 41:1–22, 1976.

[345] J.S. Huxley. *Problems of relative growth.* Methuen & Co., London, 1932.

[346] H.L. Ibsen. Prenatal growth in guinea-pigs with special reference to environmental factors affecting weight at birth. *J. Exp. Zool.*, 51:51–91, 1928.

[347] T.D. Iles and P.O. Johnson. The correlation table analysis of a sprat (*Clupea sprattus* L.) year-class to separate two groups differing in growth characteristics. *J. Cons. Int. Explor. Mer*, 27:287–303, 1962.

[348] H. Ineichen, U. Riesen-Willi, and J. Fisher. Experimental contributions to the ecology of *Chironomus* (*Diptera*) II. Influence of the photoperiod on the development of *Chironomus plumosus* in the 4th larval instar. *Oecologia (Berlin)*, 39:161–183, 1979.

[349] L. Ingle, T.R. Wood, and A.M. Banta. A study of longevity, growth, reproduction and heart rate in *Daphnia longispina* as influenced by limitations in quantity of food. *J. Exp. Zool.*, 76:325–352, 1937.

[350] J.L. Ingraham. Growth of psychrophillic bacteria. *J. Bacteriol.*, 76:75–80, 1958.

[351] R. Ivell. The biology and ecology of a brackish lagoon bivalve, *Cerastoderma glaucum* Bruguiere, in Lago Lungo, Italy. *J. Molluscan Stud.*, 45:364–382, 1979.

[352] E.A. Jackson. *Perspectives of nonlinear dynamics.*, volume I, II. Cambridge University Press, 1991.

[353] J.A. Jacquez. *Compartment analysis in biology and medicine.* Elsevier Publ. Co., 1972.

[354] M.R. James. Distribution, biomass and production of the freshwater mussel, *Hyridella menziesi* (Gray), in Lake Taupo, New Zealand. *Freshwater Biol.*, 15:307–314, 1985.

[355] M.J.W. Jansen and J.A.J. Metz. How many victims will a pitfall make? *Acta Biotheor.*, 28:98–122, 1979.

[356] G. Janssen. *On the genetical ecology of springtails.* PhD thesis, Vrije Universiteit, Amsterdam, 1985.

[357] J.W. Jeong, J. Snay, and M.M. Ataai. A mathematical model for examining growth and sporulation processes of *Bacillus subtilus. Biotechnol. Bioeng.*, 35:160–184, 1990.

[358] M. Jobling. Mathematical models of gastric emptying and the estimation of daily rates of food consumption for fish. *J. Fish Biol.*, 19:245–257, 1981.

[359] A.W.B. Johnston, J.L. Firmin, and L. Rosen. On the analysis of symbiotic genes in *Rhizobium. Symp. Soc. Gen. Microbiol.*, 42:439–455, 1988.

[360] A. Jones. Studies on egg development and larval rearing of turbot, *Scophthalmus maximus* L. and brill, *Scophthalmus rhombus* L., in the laboratory. *J. Mar. Biol. Ass. U. K.*, 52:965–986, 1972.

[361] G.P. Jones. Contribution to the biology of the redbanded perch, *Ellerkeldia huntii* (Hector), with a discussion on hermaphroditism. *J. Fish Biol.*, 17:197–207, 1980.

[362] H.G. Jones. *Plant and microclimate. A quantitative approach to environmental plant physiology.* Cambridge University Press, 1992.

[363] J.W. Jones and H.B.N. Hynes. The age and growth of *Gasterosteus aculeatus*, *Pygosteus pungitius* and *Spinachia vulgaris*, as shown by their otoliths. *J. Anim. Ecol.*, 19:59–73, 1950.

[364] E.N.G. Joosse and E. Veltkamp. Some aspects of growth, moulting and reproduction in five species of surface dwelling *Collembola. Neth. J. Zool.*, 20:315–328, 1970.

[365] D.M. Joubert. A study of pre-natal growth and development in the sheep. *J. Agric. Sci.*, 47:382–428, 1956.

[366] J. Juget, V. Goubier, and D. Barthélémy. Intrisic and extrinsic variables controlling the productivity of asexual populations of *Nais spp*, (*Naididae, Oligochaeta*). *Hydrobiologia*, 180:177–184, 1989.

[367] H. Kaiser. Populationsdynamik und Eigenschaften einzelner Individuen. *Verh. Ges. Ökol., Erlangen*, pages 25–38, 1974.

[368] H. Kaiser. The dynamics of populations as result of the properties of individual animals. *Fortschr. Zool.*, 25:109–136, 1979.

[369] J.D. Kalbfleisch and R.L. Prentice. *The statistical analysis of failure time data.* J. Wiley & Sons, Inc., 1980.

[370] G. Kapocsy. *Weissbart- und Weissflugelseeschwalbe.* Die Neue Brehm-Bücherei. A. Ziemsen-Verlag, Wittenberg Lutherstadt, 1979.

[371] W.H. Karasov. Dailey energy expenditure and the cost of activity in mammals. *Am. Zool.*, 32:238–248, 1992.

[372] V.V. Kashkin. Heat exchange of bird eggs during incubation. *Biophys.*, 6:57–63, 1961.

[373] L. Kaufman. Innere und Äussere Wachstumsfactoren. Untersuchungen an Hühnern und Tauben. *Wilhelm Roux' Arch. Entwicklungsmech. Org.*, 6:395–431, 1930.

[374] N. Kautsky. Quantitative studies on gonad cycle, fecundity, reproductive output and recruitment in a baltic *Mytilus edulis* population. *Mar. Biol.*, 67:143–160, 1982.

[375] C. Kayser and A. Heusner. Étude comparative du métabolisme énergétiques dans la série animale. *J. Physiol. (Paris)*, 56:489–524, 1964.

[376] B.D. Keller. Coexistence of sea urchins in seagrass meadows: an experimental analysis of competition and predation. *Ecology*, 64:1581–1598, 1983.

[377] C.M. Kemper. Growth and development of the Australian musid *Pseudomys novaehollandiae. Aust. J. Zool.*, 24:27–37, 1976.

[378] S.C. Kendeigh. Factors affecting the length of incubation. *Auk*, 57:499–513, 1940.

[379] M. Kennedy and P. Fitzmaurice. Some aspects of the biology of gudgeon *Gobio gobio* (L.) in Irish waters. *J. Fish Biol.*, 4:425–440, 1972.

[380] A. Kepes. Etudes cinetiques sur la galactosidepermease d'*Escherichia coli. Biochem. Biophys. Acta*, 40:70–84, 1960.

[381] K. Kersting and W. Holterman. The feeding behaviour of *Daphnia magna*, studied with the coulter counter. *Mitt. Int. Ver. Theor. Angew. Limnol.*, 18:1434–1440, 1973.

[382] A. Keve. *Der Eichelhaher.* Die Neue Brehm-Bücherei. A. Ziemsen-Verlag, Wittenberg Lutherstadt, 1985.

[383] P. Kindlmann and A.F.G. Dixon. Developmental constraints in the evolution of reproductive strategies: telescoping of generations in parthenogenetic aphids. *Funct. Ecol.*, 3:531–537, 1989.

[384] C.M. King. *The natural history of weasels & stoats.* C. Helm, London, 1989.

[385] P. Kingston. Some observations on the effects of temperature and salinity upon the growth of *Cardium edule* and *Cardium glaucum* larvae in the laboratory. *J. Mar. Biol. Ass. U. K.*, 54:309–317, 1974.

[386] F.C. Kinsky. The yearly cycle of the northern blue penguin (*Eudyptula minor novaehollandiae*) in the Wellington Harbour area. *Rec. Dom. Mus. (Wellington)*, 3:145–215, 1960.

[387] T. Kiørboe, F. Møhlenberg, and O. Nøhr. Effect of suspended bottom material on growth and energetics in *Mytilus edulis. Mar. Biol.*, 61:283–288, 1981.

[388] T.B.L. Kirkwood, R. Holliday, and R.F. Rosenberger. Stability of the cellular translation process. *Int. Rev. Cytol.*, 92:93–132, 1984.

[389] M. Klaassen, G. Slagsvold, and C. Beek. Metabolic rate and thermostability in relation to availability of yolk in hatchlings of black-legged kittiwake and domestic chicken. *Auk*, 104:787–789, 1987.

[390] M. Kleiber. Body size and metabolism. *Hilgardia*, 6:315–353, 1932.

[391] O.T. Kleiven, P. Larsson, and A. Hobæk. Sexual reproduction in *Daphnia magna* requires three stimuli. *Oikos*, 64:197–206, 1992.

[392] H.N. Kluyver. Food consumption in relation to habitat in breeding chickadees. *Auk*, 78:532–550, 1961.

[393] A.L. Koch. Overall controls on the biosynthesis of ribosomes in growing bacteria. *J. Theor. Biol.*, 28:203–231, 1970.

[394] A.L. Koch. The macroeconomics of bacterial growth. In M Fletcher and G.D. Floodgate, editors, *Bacteria in their natural environment.*, volume 16 of *Special publication of the Society for General Microbiology*, pages 1–42. Academic press, 1985.

[395] A.L. Koch. The variability and individuality of the bacterium. In F.C. Neidhardt, editor, *Escherichia coli and Salmonella typhimurium. Cellular and molecular biology.*, pages 1606–1614. American Society for Microbiology, Washington, 1987.

[396] A.L. Koch. Diffusion. the crucial process in many aspects of the biology of bacteria. *Adv. Microb. Ecol.*, 11:37–70, 1990.

[397] A.L. Koch. Quantitative aspects of cellular turnover. *Antonie van Leeuwenhoek*, 60:175–191, 1991.

[398] A.L. Koch. Biomass growth rate during the prokaryote cell cycle. *Crit. Rev. Microbiol.*, 19:17–42, 1993.

[399] A.L. Koch and M. Schaechter. A model for the statistics of the division process. *J. Gen. Microbiol.*, 29:435–454, 1962.

[400] A.C. Kohler. Variations in the growth of Atlantic cod (*Gadus morhua* L.). *J. Fish. Res. Board Can.*, 21:57–100, 1964.

[401] B.W. Kooi and S.A.L.M. Kooijman. Existence and stability of microbial prey-predator systems. submitted, 1992.

[402] B.W. Kooi and S.A.L.M. Kooijman. Many limiting behaviours in microbial food-chains. In O. Arino, M. Kimmel, and D. Axelrod, editors, *Proceedings of the 3rd conference on mathematical population dynamics.*, Biological Systems. Wuerz, 1993. to appear.

[403] S.A.L.M. Kooijman. Parametric analyses of mortality rates in bioassays. *Water Res.*, 15:107–119, 1981.

[404] S.A.L.M. Kooijman. Statistical aspects of the determination of mortality rates in bioassays. *Water Res.*, 17:749–759, 1983.

[405] S.A.L.M. Kooijman. Toxicity at population level. In J. Cairns, editor, *Multispecies toxicity testing*, pages 143–164. Pergamon Press, 1985.

[406] S.A.L.M. Kooijman. Energy budgets can explain body size relations. *J. Theor. Biol.*, 121:269–282, 1986.

[407] S.A.L.M. Kooijman. Population dynamics on the basis of budgets. In J.A.J. Metz and O. Diekmann, editors, *The dynamics of physiologically structured populations*, Springer Lecture Notes in Biomathematics., pages 266–297. Springer-Verlag, 1986.

[408] S.A.L.M. Kooijman. What the hen can tell about her egg; egg development on the basis of budgets. *Bull. Math. Biol.*, 23:163–185, 1986.

[409] S.A.L.M. Kooijman. Strategies in ecotoxicological research. *Environ. Asp. Appl. Biol.*, 17(1):11–17, 1988.

[410] S.A.L.M. Kooijman. The von Bertalanffy growth rate as a function of physiological parameters: A comparative analysis. In T.G. Hallam, L.J. Gross, and S.A. Levin, editors, *Mathematical ecology*, pages 3–45. World Scientific, Singapore, 1988.

[411] S.A.L.M. Kooijman. Biomass conversion at population level. In D.L. DeAngelis and L.J. Gross, editors, *Individual based models; an approach to populations and communities*, pages 338–358. Chapman & Hall, 1992.

[412] S.A.L.M. Kooijman, A.O. Hanstveit, and N.van der Hoeven. Research on the physiological basis of population dynamics in relation to ecotoxicology. *Water Sci. Technol.*, 19:21–37, 1987.

[413] S.A.L.M. Kooijman, A.O. Hanstveit, and H. Oldersma. Parametric analyses of population growth in bioassays. *Water Res.*, 17:749–759, 1983.

[414] S.A.L.M. Kooijman and R.J.F.van Haren. Animal energy budgets affect the kinetics of xenobiotics. *Chemosphere*, 21:681–693, 1990.

[415] S.A.L.M. Kooijman, N.van der Hoeven, and D.C.van der Werf. Population consequences of a physiological model for individuals. *Funct. Ecol.*, 3:325–336, 1989.

[416] S.A.L.M. Kooijman and J.A.J. Metz. On the dynamics of chemically stressed populations; the deduction of population consequenses from effects on individuals. *Ecotox. Environ. Saf.*, 8:254–274, 1983.

[417] S.A.L.M. Kooijman, E.B. Muller, and A.H. Stouthamer. Microbial dynamics on the basis of individual budgets. *Antonie van Leeuwenhoek*, 60:159–174, 1991.

[418] S.A.L.M. Kooijman and W.K.de Raat. Het opstellen van mathematische modellen die ten grondslag liggen aan de Ames-toets. Technical Report R 85/6, TNO, 1985.

[419] S.A.L.M. Kooijman, W.K.de Raat, and M.I. Willems. Parametric analyses of mutation rates in bioassays with cells. Technical Report P 82/15b, TNO, 1984.

[420] L.J.H. Koppes, C.L. Woldringh, and N. Nanninga. Size variations and correlation of different cell cycle events in slow-growing *Escherichia coli*. *J. Bacteriol.*, 134:423–433, 1978.

[421] H.J. Kreuzer. *Nonequilibrium thermodynamics and its statistical foundations.* Clarendon Press, Oxford, 1981.

[422] M. Kshatriya and R.W. Blake. Theoretical model of migration energetics in the blue whale, *Balaenoptera musculus*. *J. Theor. Biol.*, 133:479–498, 1988.

[423] H.E. Kubitschek. Cell growth and abrupt doubling of membrane proteins in *Escherichia coli* during the division cycle. *J. Gen. Microbiol.*, 136:599–606, 1990.

[424] D. Lack. *Ecological adaptations for breeding in birds.* Methuen & Co., London, 1968.

[425] E. Lanting. *Een wereld verdwaald in de tijd. Madagascar.* Fragment Uitgeverij, 1990.

[426] R.R. Lassiter and T.G. Hallam. Survival of the fattest: A theory for assessing acute effects of hydrophobic, reversibly acting chemicals on populations. *Ecology*, 109:411–429, 1988.

[427] D.M. Lavigne, W. Barchard, S. Innes, and N.A. Oritsland. *Pinniped bioenergetics.*, volume IV of *FAO Fisheries Series No 5*. FAO, Rome, 1981.

[428] R.M. Laws. The elephant seal (*Mirounga leonina* Linn) I. Growth and age. *RIDS Sci. Rep.*, 8:1–37, 1953.

[429] R.M. Laws. Age criteria for the African elephant, *Loxodonta a. africana* E. *Afr. Wildl.*, 4:1–37, 1966.

[430] M. LeCroy and C.T. Collins. Growth and survival of roseate and common tern chicks. *Auk*, 89:595–611, 1972.

[431] A.L. Lehninger. *Bioenergetics.* Benjamin, 1973.

[432] W.A. Lell. The relation of the volume of the amniotic fluid to the weight of the fetus at different stages of pregnancy in the rabbit. *Anat. Rec.*, 51:119–123, 1931.

[433] R. Leudeking and E.L. Piret. A kinetic study of the lactic acid fermentation. *J. Biochem. Microbiol. Technol. Eng.*, 1:393, 1959.

[434] G.N. Ling. *In search of the physical basis of life.* Plenum Press, 1984.

[435] C. Lockyer. Growth and energy budgets of large baleen whales from the southern hemisphere. In *Mammals in the seas*, volume III of *FAO Fisheries series No 5*. FAO, Rome, 1981.

[436] B.E. Logan and J.R. Hunt. Bioflocculation as a microbial response to substrate limitations. *Biotechnol. Bioeng.*, 31:91–101, 1988.

[437] B.E. Logan and D.L. Kirchman. Uptake of dissolved organics by marine bacteria as a function of fluid motion. *Mar. Biol.*, 111:175–181, 1991.

[438] H. Lohrl. *Die Tannemeise.* Die Neue Brehm-Bücherei. A. Ziemsen-Verlag, Wittenberg Lutherstadt, 1977.

[439] A. Łomnicki. *Population ecology of individuals.* Princeton University Press, 1988.

[440] B.G. Loughton and S.S. Tobe. Blood volume in the African migratory locust. *Can. J. Zool.*, 47:1333–1336, 1969.

[441] G. Louw. *Physiological animal ecology.* Longman, 1993.

[442] B.G. Lovegrove and C. Wissel. Sociality in molerats: Metabolic scaling and the role of risk sensitivity. *Oecologia (Berlin)*, 74:600–606, 1988.

[443] Y. Loya and L. Fishelson. Ecology of fish breeding in brackish water ponds near the Dead Sea (Israel). *J. Fish Biol.*, 1:261–278, 1969.

[444] A.G. Lyne. Observations on the breeding and growth of the marsupial *Pelametes nasuta* geoffroy, with notes on other bandicoots. *Aust. J. Zool.*, 12:322–339, 1964.

[445] J.W. MacArthur and W.H.T. Baillie. Metabolic activity and duration of life. I. Influence of temperature on longevity in *Daphnia magna*. *J. Exp. Zool.*, 53:221–242, 1929.

[446] R.H. MacArthur and E.O. Wilson. *The theory of island biogeography.* Princeton University Press, 1967.

[447] B.A. MacDonald and R.J. Thompson. Influence of temperature and food availability on ecological energetics of the giant scallop *Placopecten magellanicus* I. Growth rates of shell and somatic tissue. *Mar. Ecol. Prog. Ser.*, 25:279–294, 1985.

[448] E.C. Macdowell, E. Allen, and C.G. Macdowell. The prenatal growth of the mouse. *J. Gen. Physiol.*, 11:57–70, 1927.

[449] S.P. Mahoney and W. Threlfall. Notes on the eggs, embryos and chick growth of the common guillemots *Uria aalge* in Newfoundland. *Ibis*, 123:211–218, 1981.

[450] A.P. Malan and H.H. Curson. Studies in sex physiology, no. 15. Further observations on the body weight and crown-rump length of merino foetuses. *Onderstepoort J. Vet. Sci. Anim. Ind.*, 7:239–249, 1936.

[451] R.H.K. Mann. The growth and reproductive stategy of the gudgeon, *Gobio gobio* (L.), in two hard-water rivers in southern England. *J. Fish Biol.*, 17:163–176, 1980.

[452] S.C. Manolis, G.J.W. Webb, and K.E. Dempsey. Crocodile egg chemistry. In G.J.W. Webb, S.C. Manolis, and P.J. Whitehead, editors, *Wildlife Management: Crocodiles and Alligators.*, pages 445–472. Beatty, Sydney, 1987.

[453] D.A. Manuwal. The natural history of Cassin's auklet (*Ptychoramphus aleuticus*). *Condor*, 76:421–431, 1974.

[454] D.M. Maron and B.N. Ames. Revised methods for the *Salmonella* mutagenicity test. *Mutat. Res.*, 113:173–215, 1983.

[455] A.G. Marr, E.H. Nilson, and D.J. Clark. The maintenance requirement of *Escherichia coli. Ann. N. Y. Acad. Sci.*, 102:536–548, 1962.

[456] A.G. Marr, P.R. Painter, and E.H. Nilson. Growth and division of individual bacteria. *Symp. Soc. Gen. Microbiol.*, 19:237–261, 1969.

[457] M.M. Martin. *Invertebrate-microbial interactions: Ingested fungal enzymes in arthropod biology.* Explorations in chemical ecology. Comstock Publ. Associates, Ithaca, 1987.

[458] D. Masman. *The annual cycle of the kestrel Falco tinnunculus; a study in behavioural energetics.* PhD thesis, University of Groningen, The Netherlands, 1986.

[459] J. Mauchline and L.R. Fisher. *The biology of euphausids.*, volume 7 of *Adv. Mar. Biol.* Academic Press, 1969.

[460] J.E. Maunder and W. Threlfall. The breeding biology of the black-legged kittiwake in Newfoundland. *Auk*, 89:789–816, 1972.

[461] J.R.von Mayer. Über die Kräfte der unbelebten Natur. *Ann. Chem. Pharm.*, 42:233, 1842.

[462] G.M. Maynes. Growth of the Parma wallaby, *Macropus parma* Waterhouse. *Aust. J. Zool.*, 24:217–236, 1976.

[463] E. McCauley, W.W. Murdoch, and R.M. Nisbet. Growth, reproduction, and mortality of *Daphnia pulex* leydig: Life at low food. *Funct. Ecol.*, 4:505–514, 1990.

[464] P. McCullagh and J.A. Nelder. *Generalized linear models.* Monographs on statistics and applied probability. Chapman & Hall, 1983.

[465] E.H. McEwan. Growth and development of the barren-ground caribou II. Postnatal growth rates. *Can. J. Zool.*, 46:1023–1029, 1968.

[466] S.B. McGrew and M.F. Mallette. Energy of maintenance in *Escherichia coli. J. Bacteriol.*, 83:844–850, 1962.

[467] M.D. McGurk. Effects of delayed feeding and temperature on the age of irreversible starvation and on the rates of growth and mortality of pacific herring larvae. *Mar. Biol.*, 84:13–26, 1984.

[468] J.W. McMahon. Some physical factors influencing the feeding behavior of *Daphnia magna* Straus. *Can. J. Zool.*, 43:603–611, 1965.

[469] T.A. McMahon. Size and shape in biology. *Science (New York)*, 179:1201–1204, 1973.

[470] T.A. McMahon and J.T. Bonner. *On size and life.* Scientific American Library. Freeman, 1983.

[471] B.K. McNab. On the ecological significance of Bergmann's rule. *Ecology*, 52:845–854, 1971.

[472] B.K. McNab. Energy expenditure: A short history. In T.E. Tomasi and T.H. Horton, editors, *Mammalian energetics.*, pages 1–15. Comstock Publ. Assoc., Ithaca, 1992.

[473] E.M. Meijer, H.W.van Verseveld, E.G.van der Beek, and A.H. Stouthamer. Energy conservation during aerobic growth in *Paracoccus denitrificans. Arch. Microbiol.*, 112:25–34, 1977.

[474] T.H.M. Meijer. *Reproductive decisions in the kestrel Falco tinnunculus; a study in physiological ecology.* PhD thesis, University of Groningen, The Netherlands, 1988.

[475] T.H.M. Meijer, S. Daan, and M. Hall. Family planning in the kestrel (*Falco tinnuculus*): The proximate control of covariation of laying data and clutch size. *Behaviour*, 114:117–136, 1990.

[476] J.A.J. Metz and F.H.D.van Batenburg. Holling's 'hungry mantid' model for the invertebrate functional response considered as a Markov process. Part I: The full model and some of its limits. *J. Math. Biol.*, 22:209–238, 1985.

[477] J.A.J. Metz and F.H.D.van Batenburg. Holling's 'hungry mantid' model for the invertebrate functional response considered as a Markov process. Part II: Negligible handling time. *J. Math. Biol.*, 22:239–257, 1985.

[478] J.A.J. Metz and O. Diekmann. *The dynamics of physiologically structured populations.*, volume 68 of *Lecture Notes in Biomathematics.* Springer-Verlag, 1986.

[479] J.A.J. Metz and O. Diekmann. Exact finite dimensional representations of models for physiologically structured populations. I: The abstract foundations of linear chain trickery. In J.A. Goldstein, F. Kappel, and W. Schappacher, editors, *Differential equations with applications in biology, physics, and engineering.*, pages 269–289. Marcel Dekker, New York, 1991.

[480] E.A. Meuleman. Host-parasite interrelations between the freshwater pulmonate *Biomphalaria pfeifferi* and the trematode *Schistosoma mansoni. Neth. J. Zool.*, 22:355–427, 1972.

[481] H. Meyer and L. Ahlswede. Über das intrauterine Wachstum und die Körperzusammensetzung von Fohlen sowie den Nährstoffbedarf tragender Stuten. *Übersichten der Tierernährung*, 4:263–292, 1976.

[482] D.G. Meyers. Egg development of a chydorid cladoceran, *Chydorus sphaericus*, exposed to constant and alternating temperatures: significance to secondary productivity in fresh water. *Ecology*, 61:309–320, 1984.

[483] L. Michaelis and M.L. Menten. Die Kinetik der Invertinwirkung. *Biochem. Z.*, 49:333–369, 1913.

[484] H. Mikkola. *Der Bartkauz.* Die Neue Brehm-Bücherei. A. Ziemsen-Verlag, Wittenberg Lutherstadt, 1981.

[485] T.T. Milby and E.W. Henderson. The comparative growth rates of turkeys, ducks, geese and pheasants. *Poult. Sci.*, 16:155–165, 1937.

[486] J.S. Millar. Post partum reproductive characteristics of eutherian mammals. *Evolution*, 35:1149–1163, 1981.

[487] P.J. Miller. Age, growth, and reproduction of the rock goby, *Gobius paganellus* L., in the Isle of Man. *J. Mar. Biol. Ass. U. K.*, 41:737–769, 1961.

[488] P. Milton. Biology of littoral blenniid fishes on the coast of south-west England. *J. Mar. Biol. Ass. U. K.*, 63:223–237, 1983.

[489] J.M. Mitchison. The growth of single cells III *Streptococcus faecalis. Exp. Cell Res.*, 22:208–225, 1961.

[490] G.G. Mittelbach. Predation and resource patitioning in two sunfishes (*Centrarchidae*). *Ecology*, 65:499–513, 1984.

[491] J. Monod. *Recherches sur la croissance des cultures bacteriennes.* Hermann, Paris, 2nd edition, 1942.

[492] J.L. Monteith and M.H. Unsworth. *Principles of environmental physics.* E. Arnold, London, 1990.

[493] S. Moreno, P. Nurse, and P. Russell. Regulation of mitosis by cyclic accumulation of p80^{cdc25} mitotic inducer in fission yeast. *Nature (Lond.)*, 344:549–552, 1990.

[494] J.A. Morrison, C.E. Trainer, and P.L. Wright. Breeding season in elk as determined from known-age embryos. *J. Wildl. Manage.*, 23:27–34, 1959.

[495] B.S. Muir. Comparison of growth rates for native and hatchery-stocked populations of *Esox masquinongy* in Nogies Creek, Ontario. *J. Fish. Res. Board Can.*, 17:919–927, 1960.

[496] Y. Mukohata, editor. *New era of bioenergetics.* Academic Press, 1991.

[497] M.T. Murphy. Ecological aspects of the reproductive biology of eastern kingbirds: Geographic comparisons. *Ecology*, 64:914–928, 1983.

[498] A.W. Murray and M.W. Kirschner. What controls the cell cycle? *Sci. Am.*, (3):34–41, 1991.

[499] S. Nagasawa. Parasitism and diseases in chaetognats. In Q. Bone, H. Kapp, and A.C. Pierrot-Bults, editors, *The biology of chaetognats.* Oxford University Press, 1991.

[500] S. Nagasawa and T. Nemoto. Presence of bacteria in guts of marine crustaceans on their fecal pellets. *J. Plankton Res.*, 8:505–517, 1988.

[501] R.D.M. Nash. The biology of Fries' goby, *Lesueurigobius friesii* (Malm) in the Firth of Clyde, Scotland, and a comparison with other stocks. *J. Fish Biol.*, 21:69–85, 1982.

[502] S. Nee, A.F. Read, J.D. Greenwood, and P.H. Harvey. The relationship between abundance and body size in British birds. *Nature (Lond.)*, 351:312–313, 1991.

[503] F.C. Neidhardt, J.L. Ingraham, and M. Schaechter. *Physiology of the bacterial cell: A molecular approach.* Sinauer Assoc. Inc., Sunderland, Massachusetts, 1990.

[504] J.A. Nelder. The fitting of a generalization of the logistic curve. *Biometrics*, 17:89–110, 1961.

[505] B. Nelson. *The gannet.* T. & A.D.Poyser, Berkhamsted, 1978.

[506] E. Nestaas and D.I.C. Wang. Computer control of the penicillum fermentation using the filtration probe in conjunction with a structured model. *Biotechnol. Bioeng.*, 25:781–796, 1983.

[507] D.G. Nicholls. *Bioenergetics. An introduction to the chemiosmotic theory.* Academic Press, 1982.

[508] A.J. Niimi and B.G. Oliver. Distribution of polychlorinated biphenyl congeners and other hydrocarbons in whole fish and muscle among Ontario salmonids. *Environ. Sci. Technol.*, 23:83–88, 1989.

[509] B. Nisbet. *Nutrition and feeding strategies in protozoa.* Croom Helm, London, 1984.

[510] R.M. Nisbet, A. Cunningham, and W.S.C. Gurney. Endogenous metabolism and the stability of microbial prey-predator systems. *Biotechnol. Bioeng.*, 25:301–306, 1983.

[511] P.S. Nobel. *Physicochemical and environmental plant physiology.* Academic Press, 1991.

[512] F. Norrbin and U. Båmstedt. Energy contents in benthic and planktonic invertebrates of Kosterfjorden, Sweden. a comparison of energy strategies in marine organism groups. *Ophelia*, 23:47–64, 1984.

[513] G. North. Back to the RNA world and beyond. *Nature (Lond.)*, 328:18–19, 1987.

[514] T.H.Y. Nose and Y. Hiyama. Age determination and growth of yellowfin tuna, *Thunnus albacares* Bonnaterre by vertebrae. *Bull. Jpn Soc. Sci. Fish.*, 31:414–422, 1965.

[515] N. Nyholm. A mathematical model for microbial growth under limitation by conservative substrates. *Biotechnol. Bioeng.*, 18:1043–1056, 1976.

[516] N. Nyholm. Kinetics of phosphate limited algal growth under limitation by conservative substances. *Biotechnol. Bioeng.*, 19:467–492, 1977.

[517] A. Okubo. *Diffusion and ecological problems: Mathematical models.*, volume 10 of *Biomathematics.* Springer-Verlag, 1980.

[518] K. Okunuki. Denaturation and inactivation of enzyme proteins. *Adv. Emzymol. Relat. Areas Mol. Biol.*, 23:29–82, 1961.

[519] B.G. Oliver and A.J. Niimi. Trophodynamic analysis of polychlorinated biphenyl congeners and other chlorinated hydrocarbons in Lake Ontario ecosystem. *Environ. Sci. Technol.*, 22:388–397, 1988.

[520] R.V. O'Neill, D.L. DeAngelis, J.B. Waide, and T.F.H. Allen. *A hierarchical concept of ecosystems.*, volume 23 of *Monographs in population biology.* Princeton University Press, 1986.

[521] A. Opperhuizen. *Bioconcentration in fish and other distribution processes of hydrophobic chemicals in aquatic environments.* PhD thesis, University of Amsterdam, 1986.

[522] A.Y. Ota and M.R. Landry. Nucleic acids as growth rate indicators for early developmental stages of *Calanus pacificus* Brodsky. *J. Exp. Mar. Biol. Ecol.*, 80:147–160, 1984.

[523] M.J. Packard, G.C. Packard, and W.H.N. Gutzke. Calcium metabolism in embryos of the oviparous snake *Coluber constrictor. J. Exp. Biol.*, 110:99–112, 1984.

[524] M.J. Packard, G.C. Packard, J.D. Miller, M.E. Jones, and W.H.N. Gutzke. Calcium mobilization, water balance, and growth in embryos of the agamid lizard *Amphibolurus barbatus. J. Exp. Zool.*, 235:349–357, 1985.

[525] M.J. Packard, T.M. Short, G.C. Packard, and T.A. Gorell. Sources of calcium for embryonic development in eggs of the snapping turtle *Chelydra serpentina. J. Exp. Zool.*, 230:81–87, 1984.

[526] R.T. Paine. Growth and size distribution of the brachiopod *Terebratalia transversa* Sowerby. *Pac. Sci.*, 23:337–343, 1969.

[527] R.T. Paine. The measurement and application of the calorie to ecological problems. *Ann. Rev. Ecol. Syst.*, 2:145–164, 1971.

[528] P.R. Painter and A.G. Marr. Mathematics of microbial populations. *Annu. Rev. Microbiol.*, 22:519–548, 1968.

[529] J.E. Paloheimo, S.J. Crabtree, and W.D. Taylor. Growth model for *Daphnia. Can. J. Fish. Aquat. Sci.*, 39:598–606, 1982.

[530] J.R. Parks. *A theory of feeding and growth of animals.* Springer-Verlag, 1982.

[531] G.D. Parry. The influence of the cost of growth on ectotherm metabolism. *J. Theor. Biol.*, 101:453–477, 1983.

[532] S.J. Parulekar, G.B. Semones, M.J. Rolf, J.C. Lievense, and H.C. Lim. Induction and elimination of oscillations in continuous cultures of *Saccharomyces cerevisiae. Biotechnol. Bioeng.*, 28:700–710, 1986.

[533] C.S. Patlak. Energy expenditure by active transport mechanisms. *Biophys. J.*, 1:419–427, 1961.

[534] M.R. Patterson. A chemical engineering view of cnidarian symbioses. *Am. Zool.*, 32:566–582, 1993.

[535] D. Pauly and P. Martosubroto. The population dynamics of *Nemipterus marginatus* (Cuvier & Val.) off Western Kalimantan, South China Sea. *J. Fish Biol.*, 17:263–273, 1980.

[536] P.R. Payne and E.F. Wheeler. Growth of the foetus. *Nature (Lond.)*, 215:849–850, 1967.

[537] R. Pearl. The growth of populations. *Q. Rev. Biol.*, 2:532–548, 1927.

[538] D.L. Penry and P.A. Jumars. Chemical reactor analysis and optimal digestion. *Bioscience*, 36:310–315, 1986.

[539] D.L. Penry and P.A. Jumars. Modeling animal guts as chemical reactors. *Am. Nat.*, 129:69–96, 1987.

[540] J.A. Percy and F.J. Fife. The biochemical composition and energy content of Arctic marine macrozooplankton. *Arctic*, 34:307–313, 1981.

[541] B. Peretz and L. Adkins. An index of age when birthdate is unknown in *Aplysia californica*: Shell size and growth in long-term maricultured animals. *Biol. Bull. (Woods Hole)*, 162:333–344, 1982.

[542] R.H. Peters. *The ecological implications of body size.* Cambridge University Press, 1983.

[543] T.N. Pettit, G.S. Grant, G.C. Whittow, H. Rahn, and C.V. Paganelli. Respiratory gas exchange and growth of white tern embryos. *Condor*, 83:355–361, 1981.

[544] T.N. Pettit, G.S. Grant, G.C. Whittow, H. Rahn, and C.V. Paganelli. Embryonic oxygen consumption and growth of laysan and black-footed albatross. *Am. J. Physiol.*, 242:121–128, 1982.

[545] T.N. Pettit, G.C. Whittow, and G.S. Grant. Caloric content and energetic budget of tropical seabird eggs. In G.C. Whittow and H. Rahn, editors, *Seabird energetics.*, pages 113–138. Plenum Press, New York, 1984.

[546] T.N. Pettit and G.S. Whittow. Embryonic respiration and growth in two species of noddy terns. *Physiol. Zool.*, 56:455–464, 1983.

[547] H.G. Petzold. *Die Anakondas.* Die Neue Brehm-Bücherei. A. Ziemsen-Verlag, Wittenberg Lutherstadt, 1984.

[548] J. Phillipson. Bioenergetic options and phylogeny. In C.R. Townsend and P. Calow, editors, *Physiological ecology.* Blackwell Sci. Publ., 1981.

[549] E.R. Pianka. On r and K selection. *Am. Nat.*, 104:592–597, 1970.

[550] E.R. Pianka. *Evolutionary ecology.* Harper & Row Publ. Inc., 1978.

[551] E.R. Pianka. *Ecology and natural history of desert lizards; Analyses of the ecological niche of community structure.* Princeton University Press, 1986.

[552] T. Piersma. Estimating energy reserves of great crested grebes *Podiceps cristatus* on the basis of body dimensions. *Ardea*, 72:119–126, 1984.

[553] H. Pieters, J.H. Kluytmans, W. Zurburg, and D.I. Zandee. The influence of seasonal changes on energy metabolism in *Mytilus edulis* (L.) I. Growth rate and biochemical composition in relation to environmental parameters and spawning. In E. Naylor and R.G. Hartnoll, editors, *Cyclic phenomena in marine plants and animals.*, pages 285–292. Pergamon Press, 1979.

[554] J. Pilarska. Eco-physiological studies on *Brachionus rubens* ehrbg (*Rotatoria*) I. Food selectivity and feeding rate. *Pol. Arch. Hydrobiol.*, 24:319–328, 1977.

[555] E.M.del Pino. Marsupial frogs. *Sci. Am.*, (5):76–84, 1989.

[556] S.J. Pirt. The maintenance energy of bacteria in growing cultures. *Proc. R. Soc. Lond. B Biol. Sci.*, 163:224–231, 1965.

[557] S.J. Pirt. *Principles of microbe and cell cultivation.* Blackwell Sci. Publ., Oxford, 1975.

[558] S.J. Pirt and D.S. Callow. Studies of the growth of *Penicillium chrysogenum* in continuous flow culture with reference to penicillin production. *J. Appl. Bacteriol.*, 23:87–98, 1960.

[559] G. Pohle and M. Telford. Post-larval growth of *Dissodactylus primitivus* Bouvier, 1917 (*Brachyura*: *Pinnotheridae*) under laboratory conditions. *Biol. Bull. (Woods Hole)*, 163:211–224, 1982.

[560] J.S. Poindexter. Oligotrophy: Fast and famine existence. *Adv. Microb. Ecol.*, 5:63–89, 1981.

[561] D.E. Pomeroy. Some aspects of the ecology of the land snail, *Helicella virgata*, in South Australia. *Aust. J. Zool.*, 17:495–514, 1969.

[562] R.W. Pomeroy. Infertility and neonatal mortality in the sow. III. Neonatal mortality and foetal development. *J. Agric. Sci.*, 54:31–56, 1960.

[563] W.E. Poole. Breeding biology and current status of the grey kangaroo, *Macropus fuliginosus fuliginosus*, of Kangaroo Island, South Australia. *Aust. J. Zool.*, 24:169–187, 1976.

[564] D. Poppe and B. Vos. *De buizerd.* Kosmos Vogelmonografieën. Kosmos, Amsterdam, 1982.

[565] K.G. Porter, J. Gerritsen, and J.D. Orcutt. The effect of food concentration on swimming patterns, feeding behavior, ingestion, assimilation and respiration by *Daphnia*. *Limnol. Oceanogr.*, 27:935–949, 1982.

[566] K.G. Porter, M.L. Pace, and J.F. Battey. Ciliate protozoans as links in freshwater planktonic food chains. *Nature (Lond.)*, 277:563–564, 1979.

[567] R.K. Porter and M.D. Brand. Body mass dependence of H^+ leak in mitochondria and its relevance to metabolic rate. *Nature (Lond.)*, 362:628–630, 1993.

[568] D.M. Prescott. Relations between cell growth and cell division. In D. Rudnick, editor, *Rythmic and synthetic processes in growth.*, pages 59–74. Princeton University Press, 1957.

[569] H.H. Prince, P.B. Siegel, and G.W. Cornwell. Embryonic growth of mallard and pekin ducks. *Growth*, 32:225–233, 1968.

[570] W.G. Pritchard. Scaling in the animal kingdom. *Bull. Math. Biol.*, 55:111–129, 1993.

[571] D.R. Prothero and W.A. Berggren, editors. *Eocene-Oligocene climatic and biotic evolution.* Princeton University Press, 1992.

[572] T. Prus. Calorific value of animals as an element of bioenergetical investigations. *Pol. Arch. Hydrobiol.*, 17:183–199, 1970.

[573] D.M. Purdy and H.H. Hillemann. Prenatal growth in the golden hamster (*Cricetus auratus*). *Anat. Rec.*, 106:591–597, 1950.

[574] A. Pütter. Studien über physiologische Ähnlichkeit. VI Wachstumsähnlichkeiten. *Arch. Gesamte Physiol. Mench. Tiere*, 180:298–340, 1920.

[575] P.A. Racey and S.M. Swift. Variations in gestation length in a colony of pipistrelle bats (*Pipistrellus pipistrellus*) from year to year. *J. Reprod. Fertil.*, 61:123–129, 1981.

[576] U. Rahm. *Die Afrikanische Wurzelratte.* Die Neue Brehm-Bücherei. A. Ziemsen-Verlag, Wittenberg Lutherstadt, 1980.

[577] H. Rahn and A. Ar. The avian egg: incubation time and water loss. *Condor*, 76:147–152, 1974.

[578] H. Rahn, A. Ar, and C.V. Paganelli. How bird eggs breathe. *Sci. Am.*, (2):38–47, 1979.

[579] H. Rahn and C.V. Paganelli. Gas fluxes in avian eggs: Driving forces and the pathway for exchange. *Comp. Biochem. Physiol.*, 95A:1–15, 1990.

[580] H. Rahn, C.V. Paganelli, and A. Ar. Relation of avian egg weight to body weight. *Auk*, 92:750–765, 1975.

[581] J.E. Randall, G.R. Allen, and R.C. Steen. *Fishes of the great barrier reef and coral sea.* University of Hawai Press, Honolulu, 1990.

[582] C. Ratledge. Biotechnology as applied to the oils and fats industry. *Fette Seifen Anstrichm.*, 86:379–389, 1984.

[583] C. Ratsak, S.A.L.M. Kooijman, and B.W. Kooi. Modelling the growth of an oligochaete on activated sludge. *Water Res.*, 27:739–747, 1992.

[584] D.M. Raup. Geometric analysis of shell coiling: General problems. *J. Paleontol.*, 40:1178–1190, 1966.

[585] D.M. Raup. Geometric analysis of shell coiling: Coiling in ammonoids. *J. Paleontol.*, 41:43–65, 1967.

[586] P.J. Reay. Some aspects of the biology of the sandeel, *Ammodytes tobianus* L., in Langstone Harbour, Hampshire. *J. Mar. Biol. Ass. U. K.*, 53:325–346, 1973.

[587] A.C. Redfield. The biological control of chemical factors in the environment. *American Scientist*, 46:205–221, 1958.

[588] M.R. Reeve. The biology of Chaetognatha I. Quantitative aspects of growth and egg production in *Sagitta hispida.* In J.H. Steele, editor, *Marine food chains.* Univ. Calif. Press, Berkeley, 1970.

[589] M.R. Reeve and L.D. Baker. Production of two planktonic carnivores (Chaetognath and Ctenophore) in south Florida inshore waters. *Fish. Bull. (Dublin)*, 73:238–248, 1975.

[590] M.J. Reiss. *The allometry of growth and reproduction.* Cambridge University Press, 1989.

[591] H. Remmert. *Arctic animal ecology.* Springer-Verlag, 1980.

[592] F.J. Richards. A flexible growth function for empirical use. *J. Exp. Bot.*, 10:290–300, 1959.

[593] S.W. Richards, D. Merriman, and L.H. Calhoun. Studies on the marine resources of southern New England IX. The biology of the little skate, *Raja erinacea* Mitchill. *Bull. Bingham Oceanogr. Collect. Yale Univ.*, 18:5–67, 1963.

[594] D. Richardson. *The vanishing lichens: Their history, biology and importance.* David and Charles, Newton Abbot, 1975.

[595] S. Richman. The transformation of energy by *Daphnia pulex. Ecol. Monogr.*, 28:273–291, 1958.

[596] R.E. Ricklefs. Patterns of growth in birds. *Ibis*, 110:419–451, 1968.

[597] R.E. Ricklefs. Patterns of growth in birds III. growth and development of the cactus wren. *Condor*, 77:34–45, 1975.

[598] R.E. Ricklefs. Adaptation, constraint, and compromise in avian postnatal development. *Biol. Rev. Camb. Philos. Soc.*, 54:269–290, 1979.

[599] R.E. Ricklefs, S. White, and J. Cullen. Postnatal development of Leach's storm-petrel. *Auk*, 97:768–781, 1980.

[600] U. Riebesell, D.A. Wolf-Gladrow, and V. Smetacek. Carbon dioxide limitation of marine phytoplankton growth rates. *Nature (Lond.)*, 361:249–251, 1993.

[601] F.H. Rigler. Zooplankton. In W.T. Edmondson, editor, *A manual on methods for the assessment of secondary productivity in fresh waters.*, number 17 in IBP handbook, pages 228–255. Bartholomew Press, Dorking, 1971.

[602] H.U. Riisgård, E. Bjornestad, and F. Møhlenberg. Accumulation of cadmium in the mussel *Mytilus edulis*: Kinetics and importance of uptake via food and sea water. *Mar. Biol.*, 96:349–353, 1987.

[603] H.U. Riisgård and F. Møhlenberg. An improved automatic recording apparatus for determining the filtration rate of *Mytilus edulis* as a function of size and algal concentration. *Mar. Biol.*, 66:259–265, 1979.

[604] H.U. Riisgård and A. Randløv. Energy budgets, growth and filtration rates in *Mytilus edulis* at different algal concentrations. *Mar. Biol.*, 61:227–234, 1981.

[605] C.T. Robbins and A.N. Moen. Uterine composition and growth in pregnant white-tailed deer. *J. Wildl. Manage.*, 39:684–691, 1975.

[606] J.R. Robertson and G.W. Salt. Responses in growth, mortality, and reproduction to variable food levels by the rotifer, *Asplanchna girodi*. *Ecology*, 62:1585–1596, 1981.

[607] P.G. Rodhouse, C.M. Roden, G.M. Burnell, M.P. Hensey, T. McMahon, B. Ottway, and T.H. Ryan. Food resource, gametogenesis and growth of *Mytilus edulis* on the shore and in suspended culture: Killary Harbour, Ireland. *Ecology*, 64:513–529, 1984.

[608] J.A. Roels. *Energetics and kinetics in biotechnology.* Elsevier Biomedical Press, 1983.

[609] A.L. Romanov. *The Avian Embryo.* MacMillan Publ. Co., New York, 1960.

[610] C. Romijn and W. Lokhorst. Foetal respiration in the hen. *Physiol. Zool.*, 2:187–197, 1951.

[611] A.de Roos. Numerical methods for structured population models: The escalator boxcar train. *Num. Meth. Part. Diff. Eq.*, 4:173–195, 1988.

[612] M.R. Rose. Laboratory evolution of postponed senescence in *Drosophila melanogaster*. *Evolution*, 38:1004–1010, 1984.

[613] J.L. Roseberry and W.D. Klimstra. Annual weight cycles in male and female bobwhite quail. *Auk*, 88:116–123, 1971.

[614] M. Rubner. Über den Einfluss der Körpergrösse auf Stoff- und Kraftwechsel. *Z. Biol.*, 19:535–562, 1883.

[615] W. Rudolph. *Die Hausenten.* Die Neue Brehm-Bücherei. A. Ziemsen-Verlag, Wittenberg Lutherstadt, 1978.

[616] M.J.S. Rudwick. The growth and form of brachiopod shells. *Geol. Mag.*, 96:1–24, 1959.

[617] M.J.S. Rudwick. Some analytic methods in the study of ontogeny in fossils with accretionary skeletons. *Paleont. Soc., Mem.*, 2:35–59, 1968.

[618] D. Ruelle. *Elements of differentiable dynamics and bifurcation theory.* Academic Press, 1989.

[619] P.C.de Ruiter and G. Ernsting. Effects of ration on energy allocation in a carabid beetle. *Funct. Ecol.*, 1:109–116, 1987.

[620] R.W. Russel and F.A.P.C. Gobas. Calibration of the freshwater mussel *Elliptio complanata*, for quantitative monitoring of hexachlorobenzene and octachlorostyrene in aquatic systems. *Bull. Environ. Contam. Toxicol.*, 43:576–582, 1989.

[621] R.L. Rusting. Why do we age? *Sci. Am.*, (12):86–95, 1992.

[622] M. Rutgers. *Control and thermodynamics of microbial growth.* PhD thesis, University of Amsterdam, 1990.

[623] M. Rutgers, M.J. Teixeira de Mattos, P.W. Postma, and K.van Dam. Establishment of the steady state in glucose-limited chemostat cultures of *Kleibsiella pneumoniae*. *J. Gen. Microbiol.*, 133:445–453, 1987.

[624] I. Ružić. Two-compartment model of radionuclide accumulation into marine organisms. I. Accumulation from a medium of constant activity. *Mar. Biol.*, 15:105–112, 1972.

[625] T.L. Saaty. *Elements of queueing theory with applications.* Dover, New York, 1961.

[626] S.I. Sandler and H. Orbey. On the thermodynamics of microbial growth processes. *Biotechnol. Bioeng.*, 38:697–718, 1991.

[627] J.R. Sargent. Marine wax esters. *Sci. Prog.*, 65:437–458, 1978.

[628] Sarrus and Rameaux. Mémoire adressé à l'Académie Royale. *Bull. Acad. R. Med.*, 3:1094–1100, 1839.

[629] J. Sarvala. Effect of temperature on the duration of egg, nauplius and copepodite development of some freshwater benthic copepoda. *Freshwater Biol.*, 9:515–534, 1979.

[630] P.T. Saunders. *An introduction to catastrophe theory.* Cambridge University Press, 1980.

[631] H. Schatzmann. *Anaerobes Wachstum von Saccharomyces cervisiae: Regulatorische Aspekte des glycolytischen und respirativen Stoffwechsels.* Diss. eth 5504, ETH, Zürich, 1975.

[632] H.G. Schegel. *Algemeine Mikrobiologie.* Thieme, 1981.

[633] H. Scheufler and A. Stiefel. *Der Kampflaufer.* Die Neue Brehm-Bücherei. A. Ziemsen-Verlag, Wittenberg Lutherstadt, 1985.

[634] D.E. Schindel. Unoccupied morphospace and the coiled geometry of gastropods: Architectural constraint or geometric covariance? In R.M. Ross and W.D. Allmon, editors, *Causes of evolution. A Paleontological perspective.* The University of Chicago Press, 1990.

[635] D.W. Schindler. Feeding, assimilation and respiration rates of *Daphnia magna* under various environmental conditions and their relation to production estimates. *J. Anim. Ecol.*, 37:369–385, 1968.

[636] I.I. Schmalhausen and E. Syngajewskaja. Studien über Wachstum und Differenzierung. I. Die individuelle Wachtumskurve von *Paramaecium caudatum. Roux's Arch. Dev. Biol.*, 105:711–717, 1925.

[637] U. Schmidt. *Vampirfledermäuse.* Die Neue Brehm-Bücherei. A. Ziemsen-Verlag, Wittenberg, 1978.

[638] K. Schmidt-Nielsen. *Scaling: Why is animal size so important?* Cambridge University Press, 1984.

[639] R. Schneider. Polychlorinated biphenyls (PCBs) in cod tissues from the Western Baltic: Significance of equilibrium partitioning and lipid composition in the bioaccumulation of lipophilic pollutants in gill-breathing animals. *Meeresforschung*, 29:69–79, 1982.

[640] S.A. Schoenberg, A.E. Maccubbin, and R.E. Hodson. Cellulose digestion by freshwater microcrustacea. *Limnol. Oceanogr.*, 29:1132–1136, 1984.

[641] P.F. Scholander. Evolution of climatic adaptation in homeotherms. *Evolution*, 9:15–26, 1955.

[642] M. Schonfeld. *Der Fitislaubsanger.* Die Neue Brehm-Bücherei. A. Ziemsen-Verlag, Wittenberg Lutherstadt, 1982.

[643] S. Schonn. *Der Sperlingskauz.* Die Neue Brehm-Bücherei. A. Ziemsen-Verlag, Wittenberg Lutherstadt, 1980.

[644] R.M. Schoolfield, P.J.H. Sharpe, and C.E. Magnuson. Non-linear regression of biological temperature-dependent rate models based on absolute reaction-rate theory. *J. Theor. Biol.*, 88:719–731, 1981.

[645] K.L. Schulze and R.S. Lipe. Relationship between substrate concentration, growth rate, and respiration rate of *Escherichia coli* in continuous culture. *Arch. Mikrobiol.*, 48:1–20, 1964.

[646] W.A. Searcy. Optimum body sizes at different ambient temperatures: an energetics explanation of Bergmann's rule. *J. Theor. Biol.*, 83:579–593, 1980.

[647] L.A. Segel. *Modeling dynamic phenomena in molecular and cellular biology.* Cambridge University Press, 1984.

[648] H.P. Senn. *Kinetik und Regulation des Zuckerabbaus von Escherichia coli ML 30 bei tiefen Zucker Konzentrationen.* PhD thesis, Techn. Hochschule Zurich., 1989.

[649] M. Shafi and P.S. Maitland. The age and growth of perch (*Perca fluviatilis* L.) in two Scottish lochs. *J. Fish Biol.*, 3:39–57, 1971.

[650] J.A. Shapiro. Bacteria as multicellular organisms. *Sci. Am.*, (6):62–69, 1988.

[651] P.J.H. Sharpe and D.W. DeMichele. Reaction kinetics of poikilotherm development. *J. Theor. Biol.*, 64:649–670, 1977.

[652] J.E. Shelbourne. A predator-prey relationship for plaice larvae feeding on *Oikopleura*. *J. Mar. Biol. Ass. U. K.*, 42:243–252, 1962.

[653] J.C. Sherris, N.W. Preston, and J.G. Shoesmith. The influence of oxygen and arginine on the motility of a strain of *Pseudomonas sp*. *J. Gen. Microbiol.*, 16:86–96, 1957.

[654] J.M. Shick. *A functional biology of sea anemones.* Chapman & Hall, 1991.

[655] R. Shine and E.L. Charnov. Patterns of survival, growth, and maturation in snakes and lizards. *Am. Nat.*, 139:1257–1269, 1992.

[656] R.V. Short. Species differences in reproductive mechanisms. In C.R. Austin and R.V. Short, editors, *Reproductive fitness.*, volume 4 of *Reproduction in mammals.*, pages 24–61. Cambridge University Press, 1984.

[657] L. Sigmund. Die Postembryonale Entwicklung der Wasserralle. *Sylvia*, 15:85–118, 1958.

[658] G.G. Simpson. *Penguins. Past and present, here and there.* Yale University Press, 1976.

[659] M.R. Simpson and S. Boutin. Muskrat life history: a comparison of a northern and southern population. *Ecography*, 16:5–10, 1993.

[660] B. Sinervo. The evolution of maternal investment in lizard. An experimental and comparative analysis of egg size and its effect on offspring performance. *Evolution*, 44:279–294, 1990.

[661] J.W. Sinko and W. Streifer. A model for populations reproducing by fission. *Ecology*, 52:330–335, 1967.

[662] S. Sjöberg. Zooplankton feeding and queueing theory. *Ecol. Modell.*, 10:215–225, 1980.

[663] J.G. Skellam. The formulation and interpretation of mathematical models of diffusionary processes in population biology. In M.S. Bartlett and R.W. Hiorns, editors, *The mathematical theory of the dynamics of biological populations.*, pages 63–85. Academic Press, 1973.

[664] W. Slob and C. Janse. A quantitative method to evaluate the quality of interrupted animal cultures in aging studies. *Mech. Ageing Dev.*, 42:275–290, 1988.

[665] L.B. Slobodkin. Population dynamics in *Daphnia obtusa* Kurz. *Ecol. Monogr.*, 24:69–88, 1954.

[666] J.F. Sluiters. *Parasite-host relationship of the avian schistosome Trichobilharzia ocellata and the hermaphrodite gastropod Lymnea stagnalis.* PhD thesis, Vrije Universiteit, Amsterdam., 1983.

[667] G.L. Small. *The blue whale.* Columbia Univer. Press, N.Y., 1971.

[668] F.A. Smith and N.A. Walker. Photosynthesis by aquatic plants: effects of unstirred layers in relation to assimilation of CO_2 and HCO_3^- and to carbon isotopic discrimination. *New Phytol.*, 86:245–259, 1980.

[669] M.L. Smith, J.N. Bruhn, and J.B. Anderson. The fungus *Armillaria bulbosa* is among the largest and oldest living organsism. *Nature (Lond.)*, 356:428–431, 1992.

[670] S. Smith. Early development and hatching. In M.E. Brown, editor, *The physiology of fishes*, volume 1, pages 323–359. Academic Press, 1957.

[671] S.M. Smith. *The black-capped chickadee. Behavioral ecology and natural history.* Comstock Publishing Associates, Ithaca, 1991.

[672] O. Snell. Die Abhängigkeit des Hirngewichtes von dem Körpergewicht und den geistigen Fähigkeiten. *Arch. Psychiat. Nervenkr.*, 23:436–446, 1891.

[673] D.W. Snow. The natural history of the oilbird, *Steatornis caripensis* in Trinidad, W.I. part 1. General behaviour and breeding habits. *Zoologica (N. Y.)*, 46:27–48, 1961.

[674] G.M. Southward. A method of calculating body lengths from otolith measurements for pacific halibut and its application to Portlock-Albatross Grounds data between 1935 and 1957. *J. Fish. Res. Board Can.*, 19:339–362, 1962.

[675] A.L. Spaans. On the feeding ecology of the herring gull *Larus argentatus* Pont. in the northern part of the Netherlands. *Ardea*, 59:75–188, 1971.

[676] G. Spitzer. Jahreszeitliche Aspekte der Biologie der Bartmeise (*Panurus biarmicus*). *J. Ornithol.*, 11:241–275, 1972.

[677] J.R. Spotila and E.A. Standora. Energy budgets of ectothermic vertebrates. *Am. Zool.*, 25:973–986, 1985.

[678] J.B. Sprague. Measurement of pollutant toxicity to fish I. Bioassay methods for acute toxicity. *Water Res.*, 3:793–821, 1969.

[679] M. Sprung. Physiological energetics of mussel larvae (*Mytilus edulis*) I. Shell growth and biomass. *Mar. Ecol. Prog. Ser.*, 17:283–293, 1984.

[680] M.W. Stanier, L.E. Mount, and J. Bligh. *Energy balance and temperature regulation.* Cambridge University Press, 1984.

[681] R.S. Stemberger and J.J. Gilbert. Body size, food concentration, and population growth in planktonic rotifers. *Ecology*, 66:1151–1159, 1985.

[682] M.N. Stephens. The otter report. The Univ. Fed. Animal Welfare, Herts., 1957.

[683] J.D. Stevens. Vertebral rings as a means of age determination in the blue shark (*Prionace glauca* L.). *J. Mar. Biol. Ass. U. K.*, pages 657–665, 1975.

[684] J.R. Stewart. Placental structure and nutritional provision to embryos in predominantly lecithotrophic viviparous reptiles. *Am. Zool.*, 32:303–312, 1992.

[685] B. Stonehouse. The emperor penguin I. Breeding behaviour and development. F.I.D.S. Scientific Reports 6, 1953.

[686] B. Stonehouse. The brown skua of South Georgia. F.I.D.S. Scientific Reports 14, 1956.

[687] B. Stonehouse. The king penguin of South Georgia I. Breeding behaviour and development. F.I.D.S. Scientific Reports 23, 1960.

[688] J.M. Stotsenburg. The growth of the fetus of the albino rat from the thirteenth to the twenty-second day of gestation. *Anat. Rec.*, 9:667–682, 1915.

[689] A.H. Stouthamer. Metabolic pathways in *Paracoccus denitrificans* and closely related bacteria in relation to the phylogeny of prokaryotes. *Antonie van Leeuwenhoek*, 61:1–33, 1992.

[690] A.H. Stouthamer, B.A. Bulthuis, and H.W. van Verseveld. Energetics of growth at low growth rates and its relevance for the maintenance concept. In R.K. Poole, M.J. Bazin, and C.W. Keevil, editors, *Microbial growth dynamics*, pages 85–102. IRL Press, Oxford, 1990.

[691] A.H. Stouthamer and S.A.L.M. Kooijman. Why it pays for bacteria to delete disused DNA and to maintain megaplasmids. *Antonie van Leeuwenhoek*, 63:39–43, 1993.

[692] N.M.van Straalen. Production and biomass turnover in stationary stage-structured populations. *J. Theor. Biol.*, 113:331–352, 1985.

[693] R. Strahan, editor. *The complete book of Australian mammals.* Angus and Robertson Publ., London, 1983.

[694] R.R. Strathmann and M.F. Strathmann. The relationship between adult size and brooding in marine invertebrates. *Am. Nat.*, 119:91–1011, 1982.

[695] J.R. Strickler. Calanoid copepods, feeding currents and the role of gravity. *Science (New York)*, 218:158–160, 1982.

[696] T. Strömgren and C. Cary. Growth in length of *Mytilus edulis* L. fed on different algal diets. *J. Mar. Biol. Ass. U. K.*, 76:23–34, 1984.

[697] W.C. Summers. Age and growth of *Loligo pealei*, a population study of the common Atlantic coast squid. *Biol. Bull. (Woods Hole)*, pages 189–201, 1971.

[698] D.W. Sutcliffe, T.R. Carrick, and L.G. Willoughby. Effects of diet, body size, age and temperature on growth rates in the amphipod *Gammarus pulex*. *Freshwater Biol.*, 11:183–214, 1981.

[699] A. Suwanto and S. Kaplan. Physical and genetic mapping of the *Rhodobacter sphaeroides* 2.4.1. genome: Presence of two unique circular chromosomes. *J. Bacteriol.*, 171:5850–5859, 1989.

[700] C. Swennen, M.F. Leopold, and M. Stock. Notes on growth and behaviour of the Americam razor clam *Ensis directus* in the Wadden Sea and the predation on it by birds. *Helgol. Wiss. Meeresunters.*, 39:255–261, 1985.

[701] E. Syngajewskaja. The individual growth of protozoa: *Blepharisma lateritia* and *Actinophrys sp.* *Trav. de l'Inst. Zool. Biol. Acad. Sci. Ukr.*, 8:151–157, 1935.

[702] W.M. Tattersall and E.M. Sheppard. Observations on the asteriod genus *Luidia*. In *James Johnstone Memorial Volume.* Liverpool University Press, 1934.

[703] C.R. Taylor, K. Schmidt-Nielsen, and J.L. Raab. Scaling of energetic costs of running to body size in mammals. *Am. J. Physiol.*, 219:1104–1107, 1970.

[704] J.M. Taylor and B.E. Horner. Sexual maturation in the Australian rodent *Rattus fuscipes assimilis*. *Aust. J. Zool.*, 19:1–17, 1971.

[705] R.H. Taylor. Growth of adelie penguin (*Pygoscelis adeliae* Hombron and Jacquinot) chicks. *N. Z. J. Sci.*, 5:191–197, 1962.

[706] W.D. Taylor. Growth responses of ciliate protozoa to the abundance of their bacterial prey. *Microb. Ecol.*, 4:207–214, 1978.

[707] D.W. Tempest and O.M. Neijssel. The states of Y_{atp} and maintenance energy as biologically interpretable phenomena. *Annu. Rev. Microbiol.*, 38:459–486, 1984.

[708] H. Tennekes. *De wetten van de vliegkunst; over stijgen, dalen, vliegen en zweven.* Aramith uitgevers, Bloemendaal, 1992.

[709] A.J. Tessier and C.E. Goulden. Cladoceran juvenile growth. *Limnol. Oceanogr.*, 32:680–685, 1987.

[710] A.J. Tessier, L.L. Henry, C.E. Goulden, and W.W. Durand. Starvation in *Daphnia*: Energy reserves and reproductive allocation. *Limnol. Oceanogr.*, 28:489–496, 1983.

[711] H.R. Thieme. Well-posedness of physiologically structured population models for *Daphnia magna*. *J. Math. Biol.*, 26:299–317, 1988.

[712] T.W. Thomann and J.A. Mueller. *Principles of surface water quality modeling and control.* Harper & Row Publ. Inc., 1987.

[713] D'Archy W. Thompson. *Growth and form.*, volume I & II. Cambridge University Press, 1917.

[714] J.M.T. Thompson and H.B. Stewart. *Nonlinear dynamics and chaos.* J. Wiley & Sons, Inc., 1986.

[715] K.S. Thompson. *Living fossil: The story of the coelacanth.* Hutchinson Radius, London, 1991.

[716] M.B. Thompson. Patterns of metabolism in embryonic reptiles. *Respir. Physiol.*, 76:243–256, 1989.

[717] S.T. Threlkeld. Starvation and the size structure of zooplankton communities. *Freshwater Biol.*, 6:489–496, 1976.

[718] E.V. Thuesen. The tetrodotoxin venom of chaetognaths. In Q. Bone, H. Kapp, and A.C. Pierrot-Bults, editors, *The biology of chaetognaths.* Oxford University Press, 1991.

[719] R.R. Tice and R.B. Setlow. DNA repair and replication in aging organisms and cells. In C.E. Finch and E.L. Schneider, editors, *Handbook of the biology of aging.*, pages 173–224. Van Nostrand, New York, 1985.

[720] W.L.N. Tickell. *The biology of the great albatrosses, Diomedea exulans and Diomedea epomophora.*, pages 1–55. Antarctic Research series No 12. Am. Geophysical Union., 1968.

[721] K. Tiews. Biologische Untersuchungen am Roten Thun (*Thunnus thynnus* [Linnaeus]) in der Nordsee. *Ber. Dtsch. Wiss. Komm. Meeresforsch.*, 14:192–220, 1957.

[722] H. Topiwala and C.G. Sinclair. Temperature relationship in continuous culture. *Biotechnol. Bioeng.*, 13:795–813, 1971.

[723] N.R. Towers, J.K. Raison, G.M. Kellerman, and A.W. Linnane. Effects of temperature-induced phase changes in membranes on protein synthesis by bound ribosomes. *Biochim. Biophys. Acta*, 287:301–311, 1972.

[724] A.P.J. Trinci. A kinetic study of the growth of *Aspergillus nidulans* and other fungi. *J. Gen. Microbiol.*, 57:11–24, 1969.

[725] A.P.J. Trinci, G.D. Robson, M.G. Wiebe, B. Cunliffe, and T.W. Naylor. Growth and morphology of *Fusarium graminearum* and other fungi in batch and continuous culture. In R.K. Poole, M.J. Bazin, and C.W. Keevil, editors, *Microbial growth dynamics.*, pages 17–38. IRL Press, Oxford, 1990.

[726] F.J. Trueba. *A morphometric analysis of Escherichia coli and other rod-shaped bacteria.* PhD thesis, University of Amsterdam, 1981.

[727] A.W.H. Turnpenny, R.N. Bamber, and P.A. Henderson. Biology of the sand-smelt (*Atherina presbyter* Valenciennes) around Fawlay power station. *J. Fish Biol.*, 18:417–427, 1981.

[728] H. Tyndale-Biscoe. *Life of marsupials.* E. Arnold, London, 1973.

[729] D.E. Ullrey, J.I. Sprague, D.E. Becker, and E.R. Miller. Growth of the swine fetus. *J. Anim. Sci.*, 24:711–717, 1965.

[730] E. Ursin. A mathematical model of some aspects of fish growth, respiration, and mortality. *J. Fish. Res. Board Can.*, 24:2355–2453, 1967.

[731] K. Uvnäs-Moberg. The gastrointestinal tract in growth and reproduction. *Sci. Am.*, (7):60–65, 1989.

[732] H.A. Vanderploeg, G.-A. Paffenhöfer, and J.R. Liebig. Concentration-variable interactions between calanoid copepods and particles of different food quality: Observations and hypothesis. In R.N. Hughes, editor, *Behavioural mechanisms of food selection.*, volume G 20 of *NATO ASI*, pages 595–613. Springer-Verlag, 1990.

[733] L.J. Verme. Effects of nutrition on growth of white-tailed deer fawns. *Trans. N. Am. Widl. Nat. Resour. Conf.*, 28:431–443, 1963.

[734] K. Vermeer. The importance of plankton to Cassin's auklets during breeding. *J. Plankton Res.*, 3:315–329, 1981.

[735] H.W.van Verseveld, M. Braster, F. Boogerd, B. Chance, and A.H. Stouthamer. Energetic aspects of growth of *Paracoccus denitrificans*: oxygen-limitation and shift from anaerobic nitrate-limitation to aerobic succinate-limitation. *Arch. Microbiol.*, 135:229–236, 1983.

[736] H.W.van Verseveld and A.H. Stouthamer. Oxidative phosphorylation in *Micrococcus denitrificans*; calculation of the P/O ration in growing cells. *Arch. Microbiol.*, 107:241–247, 1976.

[737] H.W.van Verseveld and A.H. Stouthamer. Growth yields and the efficiency of oxydative phosphorylation during autotrophic growth of *Paracoccus denitrificans* on methanol and formate. *Arch. Microbiol.*, 118:21–26, 1978.

[738] H.W.van Verseveld and A.H. Stouthamer. Two-(carbon) substrate-limited growth of *Paracoccus denitrificans* on mannitol and formate. *FEMS Microbiol. Lett.*, 7:207–211, 1980.

[739] J. Vidal and T.E. Whitledge. Rates of metabolism of planktonic crustaceans as related to body weight and temperature of habitat. *J. Plankton Res.*, 4:77–84, 1982.

[740] J. Vijverberg. Effect of temperature in laboratory studies on development and growth of *Cladocera* and *Copepoda* from Tjeukemeer, the Netherlands. *Freshwater Biol.*, 10:317–340, 1980.

[741] C.M. Vleck, D.F. Hoyt, and D. Vleck. Metabolism of embryonic embryos: patterns in altricial and precocial birds. *Physiol. Zool.*, 52:363–377, 1979.

[742] C.M. Vleck, D. Vleck, and D.F. Hoyt. Patterns of metabolism and growth in avian embryos. *Am. Zool.*, 20:405–416, 1980.

[743] D. Vleck, C.M. Vleck, and R.S. Seymour. Energetics of embryonic development in the megapode birds, mallee fowl *Leipoa ocellata* and brush turkey *Alectura lathami*. *Physiol. Zool.*, 57:444–456, 1984.

[744] N.J. Volkman and W. Trivelpiece. Growth in pygoscelid penguin chicks. *J. Zool. Lond.*, 191:521–530, 1980.

[745] W.J. Voorn and A.L. Koch. Characterization of the stable size distribution of cultured cells by moments. In J.A.J. Metz and O. Diekmann, editors, *The dynamics of physiologically structured populations*, volume 68 of *Springer Lecture Notes in Biomathematics.*, pages 430–440. Springer-Verlag, 1986.

[746] C.H. Walker. Kinetic models to predict bioaccumulation of pollutants. *Funct. Ecol.*, 4:69–79, 1990.

[747] E.P. Walker. *Mammals of the world.* J. Hopkins University Press, Baltimore, 1975.

[748] N. Walworth, S. Davey, and D. Beach. Fission yeast chk1 protein kinase links the rad checkpoint pathway to cdc2. *Nature (Lond.)*, 363:368–371, 1993.

[749] J. Warham. The crested penguins. In B. Stonehouse, editor, *The biology of penguins.*, pages 189–269. MacMillan Publ. Co., 1975.

[750] B.L. Warwick. Prenatal growth of swine. *J. Morphol.*, 46:59–84, 1928.

[751] I. Watanabe and S. Okada. Effects of temperature on growth rate of cultured mammalian cells (l5178y). *J. Cell Biol.*, 32:309–323, 1967.

[752] L. Watson. *Whales of the world.* Hutchinson & Co. Ltd, London, 1981.

[753] E. Watts and S. Young. Components of Daphnia feeding behaviour. *J. Plankton Res.*, 2:203–212, 1980.

[754] H. Wawrzyniak and G. Sohns. *Die Bartmeise.*, volume 553 of *Die Neue Brehm-Bücherei.* A. Ziemsen-Verlag, Wittenberg Lutherstadt, 1986.

[755] G.J.W. Webb, D. Choqeunot, and P.J. Whitehead. Nests, eggs, and embryonic development of *Carettochelys insculpta* (*Chelonia: Carettochelidae*) from Northern Australia. *J. Zool. Lond.*, B1:521–550, 1986.

[756] G.J.W. Webb, S.C. Manolis, K.E. Dempsey, and P.J. Whitehead. Crocodilian eggs: a functional overview. In G.J.W. Webb, S.C. Manolis, and P.J. Whitehead, editors, *Wildlife Management: Crocodiles and alligators.*, pages 417–422. Beatty, Sydney, 1987.

[757] W. Weibull. A statistical distribution of wide applicability. *J. Appl. Mech.*, 18:293–297, 1951.

[758] A.P. Weinbach. The human growth curve: I. prenatal. *Growth*, 5:217–233, 1941.

[759] J. Weiner. Physiological limits to sustainable energy budgets in birds and mammals: Ecological implications. *Trends Ecol. & Evol.*, 7:384–388, 1992.

[760] E. Weitnauer-Rudin. Mein Vogel. aus dem Leben der Mauerseglers *Apus apus.* Basellandschaftlicher Natur- und Vogelschutzverband, Liestal., 1983.

[761] M.J. Wells. Cephalopods do it differently. *New Sci.*, 3:333–337, 1983.

[762] D.F. Werschkul. Nestling mortality and the adaptive significance of early locomotion in the little blue heron. *Auk*, 96:116–130, 1979.

[763] I.C. West. *The biochemistry of membrane transport.* Chapman & Hall, 1983.

[764] K. Westerterp. The energy budget of the nestling starling *Sturnus vulgaris*, a field study. *Ardea*, 61:137–158, 1973.

[765] P.J. Whitehead. Respiration of *Crocodylus johnstoni* embryos. In G.J.W. Webb, S.C. Manolis, and P.J. Whitehead, editors, *Wildlife Management: Crocodiles and Alligators.*, pages 473–497. Beatty, Sydney, 1987.

[766] P.J. Whitehead, G.J.W. Webb, and R.S. Seymour. Effect of incubation temperature on development of *Crocodylus johnstoni* embryos. *Physiol. Zool.*, 63:949–964, 1990.

[767] J. Widdows, P. Fieth, and C.M. Worral. Relationship between seston, available food and feeding activity in the common mussel *Mytilus edulis. Mar. Biol.*, 50:195–207, 1979.

[768] H. Wijnandts. Ecological energetics of the long-eared owl (*Asio otus*). *Ardea*, 72:1–92, 1984.

[769] M. Wilbrink, M. Treskes, T.A.de Vlieger, and N.P.E. Vermeulen. Comparative toxicokinetics of 2,2'- and 4,4'- dichlorobiphenyls in the pond snail *Lymnaea stagnalis* (L.). *Arch. Environ. Contam. Toxicol.*, 19:69–79, 1989.

[770] H.M. Wilbur. Interactions of food level and population density in *Rana sylvatica. Ecology*, 58:206–209, 1977.

[771] D.I. Williamson. *Larvae and evolution.* Chapman & Hall, 1992.

[772] P. Williamson and M.A. Kendall. Population age structure and growth of the trochid *Monodonta lineata* determined from shell rings. *J. Mar. Biol. Ass. U. K.*, 61:1011–1026, 1981.

[773] D.B. Wingate. First successful hand-rearing of an abondoned Bermuda petrel chick. *Ibis*, 114:97–101, 1972.

[774] J.E. Winter. The filtration rate of *Mytilus edulis* and its dependence on algal concentrations, measured by a continuous automatic recording apparatus. *Mar. Biol.*, 22:317–328, 1973.

[775] L.M. Winters and G. Feuffel. Studies on the physiology of reproduction in the sheep. IV. fetal development. *Technical Bulletin Minnesota Agricultural Experiment Station*, 118:1–20, 1936.

[776] L.M. Winters, W.W. Green, and R.E. Comstock. Prenatal development of the bovine. *Technical Bulletin Minnesota Agricultural Experiment Station*, 151:1–50, 1942.

[777] P.C. Withers. *Comparative animal physiology.* Saunders College Publ., 1992.

[778] P.C. Withers and J.U.M. Jarvis. The effect of huddling on thermoregulation and oxygen consumption for the naked mole-rat. *Comp. Biochem. Physiol.*, 66:215–219, 1980.

[779] M. Witten. A return to time, cells, systems, and aging III: Gompertzian models of biological aging and some possible roles for critical elements. *Mech. Ageing Dev.*, 32:141–177, 1985.

[780] R.J. Wootton. *Ecology of teleost fishes.* Chapman & Hall, 1990.

[781] J.P. Wourms. Viviparity: The maternal-fetal relationship in fishes. *Am. Zool.*, 21:473–515, 1981.

[782] J.P. Wourms and J. Lombardi. Reflections on the evolution of piscine viviparity. *Am. Zool.*, 32:276–293, 1992.

[783] J.R. Wright and R.G. Hartnoll. An energy budget for a population of the limpet *Patella vulgata*. *J. Mar. Biol. Ass. U. K.*, 61:627–646, 1981.

[784] I. Wyllie. *The Cuckoo.* B.T. Batsford Ltd, London, 1981.

[785] A. Ykema. *Lipid production in the oleaginous yeast Apiotrichum curvantum.* PhD thesis, Vrije Universiteit, Amsterdam, 1989.

[786] P. Yodzis. *Introduction to theoretical ecology.* Harper & Row Publ. Inc., 1989.

[787] E. Zeuthen. *Body size and metabolic rate in the animal kingdom.* H. Hagerup, Copenhagen, 1947.

[788] C. Zonneveld and S.A.L.M. Kooijman. The application of a dynamic energy budget model to *Lymnaea stagnalis*. *Funct. Ecol.*, 3:269–278, 1989.

[789] C. Zonneveld and S.A.L.M. Kooijman. Body temperature affects the shape of avian growth curves. 1993. submitted.

[790] C. Zonneveld and S.A.L.M. Kooijman. Comparative kinetics of embryo development. *Bull. Math. Biol.*, 55:609–635, 1993.

Glossary

allometry The group of analyses based on a linear relationship between the logarithm of some physiological or ecological variable and the logarithm of the body weight of individuals

altricial A mode of development where the neonate is still in an early stage of development and requires attention from the parents. Typical altricial birds and mammals are naked and blind at birth. The opposite of altricial is precocial

anabolism The collection of biochemical processes involved in the synthesis of structural body mass

Arrhenius temperature The value of the slope of the linear graph one gets if the logarithm of a physiological rate is plotted against the inverse absolute temperature. It has dimension temperature, but it does not relate to a temperature that exists at a site

aspect ratio The dimensionless ratio between the length and the diameter of an object with the shape of a cylinder (filaments, rods). The length of rods includes both hemispheres

ATP Adenosine triphosphate is a chemical compound that is used by all cells to store or retrieve energy via hydrolysis of one or two phosphate bonds

Bernoulli equation A differential equation of the type $\frac{d}{dx}y + f(x)y = g(x)y^a$, where a is any real number and f and g arbitrary functions of x. Bernoulli found a solution technique for this type of equation

catabolism The collection of biochemical processes involved in the decomposition of compounds for the generation of energy and/or source material for anabolic processes

chemical potential The change in the total free energy of a mixture of compounds per mole of substance when an infinitesimal amount of a substance is added, while temperature, pressure and all other compounds are constant

coefficient of variation The dimensionless ratio of the (sample) standard deviation and the mean. It is a useful measure for the scatter of realizations of a random variable that has a natural origin. The measure is useless for temperatures measured in degrees Celsius, for example

combustion reference In this frame of reference, the chemical potentials of H_2O, HCO_3^-, NH_4^+, H^+ and O_2 are taken to be 0. The chemical potentials of organic compounds in the standard thermodynamical frame of reference (pH=7, 298 K, unit molarity) are corrected for this setting by equating the dissipation free energy in both frames of reference, when the compound is fully oxidized. The chemical potential of compound $CH_xO_yN_z$ in the combustion frame of reference is expressed in the standard frame of reference as $\tilde{\mu}_{CH_xO_yN_z} = \tilde{\mu}^\circ_{CH_xO_yN_z} + \frac{1}{2}(2 - x + 3z)\tilde{\mu}^\circ_{H_2O} - \tilde{\mu}^\circ_{HCO_3^-} - (1 - z)\tilde{\mu}^\circ_{H^+} - z\tilde{\mu}^\circ_{NH_4^+} + \frac{1}{4}(4 + x - 2y - 3z)\tilde{\mu}^\circ_{O_2}$

compound parameter A function of original parameters. It is usually a simple product and/or ratio

cubic spline function A function consisting of a number of third degree polynomials glued together in a smooth way for adjacent intervals of the argument. This is done by requiring that polynomials that meet at a particular argument value x_i, have the same value y_i, and the same first two derivatives at that point. The points x_i, y_i, for $i = 1, 2, \cdots, n$ with $n \leq 4$, are considered as the parameters of the cubic spline. For descriptive purposes, splines have the advantage over higher order polynomials because their global behaviour is much less influenced by local behaviour

DEB Initials of the Dynamic Energy Budget model or theory, that is discussed in this book. The term 'dynamic' refers to the contrast with the frequently used static energy budget models, where the specifications of the individual do not change explicitly in time

ectotherm An organism that is not an endotherm

eigenvalue If a special vector, an eigenvector, is multiplied by a square matrix, the result is the same as multiplying that vector by a scalar value, known as the eigenvalue. Each square matrix has a number of different independent eigenvectors. This number equals the number of rows (or columns). Each eigenvector has its own eigenvalue, but some of the eigenvalues may be equal

endotherm An animal that usually keeps its body temperature within a narrow range by producing heat. Birds and mammals do this for most of time that they are active. Some other species (insects, tuna fish) have endothermic tendencies

enthalpy Heat content with dimension energy mole^{-1}. The enthalpy of a system increases by an amount equal to the energy supplied as heat if the temperature and pressure do not change

entropy The cumulative ratio of heat capacity and temperature of a body when its temperature is gradually increased from zero (absolute) temperature to the temperature of observation. Its dimension is energy$\times$(temperature mole)$^{-1}$. The equivalent definition of the ratio of enthalpy minus free energy and temperature is more useful in biological applications

estimation The use of measurements to assign values to one or more parameters of a model. This is usually done in some formalized manner that allows evaluation of the uncertainty of the result

expectation The theoretical mean of a function of a random variable. For a function g of a random variable $\underline{x}$ with probability density $\phi_{\underline{x}}$, its formal definition is $\mathcal{E}g(\underline{x}) \equiv \int_x g(x)\phi_{\underline{x}}(x)\,dx$. For $g(\underline{x}) = \underline{x}$, the expectation of $\underline{x}$ is the theoretical mean

exponential distribution The random variable $\underline{t}$ is exponentially distributed with parameter $\dot{r}$ if the probability density is $\phi_{\underline{t}}(t) = \dot{r}\exp\{-\dot{r}t\}$. The mean of $\underline{t}$ equals $\dot{r}^{-1}$

filament An organism with the shape of a cylinder that grows in length only. The aspect ratio is so small that the caps can be neglected in its energetics

first order process A process that can be described by a differential equation where the change of a quantity is linear in the quantity itself

flux An amount of mass or energy per unit of time. An energy flux is physically known as a power

free energy The maximum amount of energy of a system that is potentially available for 'work'. In biological systems, this 'work' usually consists of driving chemical reactions against the direction of their thermodynamic decay

functional response The ingestion rate of an organism as a function of food density

growth Increase in structural body mass, measured as an increase in volume in most organisms. I do not include anabolic processes that are part of maintenance

hazard rate The probability per time increment that death strikes at a certain age, given survival up to that age

heat capacity The mole-specific amount of heat absorbed by a substance to increase one Kelvin in temperature. Heat capacity typically depends on temperature and has dimension energy mole^{-1}

heterotroph An organism that uses organic compounds as a source of energy

homeostasis The ability of most organisms to keep the chemical composition of their body constant, despite changes in the chemical composition of the environment

isomorph An organism that does not change its shape during growth

large number law The strong law of large numbers states that the difference between the mean of a set of random variables and its theoretical mean is small, with an overwhelming probability, given that the set is large enough

maintenance A rather vague term denoting the collection of energy demanding processes that life seems to require to keep going, excluding all production processes. I also exclude heat production in endotherms

mass action law The law that states that the meeting frequency of two types of particles is proportional to the product of their densities, i.e. number of particles per unit of volume

NADPH Nicotinamide adenine dinucleotide phosphate is a chemical compound that is used by all cells to accept pairs of electrons

parameter A quantity in a model that describes the behaviour of state variables. It is usually assumed to be a constant

parthenogenesis The mode of reproduction where females produce eggs that hatch into new females without the interference of males

partition coefficient The ratio of the equilibrium concentrations of a compound dissolved in two immiscible solvents, which is taken to be independent of the actual concentrations. The concentrations are here expressed per unit of weight of solvent (not per unit of volume or per mole of solvent)

phylum A taxon that collects organisms with the same body plan

Poisson distribution A random variable $\underline{X}$ is Poisson distributed with parameter (mean) λ if $\text{Prob}\{\underline{X} = x\} = \frac{\lambda^x}{x!}\exp\{-\lambda\}$. If intervals between independent events are exponentially distributed, the number of events in a fixed time period will be Poisson distributed

polynomial A polynomial of degree n of argument x is a function of the type $\sum_{i=0}^{n} c_i x^i$, where $c_0, c_0, ., c_n$ are fixed coefficients

precocial A mode of development where the neonate is in an advanced state of development and usually does not require attention from the parents. Typical precocial birds and mammals have feathers or hair and gather food by themselves. The opposite of precocial is altricial

probability density function A non-negative function, here called ϕ, belonging to a continuous random variable, $\underline{x}$ for instance, with the property that $\int_{x_1}^{x_2} \phi_{\underline{x}}(x)\, dx = \text{Prob}\{x_1 < \underline{x} < x_2\}$

prokaryote An organism that does not have a nucleus, i.e. an eubacterium or archaebacterium

reduction degree A property of a molecule. Its value equals the sum of the valences of the atoms minus the electrical charge

relaxation time A characteristic time that indicates how long a dynamic system requires to return to its equilibrium after perturbation. It is a compound parameter with dimension time standing for the first term of the Taylor expansion of the differential equation that describes the dynamics of the system, evaluated in its equilibrium

respiration quotient The ratio between carbon dioxide production and oxygen consumption, expressed on a molar basis

rod A bacterium with the shape of a croquette or sausage, that grows in length only, at a certain substrate density. It is here idealized by a cylinder with hemispheres at both ends

state variable A variable which determines, together with other state variables, the behaviour of a system. The crux of the concept is that the collection of state variables together with the input, determine the behaviour of the system completely

survivor function A rather misleading term standing for the probability that a given random variable exceeds a specified value. All random variables have a survivor function, even those without any connection to life span. It equals one minus the distribution function. The term is sometimes synonymous with upper tail probability

taxon A systematic unit, which is used in the classification of organisms. It can be species, genus, family, order, class, phylum, kingdom

Taylor expansion The approximation of a function by a polynomial of a certain degree that is thought to be accurate for argument values around a specified value. The coefficients of the polynomial are obtained by equating the function value and its first n derivatives at the specified value, to that of the n degree polynomial

volumetric length The cubic root of the volume of an object. It has dimension length

weighted sum The sum of terms that are multiplied with weight coefficients before addition. If the terms do not have the same dimension, the dimensions of the different weight coefficients convert the dimensions of weighted terms to the same dimension

zero-th order process A process that can be described by a differential equation where the change of a quantity is constant

zooplankter An individual belonging to the zooplankton, i.e. a group of usually small aquatic animals that live in free suspension and do not actively move far in the horizontal direction

Notation and symbols

Some readers will be annoyed by the notation, which sometimes differs from the one usual in a particular specialization. One problem is that conventions in e.g. microbiology differ from those in ecology, so not all conventions can be observed at the same time. The symbol D, for example, is used by microbiologists for the dilution rate in chemostats, but by chemists for diffusivity. Another problem is that most literature does not distinguish structural biomass from energy reserves, which both contribute to e.g. dry weight. So the conventional symbols actually differ in meaning from the ones used here. For the sake of consistency, I even found it necessary to deviate sometimes from the notation I have used myself in earlier papers. Originally I thought e.g. that it was possible to measure the size of an individual in several more or less equivalent ways, such as volume or wet weight. Now I see that this theory requires volume, which urges the use of V rather than W. Few texts deal with such a broad spectrum of phenomena as this book. A consequence is that any symbol table is soon exhausted if one carelessly assigns new symbols to all kinds of variables that show up. A voluminous literature on population dynamics exists, where it is standard to use the symbol l for survival probability. This works well as long as one does not want to use lengths in the same text!

The following conventions are used to reduce this problem and to aid memory.

- Analogous to the tradition in chemistry, quantities which are expressed per unit of biovolume have square brackets, $[\,]$. Quantities per unit of biosurface area have braces, $\{\,\}$. Quantities per unit of wet weight have angles, $\langle\,\rangle_w$. This notation is chosen to stress that these symbols refer to relative quantities, rather than absolute ones. They do not indicate concentrations in the chemical sense, because most of the compounds concerned are not soluble. Parentheses, square brackets and braces around numbers refer to equations, references and pages respectively. Parentheses around numbers behind other numbers are standard deviations of estimated parameter values.

- Rates have dots, which merely indicate the dimension 'per time'. Unless indicated by an index or argument t, these rates are assumed to be constant in time. Dots, brackets and braces are introduced to have an easy test for some dimensions and to reduce the number of different symbols for related variables. If time has been scaled, i.e. the time unit is some particular value making scaled time dimensionless, the dot has been removed from the rate that is expressed in scaled time.

- Sometimes, an expression between parentheses has an index '+'. This means: take the maximum of 0 and that expression, so $(x - y)_+ \equiv \max\{0, x - y\}$. The symbol

'≡' means 'is per definition'. It is just another way of writing, you are not supposed to understand that the equality is true.

- Although the mathematical standard for notation should generally be preferred over that of any computer language, I make one exception: the logic boolean, e.g. $(x < x_s)$. It always comes with parentheses and stands for the number 1 if true or the number 0 if false. It appears as part of an expression. Simple rules apply, such as $(x \leq x_s)(x \geq x_s) = (x = x_s)$, or $\int_{x1=-\infty}^{x} (x_1 = x_s)\, dx_1/dx = (x \geq x_s)$ and $\int_{x1=-\infty}^{x} (x \geq x_s)\, dx_1 = (x - x_s)_+$. The usual notation for the logic boolean is the Heavyside function, which I consider clumsy, compared with this APL derived notation.

- The symbol $*$ as index is used to indicate that several other symbols can be substituted. It is known as 'wildcard' in computer science.

- Random variables are underscored. The notation $\underline{x}|\underline{x} > x$ means: the random variable $\underline{x}$ given that it is larger than the value x. It can occur in expressions for the probability, Prob{}, or for the probability density function, $\phi()$.

- Vectors and matrices are printed in bold face.

- The SI system is used to present units of measurements. My experience is that most American readers are unfamiliar with the symbol 'a' for year.

- The following operators occur:

$\frac{d}{dt}X\|_{t_1}$	derivative of X with respect to t evaluated at $t = t_1$
$\frac{\partial}{\partial t}X\|_{t_1}$	partial derivative of X with respect to t evaluated at $t = t_1$
$\mathcal{E}g(\underline{x})$	expectation of a function g of the random variable $\underline{x}$
var $\underline{x}$	variance of the random variable $\underline{x}$: $\mathcal{E}(\underline{x} - \mathcal{E}\underline{x})^2$
cv $\underline{x}$	coefficient of variation of the random variable $\underline{x}$: $\sqrt{\text{var }\underline{x}}/\mathcal{E}\underline{x}$
cov $(\underline{x}, \underline{y})$	covariance between the random variables $\underline{x}$ and $\underline{y}$: $\mathcal{E}(\underline{x} - \mathcal{E}\underline{x})(\underline{y} - \mathcal{E}\underline{y})$
cor $(\underline{x}, \underline{y})$	correlation between $\underline{x}$ and $\underline{y}$: cov $(\underline{x}, \underline{y})/\sqrt{\text{var }\underline{x}\text{ var }\underline{y}}$
$\mathbf{x}^T$	the transpose of vector or matrix $\mathbf{x}$

In the description of the dimensions in the list of symbols, the following symbols are used:

–	no dimension	L	length (of individual)	e	energy ($\equiv ml^2t^{-2}$)
#	number	l	length (of environment)	T	temperature
t	time	m	mass		

These dimension symbols just stand for an abbreviation of the dimension, and differ in meaning from symbols in the symbol column. A difference between the dimensions l and L is that the latter involves an arbitrary choice of the length to be measured (e.g. including or excluding a tail). The morph interferes with the choice. The dimensions differ because the sum of lengths of objects for which l and L apply, does not have any useful meaning. The list below does not include symbols that are used in a brief description only. The page number refers to the page where the symbol is introduced.

symbol	dimension	page	interpretation
a	t	{82}	age, i.e. time since gametogenesis
a_b	t	{82}	age at birth (hatching), i.e. end of embryonic stage
a_p	t	{82}	age at puberty, i.e. end of juvenile stage
$a_\dagger$	t	{107}	age at death (life span)
$\dot{A}$	$e\,t^{-1}$	{72}	assimilation rate
$\{\dot{A}_m\}$	$e\,L^{-2}t^{-1}$	{72}	surface area-specific maximum assimilation rate
$[\dot{A}_m]$	$e\,L^{-3}t^{-1}$	{93}	volume-specific maximum assim. rate: $\{\dot{A}_m\}V_d^{-1/3}$
$B_x(a,b)$	-	{92}	incomplete beta function
c	$m\,l^{-3}$	{256}	concentration of xenobiotic compound in the environment
c_d	$m\,l^{-3}$	{267}	concentration of xenobiotic compound in the water (dissolved)
c_{L50}	$m\,l^{-3}$	{274}	value of c that results in 50% mortality: LC50
c_x	$m\,l^{-3}$	{267}	concentration of xenobiotic compound in food
$\dot{C}$	$e\,t^{-1}$	{72}	rate of reserve energy utilization (catabolic rate)
d_m	-	{21}	shape (morph) coefficient: $V^{1/3}L^{-1}$
d_Q	L^3e^{-1}	{107}	amount of damage inducing compound per utilized energy unit
$[d_w]$	$m\,L^{-3}$	{21}	volume-specific weight (density): $W_wV^{-1} \equiv [W_w]$
$[d_{de}]$	$m\,L^{-3}$	{34}	reserve-specific dry weight times the max. energy density $[E_m]$
$[d_{dv}]$	$m\,L^{-3}$	{34}	structural volume-specific dry weight
$[d_{me}]$	$\#L^{-3}$	{37}	number of C atoms per unit of reserve energy volume $E[E_m]^{-1}$
$[d_{mv}]$	$\#L^{-3}$	{37}	number of C atoms per unit of structural body volume V
d_{mx}	$\#l^{-3}$	{37}	conversion coefficient from volume to C-mole of substrate
$[d_{P*}]$	$\#L^{-3}$	{190}	C-moles of product per energy volume associated with energy flux *
$[d_{we}]$	$m\,L^{-3}$	{34}	reserve-specific wet weight times the max. energy density $[E_m]$
$[d_{wv}]$	$m\,L^{-3}$	{34}	structural volume-specific wet weight
$\dot{D}$	l^2t^{-1}	{142}	diffusivity
e	-	{92}	scaled energy density: $[E]\,[E_m]^{-1}$
e_0	-	{92}	scaled energy costs for one egg/foetus: $E_0([E_m]V_m)^{-1}$
e_b	-	{92}	scaled energy density at birth
$e_{\dot{R}}$	-	{147}	scaled energy allocated to reproduction: $E_{\dot{R}}([E_m]V_m)^{-1}$
E	e	{72}	non-allocated energy in reserve
$[E]$	$e\,L^{-3}$	{73}	energy density: $E\,V^{-1}$
E_0	e	{84}	energy costs for one egg/foetus
$[E_b]$	$e\,L^{-3}$	{84}	energy density at birth
E_f	$e\,l^{-3}$	{248}	metabolizable energy density in the gut
E_g	$e\,l^{-3}$	{248}	absorbable energy density in the gut
$[E_m]$	$e\,L^{-3}$	{73}	maximum energy density
$E_{\dot{R}}$	e	{34}	energy in reserve with allocation reproduction
f	-	{63}	scaled functional response: $f = \frac{X}{K+X} = \frac{x}{1+x}$
f_f	-	{248}	scaled functional response of digestion in the gut: $\frac{E_f}{K_f+E_f}$
f_g	-	{248}	scaled functional response of absorption in the gut: $\frac{E_g}{K_g+E_g}$
$\dot{F}$	l^3t^{-1}	{64}	filtering rate
$\dot{F}_m$	l^3t^{-1}	{66}	maximum filtering rate

g	-	{81}	energy investment ratio: $\frac{[G]}{\kappa[E_m]}$
g_A	-	{96}	wall investment ratio: $\frac{[G_A]}{\kappa[E_m]}$
$[G]$	$e\,L^{-3}$	{80}	volume-specific costs for growth
$[G_A]$	$e\,L^{-3}$	{96}	surface area-specific costs for bacterial cell wall times $V_d^{-1/3}$
$[G_V]$	$e\,L^{-3}$	{96}	volume-specific costs for bacterial cytoplasm
$\dot{h}$	t^{-1}	{107}	hazard rate
$\dot{H}$	$e\,t^{-1}$	{79}	heating rate in endotherms
$\{\dot{H}\}$	$e\,L^{-2}t^{-1}$	{79}	surface area-specific heating rate: $\dot{H}\,V^{-2/3}$
ΔH	$e\#^{-1}$	{203}	free energy that dissipates per mole of consumed substrate
$\imath_m$	L^3	{115}	scaled maximum ingestion rate: $\{\dot{I}_m\}V_m^{2/3}\dot{m}^{-1}$
$\dot{I}$	L^3t^{-1}	{63}	ingestion rate: volume of food per time
$\{\dot{I}_m\}$	$L^3L^{-2}t^{-1}$	{66}	surface area-specific max ingestion rate
$[\dot{I}_m]$	$L^3L^{-3}t^{-1}$	{67}	volume-specific maximum ingestion rate: $\{\dot{I}_m\}V_d^{-1/3}$
$\dot{k}$	t^{-1}	{194}	rate of macro-chemical reaction
$\dot{k}_a$	t^{-1}	{256}	xenobiotic elimination rate
$\dot{k}_e$	t^{-1}	{248}	decay rate of enzyme activity
$\dot{k}_f$	$e\,m^{-1}t^{-1}$	{248}	digestion rate constant
$\dot{k}_g$	$e\,m^{-1}t^{-1}$	{248}	absorption rate constant in the gut
$\dot{k}_\dagger$	$l^3m^{-1}t^{-1}$	{268}	killing rate by xenobiotic compound
$\dot{k}_{ad}$	$L\,t^{-1}$	{268}	xenobiotic conductance
$[\dot{k}_{da}]$	$l^3L^{-3}t^{-1}$	{257}	volume-specific xenobiotic uptake rate from water
$\{\dot{k}_{da}\}$	$l^3L^{-2}t^{-1}$	{268}	surface area-specific xenobiotic uptake rate from water
$\{\dot{k}_{xa}\}$	$l^3L^{-2}t^{-1}$	{277}	surface area-specific xenobiotic uptake rate from food
K	$L^3l^{-3 \text{ or } 2}$	{63}	saturation coefficient
K_f	$e\,l^{-3}$	{249}	saturation coefficient of enzymatic digestion
K_g	$e\,l^{-3}$	{248}	saturation coefficient of absorption
l	-	{92}	scaled body length: $(V\,V_m^{-1})^{1/3}$
l_b	-	{92}	scaled body length at birth: $(V_bV_m^{-1})^{1/3}$
l_d	-	{103}	scaled cell length at division: $(V_dV_m^{-1})^{1/3} = \dot{m}g/\dot{\nu}$
l_h	-	{118}	scaled heating length: $(V_hV_m^{-1})^{1/3}$
l_p	-	{98}	scaled body length at puberty: $(V_pV_m^{-1})^{1/3}$
L	L	{21}	length: $V^{1/3}d_m^{-1}$
L_b	L	{148}	length at birth: $V_b^{1/3}d_m^{-1}$
L_d	L	{33}	length at cell division
L_p	L	{148}	length at puberty: $V_p^{1/3}d_m^{-1}$
L_m	L	{148}	maximum length: $V_m^{1/3}d_m^{-1}$
L_λ	L	{248}	length of a slice of gut contents
L_ϕ	L	{248}	diameter (cross section) of gut
$\dot{m}$	t^{-1}	{80}	maintenance rate coefficient: $[\dot{M}]\,[G]^{-1}$
$\dot{M}$	$e\,t^{-1}$	{78}	maintenance rate
$\dot{M}_d$	$e\,t^{-1}$	{99}	energy costs for maintaining maturity
$[\dot{M}]$	$e\,L^{-3}t^{-1}$	{78}	volume-specific maintenance rate: $\dot{M}/V$
$\mathcal{M}(V)$	-	{30}	shape (morph) correction function: $\frac{\text{real surface area}}{\text{isomorphic surface area}}$
$n_{*_1*_2}$	#	{192}	number of atoms of element $*_1$ present in compound $*_2$

$\mathbf{n}$	#	{193}	matrix of numbers $n_{*_1 *_2}$ that relate to organic compounds
$n(a)\,da$	#	{170}	number of individuals of age in the interval $(a, a+da)$
N	#	{170}	(total) number of individuals: $\int_a n(a)\,da$
$\dot{p}$	t^{-1}	{160}	individual-specific predation probability rate
$\dot{p}_a$	t^{-1}	{111}	aging rate for unicellulars; compound parameter: $d_Q \frac{[G]}{\kappa} \frac{\dot{\nu}+\dot{m}}{g+1}$
$\ddot{p}_a$	t^{-2}	{107}	aging acceleration; compound parameter $\propto d_Q[G]/\kappa$
$\dot{p}_i$	t^{-1}	{152}	aging rate for imagos; compound parameter: $(\frac{1}{6}\ddot{p}_a \dot{m} f/l)^{1/3}$
$\dot{p}_m$	t^{-1}	{190}	max. throughput rate in a chemostat without complete washout
P	$\# l^{-3}$	{190}	(microbial) product density on the basis of moles
P_{ea}	-	{267}	partition coefficient for energy reserves/aqueous fraction
P_{wa}	-	{267}	partition coefficient for total body mass/aqueous fraction
P_{xd}	-	{271}	partition coefficient for food/water (dissolved fraction)
$[Q]$	$m\,L^{-3}$	{107}	density of damage inducing or xenobiotic compounds
$[Q]_0$	$m\,L^{-3}$	{277}	maximum density of xenobiotic compound that has no effects
$\langle Q \rangle_w$	$m\,m^{-1}$	{268}	quantity of xenobiotic compound per unit of wet weight: $[Q]/[W_w]$
q	-	{100}	survival probability of the embryonic stage
$q(c,t)$	-	{275}	survival probability to a toxic compound
$\dot{R}$	$\# t^{-1}$	{100}	reproduction rate, i.e. number of eggs or young per time
$\dot{R}_m$	$\# t^{-1}$	{101}	max reproduction rate
$\{\dot{S}_g\}$	$m\,L^{-2}t^{-1}$	{248}	surface area-specific secretion rate of enzyme by the gut wall
t	t	{25}	time
t_g	t	{69}	gut residence time
t_d	t	{96}	inter division period
t_D	t	{103}	DNA duplication time
t_E	t	{191}	time parameter for energy reserves: $[d_{me}](d_{mx}[\dot{I}_m])^{-1}$
$t_{\dot{R}}$	t	{268}	time at spawning
t_s	t	{69}	mean stomach residence time
t_{X_1}	t	{191}	time parameter for structural biomass: $[d_{mv}](d_{mx}[\dot{I}_m])^{-1}$
$t_{*_1 *_2}$	t	{194}	time parameter for compound $*_1$ associated with energy flux $*_2$
t_{H*}	t	{204}	time parameter for dissipating heat associated with energy flux $*$
$\mathbf{t}_M$	t	{194}	matrix of time parameters for 'minerals'
$\mathbf{t}_D$	t	{194}	matrix of time parameters for organic compounds
T	T	{44}	temperature
T_A	T	{44}	Arrhenius temperature
T_b	T	{79}	body temperature
T_e	T_e	{79}	environmental temperature
$\mathbf{u}$	$\#^{-1}$	{193}	inverse of matrix of $n_{*_1 *_2}$ for 'minerals'
$\dot{v}$	$L\,t^{-1}$	{74}	energy conductance: $\{\dot{A}_m\}\,[E_m]^{-1}$
$\dot{\nu}$	t^{-1}	{93}	specific-energy conductance: $\{\dot{A}_m\}V_d^{-1/3}[E_m]^{-1} = [\dot{A}_m][E_m]^{-1}$
V	L^3	{21}	structural body volume
V_b	L^3	{97}	body volume at birth (transition embryo/juvenile)
V_d	L^3	{33}	cell volume at division
V_h	L^3	{81}	the volume reduction for endotherms due to heating: $\{\dot{H}\}^3[\dot{M}]^{-3}$
V_m	L^3	{81}	maximum body volume: $(\kappa\{\dot{A}_m\})^3[\dot{M}]^{-3} = (\dot{v}/\dot{m}g)^3$
V_p	L^3	{97}	body volume at puberty (transition juvenile/adult)
V_∞	L^3	{94}	ultimate body volume

$\mathcal{V}$	L^3	{222}	maximum body volume compared to reference: $z^3 V_{m,1}$
w	$m\#^{-1}$	{37}	molar weight
W_1	$\#l^{-3}$	{191}	(total) biomass density on the basis of C-moles
W_d	m	{34}	dry weight of (total) biomass
W_w	m	{21}	wet weight of (total) biomass
x	-	{161}	scaled biovolume density: XK^{-1}
X	$L^3 l^{-3 \text{ or } 2}$	{63}	biovolume density, usually food
$[X_{gm}]$	$L^3 L^{-3}$	{70}	maximum volume-specific capacity of the gut for food
$[X_{sm}]$	$L^3 L^{-3}$	{69}	maximum volume-specific capacity of the stomach for food
y	-	{124}	scaled yield factor: $Y\{\dot{I}_m\}\dot{v}^{-1}$
Y	$L^3 L^{-3}$	{161}	yield factor
Y_i	$L^3 L^{-3}$	{124}	instantaneous yield factor
Y_n	$L^3 L^{-3}$	{124}	non-instantaneous yield factor
${}_mY_{W_1}$	$\#\#^{-1}$	{191}	yield factor from substrate to biomass on the basis of C-moles
${}_mY_P$	$\#\#^{-1}$	{191}	yield factor from substrate to product on the basis of C-moles
${}_mY_*$	$\#\#^{-1}$	{192}	yield factor from substrate to compound $*$
$\mathbf{Y}_M$	$\#\#^{-1}$	{193}	matrix of yield factors from substrate to 'minerals'
$\mathbf{Y}_D$	$\#\#^{-1}$	{193}	matrix of yield factors from substrate to organic compounds
z	-	{218}	zoom factor to compare body sizes
$\dot{\gamma}$	t^{-1}	{81}	von Bertalanffy growth rate: $(3/\dot{m} + 3fV_m^{1/3}/\dot{v})^{-1} = \dot{m}g/3(f+g)$
$\Gamma(x)$	-	{152}	gamma function
δ	-	{33}	aspect ratio
κ	-	{53}	fraction of utilized energy spent on maintenance plus growth
$\kappa_{*_1 *_2}$	-	{203}	fraction of energy of flux $*_2$ that is fixed in compound $*_1$
$\dot{\mu}$	t^{-1}	{170}	individual-specific population growth rate
$\dot{\mu}_m$	t^{-1}	{166}	(net) maximum individual-specific population growth rate
$\dot{\mu}_m^\circ$	t^{-1}	{166}	gross maximum individual-specific population growth rate
$\tilde{\mu}_*$	$e\#^{-1}$	{202}	chemical potential of compound $*$
$\tilde{\mu}_E$	$e\#^{-1}$	{202}	free energy per C-mole of energy reserves: $[E_m]/[d_{me}]$
$\tilde{\boldsymbol{\mu}}_M$	$e\#^{-1}$	{202}	vector of free energies of 'minerals'
$\tilde{\boldsymbol{\mu}}_D$	$e\#^{-1}$	{202}	vector of free energies of organic compounds, determined by DEB
$\phi_{\underline{x}}(x)dx$	-	{170}	probability density of $\underline{x}$ evaluated in x

Author index

[0]Figures that are printed in bold type refer to bibliography entries, not to pages

Taxonomic index

Subject index